Powdery Mildew Disease of Crucifers: Biology, Ecology and Disease Management

Govind Singh Saharan • Naresh K. Mehta
Prabhu Dayal Meena

Powdery Mildew Disease of Crucifers: Biology, Ecology and Disease Management

Springer

Govind Singh Saharan
Department of Plant Pathology
CCS Haryana Agricultural University
Hisar, Haryana, India

Naresh K. Mehta
Department of Plant Pathology
CCS Haryana Agricultural University
Hisar, Haryana, India

Prabhu Dayal Meena
Crop Protection Unit, ICAR-Directorate
of Rapeseed-Mustard Research
Bharatpur, Rajasthan, India

ISBN 978-981-13-9852-0 ISBN 978-981-13-9853-7 (eBook)
https://doi.org/10.1007/978-981-13-9853-7

This Springer imprint is published by the registered company Springer Nature Singapore Pte Ltd.
The registered company address is: 152 Beach Road, #21-01/04 Gateway East, Singapore 189721, Singapore

Foreword

Crucifers are crops grown all over the world in temperate, cool temperate, continental, and sub-tropical regions. Economically important crucifers include oil yielding, vegetable, fodder, and horticultural *Brassica* crops. Several weeds also belong to family Cruciferae. The crucifer *Brassica* vegetables constitute major source of vitamins, fiber, minerals, and proteins in human diet, while *Brassica* oilseeds are major source of quality vegetable oil and cake for animal feed. The demand for *Brassica* vegetables and oil is consistently increasing every year all over the globe. Crucifer crops are threatened by several biotic and abiotic stresses under variable and changing climatic conditions wherever these are cultivated. Out of biotic stresses, powdery mildew (belonging to Erysiphales) is the most widespread and devastating disease causing yield losses both quantitatively and qualitatively. It has been reported that it is capable of causing up to 90 percent loss in oil quality and up to 7 percent in quantity. Powdery mildew being a favorable host-pathogen system has been largely exploited as model for basic research on host-parasite interactions, developmental morphology, cytology, and molecular biology to detect effective proteins/genes governing different biological functions. *Arabidopsis thaliana* has been widely used as a tool for molecular and genetic studies. This book *Powdery Mildew Disease of Crucifers: Biology, Ecology and Disease Management* is a comprehensive treatise on important disease of crucifers encompassing most of the published information. The information in this book has been arranged in 11 different chapters with appropriate headings and subheadings. Photographs, graphs, figures, tables, and references stimulate interest and better comprehension of the description on the disease. This book provides much needed background and current information projecting future priorities, areas of research, and methodologies. It ensures its place as a central document necessary for the Brassicalogists of the world for further investigations on this host parasite system.

The book has been crafted as the most useful document with a wide range of logically organized and easily accessible information. The authors have already contributed books on Sclerotinia diseases, White rust disease, *Alternaria* diseases, and downy mildew diseases of crucifers published by Springer. I congratulate Drs. G. S. Saharan, Naresh Mehta, and P. D. Meena for bringing out this publication which is an addition to the series and outcome of their lifelong professional interest and expertise. I am sure it will be useful for researchers, teachers, students, extension experts, industrialists, and farmers.

S.S. Chahal

Honorary Professor
Panjab University
Chandigarh, India

S. S. Chahal

Former Vice-Chancellor MPUA&T
Udaipur, Rajasthan, India

Preface

Powdery mildews are one of the world's most frequently encountered pathogenic fungi causing quantitative and qualitative yield losses in all kinds of annual and perennial, horticultural and ornamental, cash, and industrial crops, forest trees, shrubs, grasses, and all kinds of vegetation in tropical, sub-tropical, and temperate regions of the world. It is fourth most widespread and devastating disease on cruciferous crops causing yield losses up to 90 percent with loss in oil quality and up to 7 percent in rapeseed-mustard quantity. Powdery mildews are often very conspicuous owing to their profuse production of conidia on the host surface in the form of white granular coating giving them their common name. Powdery mildews are also favorable host-pathogen system as model for basic research on host-parasite interactions, developmental morphology, cytology, and molecular biology to dissect effector proteins/genes governing different biological functions. This book *Powdery Mildew Disease of Crucifers: Biology, Ecology and Disease Management* is a comprehensive treatise on the fourth most important disease of crucifers encompassing all the published information which will be useful for researchers, teachers, students, extension experts, industrialists, and farmers. The information has been arranged in 11 chapters with appropriate headings and subheadings, illustrations, photographs, graphs, figures, tables, histogram, colored plates of micrographs, electron micrographs, and flow charts for effective and stimulating comprehension by the readers. The different chapters of the book include detailed information on the status of disease and pathogen; the disease, its distribution, symptomatology, host range, yield losses, and disease assessment; pathogen, its taxonomy, morphology, phylogeny, variability, sporulation, survival, and perpetuation; spore germination, infection, pathogenesis, disease cycle, epidemiology, forecasting, and fine structures; mechanisms of host resistance, biochemical, histological, genetic, and molecular including cloning and mapping of R genes; sources of resistance, disease resistance breeding strategies, and genetics of host-parasite interactions; disease management through cultural, chemical, biological, host resistance, and integrated approach; and standardized reproducible techniques. Although we have taken every care to seek permission from the authors/publishers to include their valuable contributions in this book, nevertheless inadvertently for any error, we humbly request to excuse us. All sources have been duly acknowledged. We owe the responsibility for any error or omission and are open to include your suggestions in the revised editions.

Hisar, Haryana, India — Govind Singh Saharan
Hisar, Haryana, India — Naresh K. Mehta
Bharatpur, Rajasthan, India — Prabhu Dayal Meena

Acknowledgments

Authors are very grateful to the following persons/scientists/publishers/societies/journals/institutes/websites and all others whose valuable materials such as photographs (macroscopic, microscopic, electron micrographs, scanning electron micrographs), drawings, figures, histograms, graphs, tables, flow charts, etc. have been used through reproduction in the present document. The details of the author(s)/source(s) from where the material was adopted can be obtained from the reference section of the book.

A. **Persons/Scientists**

Acevedo-Garcia Johanna
Adam L.
Agrawal Ranjana
Ali Sajad
Alkooranee Jawadayn Talib
Allen R.L.
Ausubel F.
Barbetti M.
Braun U.
Chattopadhyay C.
Choi Hyong Woo
Ciola Victoria Lyn
Cook R.T.A.
Dange S.R.S.
Desai A.G.
Frye Catherine A.
Gollner K.
Griffiths Phillip D.
Hirata T.
Hoewing Timo
Innes Roger W.
Koch E.

Kohire O.D.
Kuhn Hannah
Kumar A.
Lipka Ulrike
Liu Simu
Lomate C.B.
Mert-Turk Figen
Micali C.O.
Pandey S.P.
Panstruga Ralph
Petkova M.
Plotnikova Julia M.
Quentin Michaël
Ramonell K.M.
Reuber T.L.
Slusarenko A.J.
Somerville S.
Somssich I.
Uloth M.B.
Vellios Evangelos
Vogel John P.
Weis Corina
Weßling Ralf
Xiao Shunyuan
Yarwood C.E.

B. **Journals**

Agricultural Science and *Technology*
Botanica Helvetica
CBS Biodiversity Series
Cellular Microbiology
Crop Protection
Current Opinion in Plant Biology
Frontiers in Plant Science
International Journal of Advanced Research
International Journal of Plant Protection
Journal of Experimental Botany
Journal of Oilseed Brassica
Journal of Plant Diseases and Protection
Mausam
Molecular Plant-Microbe Interactions
Molecular Plant Pathology
Mycological Research
Mycological Society of America
Mycopathologia

New Phytologist
Nova Hedwigia
Plant Breeding
The Plant Cell
Plant Disease Research
Plant Methods
Plant Pathology
Plant Pathology Journal
PLOS One
Proceedings of the National Academy of Sciences of the United States of America
Scientific Reports
The Arabidopsis Book
The Plant Journal
Trends in Biosciences Journal
World Applied Sciences Journal

Websites

http://www.ncbi.nl.nih.gov/
www.genevestigator.com
https://www.genevestigator.ethz.ch/andATGenExpress (http://www.arabidopsis.org/info/ expression/ATGenExpress.jsp)

Publishers

Academic Journals
Blackwell Science Ltd
CABI
CSIRO Publishing
Elsevier
John Wiley & Sons Inc.
Kluwer Academic Publishers
Oxford Academic
Springer
Taylor & Francis Group

Institutions

Agriculture and Agri-Food Canada, Saskatoon Research and Development Centre, Saskatoon, Canada
Canadian Phytopathological Society
CCS Haryana Agricultural University, Hisar, India
Global Council for Innovation in Rapeseed and Canola (GCIRC)
ICAR-Directorate of Rapeseed-Mustard Research, Bharatpur, India
Indian Council of Agricultural Research, India
Indian Phytopathological Society
Indian Society of Mycology and Plant Pathology
International Development Research Centre, Ottawa, Ontario, Canada

Mycological Society of America
University of Arizona, School of Plant Sciences, USA
Society for Rapeseed-Mustard Research
The American Phytopathological Society
The Australasian Plant Pathology Society
The British Society for Plant Pathology
The Korean Society of Plant Pathology

Databases

MycoBank, International Mycological Association
Systematic Mycology and Microbiology Laboratory Fungal Database, US Department of Agriculture

Contents

About the Authors

Prof. (Dr.) Govind Singh Saharan, former Professor and Head, Department of Plant Pathology, retired from the active service in 2002. He did his B.Sc. Agriculture (1965) and M.Sc. Agriculture (1967) from SKN College of Agriculture, Jobner, University of Udaipur, and Ph.D. (1977) from Himachal Pradesh University, Palampur, India. He served as Lecturer (1967–1976) and Assistant Professor (1976–1980) at HPKV, Palampur, and as Associate Professor (1980–1988), Professor (1988–2002), and Professor and Head (2002) at the Department of Plant Pathology, CCS Haryana Agricultural University, Hisar. He has been a Visiting Professor at the Department of Plant Sciences, University of Alberta, Edmonton, Canada (1991 and 1994); Agriculture and Agri-Food Canada, Saskatoon Research and Development Centre, Saskatoon, Canada (1991, 1994, 1997); and Rothamsted Research, IACR, Harpenden, UK (1994 and 1997).

Dr. Saharan has more than 250 research publications in journals of national and international repute. He has been Editor of books, such as *Diseases of Oilseed Crops*, *Annual Review of Plant Pathology*, *Phytopathological Techniques*, *Plant Pathology at a Glance*, and *Plant Pathological Research Problems and Progress*, and Author of books such as *Diseases of Oilseed Crops* (in Hindi); *Sclerotinia Diseases of Crop Plants: Biology, Ecology and Disease Management*; *White Rust of Crucifers: Biology, Ecology and Management*; *Alternaria Diseases of Crucifers: Biology, Ecology and Disease Management*; and *Downy Mildew Disease of Crucifers: Biology, Ecology and Disease Management.* He has authored monographs on white rust, *Alternaria* blight, and downy mildew diseases of rapeseed-mustard, including 5 bibliographies and 40 review articles in books. He is on the Panel of Experts of State Agricultural Universities, Indian Council of Agricultural Research, Central Scientific and Industrial Research, University Grant Commission, and Department of Biotechnology in India. He has contributed immensely in the preparation and release of Crop Protection Compendium (2002), CAB International, UK. He has guided three M.Sc. and eight Ph.D. students who are well-established scientists at different universities and research organizations in India and abroad.

Dr. Saharan has conducted research in diverse fields of plant pathology including standardization of artificial inoculation techniques, identification of sources of resistance, determination of pathogenic variability, genetics of host-parasite interaction, epidemiology, and management of several diseases. He has been President (North Zone) of the Indian Phytopathological Society (2001), Editor-in-Chief of the *Journal of Mycology and Plant Pathology* (1999–2000) and *Journal of Oilseed Brassica* (2012 to date), and President of the Indian Society of Mycology and Plant Pathology (2009). He has also played a major role in the organization of the global and Asian congress by the leading Phytopathological Societies of India. He has been Member of QRT, ICAR, New Delhi, for the Soybean (2010) and Rapeseed-Mustard (2015). He has been awarded with Y. L. Nene Outstanding Plant Pathology Teacher Award (2015) by the Indian Society of Mycology and Plant Pathology, Udaipur, India. He has been bestowed with Life Time Achievement Award (2017) for his outstanding research leadership and expertise in Oilseed Brassica Research by the Society for Rapeseed-Mustard Research, Bharatpur, India.

Prof. (Dr.) Naresh Kumar Mehta completed his B.Sc. Agriculture (Hons.) in 1978 and M.Sc. Plant Pathology in 1980 from Haryana Agricultural University, Hisar. He obtained his Ph.D. degree in 1993 in Plant Pathology from CCS HAU, Hisar, under the guidance of Dr. G. S. Saharan. During his study, he attained his first position in elective Plant Protection at Bachelor's degree level. He is the recipient of Excel Industries Ltd., Bombay, India, Award as Research Fellowship for Master's degree program and "Senior Research Fellowship" (SRF) Award to pursue his Ph.D. program by the Council of Scientific and Industrial Research (CSIR), New Delhi. He was awarded Ms. Manju Utereja Memorial Gold Medal for best Ph.D. thesis (1993–1994).

Dr. Mehta joined as Assistant Scientist (Plant Pathology) in 1981, Scientist/Associate Professor in 1994, and Professor in 2002 at CCS, HAU, Hisar. He was Co-Principle Investigator in the scheme "Pathogenic variability and epidemiology of Alternaria Brassicae" funded by ICAR, New Delhi.

He has been teaching Plant Pathology courses to undergraduate and postgraduate students. He has guided five M.Sc. (Plant Pathology) and two Ph.D. students and was Member of many students' advisory committees. He is the Recipient of Best Poster Paper Award for the year 2005 by the INSOPP and Indian Phytopathological Society (NZ). His students have been awarded P.R. Verma M. Sc. Student Thesis Award for the year 2009 by the Indian Society of Mycology and Plant Pathology and M.J. Narasimhan Academic Awards (NZ) by Indian Phytopathological Society, New Delhi, for the year 2010.

He has conducted research in diverse fields of plant pathology covering pathogenic variability, genetics of host-pathogen interaction, epidemiological studies, identification of resistant sources, biochemical/genetic basis for resistance, residual analysis of fungicides, and disease management.

Dr. Mehta was Member of Expert Committee, UGC, New Delhi, for 12 B status for The Gandhigram Rural Institute-Deemed University, Gandhigram, Tamil Nadu.

Dr. Mehta is one of the editors of the book entitled *Diseases of Oilseed Crops* published by Indus

Publishing Co., New Delhi, and one of the authors of three books, i.e., *Sclerotinia Diseases of Crop Plants: Biology, Ecology and Disease Management*, *Alternaria Diseases of Crucifers: Biology, Ecology and Disease Management*, and *Downy Mildew Disease of Crucifers: Biology, Ecology and Disease Management* published by Springer. He has published more than 100 research papers in the journals of national and international repute. In addition, 11 review articles, 20 book chapters, 10 practical manuals, 26 lead lectures in the conferences, 91 research paper presentations in the conferences, 35 popular articles, and 13 radio/TV talks are to his credit.

Dr. Mehta is Fellow of the Indian Phytopathological Society (FPSI), New Delhi; the Indian Society of Plant Pathologist (FINSOPP), Ludhiana; and the Indian Society of Mycology and Plant Pathology (FISMPP), Udaipur. He has been on the editorial board of the Indian Phytopathological Society (2012–2013; 2017–2019), Councilor (North Zone) of ISMPP (2005, 2011), Member of the editorial board (2012–2014), and Editor-in-Chief during 2014. He is also Member of the editorial board of the Indian Society of Plant Pathologist, Ludhiana (2017–2018).

Dr. Mehta has been a Visiting Scientist at the University of Alberta, Edmonton, Canada, in 1999 and as a FAO Fellow and presented a research paper in 8th International Congress of Plant Pathology at Christchurch, New Zealand, in 2002. He was invited to deliver lectures in the 9th International Congress of Plant Pathology at Torino, Italy, in 2008, and in the 5th International Conference on Plant Pathology "Plant Pathology in the Globalized Era," New Delhi, in 2009. He has delivered lead lecture in the 3rd Global Conference "Plant Pathology for Food Security" during 2012 with several lead lectures in the national conferences held from time to time.

Dr. Prabhu Dayal Meena is working as Principal Scientist (Plant Pathology) at the ICAR-Directorate of Rapeseed-Mustard Research, Bharatpur, Rajasthan, India. He started his carrier in the Indian Council of Agricultural Research in 1989 as a Senior Technical Assistant at Central Soil and Water Conservation Research and Training Institute, Dehradun. He obtained his B.Sc. Agriculture (1987) from the University of Rajasthan, Jaipur; M.Sc. Agriculture and Plant Pathology (1997) from Rajasthan Agricultural University, Bikaner; and Ph.D. in Botany (2005) from the University of Rajasthan, Jaipur. He has developed garlic bulb aqueous extract (2 % w/v) as botanical product for control of *Sclerotinia* rot and *Alternaria* blight diseases of mustard. He has identified, namely, white rust resistance genotypes NRCDR 515, DRMR 2019, DRMR 2035 and involved in the development of NRCDR-02, NRCHB-506, NRCHB-101, NRCDR-601 cultivars of *Brassica juncea*, and NRCYS-05-2 of *B. rapa* ssp. yellow sarson.He developed weather-based forecasting models for rapeseed-mustard diseases. He has also developed screening methods for different diseases of oilseed Brassica. Dr. Meena published more than 85 research papers, 5 reviews, and 13 book chapters in international and national reputed refereed journals and is also one of the authors of the 6 books entitled *Principles of Plant Breeding, White rust of crucifers: Biology, Ecology and Management; Alternaria blight of crucifers: Biology, Ecology and Management; Downy Mildew Disease of Crucifers: Biology, Ecology and Disease Management; Brassica Oilseeds Breeding and Management; and Climate Change and Sustainable Agriculture*. He has been a Member of Monitoring Team for All India Coordinated Research Project on Rapeseed Mustard during 2004–2019.

Dr. Meena honored as a Fellow of the Indian Society of Mycology and Plant Pathology and Fellow of the Plant Protection Association of India and has also been awarded with Dr. P.R. Kumar Outstanding Brassica Scientist Award in 2011, Brassica Gold Medal 2019 by the Society for Rapeseed-Mustard Research. He has served as Councilor of the Plant Protection Association of India, the Founder Secretary of the Society for

Rapeseed-Mustard Research since 2008, and Managing Editor (2012–2019) for the *Journal of Oilseed Brassica*.

Dr. Meena has been a Principal Investigator (PI) and Co-PI for ICAR-Outreach Programme on Diagnosis and Management of Leaf Spot Diseases in Field and Horticultural Crops (2009–2013), ICAR-Network Project on Transgenics in Crops (Functional Genomics Component for *Alternaria* and Drought), and National Network for Management of *Alternaria* blight in *Brassica juncea* and Vegetable Crops (2004–2008), All India Coordinated Research Project on Rapeseed-Mustard (2017 to continue).

He undertook 3 months of research attachment training (2007) at the Rothamsted Research, Harpenden, UK, under Indo-UK Collaborative Research on Oilseed Brassica crops. He has supervised nine M.Sc. students and co-supervised one Ph.D. student.

Abbreviations

%	Percent
/	Per
@	At the rate of
~	Tilde
<	Less than
=	Is equal to
>	More than
≥	Greater than or equal to
μl	Microliter
μm	Micrometer
μmol	Micromoles
4M13G	4-methoxy-indol-3-ylmethyl-glucosinolate
a.i.	Active ingredient
AAA- ATpase	ATpase associated with diverse cellular activities
ABA	Abscisic acid
ABC	ATP-binding cassette
ADFs	Actin depolymerizing factors
AICRPRM	All India Coordinated Research Project on Rapeseed and Mustard
ALD	Aminotransferase AGD2-like defense response protein
AM fungi	Arbuscular mycorrhizal fungi
ANN	Artificial neural network
ARF-GAP	ARF-GTPase-activating protein
ARF-GEF	ADP ribosylation factor-GTP exchange factor
At	*Arabidopsis thaliana*
At STP	*Arabidopsis* sugar transport protein
ATAF	*Arabidopsis thaliana* activating factor
ATG	Autophagy-related gene
ATL	*Arabidopsis* toxicos en levadura
AUDPC	Area under the disease progress curve
Avp. E	Evaporation evening

Avp. M	Evaporation morning
Avr	Avirulence
BC	Backcross
BC/Ratio	Benefit-cost ratio
BDM	2,3-butanedione monoxime
BEC	Bgh effector candidate
Bgh	*Blumeria graminis* f. sp. *hordei*
bHLH	Basic helix-loop-helix
BI-1	Bax inhibitor -1-(endoplasmic reticulum-resident cell death suppressor)
BjNPR	*Brassica juncea* NPR
CA	Constitutively activated
CAM	Calmodulin
CAPS	Cleaved amplified polymorphic sequence
CCaMKs	Calcium/calmodulin-dependent protein kinases
CD	Critical difference
CDC	Cell division control protein
cDNA	Complementary DNA
CDPK	Calcium-dependent protein kinase
CEP	Constitutive expression of protein
CEP	Cysteine endopeptidase
CERK	Chitin elicitor receptor kinase gene
CESA	Cellulose synthase
CEV	Constitutive expression of VSP
CF	Culture filtrate
CFU	Colony-forming unit
CH_4	Methane
CHI	Chitinase
CHIP	Chromatin immunoprecipitation
cm	Centimeter
CML	Calmodulin-like
CO_2	Carbon dioxide
COI	Coronation-insensitive protein
Conc.	Concentration
CPR	Constitutive expression of PR genes
CV	Coefficient of variation
cv.	Cultivar
cvs.	Cultivars
CWAs	Cell wall apposition
CY	Cytochrome
Cys EPs	Cysteine endopeptidases
DA	Ubiquitin receptor (DA is Chinese for large)
dai	Days after inoculation
DAMPs	Damage-associated molecular patterns
DAPG	2-4-Diacetyl-phloroglucinol

DAR	DA-related
Dec.	December
DEL1	Loss of function mutation of DP-E2F-like 1
diam	Diameter
dpi	Days post-inoculation
DR	Disease reaction
DSI	Disease severity index
e.g.	For example
EC	Emulsifiable concentrate
EDR	Enhanced disease resistance
eds	Enhanced disease susceptibility
EHM	Extra-haustorial membrane
EIN	Ethylene-insensitive
ER	Endoplasmic reticulum
ERF	Ethylene response factor
ET	Ethylene
et al	*et* alia
ETI	Effector-triggered immunity
ETL	*Economic threshold level*
ETR	Encoding a transmembrane protein kinase with a LRR domain
f. sp.	Fungal forma specialis
F1 GHs	Family-1 glycoside hydrolases
FLS	Flagellin-triggered signaling
FS	Foliar spray
FYM	Farmyard manure
GAAP	Golgi anti-apoptotic protein
GAPDH	Glyceraldehyde 3-phosphate dehydrogenase
GBP	Glutamate-binding protein
Gc	*Golovinomyces* (Syn. *Erysiphe*) *cichoracearum*
GFP	Green fluorescent protein
GHGs	Greenhouse gases
gm	Gram
Go	*Golovinomyces* (Syn. *Erysiphe*) *orontii*
GRX/ROX	TGA-interacting glutaredoxin
GSL	Glucosinolate
GSL	Glucan synthase-like
GSNOR	S-nitrosoglutathione reductase
Gy	Gyro
ha	Hectare
hpi	Hours post-inoculation
hr	Hours
HR	Hypersensitive response
HS	Highly susceptible
i.e.	That is
IAA	Indole-3-acetic acid

IAN	Indole-3-acetonitrile
IAOx	Indole-3-acetaldoxime
ICM	Integrated crop management
ICN	International Code of Nomenclature for algae, fungi, and plants
ICS	Isochorismate synthase
IDM	Integrated disease management
INR	Indian Rupees
IPM	Integrated pest management
IR	Induced resistance
ISR	Induced systemic resistance
ITS	Internal transcribed spacer
JA	Jasmonic acid
JAZ	Jasmonate ZIM-domain protein
kb	Kilobyte
KDEL-CysEPs	C-terminal KDL endoplasmic reticulum retention signal with cysteine endopeptidases from Castor bean
KEG	Keep on going
km	Kilometer
L.s.d.	Least significant difference
LFG	Life guard proteins
LIN	Lesion initiation
LM	Light microscopy
LYK	Lysine motif receptor-like kinase
LysM RLK	Receptor-like kinase gene
M	Mutant
MAPKK	Mitogen-activated protein kinase kinase
MAMPs	Microbe-associated molecular patterns
MAPE	Mean absolute percentage error
MAPK	Mitogen-activated protein kinase
MAPs	Microtubule-associated proteins
Max.	Maximum
Mbp	Genome size megabase pair
MGH	Massachusetts General Hospital isolate of powdery mildew
min	Minute
Min.	Minimum
ml	Milliliter
MLO	Mildew resistance locus
MLOs	Mycoplasma-like organisms
MLP	Multilayer perception
mm	Millimeter
MR	Moderately resistant
MS	Moderately susceptible
MT	Microtubule
MYB	Myeloblastosis
N/A	Not available

N_2O	Nitrous oxide
NB-LRRs	Nucleotide binding site-leucine-rich repeats
NDR	Non-race-specific disease resistance
NEM	*N*-ethylmaleimide
ng	*Nanogram*
NHR	Nonhost resistance
NNM	Neural network models
Nov.	November
NPR	Non-expression of PR genes
NS	Nonsignificant
^{o}C	Degree Celsius
Oct.	October
OECs	Go effector candidates
ORA	Octadecanoid-responsive *Arabidopsis*
OXLP	Oxalate oxidase-like protein
PAD	Phytoalexin-deficient
PAMPs	Pathogen-associated molecular patterns
PAPP	Phytochrome-associated protein phosphatase
PCD	Programmed cell death
PCR	Polymerase chain reaction
PDF	Plant defensin factor
PDI	Per cent disease intensity
PDI	Percentage disease incidence
PDR	Pleiotropic drug resistance
PEN	Penetration gene
PLT	Pyoluteorin
PM	Powdery mildew
PMR	Powdery mildew-resistant
PR	Pathogenesis-related
PR proteins	*Pathogenesis-related* (*PR*) *proteins*
PRRs	Pattern recognition receptors
PTI	Pattern-triggered immunity
PUB	Peptide-N-glycanase/UBA- or UBX-containing protein
PUX	Plant UBX domain-containing protein
Pv.	Pathovar
q	Quintal
qPCR	Quantitative PCR
R	Resistant
R^2	Coefficient of determination
Rab	Ras-related to brain
RAPD	Random amplification of polymorphic DNA
RAR	Encoding a protein with two zinc finger-like domains required for accumulation of many proteins
RBF	Radial basis function
rDNA	Ribosomal DNA

repetitive DNA	Repeated *sequences*
RFLP	Restriction fragment length polymorphism
RFP	Red fluorescent protein
R-genes	*Resistance genes*
RH	Relative humidity
RHE	Relative humidity evening
RHM	Relative humidity morning
RLCKs	Receptor-like cytoplasmic kinases
RLK	Receptor-like kinase
RLK	Receptor-like kinase gene
RNAi	RNA interference
ROP	Rho (RAS homologous) of plants
ROS	*Reactive oxygen species*
RPW	Resistance to powdery mildew
RRTF	Redox-responsive transcription factors
Rs.	Rupees
S	Susceptible
s. str.	Sub-strain
SA	*Salicylic acid*
SAE	Sinapic acid esters
SAG	Senescence-associated gene
SAR	Systemic acquired resistance
SAS	Statistical Analysis System
SEM	Scanning electron microscopy
SEm	Standard error
Si	Silicon
SID	Salicylic acid induction-deficient
SNAP	Soluble N-ethylmaleimide-sensitive factor adaptor protein
SNAP	Synaptosomal-associated protein
SNAREs	Soluble N-ethylmaleimide–sensitive factor attachment protein receptors
SR	Signal-responsive
ssh	Sunshine hours
SSLP	Simple sequence length polymorphism
ST	Seed treatment
START	Steroidogenic acute regulatory protein-related lipid transfer
syn.	Synonyms
SYPs	Syntaxin of plants
T.	Temperature
TCP	Teosinte branched/cycloidea /proliferating cell factor
T-DNA	Transferred DNA
TF	Transcription factor
TGA	Transcription factors (binding to a TGACG nucleotide motif)
TGN	*trans*-Golgi network
TH	*Trichoderma harzianum*

TIGS	Transient-induced gene silencing
TIR	Toll/interleukin-1 resistance
TMBIM	Transmembrane Bax inhibitor motif
UBC2	Ubiquitin-conjugating enzyme 2
UBQ5	Ubiquitin 5
UBX	Ubiquitin regulatory X
UCSC1	An isolate of *E. cichoracearum*
UEA1	An isolate of *E. cruciferarum*
UK	United Kingdom
USA	United States of America
VAMPs	Vesicle-associated membrane proteins
viz.	In other words
w/v	Weight/volume
wai	Weeks after inoculation
WI	Weather indices
WP	Wettable powder
WRKY	Transcription factor containing a highly conserved WRKY domain
WT	Wild type

List of Figures

List of Plates

List of Tables

Chapter 1
Powdery Mildew Perspective

1.1 Introduction

Powdery mildews (*Erysiphales*) are a very large group of biotrophic pathogens which grow principally on the aerial parts of angiosperms including crucifers and cause qualitative and quantitative damage on a wide variety of crops. Throughout, more than 250 years since Linnaeus first gave a scientific name to a powdery mildew, they have figured prominently in the history of plant pathology. The white, floury superficial mycelium, conidiophores, conidia, and pinhead dark brown chasmothecia (cleistothecia) of powdery mildews make them among the most easily recognized and best known plant pathogens. Their luxuriant growth, and development in dry weather, the high water content of their large turgid airborne conidia, the compatibility with their hosts, and their vulnerability to chemical control make them distinct from other plant pathogens. The taxonomy and identification of powdery mildews are based on the characteristics of teleomorph (cleistothecial appendages, number of asci per cleistothecium) and anamorphs morphology of conidial stages on host including host range data, etc. With the use of scanning electron microscopy (SEM), light microscopy (LM), host range, and phylogenetic analysis, identification keys have been constructed for different powdery mildews genera and species infecting different crops including crucifers. On crucifers, earlier powdery mildew pathogenic species were recorded as *Erysiphe communis*, *E. cichoracearum,* and *E. polygoni*. Now, with latest authentic nomenclature, and identification criteria, major powdery mildew of crucifers is caused by *E. cruciferarum* under field conditions all over the world. However, on *Arabidopsis* under artificial inoculation conditions, non-adapted species of powdery mildew (not reported under natural conditions) like *Golovinomyces* (syn. *Erysiphe*) *cichoracearum*, *G. orontii,* and *Oidium neolycopersici* have been used for molecular and genetical studies. Powdery mildews can take epidemic form under field conditions within a very short period of time if congenial weather conditions prevail after landing of primary inoculum on the host surface. Powdery mildews have been managed through cultural, chemical, biological, and host resistance on different crops.

G. S. Saharan et al., *Powdery Mildew Disease of Crucifers: Biology, Ecology and Disease Management*, https://doi.org/10.1007/978-981-13-9853-7_1

1.2 Status of Powdery Mildews

The powdery mildews constitute a very large group of biotrophic fungi belonging to the *Ascomycotina* division and Erysiphaceae family of *Mycota* phylum. The members of this group of fungi are obligate parasites causing disease in various angiosperms. In fact, most of these are mainly pathogenic to the dicotyledons, with exception of *Erysiphe graminis* synonym *Blumeria graminis* which is pathogenic on monocotyledons on the members of the grass family. Initially, various common names have been applied to this group of fungi, as mildew, white mildew, blight, and powdery mildew. The latter name probably has a significance based on the kind of symptoms these group of fungi cause on various host plants. The diseased plants can be recognized from a distance on the basis of white, cottony, powdery growth of conidia and conidiophores of the fungus on the infected leaves, stems, buds, inflorescence, and fruits of field crops, fruit crops, ornamental plants, shrubs, forest trees, and several weeds.

The powdery mildew fungi are characterized by the possession of two distinct fruiting stages. One, the asexual or conidial stage produced on conidiophores resulting in the formation of a large number of white conidia on the diseased host plants which are easily distributed by air currents to act as a resource of secondary inoculum. Later in the season, second stage of fruiting bodies is produced in the form of sexual spores, ascospores inside an ascus, and ascocarp (perithecia, cleistothecia, chasmothecia) called perfect stage. These fruiting bodies are pinhead sized, small, and more or less globose at first and have yellow brown, dark brown, brown, or black structures without differentiated openings for the escape of the asci and ascospores. These ascocarps (perithecia or cleistothecia or chasmothecia) are provided with characteristic appendages which are outgrowth of outer layer of cells of the perithecial wall. Appendages are filamentous, branched, and bulbous at base and have curved or pointed tips, and their characteristics have a great value in classification and identification of Erysiphaceae fungal genera. The asci are sac-like structures, more or less oval in shape, and at maturity contain two to eight ascospores depending upon the fungal species. This stage serves as a source of survival of pathogen called over summering or overwintering. Several classical publications have explained powdery mildew host pathosystem with details of infection process; pathogenesis; morphological features like production of hyphae, mycelium, haustoria, conidia, and conidiophores; asexual and sexual reproduction; development of perithecia; nuclear processes; biologic specialization; economic importance; pathological effects; collection; preservation; and cultivation of powdery mildews in laboratory conditions including taxonomy, classification, and nomenclature (Smith 1900; Salmon 1900, 1903a, b, 1904a, b, c, 1906; de Bary 1863, 1871; Harper 1895, 1896, 1905; Galloway 1895; Berkeley 1847; Coudarc 1893; Neger 1902, 1903; Marchal 1902; Reed 1905, 1907, 1908, 1909, 1912, 1913; Steiner 1908; Voglino 1905; Stewart 1910; Melhus 1912). Yarwood (1957) wrote a very comprehensive review has covered powdery mildew's morphology, taxonomy, and nomenclature, host range, economic importance, symptomatology, and life history including inoculation, infection, sporulation, dissemination, artificial inoculation, occurrence of perithecia, perithecium formation, sexuality, functions of appendages, maturation of perithecia, overwintering, bud infections, diurnal cycle, and obligate parasitism in addition to biologic specialization, host–parasite relationship, and epidemiological factors like moisture, temperature, light, atmospheric pressure, smoke, air circulation, and soil fertility. The control measures like exclusion, eradication, protection, biological control, resistant varieties, and inheritance of resistance determined by dominant or recessive genes of the pathogen.

The taxonomic studies in genus *Erysiphe* with respect to generic delimitation and position in the system of the Erysiphaceae have been described in detail by Braun (1981). Later, a very comprehensive monograph on powdery mildew, during 1987, described very nicely the details of powdery mildews with illustrations. The latest taxonomic manual of the *Erysiphales* (powdery mildews) by Braun (2012) covers the taxonomy of all powdery mildew fungi along with phylogenetic relationships, conidial germination, conidia as viewed by scanning electron microscopy (SEM), fossil powdery mildews, and holomorph classification. The treatment of the *Erysiphales*, its tribes, and genera is based on recent molecular phylogenetic classifications. A key to the genera (and sections), based on teleomorph and anamorph characters, is provided, supplemented by a key solely using anamorph features. Keys to the species are to be found under the particular genera. A special tabular key to species based on host families and genera completes the tools for identification of powdery mildew taxa. In total, 873 powdery mildew species are described and illustrated in 853 figures (plates). The additional data are given for the particular species and sub-specific taxa such as bibliographic data, synonyms, references to descriptions, and illustrations in literature, full descriptions, type details, host range, distribution, and notes. More than 236 taxonomic novelties are introduced, comprising the new genus *Takamatsuella*, 55 new species, 4 new varieties, 6 new names, and 170 new combinations. A list of excluded and doubtful taxa with notes and their current status is attached, followed by a list of references and a glossary. In total, more than 873 powdery mildew species have been reported to cause infection to nearly 10,000 plant species all over the world (Braun 1981, 1987, 2012; Kuhn et al. 2016). Powdery mildews are considered as one of the top ten fungal pathogens in molecular plant pathology studies (Dean et al. 2012).

1.3 Powdery Mildew of Crucifers

Since 1886 (Riley 1886), the term ‘powdery mildew’ has been used for the members of *Erysiphales* and term ‘downy mildew’ to the *Peronosporales*. Powdery mildews were recognized and named by Linnaeus as early as 1753 (Salmon 1900). The historical account of powdery mildew research between 1753 and 1900 has been reviewed by Salmon (Salmon 1900). Several classical monographic treatments to powdery mildews have been published from time to time to update our knowledge on this important group of pathogenic fungi which have been strengthened by various workers (Smith 1900; Reed 1913; Klika 1924; Jorstad 1925; Skoric 1927; Sawada 1927; Jaczewski 1927; Brundza 1933; Blumer 1933, 1967; Homma 1937; Golovin 1956; Yarwood 1957, 1978; Moseman 1966; Junell 1967; Hirata 1966, 1976; Spencer 1978; Zheng and Chen 1980; Zheng 1985; Braun 1981, 1987, 1995, 2012; Boesewinkel 1980; Ellis and Ellis 1997, 1998; Cook et al. 1997; Braun et al. 2002; Braun and Cook 2012; Belanger et al. 2002; Glawe 2008; Micali et al. 2008; Kirk et al. 2008; Kuhn et al. 2016). A splendid overview of *Erysiphales* has been provided by Braun et al. (2002) with scanning electron micrographs as well as numerous, detailed line drawings of micromorphological features. Crucifer's powdery mildew has gained its importance after 1980 (Saharan and Kaushik 1981) especially on oil yielding and vegetable crops. Most of the earlier reports are of mycological interest rather than pathological aspect of the disease. In epidemic conditions, the disease may cause yield losses ranging from 10% to 90% in oilseed crops (Saharan and Sheoran 1985; Hare 1994; Hingole 1995; Kohire et al. 2008; Mehta et al. 2008). The disease is distributed all over the world wherever crucifer's crops are grown. However, its nature and amount of losses have been assessed only from few countries, viz. India (Saharan and Sheoran 1985; Mehta et al. 2008), France (Penaud 1998), and Australia (Uloth et al. 2017). Powdery mildew of crucifers has received less attention of researchers because of their preoccupation with other diseases (white rust, *Alternaria* blight, downy mildew, *Sclerotinia* rot, wilts, and root rots) since their occurrence is early in the growth period of crucifers and damage is visible at the first sight. Secondly, powdery mildew appears at later stages of plant growth when temperature ranges between 20 and 25 °C, and its damage is assessed only after the harvest of the crop. Once the host gets infection, its spread is very fast under congenial environmental conditions, and it takes epidemic form very soon (Saharan et al. 2005, 2014, 2016, 2017; Saharan and Mehta 2008; Meena et al. 2014). Under natural conditions, powdery mildew of cultivated and wild crucifers is caused by *Erysiphe cruciferarum,* whereas under controlled artificial inoculation conditions, four species of powdery mildew pathogen, viz. *E. cruciferarum*, *Golovinomyces* (syn. *Erysiphe*) *cichoracearum*, *G. orontii,* and the tomato powdery mildew pathogen *Oidium neolycopersici*, have been used for molecular studies using *Arabidopsis thaliana* under powdery mildew host pathosystem.

1.4 Economic Importance

It is a common belief that powdery mildews cause less spectacular losses than those caused by several other important groups of plant pathogens, such as MLOs, viruses, downy mildews, rusts, smuts, and root-rotting fungi. Some characteristics of powdery mildew infections which may account for this are their lack of systemic infections, lack of root infections, relatively slow increase during the season, infrequent death of the host, and ease and efficiency of control methods. Powdery mildews usually cause chronic infections which appear in moderate amounts every year, in contrast to such diseases as late blight of potatoes and downy mildew of hops which induce little if any losses during most years but bring about heavy losses in occasional years. The most widespread and disastrous losses attributable to a powdery mildew were on grapes in France, where the yield of wine decreased from about 45 million hectoliters in 1850, in the early stages of the epidemic, to about 10 million hectoliters in 1854 at about the height of the epidemic (Gaumann 1945). Other records of losses resulting from powdery mildew are killing of barley up to 42% reduction in yield of barley (Jensen 1951), gooseberry crop rendered worthless, 61–71% reduction in yield of hops, 33–90% reduction in yield of grapes, up to 80% reduction in yield of peaches, 40% reduction in yield of clover, and 75% reduction in yield of cucumbers (Yarwood 1957). In crucifer's powdery mildew, it has been reported to cause 10–90% reduction in seed yield of *Brassica juncea* along with reduction of oil content up to 7.00% (Samudre 1994; Hare 1994; Hingole 1995; Saharan and Sheoran 1985; Singh and Singh 2003; Kohire et al. 2008; Mehta et al. 2008) under Indian conditions. In France, reduction in seed yield of *B. napus* up to 10–30% has been recorded (Penaud 1998). Other aspects of damage from powdery mildews are their action as allergens (Anderson and Mason 1941) and the injurious effect on silkworms of mildewed mulberry leaves used as feed (Ressencourt 1927).

1.5 Concepts for Names in Powdery Mildews

A new updated taxonomic monograph of the powdery mildews has recently been published (Braun and Cook 2012). Within this group of obligate plant pathogens, clear connections between anamorph and teleomorph genera (e.g. *Blumeria* with *Oidium s. str.*, *Erysiphe* with *Pseudoidium*, *Golovinomyces* with *Euoidium*) are evident and proven by means of morphology and molecular sequence analyses. All anamorph-typified genera are younger than the corresponding teleomorph-typified genera (except for *Oidium*), and hence will be younger facultative synonyms in future, but nevertheless they will remain legitimate and valid. Anamorph genera play an important role in the taxonomy of powdery mildews and reflect phylogenetic relations within this fungal group. Indeed they provided crucial evidence for the recent reclassification of all the holomorph genera. On the other hand, at species level, anamorph species (unlike the anamorph genera) and particularly the conidial stages of powdery mildew species are morphologically often poorly differentiated and of little diagnostic value. Therefore, teleomorphs traditionally prevail in the taxonomy at species level. Hence, in all cases, it is proposed to give preference to teleomorph-typified names when they are threatened by anamorph names. There is only a single generic problem in powdery mildews, viz. the anamorph genus *Oidium* Link 1824, with its type species *Oidium monilioides*, which is the anamorph of *Blumeria graminis*, the type species of the teleomorph genus *Blumeria* Golovin ex Speer 1974. Hence, *Oidium* would be an older name for *Blumeria*, and "*Oidium graminis*" would be the correct name for the powdery mildew of grasses and cereals in future; this is, of course, unacceptable, and *Blumeria* will be proposed as the accepted generic name for this taxon. Most powdery mildew anamorphs are morphologically poorly differentiated at species level, and it is often difficult to truly distinguish separate species in the absence of the teleomorph. However, their relations to teleomorphic genera are almost always clear. Host switches often occur in glass houses, and also in nature, usually connected with anamorph growth but lacking the teleomorph. Even results of molecular sequence analyses are often not helpful here due to a lack of data from other specimens for comparison or other problems. Hence, descriptions of anamorph-typified taxa should be avoided, also in future, but when new descriptions are intended, that should only be based on striking morphological differences combined, if possible, with molecular data, and the taxa concerned should preferably be assigned to the existing anamorph genera. That can also be used in future as it remains legitimate, valid, and available, as already explained. Descriptions of anamorph-typified new species in *Erysiphe*, *Golovinomyces*, *Neoerysiphe,* and other teleomorph-typified genera are in future of course also valid, and in accordance with the *Code*, but that should only be proposed in absolutely clear and molecularly proven cases.

Recently, found powdery mildew anamorph on *Solanum betaceum* (tamarillo or tree tomato) in India is a striking example. This host is phylogenetically closely related to *S. lycopersicum* (tomato), and the anamorph found on tree tomato is morphologically indistinguishable from *Pseudoidium neolycopersici* (syn. *Oidium neolycopersici*) on tomato (Baiswar et al. 2009). Nevertheless, this powdery mildew disease was only recorded as *Oidium* sp. and not as *O. neolycopersici* because reviewers refused the latter denomination without inoculation results and/or molecular analyses. Therefore, cross inoculation tests were later carried out, and the tree tomato powdery mildew was subjected to molecular examinations based on amplification of the rDNA ITS region, including the 5.8S rDNA, but, unfortunately, these new results also failed to elucidate its taxonomy. The powdery mildew on *Solanum betaceum* was unable to infect tomato, and several other species of *Solanum*, but the sequence derived from this powdery mildew differed only in one base pair from that of *Pseudoidium neolycopersici*. Is the tree tomato powdery mildew conspecific with the latter species and only a special form? Or Is it a separate species, morphologically indistinguishable from *P. neolycopersici*, but biologically distinguished, and genetically distinct in one base pair in rDNA ITS sequence data? A final answer cannot yet be given. Incidentally, in this case, a study of the morphology of this pathogen would now allow it to be referred to the morphospecies *Pseudoidium lycopersici* as listed in the updated monograph (Braun and Cook 2012). Now, it is apparent from the above the anamorphic genus *Oidium s. str.* belongs solely to *Blumeria* (Braun 2012).

1.6 Discontinuation of Dual Nomenclature of Pleomorphic Fungi

The symposium 'One fungus = which name' held in Amsterdam in 2012 addressed the drastic changes in the naming of pleomorphic fungi adopted by the 18th International Botanical Congress at Melbourne in 2011. Possible solutions and ways to face resulting problems were suggested. The fundamental change is that under the new rules, fungi in future will be treated nomenclaturally, like plants, and all other groups of organisms ruled by the International Code of Nomenclature for algae, fungi, and plants (ICN), i.e. with one correct name for each species. Numerous discussions and statements during the symposium reflected widespread anxieties that these rules could negatively influence taxonomic work on pleomorphic fungi. However, they are groundless, being based on misunderstandings and confusion of nomenclature and taxonomy. With pleomorphic fungi, taxonomists will in future have to answer the question whether different morphs can represent one fungus (taxon), but this remains a taxonomic decision and has nothing to do with nomenclature. Furthermore, the ICN does not and cannot rule on how this decision is made. Thus, it cannot provide rules based solely on methods involving morphology in vivo or in vitro, molecular analyses, physiological and biochemical data, inoculation experiments in pathogenic groups, or any other methods or their combinations. It is up to the taxonomist to select appropriate methods and to decide which data are sufficient to introduce new taxa. Some future problems, and strategies around the application of anamorph- and teleomorph-typified taxon names (genera and species), have been discussed, using the recent monograph on powdery mildews (*Erysiphales*) as an example (Braun 2012).

1.7 Powdery Mildew as Biocontrol Agent

Due to importance in determining agricultural productivity, the effects of plant pathogens on survival, growth, and reproduction of crop plants have been widely reported (Williamson and Smith 1986). Less studied are the effects of plant pathogens on wild plant populations (Jarosz and Davelos 1995), which can range from a mild depression in individual plant growth rate to widespread mortality, and population declines. The literature also reveals that infection rates and impacts of plant pathogens on agricultural or wild plant populations can vary with environmental conditions, often being more severe in moist environments or during wet years (Jarosz and Davelos 1995). Because herbivores and pathogens may exert top-down control on plant populations, escape from population regulation by herbivores and pathogens can contribute to the dramatic success of invasive plants in novel habitats (Carpenter and Cappuccino 2005). The introduction of biocontrol agents to control invasive plants that have otherwise escaped attack is an approach that has succeeded in some circumstances (Culliney 2005).

Garlic mustard (*Alliaria petiolata*) is an invasive biennial that negatively impacts plant and animal communities throughout North America and lacks significant herbivory in its invasive range. Throughout Ohio, many garlic mustard populations support the powdery mildew fungus *Erysiphe cruciferarum*, although disease incidence varies among populations and environments. Effects of infection on plant growth, as well as both plant and fungal responses to drought and light conditions, were examined on greenhouse-grown, first-year garlic mustard plants. Also, the effects of the fungus on plant growth and fitness were studied in a naturally growing population of second-year plants in the field. Powdery mildew significantly reduced growth of first-year plants in the greenhouse, eventually causing complete mortality. Simulated drought slowed both plant growth and disease development, independent of light conditions. In the field, plants with less disease incidence after their first year grew taller during their second year, producing significantly more siliquae and twice as many seeds as heavily diseased plants did. Seed germination rates did not differ between plants with different levels of disease severity. Consistent reductions in survival, growth, and fitness caused by fungal infection may reduce populations of garlic mustard. These effects may be more evident in moist sites that favour fungal development (Enright and Cipollini 2007).

1.8 Hyperparasites of Powdery Mildews

Hyperparasitism of powdery mildew has been commonly observed during asexual reproduction phase of different powdery mildew genera. The hyphae and conidia of powdery mildews are frequently themselves parasitized by another fungus like *Ampelomyces quisqualis* and *Cicinnobolus cesatii*. This can be detected on the powdery mildew-infected host plants only after constant observation of powdery mildew symptoms with the changing colour and pattern of powdery mildew patches. The hyperparasitized powdery mildew growth colour changes slowly from white to dark brown with the increasing intensity of hyperparasite on its host powdery mildew. The pattern of powdery mildew symptoms changes from circular to irregular patches. The hyperparasitism of powdery mildew by *A. quisqualis* can be detected by its golden-brown pycnidia which appear as net shopping bags under the microscope. This net-like surface ornamentation could lead these pycnidia to be confused with the cleistothecia or chasmothecia of the host powdery mildew, but these are variably ovoid and produce large numbers of small, smooth, bean-shaped conidia. The hyperparasite *C. cesatii* covers the entire powdery mildew growth and is visible with dark brown growth of its pycnidial formation. In early stages of development, the small hyphae of *Cicinnobolus* can be seen ramifying throughout the mycelium and conidiophores of the host powdery mildew fungus. In advanced stage of development of the *Cicinnobolus* mycelium in the invaded conidiophores results in the formation of pycnidia. The pycnidia are globose to pyriform pseudo-parenchymatous, thin-walled, with a definite beak, ostiolate, and olivaceous brown in colour and give rise to numerous hyaline spores in cirrhi. These are embedded in short-celled geniculate, profusely branched mycelial mat and scattered clustered in groups on the powdery mildew growth. The pycnidia measure 68.47–27.68 × 37.35–27.46 (47.54 × 34.97) μm. Conidia are formed from short outgrowth of the cells lining the inner wall of the pycnidium. The conidia are oval, hyaline, single-celled, continuous, and thin-walled and measure 9.19–5.67 × 4.55–3.42 (7.47 × 3.99) μm. From the severely hyperparasitized powdery mildew host, rarely powdery mildew conidia can be detected (Saharan 1976). *Trichothecium roseum* has also shown to possess protective value against powder mildews. However, these hyperparasitic fungi need high relative humidity for their growth and splashing rains for dispersal under natural conditions.

1.9 Impact of Climate Change on Powdery Mildews

Climate change has become a subject of discussion at every platform in scientific gatherings. Its impact on occurrence and severity of diseases including powdery mildews have been analyzed on the basis of historical data of weather variables like temperature, humidity, rainfall, drought period, and wind velocity. These variables have a great bearing on the severity of powdery mildew of crucifers. Powdery mildew of crucifers has gained its importance as major disease after 1980 because of rise in temperature and low relative humidity (Saharan 1980; Saharan and Kaushik 1981) prevailing during these years. In future, the situation may aggravate with increased emission of carbon dioxide (CO_2) and other greenhouse gases (GHGs) predominantly methane CH_4 and nitrous oxide (N_2O) which are described as the main agents causing increase in global temperature. Powdery mildew of crucifers has attracted attention of researchers from number of countries as evident from the reports of its geographical distribution all over the world with a very large host range (see Sects. 2.3 and 2.4, Chap. 2).

1.10 Genetical and Molecular Mechanisms of Crucifer's Powdery Mildew Pathogenesis and Host Resistance

There are a number of events during crucifer's powdery mildew pathogenesis right from the time of host–pathogen contact after landing of pathogen inoculum (mycelium, hyphae, conidia, ascospores) on the surface of host plant (leaf, stem, siliqua). The different events include penetration of host surface, infection of host tissues, and proliferation of host tissues by the pathogen and reproduction/multiplication, as well as expression of pathogen's vegetative and reproductive structures on host surface in the form of symptoms. During pre- and post-penetration stages of powdery mildew pathogenesis, there are induction, and synthesis of several genes, proteins, enzymes, and effector factors which control the mechanism of powdery mildew pathogenesis in crucifers (Table 1.1). These genes modulate the infection process with the suppression of basal defense system of the host, increase powdery mildew pathogen's penetration efficiency, support pathogen penetration and further development, exhibit enhanced susceptibility through enhanced fungal growth, and reproduction. During powdery mildew pathogenesis, variety of molecules (phytohormones) are synthesized which act as signal molecules for pathogenesis. Salicylic acid signaling pathways are induced during early stages of pathogen penetration and

Table 1.1 Genetical and molecular mechanisms of crucifer's powdery mildew pathogenesis

Genes and molecules controlling powdery mildew pathogenesis[a]	Mechanism of action/function
Mildew resistance locus (*MLO2, MLO6, MLO12*)	Modulate infection process and suppress bassal defense
Plant UBX domain-containing protein (*PUX*) 2	Fungal growth and reproduction
Bax inhibitors (B1-1)	Support powdery mildew penetration and development
Lifeguard (LEG) proteins	-do-
Penetration gene (*PEN1*), vesicle-associated membrane protein (VAMP)	Increase powdery mildew penetration
Non-expresser of PR proteins (npr1), phytoalexin-deficient (pad 4) (mutants)	Exhibit enhanced S
nah G (transgenic)	-do-
At receptor-like cytoplasmic kinases (rlck) VIA3 mutant	Enhanced fungal growth
Salicylic acid (SA) signaling pathways	Early stages of penetration and infection
Jasmonic acid/ethylene (JA/ET) signaling pathways	Later stages of pathogenesis

[a]See full form from list of abbreviations

infection. At later stages of powdery mildew pathogenesis, JA/ET signaling pathways are in operation. Detailed description on mechanisms of action and function of genes identified for powdery mildew pathogen's pathogenesis is given in Chap. 2.

An array of r genes have been isolated, identified, cloned, and mapped governing mechanisms of host resistance under artificial inoculation conditions using powdery mildew *Arabidopsis* host pathogen (Ec, Gc, Go) interactions (Table 1.2). The mechanism of action, role, and function of these r genes, PR proteins, elicitors, enzymes, effector genes, regulators, mutants, transcription factors, interacting molecules, and metabolites has been characterized for better perception of disease management strategies. These r genes have multi-functions to confer resistance against powdery mildew of crucifers through synthesis of callose to act as physical barrier, to enhance penetration R, to confer broad-spectrum R through altered cell wall composition (post-penetration R), HR, and SHL, activation of SAR by providing basal R, induction of PCD, elicitation of defense, synthesis of camalexin, regulation of glucosinolates, activation of immune signaling system, and overexpression of defense-related genes. At pre-penetration R, SA signaling pathways are in operation, while at post-penetration R, JA/AT signaling pathways are most common. Detailed descriptions on mechanisms, role, function, and interacting molecules have been given in Chap. 7 (Kuhn et al. 2016; Micali et al. 2008).

Table 1.2 Genetical and molecular mechanisms of crucifer's host resistance to powdery mildews

Genes and molecules controlling host resistance[a]	Mechanism of action/function
PMR4	Callose synthesis a physical barrier
AtL31, RABA4, overexpression	Enhanced penetration R
MLO genes (*mlo2, mlo6, mlo12*) Triple mutant	Confer broad spectrum R through altered cell wall composition (post-penetration R)
RPW8.1, RPW8.2	R through HR or SHL
AtNPR1, AtNPR2	Key regulators of SAR
BjNPR1	Activates SAR to confer broad-spectrum R
AtMLO2, AtMLO6, AtMLO12	Provide basal R
AtROP-regulated *AtRLCK VIA3*	Provide basal R
CPR5	Control R through PCD
Cyp 83 a-1-3 mutant	High levels of camalexin
PAD3, WRKY33	High levels of camalexin for R
SR 1	Regulates EIN 3 and NDR1 expression
Chito-octamer	Elicitor of plant defenses
PEN genes	NHR at preinvasion stage
EDS1, PAD4, SAG101	NHR at post-invasion stage
Glucanases, chitonases, thaumatins, defensins (PR proteins)	Play role in PM resistance and have antifungal characteristics
WRKY transcription factors (*WRKY70*)	Confer R to PM
Overexpression of r genes, *PMR*, MLO, *PEN, EDR, MAPK, MAPK 65-3, NPR1, PAD3, PAD4, ED5, SNARE, RLCKs, KDL*	Confer R to PM
AtCEP1	Provide basal R
KDEL CysEP CEP1	Provide R during post-penetration
MYB 51 TF	Regulates glucosinolate biosynthesis genes
wrky18, wrky40 mutants	Enhanced camalexin
PAMPs	Activate immune responses
PRRs	Activate immune signaling
SA signaling pathways	Pre-penetration R
JA/ET signaling pathways	Post-penetration R
CERK1	Contributes to basal R
Soluble carbohydrate elicitor	Decreases fungal growth

[a]See full form from list of abbreviations

1.11 Exploitation of Non-host Resistance (NHR) to Powdery Mildews

Disease resistance shown by an entire plant species to all genetic variants of non-adapted pathogen species is the most common form of plant immunity known as non-host resistance (NHR). This type of R is very strong and broad-based basal R. At present, two models of NHR are in operation: the first is the absence of adapted pathogen effectors which is durable and basal R. The second is with presence of stacks of multiple R-genes encoding proteins of the NBS–LRR which recognize a number of pathogen-derived avirulence (AVR) genes encoding effector proteins for durable R. The stability of NHR is the consequence of mechanisms comprising of both constitutive (rigid host cell wall, toxic phytoanticipins) and inducible (PRPs, MAP kinase signaling, ROS, ethylene, ion fluxes, PR genes, protein phosphorylation, callose deposition) factors. NHR is due to ineffective pathogen effectors resulting in without suppression of *PTI* and/or *ETI*. Mechanisms of pre-powdery mildew penetration control is through the three penetration genes, viz. *PEN1*, *PEN2*, and *PEN3*, whose products limit the entry success of non-adapted powdery mildews. Other contributing components are S-nitrosoglutathione reductase (*GSNOR1*) and the plant transcription factor *ATAF1*. Post-penetration defense is through post-haustorial NHR which is controlled by enhanced susceptibility (*ESD1*) gene, phytoalexin deficient 4 (*PAD4*) gene, and senescence-associated gene 101 (*SAG101*). The phytohormone abscisic acid (ABA) is involved in NHR against powdery mildews. The use of mutants and gene-silencing approaches in crucifers may open new possibilities to discover useful genes governing mechanisms of NHR breeding strategy in crucifers to powdery mildews. Detailed description is given in Chap. 7 at Sect. 7.8.

References

Anderson VG, Mason LR (1941) Powdery mildews as allergens. Calif West Med 55:241–243

Baiswar P, Braun U, Chandra S, Ngachan SV (2009) First report of an *Oidium* sp.[*Neolycopersici*] on *Solanum betaceum* in India. Aust Plant Dis Notes 4:32–33

Belanger RB, Bushnell WR, Dik AJ, Carver TLW (2002) The powdery mildews: a comprehensive treatise. APS Press, St Paul: 292 ISBN 0–89054–291-0

Berkeley MJ (1847) Gardeners' Chronicle 7:779

Blumer S (1933) Die Erysiphaceen Mitteleuropasunter Besonderer Beriicksichtigung der Schweiz. In Beitr Krypt -Fl Schweiz 7(1):1–483

Blumer S (1967) Echte Mehltaupilze (*Erysiphaceae*). G Fischer, Jena

Boesewinkel HJ (1980) The morphology of the imperfect states of powdery mildews (Erysiphaceae). Bot Rev 46:167–224

Braun U (1981) Taxonomic studies in the genus *Erysiphe* I. Generic delimitation and position in the system of the Erysiphaceae. Nova Hedwigia 34:679–719

Braun U (1987) A monograph of the Erysiphales (Powdery mildews). Beihefte zur Nova Hedwigia 89:1–700

Braun U (1995) The powdery mildews (Erysiphales) of Europe. VEB G Fischer Verlag, Jena

Braun U (2012) The impacts of the discontinuation of dual nomenclature of pleomorphic fungi: the trivial facts, problems, and strategies. Intl Mycol Assoc- Fungus 3(1):81–86

Braun U, Cook RTA (2012) Taxonomic manual of the Erysiphales (Powdery mildews), CBS biodiversity series no. 11. CBS-KNAW Fungal Diversity Centre, Utrecht

Braun U, Cook RTA, Inman AJ, Shin HD (2002) The powdery mildews: a comprehensive treatise. In: Belanger RR, Bushnell WR, Dik AJ, Carver TLW (eds) The taxonomy of the powdery mildew fungi. American Phytopathological Society (APS Press), St Paul, pp 13–55

Brundza K (1933) Beitraige zur Kenntnis der Erysiphaceen. Litauens Jahrb Landwirtschaft Akad Litauen:108–197

Carpenter D, Cappuccino N (2005) Herbivory, time since introduction and the invasiveness of exotic plants. J Ecol 93:315–321

Cook RTA, Inman AJ, Billings C (1997) Identification and classification of powdery mildew anamorphs using light and scanning electron microscopy and host range data. Mycol Res 101:975–1002

Couderc G (1893) Surles peritheces del *Uncinula spiralis* en France etl identification del *Oidium* Americain et del *Oidium* Europeen. Comptes Rendus II 6(210):211

Culliney TW (2005) Benefits of classical biological control for managing invasive plants. Critical Rev Plant Sci 24:131–150

de Bary HA (1863) Uber die Fruchtentwicklung der Ascomyceten. Engelmann, Leipzig

de Bary HA (1871) Beitr Zur Morphol Und Physiol Der Pilze I, XIII And XIV 23-75, 91-95 Pl 9-12 Abhandl D Senckenb Naturf Gesell 7

Dean R, Kan JALV, Pretorius ZA, Hammond-Kosack KE, Pietro AD, Spanu PD, Rudd JJ, Dickman M, Kahmann R, Ellis J, Foster GD (2012) The top 10 fungal pathogens in molecular plant pathology. Mol Plant Pathol 13(4):414–430

Ellis MB, Ellis JP (1997) Micro fungi on land plants, an identification handbook, 2nd edn. Richmond Publishing Co Ltd Slough

Ellis MB, Ellis JP (1998) Micro fungi on miscellaneous substrates. An identification handbook ed. 2. The Richmond Publishing Co LTD, Slough

Enright SM, Cipollini D (2007) Infection of powdery mildew *Erysiphe cruciferarum* (Erysiphaceae) strongly affects growth and fitness of *Alliaria petiolata* (Brassicaceae). Am J Bot 94:1813–1820

Galloway BT (1895) Observations on the development of *Uncinula spiralis*. Bot Gazett 20:486–491

Gaumann E (1945) Cber Seuchenziige bei pflanzliche Infektionskrankheiten (concerned with epidemics of herbal infectious diseases). Exp Dermatol 1:1–12

Glawe DA (2008) The powdery mildews: a review of the world's most familiar (yet poorly known) plant pathogens. Annu Rev Phytopathol 46:27–51

Golovin PN (1956) Morphological review of the genus *Leveillula arnaud* (powdery mildew fungi, family Erysiphaceae) (in Russian). Tr VL Komarov Bot Inst Acad Sci USSR Ser II 10:193–308

Hare RM (1994) Influence of dates of sowing on powdery mildew of rapeseed mustard. M Sc. (Ag.) Thesis, Marathwada Agricultural University, Parbhani, India, pp 1–47

Harper RA (1895) Die Entwickelung des Peritheciums bei *Spherotheca castagnei*. Berichte d d Bot Gesell 13:475–481

Harper RA (1896) Uber das Verhalten der Kerne bei der Fruchtentwickelung einiger Ascomyceten. Jahr f w Botanik 29:656–686

Harper RA (1905) Sexual reproduction and the organization of the nucleus in certain mildews. *Carnegie* Instit of Washington, Publication 37:1–104

Hingole DG (1995) Estimation of yield losses in rapeseed-mustard due to powdery mildew and white rust. M.Sc. (Ag.) Thesis, Marathwada Agricultural University, Parbhani, India, pp 1–52

Hirata K (1966) Host range and geographical distribution of the powdery mildews. Faculty of Agriculture, Niigata University, Niigata

Hirata K (1976) Notes on host range and geographic distribution of the powdery mildew fungi VI distribution of the hosts of powdery mildew fungi in the families of angiosperms. Trans Mycol Soc Japan 17:35–62

Homma Y (1937) Erysiphaceae of Japan. J Fac Agri Hokkaida Univ 38:183–461

Jaczewski AA (1927) Powdery mildew fungi, 626 p [In Russian Abstr Rev Appl Mycol 7:346, 1928]

Jarosz AM, Davelos AL (1995) Effects of disease in wild plant populations and the evolution of pathogen aggressiveness. New Phytol 129:371–387

Jensen NF (1951) Powdery mildew of barley. Cornell Univ Agri Exp Sta Mere 305:39

Jorstad I (1925) The Erysiphaceen of Norway 116 p. Schriften Utgitt av det Norske Videnskaps Akademi i Oslo I Maten Naturvid

Junell L (1967) Erysiphaceae of Sweden. Symbola Botanica Upsaliensis 19:117

Kirk PM, Cannon PF, Minter DW, Stalpers JA (2008) Ainsworth Bisby's dictionary of the Fungi, 10th edn. CAB International, Wallingford. 1–655p

Klika J (1924) Monografie Cesych padli. Masarykova Akad. - Prace, 23. Praze

Kohire OD, Ahmed R, Chavan SS, Khilare VC (2008) Yield losses amongst four varieties of mustard due to powdery mildew in Maharashtra. J Phytol Res 21(2):331–332

Kuhn H, Kwaaitaal M, Kusch S, Acevedo-Garcia J, Wu H, Panstruga R (2016) Biotrophy at its best: novel findings and unsolved mysteries of the *Arabidopsis*-powdery mildew pathosystem. Arabidopsis Book:e018. https://doi.org/10.1199/tab.0184

Marchal E (1902) De la specialization du parasitisme chez *Erysiphe graminis*. Comptes Rendus 135:210–212

Meena PD, Kumar V, Rathi AS, Singh D (2014) Compendium of rapeseed-mustard diseases: identification and management. ICAR-Directorate of Rapeseed-Mustard Research, Bharatpur, India, 30p

Mehta N, Singh K, Sangwan MS (2008) Assessment of yield losses and evaluation of different varieties/ genotypes of mustard against powdery mildew in Haryana. Plant Dis Res 23(1):55–59

Melhus IE (1912) Culturing of parasitic fungi on the living host. Phytopathology 2:197–203

Micali C, Gollner K, Humphry M, Consonni C, Panstruga R (2008) The powdery mildew disease of *Arabidopsis*: a paradigm for the interaction between plants and biotrophic fungi. Arabidopsis Book 6:e0115. https://doi.org/10.1199/tab.0115

Moseman JG (1966) Genetics of powdery mildews. Ann Rev Phytopthol 4:269–290

Neger FW (1902) Beitrage zur Biologie der Erysipheen. Flora 90:221–272

Neger FW (1903) Neue Beobachtungen uber das spontane Freiwerden der Erysipheen-frucht korper. Centrb f Bakt Parasit U Infekt Abstr 2(10):570–573

Penaud A (1998) Oidium du colza, la protection du colza est maintenant possible (Rapeseed powdery mildew, protection of rapeseed is now possible). Oleoscope 50:36–38

Reed GM (1905) Infection experiments with *Erysiphe graminis* DC. Trans Wisconsin Acad Sci Arts Lett 15:135–162

Reed GM (1907) Infection experiments with the mildew on cucurbits, *Erysiphe cichoracearurm* DC. Trans Wisconsin Acad Sci Arts Lett 15:527–547

Reed GM (1908) Infection experiments with *Erysiphe cichoracearum* DC. Bull Univ Wisconsin Sci Ser 3:337–416

Reed GM (1909) The mildews of the cereals. Bull Torr Bot Club 36:353–388

Reed GM (1912) Infection experiments with the powdery mildew of wheat. Phytopathology 2:81–87

Reed GM (1913) The powdery mildews: Erysiphaceæ. Trans Am Microsc Soc 32(4):219–258

Ressencourt (1927) Recherches Sur un Oidium du Murrier (research on an *Oidium* of the Murrier). Bull Econ Indochine 80:41–62. [Abstr in Rev Appl Mycol 6: 519–520]

Riley CV (1886) The mildews of the grape-vine and an effectual remedy for *Peronospora*. Proc Am Pom Soc 5:49–54

Saharan GS (1976) Three new records of hyperparasites on powdery mildew in India. Pesticides 10:52

Saharan GS (1980) Epidemiology of powdery mildew of rape and mustard. Indian J Mycol Plant Pathol (Abstr) 10:XLVI

Saharan GS, Kaushik JC (1981) Occurrence and epidemiology of powdery mildew of *Brassica*. Indian Phytopathol 35:17–21

Saharan GS, Mehta Naresh (2008) Sclerotinia diseases of crop plants: Biology, ecology and disease management. Springer Netherlands LXII: 486p ISBN 978-1-4020-8407-2

Saharan GS, Sheoran BS (1985) Assessment of yield losses of mustard due to powdery mildew disease. Cruciferae Newslet 10:112–113

Saharan GS, Naresh M, Sangwan MS (2005) Development of disease resistance in rapeseed-mustard. In: Saharan GS, Naresh M, Sangwan MS (eds) Diseases of oilseed crops. Indus Publication Co, New Delhi, pp 561–617

Saharan GS, Verma PR, Meena PD, Kumar A (2014) White rust of crucifers: biology, ecology and disease management. Springer, New Delhi. 244p ISBN 978-81-322-1791-6

Saharan GS, Naresh M, Meena PD (2016) Alternaria blight of crucifers: biology, ecology and disease management. Springer, Singapore: 326p, ISBN 978-981-10-0019-5

Saharan GS, Mehta N, Meena PD (2017) Downy mildew disease of crucifers: biology, ecology and disease management. Springer, Singapore: 357p, ISBN 978-981-10-7499-8

Salmon ES (1900) A monograph of the Erysiphaceae. Memoirs of the Torrey Botanical Club 9:1–292

Salmon ES (1903a) Infection powers of ascospores in Erysipheae. J Bot 41(159):204–212

Salmon ES (1903b) On specialization of parasitism in the Erysiphaceae. Beihefte *zum botanischen* Centralblatt 14:261–315

Salmon ES (1904a) On specialization of parasitism in the Erysiphaceae. New Phytol 3:109

Salmon ES (1904b) Mycological notes. J Bot 42:182–186

Salmon ES (1904c) On *Erysiphe graminis* DC and its adaptive parasitism within the genus *Bromus*. Ann Mycologici 2:255–267

Salmon ES (1906) On *Oidiopsis taurica* Lev. An endophytic member of the Erysiphaceae. Ann Bot 20:187–200

Samudre RA (1994) Influence of spacing in powdery mildew disease on four cultivars of rapeseed mustard. M.Sc. (Agri.) thesis, Marathwada Agricultural University, Parbhani India, pp 1–43

Sawada K (1927) On the systematic investigation of Erysiphe in Formosa. Formosa Dept Agri Govt Res Inst Rep 24:55

Singh RB, Singh RN (2003) Management of powdery mildew of mustard. Indian Phytopathol 56(2):147–150

Skoric V (1927) Erysiphaceae croatiae Prilog fitopatoloskosistematskoj monografiji nasih pepelnica Annales pro Experimentis Foresticis, Zagreb, I: 52–118 [Abstr in Rev Appl Mycol 6: 124–125

Smith G (1900) The haustoria of the *Erysiphe*. Bot Gazette 29(3):153–184

Spencer DM (1978) The powdery mildews. Academic, New York

Steiner JA (1908) Die Specialisation der Alchemillenbewohnenden *Sphwrotheca humuli* (DC) Burr Centrall blatt fur Bakt, Parasitenkunde, und Infextious Krankheiten, Abstr II, 21:677–736

Stewart FW (1910) Notes on New York Plant Diseases I. NY Agri Exp Sta Bult 328

Uloth MB, You MP, Barbetti MJ (2017) Plant age and ambient temperature: significant drivers for powdery mildew (*Erysiphe cruciferarum*) epidemics on oilseed rape (*Brassica napus*). Plant Pathol. https://doi.org/10.1111/ppa.12740

Voglino P (1905) Contribuzione allo studio della *Phyllactinia corylea*. Nuovo Giornale Botanico Italiano 12:313–327

Williamson CJ, Smith WHM (1986) Fungicidal control of powdery mildew and its effect on yield, digestibility and chemical composition of eight forage rape cultivars. J Agric Sci Cambridge 107:385–391

Yarwood CE (1957) Powdery mildews. Bot Rev XXIII:235–301

Yarwood CE (1978) History and taxonomy of powdery mildews. In: Spencer DM (ed) The powdery mildews. Academic, London, pp 1–37

Zheng RY (1985) Genera of Erysiphales. Mycotaxon 22:209–263

Zheng R, Chen G (1980) Taxonomic studies on the genus *Erysiphe* of China I. New species and new varieties on *Caprifoloaceae*. Acta Microbiol Sinica 20(1):45–49

Chapter 2
The Disease: Powdery Mildew

2.1 Introduction

The name of the disease as powdery mildew (PM) has been coined because of its characteristic symptoms on the above-ground infected plant parts in the form of white powdery growth of the pathogens mycelium, conidiophores, and conidia. The white coating of asexual fruiting bodies is very conspicuous and catches the eyes of a person from a distance. The disease is very widespread on cultivated and wild crucifers grown or available in tropical and sub-tropical areas of the world. The disease causes very heavy yield losses to the tune of 10–90% in oil-yielding crops with 1–7% reduction in oil content of the seed. To assess the disease on different crucifers with respect to their inbuilt resistance and efficacy of management practices, disease scoring scales have been developed.

2.2 Symptomatology

The symptoms of powdery mildew infection are usually apparent in the form of signs (conidia and conidiophores as white coating visible to unaided eye) before the pathogen shows its effect on infected morphological host parts. Powdery mildew injuries appear on their hosts slowly. Powdery mildews are rarely systemic, and the growth of the pathogen appears in the form of mycelium, conidia, and conidiophores as white coating on the infected host parts which is conspicuous and visible from a distance. Infection is usually apparent within 4–5 days after inoculation. Powdery mildew disease spreads very fast and can be in epidemic form within a short period of time compared to other diseases of crucifers because of its very short period of incubation and latent time after inoculation of the pathogen under both natural and controlled conditions. The characteristic symptoms on different crucifers are as follows:

G. S. Saharan et al., *Powdery Mildew Disease of Crucifers: Biology, Ecology and Disease Management*, https://doi.org/10.1007/978-981-13-9853-7_2

2.2.1 Rapeseed–Mustard

All the above-ground parts of the plant are affected by the disease. It appears in the form of dirty white circular, floury patches on leaves, stems, and pods (Plate 2.1a A, B, C, D). With the increase in temperature, white floury patches increase in size and coalesce to cover entire plant. Severely diseased plants appear as if they are dusted with white, granular, chalk-like powdery mass (Plate 2.1a E). Such plants are poor in growth and produce less siliquae. Stems are covered with mildew growth which appears purplish or bleached brown below the fungal growth (Plate 2.1a F, G). The young green siliquae show white to dirty white circular mildew growth in the initial stage of infection (Plate 2.1a D). Later such siliquae are completely covered with white mass of mycelium and conidia. Severely diseased siliquae remain small in size and produce small-sized shriveled few seeds (Plate 2.1b H). At maturity stage of the crop, infected leaves, stem, and siliquae show minute, circular black or dark brown spherical, scattered, or concentrated bodies of perithecia, cleistothecia, or chasmothecia (Plate 2.1b I; Saharan and Kaushik 1981; Saharan 1998; Saharan et al. 2005). Rutabagas (*Brassica napus* ssp. *rapifera*)

Plate 2.1a Symptomatology of powdery mildew disease of rapeseed–mustard; (**A–B**) dirty white circular, floury patches on leaves; (**C**) on stems; (**D**) dirty white circular mildew growth on siliquae; (**E**) on pod; (**F**) whole plant infected with powdery mildew; (**G**) purplish or bleached brown stem below the fungal growth

Plate 2.1b Powdery mildew disease of rapeseed–mustard; (**H**) small-sized shriveled few seeds; (**I**) black or dark brown scattered bodies of perithecia, cleistothecia, or chasmothecia on stem; (**J**) mixed infection of *Alternaria* blight and powdery mildew; (**K**) mixed infection of white rust and powdery mildew

plants infected with powdery mildew disease typically exhibit premature senescence of foliage. In addition, this disease predisposes rutabagas to other foliage and insect infections. In severe cases, the plant canopy is reduced drastically (Shattuck and Parry 1990). Under artificial inoculation conditions, symptoms development on *B. napus* crop is greatly influenced by the growth stages of host (Plate 6.1, Chap. 6). Please see description in details in Sect. 6.5, Chap. 6.

Mixed infection of *Alternaria* blight and powdery mildew is very common on *B. juncea* leaves in the form of brown spots with concentric rings covered with dirty white floury patches (Plate 2.1b J) specially when temperature is between 20 and 25 °C. Similarly, mixed infection of white rust and powdery mildew is also common on *B. juncea* leaves showing creamy to whitish pustules covered with dirty white floury patches of powdery mildew (Plate 2.1b K). Association of these (all three) diseases may be visible depending upon weather conditions prevailing in an area coinciding with infection and development of pathogens. Under field conditions severity of powdery mildew of mustard is visible in the form of white powdery mildew growth of conidia and conidiophores on leaves, stems, siliquae, and whole plants with deformed leaves (Plate 2.1c L–M). Similar symptoms appear on siliquae of *Brassica rapa* ssp. yellow sarson (Plate 2.1c N) and on *Brassica nigra* (Plate 2.1c

Plate 2.1c Powdery mildew on leaves, stem, siliquae, and whole plant of Indian mustard (**L–M**), on siliquae of *B. rapa* ssp. yellow sarson (*N*); powdery mildew on *Brassica nigra* (**O**); severely infected field of powdery mildew (**P**)

O), and see severely infected field under natural field infection of powdery mildew of Indian mustard (Plate 2.1c P).

Dr. Martin Barbetti from Shenton Park, Western Australia, during November 2015 also collected samples of powdery mildew infection on *Brassica napus*: powdery mildew initial stage of infection on leaf (Plate 2.2a-A); powdery mildew symptoms on leaves (Plate 2.2a-B); powdery mildew infection on older leaves (Plate 2.2a-C); and reddening of leaf due to powdery mildew (Plate 2.2a-D). The symptoms of powdery mildew on stem (Plate 2.2b-E) and pods (Plate 2.2b-F); drying of the stem due to powdery mildew (Plate 2.2b-G), and drying of the pods due to powdery mildew (Plate 2.2b-H) as well were also observed at Shenton Park, Western Australia.

Plate 2.2a Powdery mildew symptoms on leaves, stem, and pods of *Brassica napus* observed at Shenton Park, Western Australia, by Dr. Martin Barbetti during November 2015. (**A**) Powdery mildew initial stage of infection on leaf; (**B**) symptoms on leaves; (**C**) powdery mildew infection on older leaves; (**D**) reddening of leaf due to powdery mildew

2.2.2 Crucifer Vegetables

On crucifer vegetables, and on radish seed crop, the disease normally appears during late in the season. The disease first appears as dusty powdery patches having mycelium, conidia, and conidiophores of the pathogen on the leaves, which are circular in the beginning, and later on coalesces to form large patches. As the disease advances, the powdery growth covers the entire leaves, stem, and siliquae/pods. Severe infection may cause premature withering of the leaves, stems, and pods (Plate 2.3). The fruiting bodies (cleistothecia) may appear on the pods of radish seed crop (Suhag and Duhan 1985).

2.2.3 Chinese Cabbage

Symptoms first appear as circular to irregular white colonies, which subsequently develop into abundant growth on both leaf surfaces (Plate 2.3). Severe infections may cause leaf distortion, withering, and premature senescence of the foliage. Chasmothecial (cleistothecial) stage has not been recorded (Jee et al. 2008) so far on Chinese cabbage.

Plate 2.2b Powdery mildew symptoms on leaves, stem, and pod of *Brassica napus* observed at Shenton Park, Western Australia, by Dr. Martin Barbetti during November 2015. (**E**) Symptoms on stem; (**F**) symptoms on pods; (**G**) drying of the stem due to powdery mildew; (**H**) drying of the pods due to powdery mildew

Plate 2.3 Chinese cabbage covered with flourish white growth of powdery mildew (Jee et al. 2008)

2.2.4 *Taramira (Eruca sativa)*

Under Indian conditions, this crop is sown early under rainfed conserved moisture situations; therefore, defoliation occurs at the time of occurrence of the disease. The symptoms can only be observed on stem and siliquae of the infected plants. On green stem and siliquae, white, powdery, isolated, circular patches of fungal growth consisting of conidia and conidiophores appear which progress and enlarge to cover entire stem and siliquae (Saharan et al. 1984).

2.2.5 *Thale Cress Weed (Arabidopsis thaliana)*

The powdery mildew symptoms appear as white star-shaped colonies on the rosette leaves of 5–6-week-old *Arabidopsis*, mostly on upper surface of leaves but rarely on the lower surface. The colonies increase in size, and coalesce, which subsequently cover the entire leaf surface (Plate 2.4). As the disease progress, infected leaves exhibit chlorotic or necrotic brown lesions and leaf distortion followed by senescence (Choi et al. 2009). The variation in powdery mildew symptoms of susceptible genotypes, Col-O (A), Do-O (B), and Sorbo (C), and resistant genotype pmr-6-3 (D) at rosette stage is visible in the form of fungal conidiophores (Plate 4.4, Chap. 4).

Plate 2.4 Powdery mildew symptoms on leaves of *Arabidopsis thaliana* (Choi et al. 2009)

Plate 2.5 (**A**) Powdery mildew symptoms on stems of *Malcolmia africana* infected with *Erysiphe cruciferarum*. (**B**) Powdery mildew symptoms on leaves of *Malcolmia africana*

2.2.6 *African Mustard (Malcolmia africana)*

On African mustard, powdery mildew symptoms appear as dense white mycelial patches ranging from 5 to 25 mm in length on stems (Plate 2.5a). Hyphae are also present on the upper surface of the leaves (Plate 2.5b; Mirzaee et al. 2010).

2.2.7 *Gold of Pleasure (Camelina sativa)*

The symptoms on *Camelina* and wild mustard plants appear as circular to irregular whitish colonies on leaves, stems, and fruits. At initial stage, the individual colonies are small, and distinct (middle of April), but later on, it increases to cover the whole leaf surface of *Camelina* (Plate 2.6a-A) and *Sinapis* siliquae (Plate 2.6a-B). The whitish colonies consist of mycelium, conidiophores, and conidia (Plate 2.6b-AB). Mature chasmothecia with appendages (Plate 2.6b-C) and asci and ascospores after rupturing of chasmothecia (Plate 2.6b-D). Chasmothecia can also be observed on the upper surface of *Camelina* leaves (Plate 2.6c-CD) and on wild mustard siliques (*Sinapis arvensis*) (Plate 2.6c-A B; Vellios et al. 2017).

Plate 2.6a (**A**) Symptoms of powdery mildew on *Camelina sativa* leaves, (**B**) on *Sinapis arvensis* siliquae

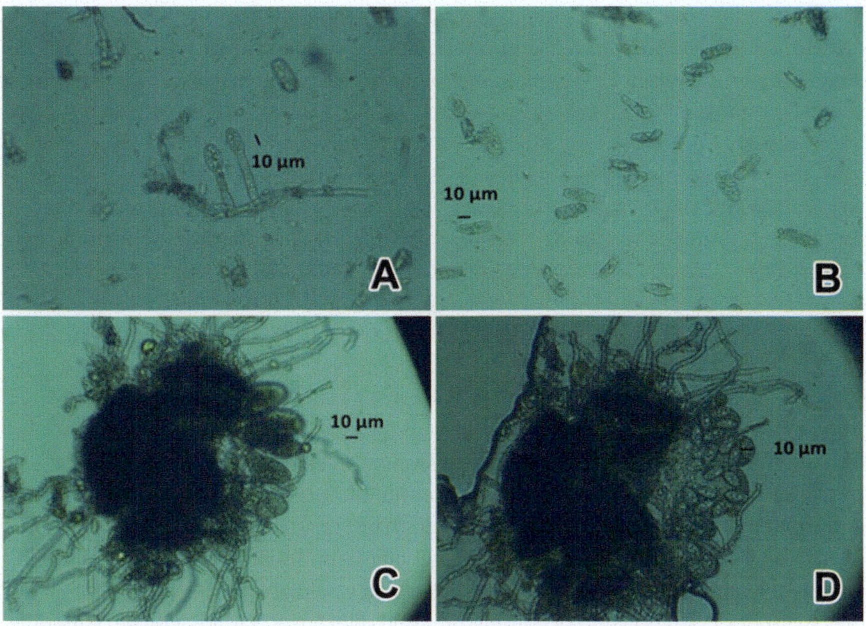

Plate 2.6b *Erysiphe cruciferarum* causing powdery mildew on *Camelina* and wild mustard. (**A**) Conidiophore with conidium, (**B**) conidia, (**C–D**) mature chasmothecia with appendages, and asci with ascospores

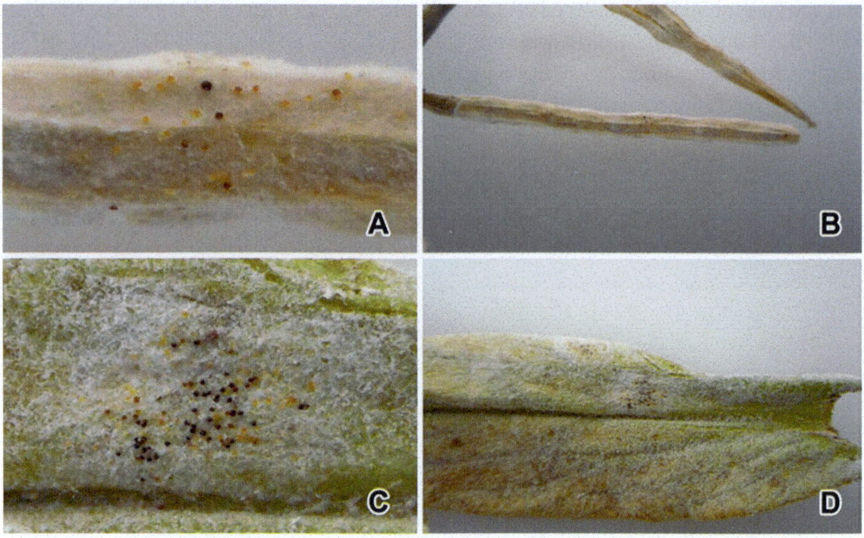

Plate 2.6c (**A**, **B**) Chasmothecia on wild mustard siliques and (**C**, **D**) *Camelina* leaves

*2.2.8 Rutabagas (***Brassica napus** *ssp.* **rapifera***)*

The first symptom of this disease is the appearance of small white spots consisting of the interwoven threads of mycelia on older leaves and stems. Under favourable conditions for the disease, the mycelia gradually expand over the surface of tissues, produce conidia, and give the appearance of a greyish powdery coating. Plants infected with this disease typically exhibit premature senescence of foliage. In addition, this disease predisposes rutabagas to other foliage and insect infections. In severe cases the plant canopy is reduced which makes the mechanical harvesting of roots when using a revolving belt-type pulling system difficult (Shattuck and Parry 1990).

2.3 Geographical Distribution

Considering the nature of the pathogen, powdery mildew is likely to infect cruciferous cultivated as well as wild plants all over the globe. However, its occurrence and severity on crucifers have been recorded from very less number of countries. Moreover, its importance as an economically devastating disease has not been realized by the crucifer's scientific community. As per published records, powdery mildew of crucifers is known to occur in Argentina (Gaetan and Madia 2004); Australia (Shivas 1989); Bulgaria (Negrean and Denchev 2000); Canada (Slopek and Peters 1988; Shattuck and Parry 1990); China (Alkooranee et al. 2015); Czech and Slovak

(Pastircakova and Pastircak 2013); Egypt (Melchers 1931); Europe (Bolay et al. 2005); Finland (Farr et al. 1989); France (Ale-Agha et al. 2008); Greece (Vellios et al. 2017); Germany (Koch and Slusarenko 1990); India (Butler 1918); Iran (Mohammadi-Doustdar 1967); Japan (Amano 1986); Korea (Shin and La 1992); Mexico (Yanez-Morales et al. 2009); New Zealand (Boesewinkel 1977); Poland (Sadowski et al. 2002); Russia (Bunkina 1991); Sweden (Junell 1967); Switzerland (Bolay 2005); Turkey (Mert-Turk et al. 2008); UK (Ellis and Ellis 1985); USA (Ellett 1966), and Vietnam (Tam et al. 2016).

2.3.1 Geographical Distribution of Rapeseed–Mustard Powdery Mildew in India

The distribution of powdery mildew of rapeseed–mustard in various parts/states of India is at large scale covering more than 17 states (Plate 2.7) with severity indices of 5–58% (Table 2.1; Meena et al. 2018). State-wise distribution of powdery mildew is given in Table 2.1. In India, Butler (1918) was probably the first person who reported a severe outbreak of powdery mildew in 1907 in the Chenab Canal Colony of Punjab province.

Powdery mildew can only progress if the conditions provided by the host plants as well as weather conditions are favourable. Previously, powdery mildew of rapeseed–mustard was a minor problem mostly occurring in Gujarat state except for mild incidences elsewhere, but now it has assumed an epidemic form in rapeseed–mustard-growing areas of India (Sharma 1979; Saharan 1980; Saharan and Kaushik 1981; Saharan and Sheoran 1985; Gohil et al. 1988; Dange et al. 2002; Dange et al. 2003; Saharan et al. 2005). During the past two decades, powdery mildew has become one of the major diseases of rapeseed–mustard in India. The appearance of the disease in the past was noticed from late January till maturity of the crop, but it was in severe form at Ranchi, Jharkhand, during December 2016. However, recently, the disease has been recorded from other oilseed *Brassica*-growing states, viz., Haryana, Central UP, MP, parts of Rajasthan, and Bihar, even in December possibly due to shortening of cold spell during the crop growth period (Chattopadhyay et al. 2011). The powdery mildew disease severity on different crops ranged from 10 to 100% during 2000 to 2017 (Table 2.2; Anonymous 2000–2017, Meena et al. 2018). The disease increased in India in an order from north (Himachal Pradesh 5.0%) to south (Gujarat 57.5%) and west (Rajasthan 22.1%) to east (Manipur 39.4%) owing to favourable environmental conditions and availability of susceptible host species *B. juncea* at large scales in contiguity.

The damage to the crop was more when infection took place in early stages of plant growth, and pods were heavily loaded with fungal spores (Saharan and Kaushik 1981). Powdery mildew disease of rapeseed–mustard has been reported from 17 states in India so far. Distribution and severity of powdery mildew disease were recorded from Gujarat (57.5%), Maharashtra (38.7%), Punjab (30.8%), Uttar

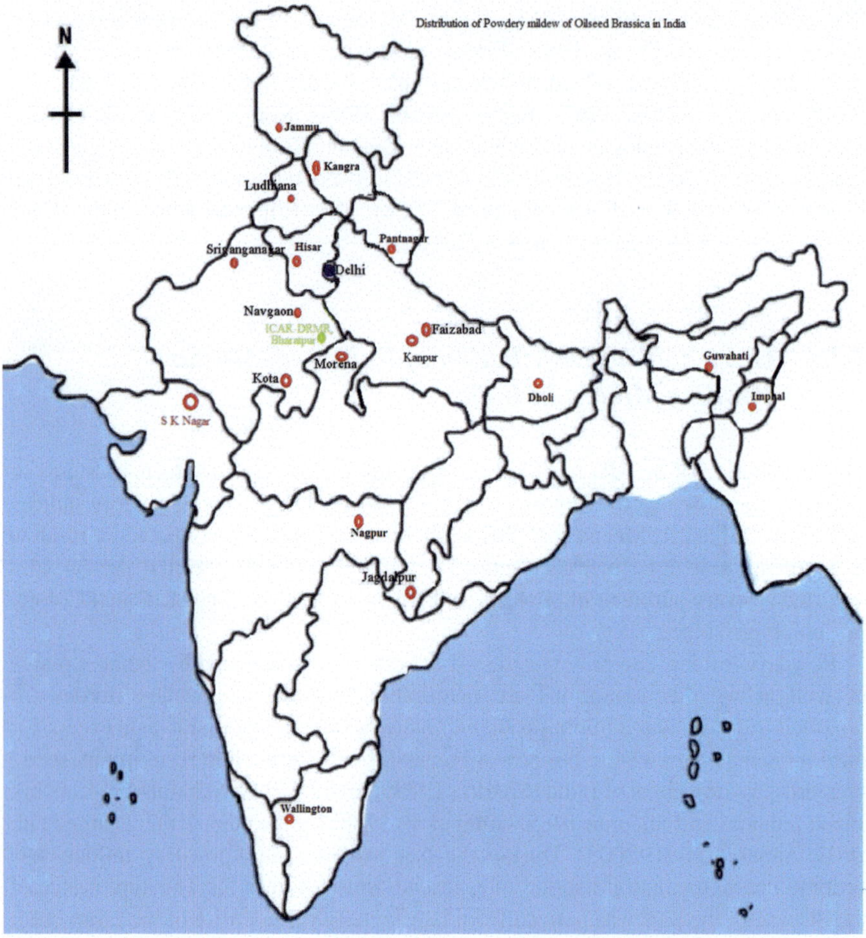

Plate 2.7 Distribution map of powdery mildew disease of rapeseed–mustard in India

Pradesh (30.1%), Manipur (39.4%), Madhya Pradesh (M.P.) (21.5%), Haryana (20.6%), Chhattisgarh (25.5%), Rajasthan (22.1%), Uttarakhand (24.5%), Delhi (20.0%), Bihar (15.5%), Jharkhand (20.0%), Assam (15.0%), TN (35.0%), Jammu and Kashmir (10.0%), Himachal Pradesh (H.P.) (5.0%), and other mustard-growing areas of the country (Tables 2.1 and 2.2; Fig. 2.1; Anonymous 2000–2017, Meena et al. 2018). The severity of the disease on oilseed *Brassicas* differed over seasons and regions as also between individual crops within a region and state. However, powdery mildew severity showed an increasing trend from north (H.P.) to south (Gujarat) with a plateau of 20–22% severity in the states like Delhi (20%), Haryana (20.6%), Jharkhand (20%), M.P. (21.5%), and Rajasthan (22.1%, Fig. 2.1) which may take a volcano shape in future.

Table 2.1 Distribution of Powdery mildew disease of rapeseed–mustard over the years in India

State	% powdery mildew severity[a]	References
Assam	15.0	Saharan (1992a, b)
Bihar	15.5	Saharan (1992a, b)
Chhattisgarh	25.5	Anonymous (2000–2017)
Delhi	20.0	Saharan (1992a, b)
Gujarat	57.5	Anonymous (2000–2017)
Haryana	20.6	Saharan (1980), Dang et al. (2000) and Mehta et al. (2008)
Himachal Pradesh	5.0	Paul (1984)
Jammu & Kashmir	10.0	Sharma (1979)
Jharkhand	20.0	Saharan (2016)
Madhya Pradesh	21.5	Anonymous (2000–2017)
Maharashtra	38.7	Vasudeva (1958), Kohire (2002), Nair et al. (2017) and Kohire et al. (2008)
Manipur	39.4	Yengkhom and Chhetry (2016, 2017)
Punjab	30.8	Anonymous (2000–2017) and Butler (1918)
Rajasthan	22.1	Anonymous (2000–2017)
Tamil Nadu	10.0	Anonymous (2000–2017)
Uttar Pradesh	30.1	Anonymous (2000–2017)
Uttarakhand	24.5	Anonymous (2000–2017)

[a]Mean disease severity over the locations and years

Table 2.2 Crucifers' powdery mildew severity (Anonymous 2000–2017)

Crucifer species	Stage	Severity (%)	Losses (%)	References
Brassica juncea	Maturity	53.1–100.0	37	Anonymous (2000–2017), Kohire et al. (2008) and Mehta et al. (2008)
B. rapa ssp. yellow sarson	Maturity	21.9–66.7		Anonymous (2000–2017)
B. rapa ssp. Toria	Maturity	11.3–66.7		Anonymous (2000–2017)
B. rapa ssp. brown sarson	Maturity	28.3		Dang et al. (2000)
B. napus	Maturity	10–100		Anonymous (2000–2017)
B. nigra	Maturity	0–10		Dang et al. (2000)
B. tournefortii	Maturity	0–5		Dang et al. (2000)
Eruca sativa	Maturity	9.2–42.9	17	Anonymous (2000–2017) and Kohire et al. (2008)

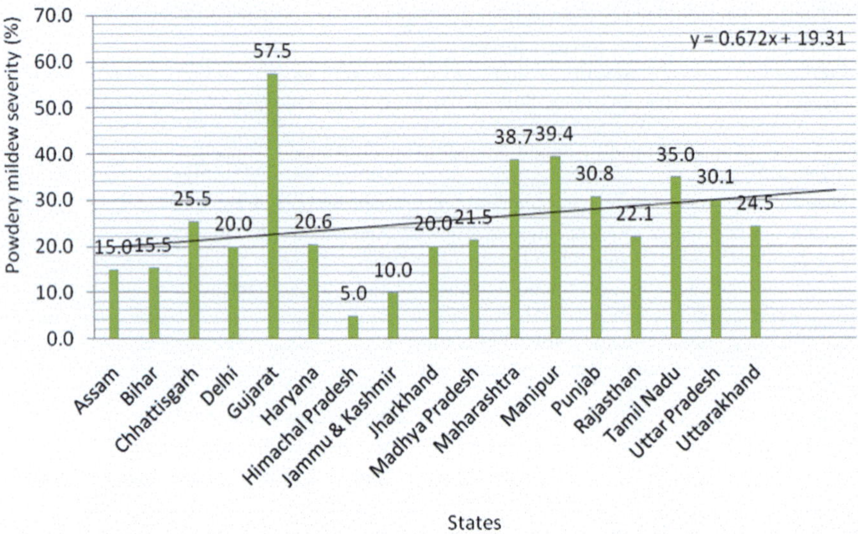

Fig. 2.1 Powdery mildew severity (%) on rapeseed–mustard at different states of India (Plateau 27.7%)

2.4 Host Range

Powdery mildew occurs world over and attacks a large number of cultivated crops along with numerous crucifer's weeds. Economically important host crops include rapeseed–mustard, cabbage, Chinese cabbage, cauliflower, broccoli, horseradish, turnip, radish, Brussels sprouts, collards, kohlrabi, kale, mustard, rape, rutabaga, swedes, turnips, and *Arabidopsis* – a model plant for molecular studies using powdery mildew pathogen for host–parasite interaction investigations (Sherf and Mac Nab 1986; Saharan et al. 2005). The inventory of hosts infected by powdery mildew pathogen along with location and years of records all over the world is given in Table 2.3.

Although the taxonomic value of host range is controversial, there are clear host limitations among *Erysiphe* spp. (Junell 1967; Yarwood 1978; Braun 1987). *Brassica* spp. have been reported as hosts for a number of *Erysiphales*: Amano (1986) identified *E. arabidis, E. cichoracearum, E. communis, E. cruciferarum, E. rorippae, Leveillula cruciferarum, L. taurica, Microsphaera alni,* various *Oidium* spp., *Sphaerotheca fuliginea, S. drabae,* and *S. macularis*. Several authors have independently reported *E. polygoni, E. arabidis, E. cruciferarum, E. orontii,* and *S. drabae* on crucifers (Spencer 1978; Plotnikova et al. 1998). In addition, one powdery mildew species may infect multiple host species. For example, *E. cichoracearum* has been narrowly defined by Braun (1987) as consisting of those powdery mildews infecting the Asteraceae, whereas others have suggested that all powdery mildews morphologically similar to *E. cichoracearum* should be regarded as members of this species regardless of their host specialization (Salmon 1900; Schmitt

Table 2.3 World records of Cruciferae powdery mildew (*E. cruciferarum* Opiz ex L. Junell)

Host species	Location	Recording year	References
Alyssum sp.	Iran	1958	Viennot-Bourgin (1958), Mohammadi-Doustdar (1967), Amano (1986) and Ershad (1995)
Alyssum hirsutum	Bulgaria	2000	Negrean and Denchev (2000)
Alyssum hirsutum	Iran	1971	Amano (1986) and Ershad (1971, 1995)
Alyssum dasycarpum	Iran	1971	Ershad (1971, 1995)
Alyssum strigosum	Iran	2000	Khodaparast et al. (2000)
Alyssum repens	Iran	1986	Amano (1986)
Alyssum alyssoides	UK	1997	Ellis and Ellis (1997)
Alliaria petiolata	Iran	1967	Mohammadi-Doustdar (1967), Amano (1986) and Ershad (1995)
Alliaria petiolata	UK	1997	Ellis and Ellis (1997)
Antirrhinum majus	UK	1997	Ellis and Ellis (1997)
Armoracia rusticana	UK	1997	Ellis and Ellis (1997)
Argemone mexicana	India	1995	Bappammal et al. (1995)
Argemone mexicana	India	1992	Sharma and Khare (1992)
Argemone sp.	Czech and Slovak	2013	Pastircakova and Pastircak (2013)
Arabidopsis thaliana	USA	1993	Karakaya et al. (1993)
Arabidopsis thaliana	Zurich, Germany	1990	Koch and Slusarenko (1990)
Arabidopsis thaliana	Korea	2008	Choi et al. (2009)
Arabidopsis thaliana	USA	1999	Adam et al. (1999) and Adam and Somerville (1996)
Arabidopsis thaliana	Europe	1997	Xiao et al. (1997) and Allen et al. (2004)
Brassica sp.	New Zealand		Cooper (2013)
Brassica sp.	Iran		Mohammadi-Doustdar (1967), Amano (1986) and Ershad (1995)
B. carinata Ethiopian mustard	Australia		Gunasinghe et al. (2013)
Brassica rapa (*B. campestris*)	New Zealand	1977	Boesewinkel (1977)
Brassica rapa (*B. campestris*)	New Zealand	2013	Cooper (2013)
Brassica rapa (*B. campestris*)	UK	1997	Ellis and Ellis (1997)
Brassica rapa var. *chinensis*	USA	1999	Adam et al. (1999)
Chinese cabbage	Suwon, Korea	2006	Jee et al. (2008)
Brassica rapa var. *pekinensis* Syn. *B. pekinensis*	Australia	2012	Gunasinghe et al. (2013)

(continued)

Table 2.3 (continued)

Host species	Location	Recording year	References
B. pekinensis	China	2014	Zhao et al. (2014)
B. pekinensis	China	2015	Alkooranee et al. (2015)
Brassica rapa var. *rapifera*	USA	1999	Adam et al. (1999)
Brassica rapa	Rajasthan, India	1967	Sankhla et al. (1967)
Brassica rapa	Ohio, USA	1965	Ellett (1966)
Brassica rapa	Iran	1986	Amano (1986) and Ershad (1995)
Brassica rapa	Haryana, India	1980	Saharan and Kaushik (1981)
Brassica rapa	Maharashtra, India	1949	Patel et al. (1949)
Brassica rapa ssp. *toria*	Haryana, India	1981	Saharan and Kaushik (1981)
Brassica rapa var. Sarson	Madhya Pradesh, India	1992	Sharma and Khare (1992)
Brassica juncea (Indian mustard)	Australia	2007	Kaur et al. (2008)
Brassica juncea	Hwascong, Korea	2011	Kim et al. (2013)
Brassica juncea	Vietnam	2014	Tam et al. (2016)
Brassica juncea	Haryana, India	1980	Saharan and Kaushik (1981)
Brassica juncea	Rajasthan, India	1967	Sankhla et al. (1967)
Brassica juncea	India	1963	Bhander et al. (1963)
Brassica juncea	Kashmir, India	1979	Sharma (1979)
Brassica juncea	India	1958	Vasudeva (1958)
Brassica juncea	USA	1999	Adam et al. (1999)
Brassica juncea var. *crispifolia*	New Zealand	2013	Cooper (2013)
Brassica nigra	Haryana, India	1980	Saharan and Kaushik (1981)
B. napobrassica	New Zealand	1977	Boesewinkel (1977)
B. napus	New Zealand	1977	Boesewinkel (1977)
B. napus	UK	1997	Ellis and Ellis (1997)
B. napus (oilseed rape)	Australia	1986	Shivas (1989)
B. napus	China	2014	Alkooranee et al. (2015)
B. napus var. *napobrassica*	Australia,	1971	Shivas (1989)
B. napus var. *napobrassica*	Ohio, USA	1965	Ellett (1966)
B. napus var. *napobrassica*	New Zealand	2013	Cooper (2013)
B. napus	Iran	1986	Amano (1986) and Ershad (1995)
B. napus	Argentina	2003	Gaetan and Madia (2004)
B. napus rutabaga	Korea	2015	Cho et al. (2016)

(continued)

Table 2.3 (continued)

Host species	Location	Recording year	References
B. napus	Alberta, Canada	1988	Slopek and Peters (1988)
B. napus ssp. *rapifera*	Ontario, Canada	1990	Reyes (1969)
B. napus	Haryana, India	1980	Saharan and Kaushik (1981)
B. napus	Turkey	2008	Mert-Turk et al. (2008)
B. napus	Poland	2002	Sadowski et al. (2002)
B. napus var. *pabularia*	USA	1999	Adam et al. (1999)
Broccoli raab	California, USA	1997	Koike and Saenz (1998)
Brassica oleracea	Iran	1967	Mohammadi-Doustdar (1967)
Brassica oleracea	USA	1965	Walker and Williams (1965)
Brassica oleracea	Japan	1902	Salmon (1902)
Brassica oleracea var. *capitata*	Western Australia	2012	Gunasinghe et al. (2013)
Brassica oleracea var. *capitata*	USA	1999	Adam et al. (1999)
Brassica oleracea var. *capitata*	New Zealand	2013	Cooper (2013)
Brassica oleracea var. *acephala*	USA	1999	Adam et al. (1999)
Brassica oleracea var. *botrytis*	USA	1999	Adam et al. (1999)
Brassica oleracea var. *botrytis*	New Zealand	1977	Boesewinkel (1977)
Brassica oleracea var. *gongylodes*	USA	1999	Adam et al. (1999)
Brassica oleracea var. *gemmiferae*	USA	1974	Dixon (1974) and Adam et al. (1999)
Brassica oleracea var. *ramosa*	New Zealand	2013	Cooper (2013)
Brassica oleracea var. *ramosa*	New Zealand	1977	Boesewinkel (1977)
Brassica tournefortii	Haryana, India	2000	Dang et al. (2000)
Cardamine debilis	New Zealand	1977	Boesewinkel (1977)
Cardamine debilis	New Zealand	2013	Cooper (2013)
Cardamine flexuosa	New Zealand	2013	Cooper (2013)
Cardamine hirsuta	New Zealand	1977	Boesewinkel (1977)
Chelidonium asiaticum	Japan	1985	Amano (1986)
Chelidonium majus	Russia	1991	Bunkina (1991)
Chelidonium majus var. *asiaticum*	Korea	2000	Shin (2000)
Camelina sativa	UK	1996	van Roessel et al. (1996)
Camelina sativa	Greece	2015	Vellios et al. (2017)

(continued)

Table 2.3 (continued)

Host species	Location	Recording year	References
Cleome hassleriana (spider flower)	France	2006	Ale-Agha et al. (2008)
Cleome hassleriana (spider flower)	Italy	2009	Garibaldi et al. (2009)
Cleome spinosa	France	2006	Ale-Agha et al. (2008)
Cleome spinosa	New Zealand	1977	Boesewinkel (1977)
Cleome spinosa	Italy	2009	Garibaldi et al. 2009
Cheiranthus cheiri	UK	1997	Ellis and Ellis (1997)
Capsella bursa-pastoris	UK	1985	Ellis and Ellis (1985)
Capsella bursa-pastoris	Jammu, India	1979	Sharma (1979)
Capsella bursa-pastoris	Massachusetts	1998	Plotnikova et al. (1998)
Capsella bursa-pastoris	Slovenia	2017	Radisek et al. (2018)
Coronopus didymus	Jammu, India	1979	Sharma (1979)
Cardaria draba	Iran	2000	Khodaparast et al. (2000) and Aeenfar (2006)
Cardaria subsp. *chalapensis*	Iran	1995	Tajik-Ghanbary (1995)
Conringia planisiligua	Iran	2000	Khodaparast et al. (2000)
Crambe sp.	Iran	2006	Kachooeian Javadi et al. (2006)
Crambe orientalis	Iran	1971	Ershad (1971, 1995), Amano (1986) and Tajik-Ghanbary (1995)
Descurainia sophia	Iran	1958	Viennot-Bourgin (1958), Mohammadi-Doustdar (1967), Amano (1986) and Ershad (1971, 1995)
Erodium moschatum	New Zealand	1977	Boesewinkel (1977)
Eruca sativa	Haryana, India	1981	Saharan et al. (1984)
Eruca sativa	Australia	2012	Gunasinghe et al. (2013)
Eruca vesicaria	Australia	2012	Gunasinghe et al. (2013)
Erysimum cheiri	UK	1997	Ellis and Ellis (1997)
Eschscholzia californica	Switzerland	2005	Bolay (2005)
Eschscholzia californica	Germany	2009	Schmidt and Scholler 2011
Eschscholzia	Czech and Slovak	2013	Pastircakova and Pastircak 2013
Fumaria sp.	Iran	1977	Mohammadi-Doustdar (1967) and Ershad (1995)
Fumaria officinalis	New Zealand	1977	Boesewinkel (1977)

(continued)

Table 2.3 (continued)

Host species	Location	Recording year	References
Glaucium	Czech and Slovak	2013	Pastircakova and Pastircak (2013)
Geranium homeanum	New Zealand	1977	Boesewinkel (1977)
Iberis amara	Jammu, India	1979	Sharma (1979)
Iberis amara	Madhya Pradesh, India	1992	Sharma and Khare (1992)
Iberis umbellata	New Zealand	1977	Boesewinkel (1977)
Iberis sp.	Iran	1965	Amano (1986); Ershad 1995
Lepidium apetalum	Korea	1992	Shin and La (1992)
Lepidium campestre	Iran	1986	Amano (1986); Ershad (1995)
Lepidium latifolium	Iran	1958	Viennot-Bourgin (1958), Mohammadi-Doustdar (1967), Amano (1986), Ershad (1995), Tajik-Ghanbary (1995) and Kachooeian Javadi et al. (2006)
Lepidium sativum	Iran	1967	Mohammadi-Doustdar (1967), Amano (1986) and Ershad (1995)
Lepidium virginicum	Himachal Pradesh, India	1984	Paul (1984)
Malcolmia africana	Iran	2010	Mirzaee et al. (2010)
Malcolmia incana	Canada	2009	Farr et al. (2009)
Malcolmia maritima	France	2009	Farr et al. (2009)
Meconopsis	Czech and Slovak	2013	Pastircakova and Pastircak (2013)
Hesperis matronalis	UK	1985	Ellis and Ellis (1985)
Papaver nudicaule	New Zealand	1977	Boesewinkel (1977)
Papaver sp.	Czech and Slovak	2013	Pastircakova and Pastircak (2013)
Papaver rhoeas	New Zealand	1977	Boesewinkel (1977)
Papaver somniferum	M.P., India	1992	Sharma and Khare (1992)
Papaver somniferum	New Zealand	1977	Boesewinkel (1977)
Raphanus sativus	Iran	1967	Mohammadi-Doustdar (1967), Amano (1986) and Ershad (1971, 1995)
Raphanus sativus	Haryana, India	1985	Suhag and Duhan (1985)
Raphanus sativus	Jammu, India	1979	Sharma (1979)
Raphanus maritimus	New Zealand	1977	Boesewinkel (1977)
Raphanus raphanistrum subsp. *maritimus*	New Zealand	2013	Cooper (2013)
Rapistrum rugosum	Iran	1967	Mohammadi-Doustdar (1967), Amano (1986), Ershad (1995), Khodaparast et al. (2000) and Tajik-Ghanbary (1995)
Rapistrum rugosum	Argentina	2000	Braun et al. (2000)
Sinapis arvensis	Greece	2015	Vellios et al. (2017)

(continued)

Table 2.3 (continued)

Host species	Location	Recording year	References
Sinapis arvensis	Iran	1971	Ershad (1971, 1995) and Amano (1986)
Sisymbrium alliaria	UK	1997	Ellis and Ellis (1997)
Sisymbrium officinale	UK	1985	Ellis and Ellis (1985)
Sisymbrium sp.	Mexico	2008	Yanez-Morales et al. (2009)
Sisymbrium irio	Argentina	2000	Braun et al. (2000)
Sisymbrium irio	Iran	1996	Niknam and Guya (1996)
Sisymbrium orientale	Iran	2000	Khodaparast et al. (2000)
Stylophorum	Czech and Slovak	2013	Pastircakova and Pastircak (2013)
Stylophorum diphyllum	Switzerland	2005	Bolay (2005)
Wasabia japonica	New Zealand	2013	Cooper (2013)

On some host species, the causal organism has been recorded earlier as *E. communis, E. cichoracearum*, and *E. polygoni* erroneously

1955; Yarwood 1978). Amano (1986) alone has reported *E. cichoracearum* on more than 55 plant families. Radisek et al. (2018) have reported from Slovenia that *Golovinomyces orontii* infected *Capsella bursa-pastoris* growing wild in Ljubljana during 2017.

2.5 Yield Losses

The powdery mildew of crucifers has been considered a minor problem in the past all over the world. However, during last two decades of the twentieth century, it has taken epidemic form all over the crucifers-growing areas of the world (Saharan and Kaushik 1981; Saharan 1980, 1984, 1992a, b, 1998; Braun 1987, 2012; Uloth et al. 2017; Saharan and Mehta 2002; Saharan et al. 2005; Meena et al. 2018).

The damage to the crop may be very severe when disease appears at early stages of plant growth since fungal growth covers entire photosynthetic green areas of the plant hindering food manufacturing mechanism of host plants. The pods heavily covered with powdery mass remain empty or produce few shriveled seeds at the base with twisted sterile tips. There is negative correlation of disease with yield components of mustard (*B. juncea*). Yield components like number of siliquae per plant, siliqua length, seeds per siliqua, 1000 seed weight, total yield, and per cent oil content are negatively correlated with the disease. There is reduction in yield of a susceptible mustard cv. (EC-126743) to the extent of 17.4% with 6.47% reduction in oil content (Table 2.4; Saharan and Sheoran 1985) under Indian conditions. Dang et al. (2000) evaluated thirty six Cruciferae genotypes belonging to different species of *Brassicas* for resistance to powdery mildew disease during three consecutive crop seasons (1994–1996) and observed on an average of 20.6% disease severity at

Table 2.4 Assessment of losses in yield of mustard infected with powdery mildew disease (Saharan and Sheoran 1985)

Yield components	Diseased	Protected[a]	Percent reduction
Number of pods/plant	653.5	693.5	**5.76**
Pod length (cm)	4.1	4.4	**6.82**
Number of seeds/pod	12.0	12.9	**6.97**
1000 grain weight (gm)	2.17	2.18	**0.82**
Total weight (g/ha)	16.5	20.0	**17.50**
Oil content (%)	43.3	46.3	**6.47**

[a]The crop was protected by spraying Karathane 0.1%

Haryana state with highest on *B. juncea* (43.2%) followed by *B. rapa* ssp. Toria (28.6%), *B. rapa* ssp. yellow sarson (23.9%), *B. napus* (10.0%), *B. nigra* (10.0%), *B. tournefortii* (5.0%), and *Eruca sativa* (15.9%) (Fig. 2.1). However, Mehta et al. (2008) reported maximum (91.0%) powdery mildew disease severity on *B. juncea* during 2007–2008 at Haryana. The highest powdery mildew severity recorded on *B. juncea* (43.2%) followed by *B. rapa* (28%) (Fig. 2.1) is very alarming since these two species are grown in wider areas of the country to meet up the edible oil demand. Recently, the powdery mildew occurrence and severity have been reported from north eastern states of India. In a survey conducted during 2014–2016, on an average 39.4% powdery mildew disease severity on two varieties of Indian mustard (Local Yella and Lamtachabi) and two varieties of rapeseed *B. rapa* ssp. yellow sarson (namely, M27 and Ragini) was observed at advanced stage in experimental site as well as at number of roaming sites located in Thoubal, Senapati, and Chandel districts of Manipur (Premlata Devi and Chhetry 2016). In southern parts of the country, rapeseed–mustard is being strategically spread as nontraditional crop to increase the area under *Brassica* crop in order towards edible oil self-sufficiency, but powdery mildew severity has been observed up to 38.7% in Nagpur and Prabhani districts of Maharashtra responsible for heavy yield losses (Nair et al. 2017). In this region, an average temperature during winter ranges from 10–15 to 25–30 °C from October to February/March which is most congenial for epidemic development of the powdery mildew disease. Based on the disease severity, and occurrence, the disease distribution map of India was prepared (Fig. 2.2) which revealed that powdery mildew disease severity on rapeseed–mustard in India was highest during 2015 (42.8%), while it was lowest during 2010 (17.0%). The increase in trend of disease severity doubled over the period which was probably one of the factors responsible for reduction in productivity of the crop in India. The zigzag trend in the severity of powdery mildew during the period from 2004 to 2017 indicated the influence of environmental conditions, which monitored the disease progression during different years (Fig. 2.3).

In France, powdery mildew of oilseed rape can cause yield losses in the range of 10–30%, a loss of up to 12q/ha potential yield in some instances (Penaud 1998). Among the fungicidal treatments, spray of Triadimefon 23 WP (0.1%) suspension thrice at fortnightly intervals, beginning at the initiation of the symptoms at 1000 l/ha resulted in lowest disease severity and highest yield (Penaud 1999). Taking it as

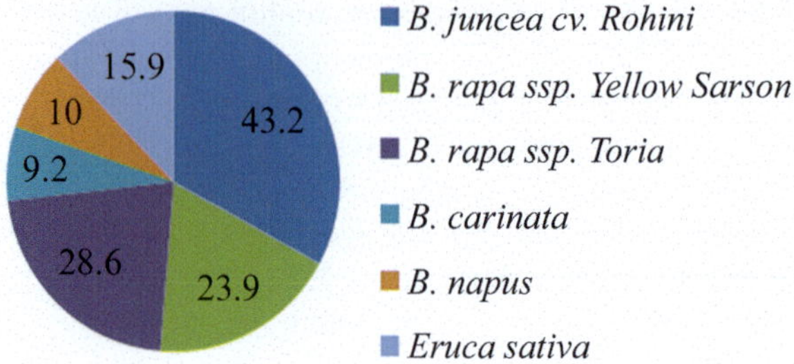

Fig. 2.2 Powdery mildew severity (%) on rapeseed–mustard species over 18 years at different locations of India

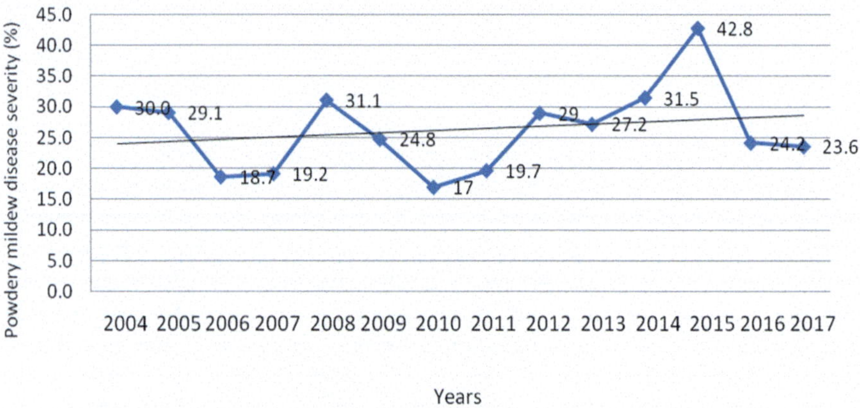

Fig. 2.3 Powdery mildew disease severity (%) over the years

a measure of the protection, 42.4% avoidable seed yield loss due to this disease and 17.5% loss in 1000 seed weight were recorded by Singh and Singh (2003). In rapeseed–mustard under Gujarat Indian conditions, early planting of mustard cvs. Indian mustard cultivars GM-1 and Varuna sown in October resulted in less severity of powdery mildew with higher grain yield than late planting in the month of November (Tables 2.5 and 2.6). Under Gujarat conditions, powdery mildew caused 24.1% yield loss (Dange et al. 2003). Earlier, Patel et al. (1995) recorded 22% loss in seed yield of mustard from Gujarat, India. Mehta et al. (2008) estimated yield losses in mustard cvs. from 10 to 29.5% with an average loss of 24.5% (Table 2.7). Hare (1994) from Maharashtra reported the losses in yields of mustard from 45 to 90% due to powdery mildew severity in cvs. Pusa Bold and Seeta. However, Kohire et al. (2008) recorded 15–40% losses in yield of mustard cvs. TM-17 and Seeta at Aurangabad with loss in oil content of 1.17% (Tables 2.8 and 2.9) in cv. Bio 903. Earlier studies from Parbhani, Maharashtra, recorded losses in the yields of rapeseed–mustard from 20 to 40% with 2–7% loss in oil content (Samudre 1994; Hingole 1995). Overall yield losses in crucifers due to powdery mildew have been estimated from 10 to 90% from India and 10 to 30% from France with loss of oil content from 1 to 7% from India (Table 2.10).

Table 2.5 Effect of different planting time on the severity of powdery mildew disease. (Dange et al. 2003)

	GM-1				Varuna			
	Disease severity (%)				Disease severity (%)			
Planting time	1994–1995	1995–1996	1996–1997	Pooled	1994–1995	1995–1996	1996–1997	Pooled
D_1 5th October	20.0 (26.53)	15.5 (23.13)	17.0 (24.01)	17.5 (24.56)	30.0 (33.17)	18.0 (25.05)	30.5 (33.44)	26.2 (30.55)
D_2 15th October	20.0 (26.36)	12.0 (20.04)	22.5 (28.07)	18.2 (24.82)	25.0 (29.89)	15.5 (23.11)	21.0 (27.21)	20.5 (26.74)
D_3 25th October	42.0 (40.37)	38.0 (38.02)	24.0 (29.24)	34.7 (35.88)	38.5 (38.30)	35.5 (36.53)	28.25 (32.06)	34.1 (35.63)
D_4 5th November	40.0 (39.20)	40.0 (39.20)	34.5 (35.95)	38.2 (38.12)	38.0 (38.02)	38.0 (38.04)	32.5 (34.65)	36.2 (36.90)
D_5 15th November	70.0 (56.79)	55.0 (47.88)	40.0 (39.16)	55.0 (47.94)	72.0 (58.08)	52.0 (46.13)	63.0 (52.56)	62.3 (52.26)
D_6 25th November	72.0 (58.13)	64.5 (58.47)	52.5 (46.42)	63.0 (52.67)	75.0 (60.00)	74.0 (59.40)	65.5 (54.03)	71.5 (57.81)
SE±	1.29	1.49	2.19	2.42	1.42	1.05	1.46	1.92
CD at 5%	3.88	4.48	6.59	7.64	4.29	3.18	4.40	6.04
CV %	6.25	8.04	12.94	9.10	6.63	5.54	7.49	6.63
Interaction (Year × Planting time)		SEm±: 1.70 CD at 5%: 4.84			SEm±: 1.33 CD at 5%: 3.78			

Table 2.6 Effect of different planting time on the grain yield of mustard due to powdery mildew disease. (Dange et al. 2003)

	GM-1				Varuna			
	Grain yield (Kg)				Grain yield (Kg)			
Planting time	1994–1995	1995–1996	1996–1997	Pooled	1994–1995	1995–1996	1996–1997	Pooled
D_1 5th October	1164.69	501.65	1466.67	1044.34	1112.06	398.86	1514.28	1008.24
D_2 15th October	1196.83	1148.57	1990.48	1445.29	1203.49	1010.55	1936.51	1383.52
D_3 25th October	1173.65	1461.90	2453.93	1696.51	1083.17	1303.41	2644.45	1677.01
D_4 5th November	898.73	1213.60	2095.24	1402.52	847.86	1050.59	2273.02	1390.49
D_5 15th November	653.33	940.00	987.30	860.21	744.45	826.49	825.40	798.70
D_6 25th November	500.32	249.09	428.57	392.66	457.44	229.96	377.78	354.96
SEm±	82.16	91.53	160.78	186.12	100.99	80.18	143.72	226.80
CD at 5%	247.62	275.85	484.53	586.44	304.36	241.63	433.13	714.59
CV %	17.65	19.92	20.48	20.50	22.34	19.96	18.02	20.23
Interaction (Year x Planting time)		SEm±: 116.87 CD at 5%: 333.05			SEm±: 111.48 CD at 5%: 317.68			

Table 2.7 Yield losses in mustard cultivars/varieties due to powdery mildew (combined basis). (Mehta et al. 2008)

Treatment	Dose (%)	Disease intensity (%) RH-30	RH-8113	RH-9304	RH-9801	Mean	Disease control (%)	Yield (g/plot) RH-30	RH-8113	RH-9304	RH-9801	Mean	Increase in yield over control (g/plot)	Yield losses (%)
Sulfex (two sprays)	0.2	6.00 (14.04)	6.66 (14.92)	4.00 (11.27)	6.66 (14.71)	5.83 (13.73)	90.96	501.7	550.0	570.0	545.0	541.6	132.99	0.0
Sulfex (one spray)	0.2	22.00 (29.30)	26.00 (30.63)	22.00) (27.94)	24.00 (29.30)	23.50) (29.30)	63.65	461.0	501.0	490.0	495.0	486.7	78.08	10.13
Water (one Spray)		56.66 (48.81)	55.53 (48.81)	51.33) (45.74)	54.00 (47.27)	54.33) (47.66)	15.76	426.0	452.7	435.0	436.0	437.4	28.74	19.24
Control (unsprayed)		68.00 (55.53)	64.00 (53.11)	62.00) (51.93)	64.00 (53.12)	64.50) (53.42)	–	406.3	423.3	402.0	403.0	408.6	0.0	24.55
Mean		38.16 (36.92)	37.99 (36.87)	34.83) (34.22)	37.16 (36.10)		–	448.7	481.7	474.2	469.8	–	–	–
C.D at 5% (P = 0.05)	Treatment – 1.35 Varieties – 1.35 Treatment x varieties – NS						Treatments – 15.71 Varieties – 15.71 Treatment x Varieties – NS							

Figures in parentheses are the angular transformed values

Table 2.8 Yield loss in mustard cultivars due to powdery mildew. (Kohire et al. 2008)

	Yield (kg/ha)						
	Aurangabad			Parbhani			
Cultivar	Protected	Unprotected	Loss %	Protected	Unprotected	Loss %	Pooled loss
Seeta	802	474	40.0	821	577	29.0	34.5
Pusa Bold	1289	855	34.0	1158	740	37.0	35.5
Bio-902	1376	876	36.0	1107	789	28.0	32.0
TM-17	1202	1014	15.0	910	748	17.0	16.0
Pooled mean	1137	861	24.0	981	732	25.0	24.7

Table 2.9 Yield and yield contributing factors of mustard influenced by powdery mildew. (Kohire et al. 2008)

	Aurangabad			Parbhani		
Cultivar	Wt. of 1000 seeds (g)	Oil content (%)	Yield (kg/ha)	Wt. of 1000 seeds (g)	Oil content (%)	Yield (kg/ha)
Seeta	1.85	37.3	688	2.65	37.3	699
Pusa Bold	3.65	37.3	1072	4.25	37.3	994
Bio-902	2.75	37.5	1108	3.15	37.5	829
TM-17	2.8	37.6	1126	3.7	37.6	998
CD at 5%	0.3	1.05	173	0.4	1.1	91
Protected	3.05	37.2	1137	3.35	37.5	981
Unprotected	2.55	37.4	1231	3.25	37.3	732
CD at 5%	0.2	0.7	99.5	1.3	0.7	126
CD at 5% (CP)	0.45	1.5	202	0.65	1.6	61

2.6 Disease Assessment

The assessment of severity of powdery mildew has been carried out using different disease scoring scales to estimate quantitative and qualitative yield loss of total oilseeds produce along with oil content in the oilseeds, to determine efficacy of control measures used, to assess extent of host resistance in different genotypes, to compare genetically modified genotypes with wild types, as well as to assess different techniques of inoculation and pathogenesis. The various scoring systems of powdery mildew host–pathosystem developed by different workers are as follows:

2.6.1 A scoring system of powdery mildew of *Arabidopsis* plants to compare disease scores for wild type and mutants or transgenics was used by Reuber et al. (1998) as:

+1: Isolated spots of infection
+2: Approximately 20% coverage of leaves
+3: Approximately 50% coverage of leaves
+4: Nearly 100% coverage of leaves

Table 2.10 Losses estimated in crucifers due to powdery mildew

Crucifers	Country	Percent reduction in Seed yield	Percent reduction in Oil content	References
Brassica juncea (mustard)	India	17.5	6.47	Saharan and Sheoran (1985)
Brassica juncea (mustard)	India	17.5	–	Singh and Singh (2003)
Brassica juncea (mustard)	India	20.0	2.0	Samudre (1994)
Brassica juncea (mustard)	India	10.0–29.5	–	Mehta et al. (2008)
Brassica juncea (mustard)	India	45.0–90.0	–	Hare (1994)
Brassica juncea (mustard)	India	24.0	–	Dange et al. (2002)
Brassica juncea (mustard)	India	40.0	7.0	Hingole (1995)
Brassica juncea (mustard)	India	22.0	–	Patel et al. (1995)
Brassica juncea (mustard)	India	15.0–40.0	1–17	Kohire et al. (2008)
Brassica napus (oilseed rape)	France	10.0–30.0	–	Penaud (1998)

Disease scores were taken at approximately 2 weeks post-infection (Tables 2.5, 2.6, 2.7, 2.8, 2.9, and 2.10).

2.6.2 Susceptibility of *Arabidopsis* plants to *E. cruciferarum* was scored by Quentin et al. (2016) in a role of R-genes *MAP65-3* by visual examination at 7, 9, and 11 days after inoculation (dai). Plants were distributed in three categories of susceptibility with 0–30%, 30–60%, and > 70% diseased leaf area.

2.6.3 Adam and Somerville (1996) and Adam et al. (1999) screened *A. thaliana* accessions for powdery mildew resistance using disease reaction (DR) scores as: 0, limited hyphal development with no conidiation (i.e. fully resistant phenotype); 1, low to moderate hyphal growth with sporadic conidiation; 2, abundant mycelial development with moderate or delayed conidiation; and 3, profuse mycelial development and conidiation (i.e. fully susceptible phenotype).

2.6.4 To assess the efficacy of chemicals against powdery mildew infection severity, Singh and Singh (2003) used 0–5 scoring scales as follows: 1, 1–10% leaf area mildewed (R); 2, 11–25% leaf area mildewed (MR); 3, 26–50% leaf area mildewed (MS); 4, 51–75% leaf area mildewed (S); and 5, over 75% leaf area mildewed (HS). Shete et al. (2008) used 0–9 grading scale; Meena et al. (2011) used 0–9 common assessment scale for *Alternaria* blight, white rust, and powdery mildew. Lomate et al. (2014) and Penaud (1999) also used a 0–9 assessment scales to assess the efficacy of chemicals against powdery mildew of crucifers.

2.6.5 Shete et al. (2008) used the scoring scale for recording the observations on incidence of powdery mildew of crucifers. Observations regarding disease severity were recorded by 0–9 grading scale, and per cent disease intensity was calculated by following formula used by Gurha and Gangal (1980):

$$\text{PDI} = \frac{\text{Summation of all categories}}{\text{Number of leaves examined} \times 9} \times 100$$

where 9 is highest rating scale

2.6.6 Saharan and Mehta (2002) and Saharan et al. (2005) used a scoring scale of 0–5 to score the disease reaction of *Brassica* host species. The detailed description of the scale is 0 = no disease; 1 = 1–10% leaf area covered with mildew growth; 2 = 11–20% leaf areas covered with mildew growth; 3 = 21–50% leaf areas covered with mildew growth; 4 = 51–75% leaf areas covered with mildew growth; and 5 = more than 75% leaf area covered with mildew growth. 0 to 1 score was considered as resistant, 2 as moderately resistant, 3 as moderately susceptible, 4 as susceptible, and 5 as highly susceptible reaction of the host. Per cent disease intensity (PDI) was calculated with the following formula:

$$\text{PDI} = \frac{\text{Sum of total numerical ratings}}{\text{No. of leaves observed} \times \text{Max number of the scale}} \times 100$$

2.6.7 Meena et al. (2011) used the scoring scales of 0–9 for rating of rapeseed–mustard entries for reaction to powdery mildew as approved by AICRPRM plant pathologists (AICRPRM Proceedings 2011) group. It is a common rating scale for assessing *Alternaria* blight, white rust, powdery mildew, and downy mildew diseases of rapeseed–mustard.

Rating of scale	Category of host reaction	Description of scale
0	Immune	No lesion
1	HR	Non-sporulating pinpoint size or small brown necrotic spots, less than 5% leaf area covered by lesion
3	R	Small roundish slightly sporulating larger brown necrotic spots, about 1–2 mm in diameter with a distinct margin or yellow halo, 5–10% leaf area covered by lesions
5	MR	Moderately sporulating, non-coalescing larger brown spots, about 2–4 mm with a distinct margin or yellow halo, 11–25% leaf area covered by the spots
7	S	Moderately sporulating, coalescing larger brown spots, about 4–5 mm in diameter, 26–50% leaf area covered by the lesions
9	HS	Profusely sporulating, rapidly coalescing brown to black spots measuring more than 6 mm diameter without margins covering more than 50% leaf area

Average severity score =

$$\frac{(N-1\times 0)+(N-2\times 1)+(N-3\times 3)+(N-4\times 5)+(N-5\times 7)+(N-6\times 9)}{\text{Number of leaf samples}\times 9}$$

Disease index =

$$\frac{(N-1\times 0)+(N-2\times 1)+(N-3\times 3)+(N-4\times 5)+(N-5\times 7)+(N-6\times 9)\times 100}{\text{Number of leaf samples}\times 9}$$

where N-1 to N-6 represent frequency of leaves in the respective score.

Note

(1): *The word spots can be read as pustules if the same scale is used for white rust ratings.*

(2): *For PM, the same rating scale will be followed ignoring the lesion/pustule characteristics.*

2.6.8 Lomate et al. (2014) proposed the scoring scales to assess disease severity by 0–9 grade scale (Mayee and Datar 1986). Per cent disease intensity was calculated by the following formula used by Wheeler (1969):

$$\text{Percent disease intensity} = \frac{\text{Sum of all numerical ratings}}{\text{Total number of leaves examined} \times \text{Maximum rating}} \times 100$$

2.6.9 Kumar et al. (2015) used a scoring scale of 0–9 for rating reaction to powdery mildew of rapeseed–mustard:

0 – (Immune) =	Nil powdery mildew area
1 – (Highly Resistant) =	less than 5% powdery mildew area
3 – (Resistant) =	5–10% powdery mildew area
5 – (Moderately Resistant) =	11–25%powdery mildew area
7 – (Susceptible) =	26–50% powdery mildew area
9 – (Highly Susceptible) =	more than 50% powdery mildew area

2.6.10 Mert-Turk et al. (2008) used 1–6 scoring scale to assess powdery mildew intensity of oilseed rape under Turkey condition to observe the influence of nitrogen and fungicide applications on quality components of the crop: 1, no powdery mildew present; 2, occurrence of some white spots on stem; 3, powdery mildew present on stem and lower leaves; 4, powdery mildew present on stems and lower and upper leaves; 5, powdery mildew present on stems, lower leaves, upper leaves, and partly pods; and 6, powdery mildew covers all the plants.

2.6.11 Area under disease progress curve (AUDPC) can be calculated with the formula given by Wilcoxson et al. (1975):

$$\text{Accordingly} = \text{A} = \sum_{i=1}^{k} 1/2\left(Si + Si - 1\right) \times d$$

where

A	=	AUDPC value
Si	=	Disease severity at the end of the week i
k	=	The number of successive evaluation of disease
d	=	Interval between two evaluations

2.6.12 Apparent infection rate can be calculated with the formula given by Vander Plank (1963):

$$r = \frac{2.3}{t_2 - t_1}\left(\log\frac{X_2}{1-X_2} - \log\frac{X_1}{1-X_1}\right)$$

where

r = Apparent infection rate
t_1 = First date for recording disease intensity
t_2 = Second date for recording disease intensity
X_1 = Disease intensity at time t_1
X_2 = Disease intensity at time t_2
2.3 = Constant value

2.6.13 Components of slow mildewing can be analyzed with respect to the following: 1, progression of disease; 2, disease intensity; 3, number of specks/leaf; 4, number of conidia/speck; and 5, incubation and latent period (days).

2.6.14 Powdery mildew assessments were made by visual score of powdery mildew severity on pods by taking a sample of 5 to 20 plants per plot using the following scale in France by Penaud (1999):

0 = Healthy
1 = Occurrence of some white spots
3 = Superficial white powdery layer
5 = White powdery layer covering the whole pods
7 = Black lesions under the white powdery layer
9 = Grey to black colour on the whole pods

The data score was analyzed using a nonparametric method based on taking the ranks of the score and analyzing these ranks instead of the original values.

2.6.15 Initially Brain (1978) used a scoring system for *E. cruciferarum* development, which was based on the stage of fungal growth reached for each conidium inoculated on host surface as referred by Manners and Hossain (1963), Manners (1966), and Purnell (1971). The fungal growth was summarized as follows:

Symbol	Fungal growth stage
–	Ungerminated
G	Germinated (germ tube length > breadth)
A	Appressorium formed
P	Primary hypha present
S	Secondary hypha present
+	Expanding young colony (largest diameter 0.4 mm)

2.6.16 A general powdery mildew severity scale was proposed by James (1971) on the basis of leaf area covered by powdery mildew growth.

2.6.17 Shattuck and Parry (1990) evaluated *B. napus* ssp. *rapifera* powdery mildew-infected plants using mid- and older-aged leaves of each plant visually on a scale of 1, 5, 25, and 50% leaf infection.

2.6.18 Uloth et al. (2016, 2017) estimated the percentage of leaf, stem, and siliquae area infected with powdery mildew and related with the development stages of *B. napus* (Table 6.2; Chap. 6). The test plants were assessed weekly, starting 9 days after inoculation (dai). The incidence of visible infection on siliquae petioles and/or siliquae was also recorded from 90 dai.

The above-mentioned different powdery mildew disease assessment scales devised by the crucifer workers are basically based on visual observations of powdery mildew symptoms at different crop growth stages grown under different agroecological conditions. There is a need to devise a standardized internationally acceptable powdery mildew disease assessment scale which can correlate per cent yield loss with per cent disease severity near to precision under geographical conditions where the crucifer crops are grown.

References

Adam L, Somerville SC (1996) Genetic characterization of five powdery mildew disease resistance loci in *Arabidopsis thaliana*. Plant J 9:341–356

Adam L, Ellwood S, Wilson I, Saenz G, Xiao S, Oliver RP, Turner JG, Somerville S (1999) Comparison of *Erysiphe cichoracearum* and *E. cruciferarum* and a survey of 360 *Arabidopsis thaliana* accessions for resistance to these two powdery mildew pathogens. Mol Plant Microbe Inter 12:1031–1043

Aeenfar H (2006) Identification of powdery mildew fungi (Erysiphaceae) and their hosts in Mashhad and Mashhad vicinity, thesis, College of Science, Tehran Noor University

Ale-Agha N, Boyle H, Braun U, Butin H, Jage H, Kummer V, Shin HD (2008) Taxonomy, host range and distribution of some powdery mildew fungi (*Erysiphales*). Schlechtendalia 17:39–54

Alkooranee JT, Liu S, Aledan TR, Yin Y, Li M (2015) First report of powdery mildew caused by *Erysiphe cruciferarum* on *Brassica napus* in China. Plant Dis 99(11):1651

Allen RL, Bittner-Eddy PD, Grenville-Briggs LJ, Meitz JC, Rehmany AP, Rose LE, Beynon JL (2004) Host parasite co-evolutionary conflict between *Arabidopsis* and downy mildew. Science 306:1957–1960

Amano K (1986) Host range and geographical distribution of the powdery mildew fungi, Tokyo

Anonymous (2000–2017) Annual progress report of All India Co-ordinated Research Project on Rapeseed-Mustard, ICAR-Directorate of Rapeseed-Mustard Research, Bharatpur, India

Anonymous (2011) Proceedings of the Group Meeting of All India Co-ordinated Research Project on Rapeseed-Mustard. ICAR-Directorate of Rapeseed-Mustard Research, Bharatpur

Bappammal M, Hosagoudar VB, Udaiyan K (1995) Powdery mildews of Tamilnadu, India. New Bot 22:81–175

Bhander DS, Thakur RN, Hussain A (1963) A new diseases of rapeseed and mustard in India. Plant Dis Reptr 47:1039

Boesewinkel HJ (1977) Identification of Erysiphaceae by conidial characteristics. Revue Mycol 41:493–507

Bolay A (2005) Les Oïdiums de Suisse (Erysiphacées). Cryptogamica Helvetica 20:3–173

Bolay A, Braun U, Delhey R, Kummer V, Piatek M, Wolczanska A (2005) *Erysiphe deutziae*-a new epidemic spread in Europe. Cryptogam Mycol 26:293–298
Brain PJ (1978) Investigations into the powdery mildew disease of swedes. PhD thesis, University of Nottingham
Braun U (1987) A monograph of the Erysiphales (Powdery mildews). Beih Nova Hedw 89:1–700
Braun U (2012) The impacts of the discontinuation of dual nomenclature of pleomorphic fungi: the trivial facts, problems, and strategies. Intl Mycol Assoc 3(1):81–86
Braun U, Kiehr M, Delhey R (2000) Some new records of powdery mildew fungi from Argentina. Sydowia 53(1):34–43
Bunkina IA (1991) *Erysiphales*. In: Azbukina ZM, (ed.), Nizshie rastenija, griby i mohoobraznye Sovetskogo Dal'nego Vostoka, Griby, Tom 2: Askomicety, Erizifal'nye, Klavicipital'nye, Gelocial'nye, 11–142 p, Leningrad [In Russian]
Butler EJ (1918) Fungi and disease in plants. Thaker Spink and co. Calcutta, p 547
Chattopadhyay C, Bhattacharya BK, Kumar V, Kumar A, Meena PD (2011) Impact of climate change on pests and diseases of oilseed *Brassica* – the scenario unfolding in India. J Oilseed Brassica 2:48–55
Cho E, Park JH, Lee SH, Lee CK, Shin HD (2016) Occurrence of powdery mildew caused by *Podosphaera pannosa* on lemon eucalyptus. For Pathol 46:264–266
Choi HO, Choi Y, Kim DS, Hwang IS, Choi DS, Kim NH, Lee DH, Shin HD, Nam J, Hwang BK (2009) First report of Powdery mildew caused by *Erysiphe cruciferarum* on *Arabidopsis thaliana* in Korea. Plant Pathol J 25(1):86–90
Cooper J (2013) Mycological Notes 25: The status of names of powdery mildews in New Zealand 1-17
Dang JK, Sangwan MS, Mehta N, Kaushik CD (2000) Multiple disease resistance against four fungal foliar diseases of rapeseed-mustard. Indian Phytopathol 53(4):455–458
Dange SR, Patel S, Mishra RL, Saxsena CM (2002) Assessment of losses in yield due to powdery mildew disease in mustard under north Gujarat conditions. J Mycol Plant Pathol 32(2):249–250
Dange SRS, Patel RL, Patel SI, Patel KK (2003) Effect of planting time on the appearance and severity of white rust and powdery mildew diseases of mustard. Indian J Agril Res 37(2):154–156
Dixon GR (1974) Field Studies of Powdery Mildew (*Erysiphe crucifearum*) on Brussels Sprouts. Plant Pathol 23(3):105–109
Ellett CW (1966) Host Range of the *Erysiphaceae* of Ohio. Ohio J Sci 66(6):570–581
Ellis MB, Ellis JP (1985) Microfungi on land plants: an identification handbook. Macmillan Pub Co., New York, p 818
Ellis MB, Ellis JP (1997) Micro-fungi on land plants. Richmond Publishing, Slough
Ershad D (1971) Contribution to the knowledge of *Erysiphaceae* of Iran. Iranian J Plant Pathol 6(3–4):50–60
Ershad D (1995) Fungi of Iran. Publication number 10, Plant Pest and Disease Research Institute, Tehran, Iran
Farr DF, Bills GF, Chamuris GP, Rossman AY (1989) Fungi on plants and plant products in the United States. APS Press, St Paul, p 1252
Farr DF, Rossman AY, Palm ME, Mc Cray EB (2009). Fungal databases systematic botany and mycology laboratory, ARS, USDA. http://nt.ars-grin.gov/fungaldatabases
Gaetan S, Madia M (2004) First report of Canola Powdery mildew caused by *Erysiphe polygoni* in Argentina. Plant Dis 88(10):1163
Garibaldi A, Bertetti D, Gullino ML (2009) Outbreak of powdery mildew caused by *Erysiphe cruciferarum* on spider flower (*Cleome hassleriana*) in Italy. Plant Dis 93:963
Gohil VP, Jani SM, Dange SRS (1988) Assessment of yield losses due to powdery mildew in North Gujarat. Indian Phytopathol 41:156–157
Gunasinghe N, You MP, Lanoiselet V, Eyres N, Barbetti MJ (2013) First report of powdery mildew caused by *Erysiphe cruciferarum* on *Brassica campestris* var. *pekinensis, B. carinata, Eruca sativa, E. vesicaria* in Australia and on *B. rapa* and *B. oleracea* var. *capitata* in Western Australia. Plant Dis 97(9):1256

Gurha SN, Gangal LK (1980) Control of powdery mildew of green gram (*Vigna radiate* L.) wilchzek. Madras Agri J 67(10):666–668
Hare RM (1994) Influence of dates of sowing on powdery mildew of rapeseed mustard. MSc (Ag.). thesis, Marathwada Agricultural University, Parbhani, pp 1–47
Hingole DG (1995) Estimation of yield losses in rapeseed-mustard due to powdery mildew and white rust. MSc (Ag.) thesis, Marathwada Agricultural University, Parbhani, India, pp 1–52
James WC (1971) An illustrated series of assessment keys for plant diseases, their preparation and usage. Can Plant Dis Surv 51:39–65
Jee HJ, Shim CK, Choi YJ, Shin HD (2008) Powdery mildew caused by *Erysiphe cruciferarum* is found for the first time on Chinese cabbage in Korea. Plant Pathol 57:777
Junell L (1967) Erysiphaceae of Sweden. Symbolae Botanicae Upsalienses 14:1–117
Kachooeian Javadi S, Abbasi M, Riahi H, Mousavi SM (2006) Study of fungal flora (Erysiphales, Ustilaginales, Uredinales) of Jajroud protected region. Environ Sci 13:41–59
Karakaya A, Gray FA, Koch DW (1993) Powdery mildew of *Brassica* spp. in Wyoming. Plant Dis 77:1063
Kaur P, Li CX, Barbetti MJ (2008) First report of powdery mildew caused by *Erysiphe cruciferarum* on *Brassica juncea* in Australia. Plant Dis 92(4):650
Khodaparast SA, Hedjaroude GA, Ershad D, Zad J, Termeh F (2000) A study on the identification of *Erysiphaceae* in Guilan province, Iran (I). Rostaniha 1:53–63
Kim JY, Kim BS, Cho SE, Shin HD (2013) First report of powdery mildew caused by *Erysiphe cruciferarum* on Indian mustard (*Brassica juncea*) in Korea. Plant Dis 97:1383
Koch E, Slusarenko AJ (1990) Fungal pathogens of *Arabidopsis thaliana* (L.) Heyhn. Bot. Helv 100:257–268
Kohire OD (2002) Epidemiology and management of powdery mildew of rapeseed-mustard. PhD thesis, Marathwada Agricultural University, Parbhani [M.S.] India, p 159
Kohire OD, Ahmed R, Chavan SS, Khilare VC (2008) Yield losses amongst four varieties of mustard due to Powdery mildew in Maharastra. J Phytol Res 21(2):331–332
Koike ST, Saenz GS (1998) First report of powdery mildew, caused by an *Oidium* sp., on poinsettia in California. Plant Dis:82: 128
Kumar S, Singh D, Yadav SP, Prasad R (2015) Studies on powdery mildew of rapeseed-mustard (*Brassica juncea* L.) caused by *Erysiphe cruciferarum* and its management. J Pure Appl Microbiol 9(2):1481–1486
Lomate CB, Mate GD, Kalaskar RR (2014) Management of powdery mildew of mustard with chemicals and biogents. Intl J Plant Prot 7(1):122–124
Manners JG (1966) Assessment of germination in the fungus spore. In: Proceedings of the 18th symposium on Colston Research Society University of Bristol, pp 165–173
Manners JG, Hossain SMM (1963) Effect of temperature and humidity on conidial germination in *Erysiphe graminis*. Trans Br Mycol Soc 46:225–234
Mayee CD, Datar VV (1986) Phytopathometry. Technical Bulletin-1 (Special bulletin-3). Marathwada Agricultural University, Parbhani, p 74
Meena PD, Awasthi RP, Godika S, Gupta JC, Kumar A, Sandhu PS, Sharma P, Rai PK, Singh YP, Rathi AS, Prasad R, Rai D, Kolte SJ (2011) Eco-friendly approaches managing major diseases of Indian mustard. World Appl Sci J 12:1192–1195
Meena PD, Mehta N, Rai PK, Saharan GS (2018) Geographical distribution of rapeseed-mustard powdery mildew disease in India. J Mycol Pl Pathol 48(3):284–302
Mehta N, Singh K, Sangwan MS (2008) Assessment of yield losses and evaluation of different varieties/ genotypes of mustard against powdery mildew in Haryana. Plant Dis Res 23(1):55–59
Melchers LE (1931) A check list of plant diseases and fungi occurring in Egypt. Trans Kansas Acad Sci 34:41–106
Mert-Turk F, Kemal Gul M, Egesel C (2008) Nitrogen and fungicide applications against *Erysiphe cruciferarum* affect quality components of oilseed rape. Mycopathologia 165(1):27–35

Mirzaee MR, Khodaparast A, Sajedi S, Javadi SB, Saberi MH (2010) *Malcolmia africana*, a new host for powdery mildew disease caused by *Erysiphe cruciferarum* in Iran. Australasian Plant Dis Notes 5:101–102

Mohammadi-Doustdar E (1967) Mycology, powdery mildew of Iran, Tehran University no. 1262

Nair B, Nartam AT, Badukale MG, Mohurle NA (2017) Gamma rays and EMS induced mutants in Indian mustard (*Brassica juncea*). J Oilseed Brassica 8(1):62–71

Negrean G, Denchev CM (2000) New records of Bulgarian parasitic fungi. Fl Medit 10:101–108

Niknam GR, Guya M (1996) Identification of Erysiphe species causing powdery mildew on plant flora in Tabriz University campus. Agril Sci 6(3–4):1–19

Pastircakova K, Pastircak M (2013) A powdery mildew (*Pseudoidium* sp.) found on *Chelidonim majus* in the Czech Republic and Slovakia. Czech Mycol 65(1):125–132

Patel MK, Kamat MN, Bhide VP (1949) Fungi of Bombay. Indian Phytopathol 02:142

Patel VA, Vaishnav KA, Dhedhi BM (1995) Chemical control of powdery mildew. Indian J Mycol Plant Pathol 25:114–115

Paul YS (1984) New host records for Indian Erysiphaceae. Indian J Mycol Plant Pathol 14:177

Penaud A (1998) Oidium du colza, la protection du colza est maintenant possible. Oléoscope 50:36–38

Penaud A (1999) Chemical control and yield losses caused by *Erysiphe cruciferarum* on oil-seed rape in France. In: Wratten N, Salisbury PA (eds) Proceedings of the 10th International Rapeseed Congress. The Regional Institute Ltd, Canberra, Australia CD-ROM Doc No. 327: 1–8

Plotnikova JM, Reuber TL, Ausubel FM, Pfister DH (1998) Powdery mildew pathogenesis of *Arabidopsis thaliana*. Mycol Soc Am 90(6):1009–1016

Premlata Devi Y, Chhetry GKN (2016) Fungal diseases associated with rapeseed-mustard in organic farms in Manipur. Intl J Agril Sci Res 6(5):93–98

Purnell TJ (1971) Environmental and physiological studies of leaf infection of swede by *Erysiphe cruciferarum*. PhD thesis, University of Leeds

Quentin M, Baures I, Hoefle C, Caillaud M-C, Allasia V, Panabieres F, Abad P, Huckelhoven R, Keller H, Favery B (2016) The *Arabidopsis* microtubule-associated protein MAP65-3 supports infection by filamentous biotrophic pathogens by down-regulating salicylic acid-dependent defences. J Exper Bot 67(6):1731–1743

Radisek S, Jakse J, Zhao T, Cho S, Shin H (2018) First report of powdery mildew of *Capsella bursa-pastoris* caused by *Golovinomyces orontii* in Slovenia. J Plant Path 100:359

Reuber TL, Plotnikova JM, Dewdney J, Rogers EE, Wood W, Ausubel FM (1998) Correlation of defense gene induction defects with powdery mildew susceptibility in *Arabidopsis* enhanced disease susceptibility mutants. The Plant J 16(4):473–485

Reyes AA (1969) The seasonal occurrence of fungal and bacterial diseases of crucifers in Ontario in 1967 and 1968. Canadian Plant Dis Surv:4956–4957

Sadowski C, Dakowska S, Lukanowski A, Jedryczka M (2002) Occurrence of fungal diseases on spring rape in Poland. In: Paul VH, Foller I (eds) Working group on integrated control in oil-seed crops. Soest, Germany: IOBC/WPRS Bulletin, pp 1–12

Saharan GS (1980) Epidemiology of powdery mildew of rape and mustard. Indian J Mycol Plant Pathol (Abstr.) 10, XLVI

Saharan GS (1984) A review of research on rapeseed-mustard pathology in India. Paper presented in the Annual Rabi Oilseed Workshop held at Jaipur, India, 6–10 August, 1984

Saharan GS (1992a) Rapeseed and Mustard. In: (Kumar D, Rai M (eds) Advances in oilseeds research, vol 1. Scientific Publishers, Jodhpur pp 155–188

Saharan GS (1992b) AICRPRM monitoring team visit of Assam, Bihar and Delhi during 1992 crop season

Saharan GS (1998) Diseases of rapeseed and mustard and their management. In: Thind TS (ed) Diseases of field crops and their management. National Agricultural Technology Information Center, Ludhiana, pp 95–114

Saharan GS (2016) ICAR AICRPRM QRT visit of Birsa Agric. Univ., Ranchi, Jharkhand during 2016 crop season

Saharan GS, Kaushik JC (1981) Occurrence and epidemiology of powdery mildew of *Brassica*. Indian Phytopathol 35:17–21

Saharan GS, Mehta N (2002) Fungal diseases of rapeseed-mustard. In: Gupta VK, Paul YS (eds) Diseases of field crops. Indus Publishing Company, New Delhi, pp 193–228

Saharan GS, Sheoran BS (1985) Assessment of yield losses of mustard due to powdery mildew disease. Cruciferae Newslt 10:112–113

Saharan GS, Kaushik CD, Kaushik JC (1984) Powdery mildew of taramira and hirankhuri. Indian J Mycol Plant Pathol 14(2):167

Saharan GS, Mehta N Sangwan MS (2005) Development of disease resistance in rapeseed-mustard. In: Saharan GS, Mehta N, Sangwan MS (eds) Diseases of oilseed crops. Indus Publication Co, New Delhi, pp 561–617

Salmon ES (1900) A monograph of the Erysiphaceae. Mem Torrey Bot Club 9:1–292

Salmon ES (1902) Supplementary notes on the Erysiphaceae. Bull Torrey Bot Club 29(5):302–316

Samudre RA (1994) Influence of spacing 1n powdery mildew disease on four cultivars of rape seed mustard. M.Sc. (Agri.). Thesis, Marathwada Agricultural University, Parbhani, India. 1-43

Sankhla HS, Dalela GG, Mathur RL (1967) Occurrence of perithecial stage of *E. polygoni* on *B. campestris* var. Sarson and *B. juncea*. Plant Dis Reptr 51:800

Schmidt A, Scholler M (2011) Studies in *Erysiphales* anamorphs (4): species on *Hydrangeaceae* and *Papaveraceae*. Mycotaxon 115:287–301

Schmitt JA Jr (1955) The host specialization of *Erysiphe cichoracearum* from Zinnia phox an cucurbits. Mycologia 47:688–701

Sharma AK (1979) Powdery mildew diseases of some crucifers from Jammu and Kashmir state. Indian J Mycol Plant Pathol 9:29–32

Sharma ND, Khare CP (1992) Morphology of anamorphs of some powdery mildew fungi of Jabalpur division. Acta Bot Indica 20:269–277

Shattuck VI, Parry R (1990) The occurrence of powdery mildew on rutabagas in southern Ontario. Canadian Plant Dis Surv 70:15–16

Sherf AF, Mac Nab AA (1986) Vegetable diseases and their control. Wiley, Toronto, 728p

Shete MH, Dake GN, Gaikwad AP, Pawar NB (2008) Chemical management of powdery mildew of mustard. J Plant Dis Sci 3(1):46–48

Shin HD (2000) Erysiphaceae of Korea. Plant Pathogens of Korea 1:1–320

Shin HD, La YJ (1992) Addition to the new records of host plants of powdery mildews in Korea. Korean J Plant Pathol 8:57–60

Shivas RG (1989) Fungal and bacterial diseases of plants in Western Australia. J Royal Soc Western Aust 72:1–62

Singh RB, Singh RN (2003) Management of mildew of mustard. Indian Phytopathol 56:147–150

Slopek SW, Peters S (1988) A survey of canola diseases in south central Alberta, 1987. Canad Plant Dis Surv 68:113–114

Spencer DM (1978) The Powdery Mildews. Academic, UK, London, p 565

Suhag LS, Duhan JC (1985) Severity of powdery mildew disease on radish seed crop in Haryana. Indian Phytopathol 38(3):549–551

Tajik-Ghanbary MA (1995) Identification of fungi belonging to Erysiphaceae in Golestan National park. Thesis, Department of Plant Protection, College of Agriculture, Tehran University, Karaj, Iran

Tam LTT, Dung PN, Liem NV (2016) First report of powdery mildew caused by *Erysiphe cruciferarum* on *Brassica juncea* in Vietnam. Plant Dis 100(4):856

Uloth MB, You MP, Barbetti MJ (2016) Cultivar resistance offers the first opportunity for effective management of the emerging powdery mildew (*Erysiphe cruciferarum*) threat to oilseed *Brassicas* in Australia. Crop Pasture Sci 67:1179–1187

Uloth MB, You MP, Barbetti MJ (2017). Plant age and ambient temperature: significant drivers for powdery mildew (*Erysiphe cruciferarum*) epidemics on oilseed rape (*Brassica napus*). Plant Pathol:1–12

van Roessel GJ, Boothh, EJ, Walker KC, Eavis RM (1996) The susceptibility of *Camelina sativa* to clubroot (*Plasmodiophora brassicae*). In: Proceedings of the international symposium on domestication, production and utilization of new crops: practical approaches, Southampton, UK, pp 8–10

Vander Plank J (1963) Plant disease epidemics and control. Academic, New York, p 349

Vasudeva RS (1958) Diseases of rape and mustard. In: Singh DP (ed) Rape and mustard. Indian Central Oilseed Committee, Hyderabad, pp 77–86

Vellios E, Karkanis A, Bilalis D (2017) Powdery mildew (*Erysiphe cruciferarum*) infection on camelina (*Camelina sativa*) under Mediterranean conditions and the role of wild mustard (*Sinapis arvensis*) as alternative host of this pathogen. Emirates J Food Agri 29(8):639–642

Viennot-Bourgin G (1958) Contribution à la connaissancedes champignons parasites de l' Iran. Annales des Epiphyties (et de Phytogenetique) N S 9:97–210

Walker JC, Williams PH (1965) The inheritance of powdery mildew resistance in cabbage. Plant Dis Reptr 49:198–201

Wheeler BEJ (1969) An introduction to plant diseases. Wiley, London

Wilcoxson RD, Skovmand B, Atif AH (1975) Evaluation of wheat cultivars for ability to retard development of stem rust. Ann Appl Biol 80:275–281

Xiao X, Ellwood S, Findlay K, Oliver RP, Turner JG (1997) Characterization of three loci controlling resistance of *Arabidopsis thaliana* accession Ms-0 to two powdery mildew diseases. Plant J 12:757–768

Yanez-Morales M, de Jesús, Braun U, Minnis AM, Tovar-Pedraza JM (2009) Some new records and new species of powdery mildew fungi from Mexico. Schlechtendalia 19:47–61

Yarwood C E (1978). History and taxonomy of powdery mildews. In: Spencer DM (ed) The powdery mildews, vol 565. Academic, London, pp 1–37

Yengkhom PD, Chhetry GKN (2016) Fungal diseases associated with rapeseed-mustard in organic farms in Manipur. Intl J Agril Sci Res (IJASR) 6(5):93–98

Yengkhom PD, Chhetry GKN (2017) Effect of traditional agronomic practices on fungal diseases of rapeseed–mustard under organic farming system in Manipur. Intl J Adv Res 5(6):1256–1260

Zhao HH, Xing HH, Liang C, Yang XY, Cho SE, Shin HD (2014) First report of powdery mildew caused by *Erysiphe cruciferarum* on Chinese cabbage in China. Plant Dis 98:421

Chapter 3
The Pathogen

3.1 Introduction

Erysiphe cruciferarum (Opiz. ex L. Junell) is the major causal organism of crucifer's powdery mildew (PM) disease (Sankhla et al. 1967; Junell 1967; Saharan and Kaushik 1981; Koch and Slusarenko 1990). The morphological characters of conidia, chasmothecia, asci, and number of asci per Chasmothecia recovered from different *Brassica* species and varieties are given in Table 3.1 (Saharan and Kaushik 1981). The collective name *E. polygoni* DC was used by Salmon (1900) for the pathogen on swede (US = rutabaga), turnip, and other crop hosts. This species was reclassified by Blumer (1933) who used the *E. communis* for those forms whose morphology was indistinctly known and where chasmothecia seldom developed. Further, revision was proposed by Junell (1967), who regarded *E. communis* as a nomen ambiguum and used *E. cruciferarum* for the powdery mildew pathogen of a wide range of cultivated and wild crucifers (Dixon 1978). Four powdery mildew species are known to be able to complete their asexual life cycle on *Arabidopsis thaliana*, viz. *E. cruciferarum* (Koch and Slusarenko 1990), *Golovinomyces* (Syn. *Erysiphe*) *cichoracearum* (Gc) isolate UCSC1 (Adam and Somerville 1996), *Golovinomyces* (Syn. *Erysiphe*) *orontii* (Go) (Plotnikova et al. 1998), and a tomato powdery mildew pathogen *Oidium neolycopersici* (Bai et al. 2008). Despite their principal capacity to colonize *Arabidopsis*, not all *Arabidopsis* ecotypes are equally susceptible to this virulent powdery mildew pathogen (Adam et al. 1999). These four powdery mildew species differ in some morphological characteristics such as the size of the conidia, the shape of appressoria and haustoria, and the number of conidiophores per colony. The distinguishing anatomical and developmental features of four powdery mildew species pathogenic to *Arabidopsis* have been given in tabular form (Table 3.2) by Micali et al. (2008). Although all the four species of powdery mildew can colonize *Arabidopsis* under controlled laboratory conditions, the fungal species' range, frequency, and intensity of infection under natural

G. S. Saharan et al., *Powdery Mildew Disease of Crucifers: Biology, Ecology and Disease Management*, https://doi.org/10.1007/978-981-13-9853-7_3

Table 3.1 Size of conidia, cleistothecia (chasmothecia) and asci (µm) of *Erysiphe cruciferarum* on different species, and varieties of *Brassica* (Saharan and Kaushik 1981)

Brassica species/ varieties	Conidia (µm)	Cleistothecia/ chasmothecia (µm)	Asci (µm)	No. of asci/ cleistothecium
B. nigra var. Matopobka	14.47 × 31.03 (12.48–16.64 × 20.80–41.60)	113.5 (83.20–137.28)	31.95 × 54.08 (24.26–37.44 × 41.60–62.40)	3–8
B. campestris var. Candle	13.72 × 36.69 (8.32–16.64 × 24.96–45.76)	106.41 (83.20–133.20)	33.57 × 54.95 (29.12–37.44 × 49.92–58.24)	3–5
B. juncea var. Blaze	14.51 × 36.64 (8.32–20.80 × 24.96–45.76)	104.41 (83.20–133.12)	34.53 × 61.98 (29.12–37.44 × 54.08–66.56)	3–6
B. juncea var. Parkash	12.48 × 33.36 (8.32–16.64 × 24.90–41.60)	108.82 (83.20–133.12)	32.69 × 52.33 (29.12–37.44 × 41.60–62.25)	3–8
B. juncea var. RC 781	14.93 × 36.44 (8.32–20.80 × 24.96–45.76)	119.14 (83.20–124.80)	31.70 × 54.75 (29.12–37.44 × 41.60–62.40)	3–6

conditions are not known as of today. According to current names in *Erysiphales* proposed for inclusion in a list of accepted names where there is an earlier anamorph-typified names available, the causal organisms of crucifer's powdery mildews are as follows.

3.2 Causal Organisms

Erysiphe cruciferarum Opiz. ex L. Junell, Sevensk. Bot. Tidskr. 61: 217 (1967). Synonyms: *Erysiphe cruciferarum* Opiz. Lotos. 5: 42 (1855), nom. inval (Art. 32). *Erysiphe pisi* var. *cruciferarum* (Opiz. ex L. Junell) Ialongo, Mycotaxon 44: 255 (1992). *Oidium mattholae* Rayss, Palestine J. Bot. Jorusalem Ser. 1: 325 (1940; 1947) ["1938–1939] is the major causal organism of powdery mildew of crucifers under natural conditions. On *Arabidopsis thaliana* four powdery mildew species are known to complete their asexual life cycle, viz. *E. cruciferarum* (Koch and Slusarenko 1990), *Golovinomyces* (Syn. *Erysiphe*) *cichoracearum* (Adam and Somerville 1996), *Golovinomyces* (syn. *Erysiphe*) *orontii* (Plotnikova et al. 1998), and *Oidium neolycopersici* (Bai et al. 2008).

The distinguishing, morphological, anatomical, and developmental features of four powdery mildew pathogens pathogenic on *Arabidopsis* are given in Table 3.2.

Table 3.2 Distinguishing, morphological, anatomical, and developmental features of four powdery mildew species pathogenic on *Arabidopsis*[a] (Micali et al. 2008; Plotnikova et al. 1998; Adam and Somerville 1996)

Development features	*G. cichoracearum* (UCSC1)	*G. orontii* (MGH)	*E. cruciferarum* (UEA1)	*O. neolycopersici*
Conidial size and appearance	26–33 μm long 16–19 μm wide Smooth surface	29–35 μm long 15–19 μm wide Mostly smooth surface	38–46 μm long 13–16 μm wide Wart-like projections	23–345 μm long 13–20 μm wide Ribbon-like projections
Number of conidia per conidiophore	3–5 in chain	7–20 in chain	1–2 in chain	Rarely more than one and up to 3[c]
Conidiophore shape	Straight, first cell curved	Straight	Bent	–
Conidial shape	Ovoid, ellipsoidal	Ovoid, ellipsoidal	Ovoid, ellipsoidal	–
Conidial size	88–160 × 9.5–11 μm	70–100 × 10 μm	88–160 × 9.5–11 μm	–
Number of germ tube per conidium	One	4–5	3–4	–
Germ tube shape	Unlobed	Twisted, bent, helicoid	Bent	–
Appressoria	Unilobed, nipple shaped	Unilobed, nipple shaped	Multilobed (2–4)	Multilobed
Haustorial shape	Ellipsoidal, oval	Spherical	Ovoid, elongated	Spherical, sac-like[c]
Hyphal branch angles	45–90 degrees	90 degrees, 5.8–6.6 μm wide	Less than 45 degrees	90 degrees[c]
Alternate hosts	Asteraceae, Cucurbitaceae	Brassicaceae, Solanaceae, Cucurbitaceae	Brassicaceae	Cucurbitaceae, Solanaceae
Conidium germination	1–2 hpi[b]	1–2 hpi	N/A	3–5 hpi[c]
Appressorium formation	6–10 hpi	~5 hpi	N/A	6–8 hpi[c]
Haustorium formation	10–14 hpi	14–16 hpi	N/A	N/A
Conidiophore formation	4 dpi[b]	5 dpi	5–6 dpi	6 dpi[c]

[a]Information was obtained from D Meyer as well as from the following publications: Adam and Somerville 1996; Plotnikova et al. 1998; Adam et al. 1999; Vogel and Somerville 2000
[b]*hpi* hours post-inoculation, *dpi* days post-inoculation, *N/A* not available
[c]Phenotype observed on tomato (Whipps et al. 1998; Jones et al. 2001)

(Micali et al. 2008; Plotnikova et al. 1998; Adam and Somerville 1996). Morphological comparisons of *E. orontii, E. cruciferarum,* and *E. cichoracearum* isolates from *A. thaliana* leaves are also presented in Table 3.2. Recently described powdery mildew isolates recovered from common sow thistle (*Sonchus oleraceus*)-

designated *Golovinomyces cichoracearum* UMSG 1 and tobacco (*Nicotiana tabacum*)-designated Gc *SICAU 1* are only partially adapted to *Arabidopsis*. These isolates show considerable host cell penetration rates but either fail to complete their life cycle (Gc *UMSG1*) or show only little sporulation (Gc *SICAV1*) on Col-0 wild-type plants (Wen et al. 2011; Zhang et al. 2015). Two species of powdery mildews, viz. *E. cichoracearum* and *E. polygoni*, are very complex and polyphagus (Salmon 1900; Junell 1967; Yarwood 1978; Braun 1987), whereas cereal powdery mildew pathogens *Blumeria graminis* f. sp. *hordei* and *B. graminis* f. sp. *tritici* are very specific on barley and wheat (Jorgensen 1988).

The systematic position of powdery mildew pathogens of crucifers is as kingdom, Fungi; phylum, *Ascomycota*; division, *Ascomycotina*; class, *Ascomycetes*; order, *Erysiphales*; family, *Erysiphaceae*; subfamily, *Erysiphoideae*; tribus, *Erysipheae*; subtribus, *Erysiphinae*; and genera, *Erysiphe* sect. *Erysiphe*; *E.* sect. *Golovinomyces*, species pathogenic on *Arabidopsis thaliana*; *Erysiphe cruciferarum*, *Golovinomyces* (syn. *Erysiphe*) *cichoracearum*, G. (syn. *Erysiphe*) *orontii*, and the tomato powdery mildew pathogen *Oidium neolycopersici*.

3.3 Classification

Linnaeus was the first person who mentioned powdery mildew under the name of *Mucor erysiphe* in species *Plantarum* (1753). Following Linnaeus, Persoon (1796, 1801) and Rebentish (1804) published the first illustrations of powdery mildews. Later, de Candolle (1805–1815), Fries (1829), and many others made additions to increase our knowledge of powdery mildews. Wallroth (1819) mentioned about systematic arrangement of powdery mildews on the basis of morphological characters rather than the host plant infected. De Schweinitz (1834) described several North American forms. Leveille (1851) divided powdery mildews into six genera on the basis of number of asci in the perithecium and the characters of the appendages. Salmon (1900) wrote the first monograph of powdery mildews and recognized 6 genera, 49 species, and 11 varieties. However, Saccardo recognized more number of species. Following Loveille, Salmon divided the *Erysiphaceae* into six genera and used the same generic names as proposed by Leveille (1851). Salmon (1900) in his monograph of *Erysiphaceae* for the first time described keys to differentiate the genera and species based on characteristics of perithecium, number of ascus per perithecium, and types of appendages on perithecial surface. Stevens (1925) in his book "*Plant Disease Fungi*" has also divided family *Erysiphaceae* into six genera, viz. *Erysiphe*, *Podosphaera*, *Sphaerotheca*, *Uncinula*, *Microsphaera,* and *Phyllactinia*. The *Erysiphales* contains one family, the *Erysiphaceae*. Kirk et al. (2008) recognized 19 genera of powdery mildew fungi. Braun and Cook (2012) reduced these to 16 plus 9 additional forms of genera to accommodate those species known only from asexual (anamorphic) states and not yet formally given names in the primary taxonomy of the family powdery mildews which occur throughout the world. Cook et al. (1997) in their detailed studies of powdery mildew anamorphs using light and scanning electron microscopy and host range have provided characteristic features of identification and classification. Previously undescribed features on the surfaces of powdery mildew conidia were revealed by the scanning electron microscope (SEM), reinforcing differences observable by light microscopy (LM). Four distinct patterns were observed on the septa, and 10 on the outer wall, thus categorizing 15 anamorph taxa. The patterns on the outer wall of conidia were frequently modified by secondary creasing. These wrinkling patterns, also observable under the LM, proved useful in identification of anamorphs especially on herbarium specimens. Scanning electron microcopy, light microscopy, and host range data were used to construct keys for the prediction of the anamorphs of all the powdery mildew genera found in the UK. The proposed new structure (Cook et al. 1997) has been compared with Braun's 1987, 1995 structure in Fig. 3.1.

3.4 Taxonomy and Nomenclature

The taxonomy, identification, and nomenclature of powdery mildews are based largely on the characteristics of the teleomorph such as the shape of the appendages on the cleistothecium or chasmothecium and whether the latter contains one or several asci. This causes problems when a powdery mildew increases its host range or geographical areas, because the teleomorph may not be seen for some years or may not be produced at all. It can then be impossible to identify its holomorph genus let alone its species. An accurate identification is essential in plant quarantine, and in locating the origin or source of inoculum. Any additional taxonomic details discernible by SEM could help to identify the pathogen more accurately especially when such information is linked with the known susceptibility of the particular powdery mildews. The identification may then be progressed beyond the extremely broadly based *Oidium* (Cook et al. 1997).

The taxonomic studies in the genus *Erysiphe* with revision of the generic delimitation, and position in the system of the *Erysiphaceae* along with key, and hypothesis about the phylogenetical relationship within the family have been described by Braun (1981).

3.4.1 Delimitation of the Genus Erysiphe

The restriction of the genus *Erysiphe* in the present sense dates back to Leveille (1851). Species with numerous asci, and mycelioid appendages, without uncinate or dichotomously branched tips, were included. This conception was recognized by Salmon (1900) in his monograph and later on by most authors of powdery mildews. The structure of the appendages played a dominant part in the taxonomy of the *Erysiphaceae*. Its taxonomic value was often overestimated. Corresponding types of appendages are present in different lines of the evolution of the powdery mildews

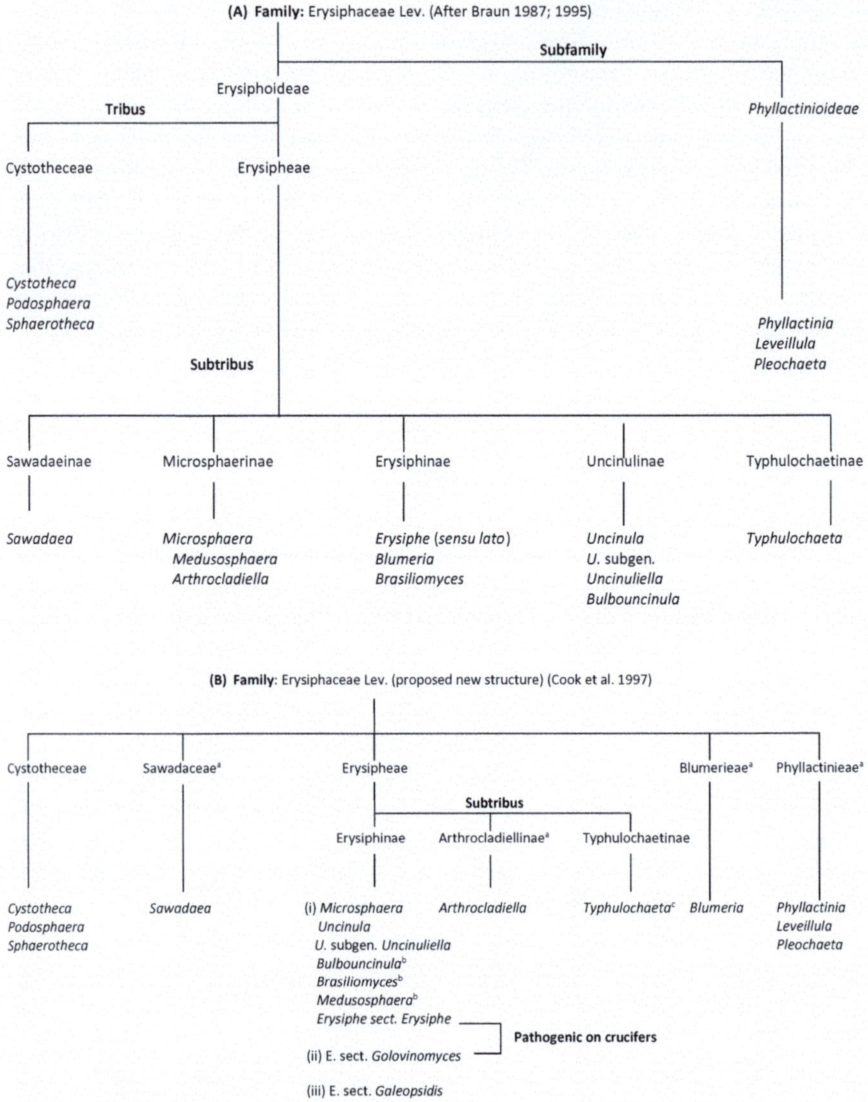

[a]Proposed new tribe/subtribe

[b]Not examined in this study. Because the anamorph is rare or unknown and is not known in Europe (Braun 1995). However, the teleomorph has as affinity with one or more of the genera in group (i) (Braun 1987)

[c]Not examined in this study because the anamorph is unknown (Braun 1987)

Fig. 3.1 Structure of the Erysiphaceae, (**a**) Braun's (1987, 1995) revisions, (**b**) proposed new structure. (Cook et al. 1997)

(*Erysiphe-Microsphaera* and *Sphaerotheca-Podosphaera*; *Erysiphe-Uncinula* and *Leveillula-Pleochaeta*). It is not always possible to place species with corresponding number of asci and an agreeing type of appendages in one genus (e.g. *Erysiphe-Leveillula*). The whole complex of features should be compared.

Braun (1978) recognized the system of Blumer (1967) and dealt with the nomenclature of the sections. The genus *Erysiphe* was divided into the following sections: 1. Sect. *Erysiphe* (type-*Erysiphe polygoni*), 2. Sect. *Golovinomyces* (type-*Erysiphe cichoracearum*), 3. Sect. *Blumeria* (type-*Erysiphe graminis*), and 4. Sect. *Trichocladia* (type – *Erysiphe tortilis*). Braun (1981) has intensively been engaged in the problem of the generic delimitation within the *Erysiphaceae*. The positions of the mentioned first two sections are as follows. The other sections have not been included here since they are not pathogenic on crucifers or beyond the scope of this book.

3.4.2 *The Position of* Erysiphe graminis *(Sect.* Blumeria*)*

Erysiphe graminis is a rather isolated species of the genus. The conidial state differs completely from the other species. The foot cells of the conidiophores are characterized by a bulbous swelling. The haustoria are digitate. The mode of their development deviates from the rest of the powdery mildews (Hirata and Kojima 1962). A coloured secondary mycelium is formed, and the structure of the fruit bodies is also somewhat different from typical *Erysiphe* (Neger 1901; Speer 1975). *E. graminis* is confined to hosts of the family Poaceae, almost the only hosts of the Monocotyledoneae. On account of the isolated position, Golovin (1958) proposed the new genus *Blumeria* for it. Speer (1975) recognized and validated *Blumeria*. *E. graminis* should be excluded from *Erysiphe*. That is an important step to attain to a more homogenous genus *Erysiphe*. Therefore, *Blumeria* is recognized as a separate genus (Braun 1981). The development of cleistothecial appendages from *Erysiphe* to *Microsphaera* has been depicted in Fig. 3.2.

3.4.3 Erysiphe *emend. nov.*

E. graminis (= *Blumeria*) and *E. trina* (= *Californiomyces*) are excluded from *Erysiphe*. The limit between *Erysiphe* and *Microsphaera* is more strictly drawn. *E. tortilis, E. trifolii*, and *E. hyperici* are transferred to *Microsphaera*. The rest of the species belongs to two sections. They are characterized as follows:

1. Sect. *Erysiphe* – conidia singly formed (*Pseudoidium* type), conidiophores short and slender, about 10–50(−80) × 5–10 μm, appressoria mostly lobed, asci (2-) 3-8-spored.
2. Sect. *Golovinomyces* – conidia in chains (*Euoidium* type), conidiophores long and broad, about 40–250 × 9–15 μm, appressoria nipple-shaped, rarely indistinct

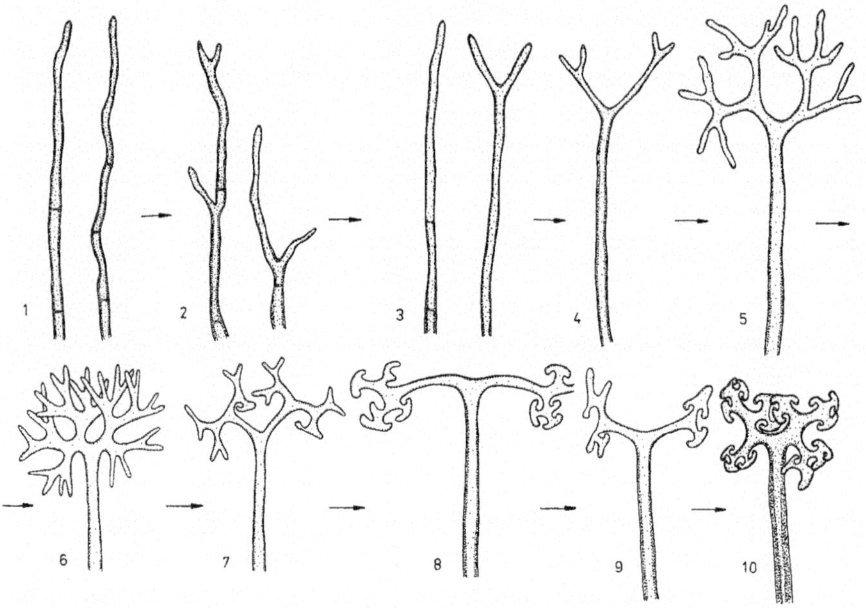

Fig. 3.2 Development of the cleistothecial appendages from *Erysiphe* to *Microsphaera*. (1) *Erysiphe* with simple appendages (*E. cruciferarum* Junell), (2) *Erysiphe* with irregularly branched appendages (*E. heraclei* DC. ex. St-Am), (3) *Microsphaera trifolii* (Grev.) U. Braun, (4) *M. baeumleri* Magn., (5) *M. diffusa* Cke. & Pk., (6) *M. grossulariae* (Wallr. Ex) Lev., (7) *M. lonicerae* (Dc. ex St. Am) Wint., (8) *M. nomurae* U. Braun, (9) *M. semitosta* Berk and Curt, (10) *M. penicillata* (Waller. Ex Fr.) Lev. (Braun 1981)

or moderately lobed, asci 2(-4)-spored. Sawada (1949, 1951, 1959) was the first who proposed to treat the two groups as separate genera. He established the new genus *Ischnochaeta* for the species with singly formed conidia (sect. *Erysiphe*). The Latin description was published by Sawada in 1959. The same proposal was done by Golovin (1958). He named the new genus *Linkomyces*. Both genera are invalid. *Erysiphe polygoni* is the lectotype of *Erysiphe* (Braun 1978). Therefore, *Ischnochaeta* and *Linkomyces* become synonyms of *Erysiphe*. Furthermore, *Linkomyces* was published without Latin description (nom. nud.). At first, Braun (1981) was inclined to distinguish the two groups at generic level. Sect. *Erysiphe* and *Golovinomyces* are, however, linked by intermediate species. *Erysiphe galeopsidis* is characterized by an unusual combination of features. The asci are (2-) 3-6(-8)-spored and the appressoria are lobed, two characteristics of section *Erysiphe*. But the conidia are produced in real chains as in sect. *Golovinomyces*. *E. galeopsidis* belongs neither to sect. *Erysiphe* nor to *Golovinomyces*. The ascospores are only developed after overwintering. This mark confirms the isolated status of the species. It should be placed in a separate section. Besides, the number of ascospores is not strictly fixed. *E. cichoracearum* forms with 3- or 4-spored asci are not unusual. *E. asperifoliorum* possesses a conidial state which clearly belongs to sect. *Golovinomyces*, whereas the number of ascospores is fairly variable (2–4). On the other hand, lobed appressoria are sometimes to be met in species of sect. *Golovinomyces* (*E. galii* or *E. cichoracearum* on *Sonchus arvensis*). Owing to these intermediate forms, a separation at generic rank is impossible. These features cannot be used to segregate genera. The discussed facts force to retain the two groups as sections and to establish a new section for *E. galeopsidis*.

***Erysiphe* DC. ex Fr. emend. nov.**
Conidial state belonging to the genus *Oidium*, mycelium ectophytic, fibrosin bodies are absent. Cleistothecia not conspicuously dorsiventral, peridium multilayered, outer layer, mostly 2–3 cells deep, composed of thick-walled cells, followed by an inner layer of thin-walled cells, peridium dark-brown to black when mature, mycelioid appendages present, simple or irregularly branched, attached between basis and equatorial zone, very rarely in the upper half, ascocarps with 2 or more asci, 2-8-spored.
Lectotype: *Erysiphe polygoni*.

3.4.4 Generic Structure of Erysiphe

Erysiphe ought to be divided into three sections:

1. Sect. Erysiphe (type: *Erysiphe polygoni*)
Conidia singly formed (*Pseudoidium* type), appressoria mostly lobed, ascospores (2-) 3-8
2. Sect. *Golovinomyces* (type: *Erysiphe cichoracearum*)
Conidia in chains (*Euoidium* type), appressoria mostly nipple-shaped, ascospores 2(-4)
3. Sect. Galeopsidis sect. nov. (type: *Erysiphe galeopsidis*)
Conidia in chains, appressoria lobed, ascospores (2-) 3-6 (-8), only developed after overwintering
Conidia catenulatis, hyphopodiis lobatis, ascus (2-) 3-6 (-8)-spores

3.4.5 Sect. Erysiphe

This section is composed of two well-characterized groups of species, which are different in the number of ascospores. Most species of the section show a considerable variability in the number of ascospores.

The asci of the species in question are mostly (2-) 3-6-spored. *E. polygoni, E. heraclei*, and allied species are members of this group. A smaller set of species possess a constantly larger number of spores (6-8, rarely 5). The limit between the two groups is not strict. There is a number of species in group 1 with usually 4-6-

spored asci (e.g. *Erysiphe liriodendri* Schw. or *E. hiratae* U. Braun). Such species form the transition between the groups. Nevertheless, the differences between both groups are striking. So, Braun (1981) considered them as subsections. It shall be a step towards a genus *Erysiphe* which is easier to be surveyed.

1. *Erysiphe* sect. *Erysiphe* subsect. *Erysiphe*: Number of ascospores variable, (2-) 3-6

The following species belong to the subsection:

Erysiphe aceriphylli Golovin and Bunkina, Bot. mat. otd. spor. rast. 14, p. 119, 1962 (on *Aceriphyllum*, Korea).

E. aquilegiae DC. ex Merat, Nouv. fl. env. Paris, ed. 2, 1, p. 132, 1821 (on Ranunculaceae).

E. betae (Vanha) Weltzien, Phytopath. Z. 47, p. 127, 1963 (on *Beta*).

E. buhrii U. Braun, Ceska Myk. 32, p. 80, 1978 (on Caryophyllaceae, Eur.-As.).

E. circaeae Junell, Sv. Bot. Tidskr. 61, p. 224, 1967 (on *Circaea*, Eur.).

E. convolvuli DC. ex St.-Am., Fl. Agen., p. 615, 1821 (on Convolvulaceae). a. var. *convolvuli* (on Convolvulus, Eur.-As.). b. var. *calystegiae* U. Braun var. nov. (on *Calystegia*, Eur.-Central As.).

Syn.: *Erysiphe convolvuli-sepii* Cast., Cat. Pl. Mars., p. 188, 1845. Diagn.: Differt a typo ascosporis (3-) 5-6, plerumque 6. Holotypus: on *Calystegia sepium* (L.) R. Br. ("Ad rheni ripas pr. Hattenheim", Fuckel, Fungi rhen., Suppl., Fasc. VIII, No. 2237. HAL).

The asci in var. *convolvuli* are (2-) 3-4(-6)-spored, in var. *calystegiae*, however, (3-) 5-6, mostly 6-spored.

E. cruciferarum **Opiz. ex Junell, 1.c., p. 217 (on Brassicaceae and other families):**

E. deserticola Speg.(?), Fung. Arg. nov. v. crit., p. 224, 1899 (on *Hoffmannseggia*, Argentina).

E. epimedii (Tai) Z. et C., Acta Mier. Sin. 20, p. 356, 1980 (on *Epimedium*, China).

E. frickii Neger(?), Ber. Deutsch. Bot. Ges. 17, p. 241, 1899 (on *Geum*, Chile).

E. heraclei DC. ex St.-Am., 1.c., p. 615 (on Apiaceae).

E. hiratae U. Braun, Feddes Repert. 92, p. 500, 1981 (on *Quercus*, Asia).

E. hommae U. Braun, 1.c., p. 501 (on Lamiaceae, Asia-N. Am.).

E. jatrophae Doidge, Bothalia 4, p. 839, 1948 (on *Jatropha* and *Acalypha*, S. Afr.).

E. knautiae Duby, Bot. Gall. 2, p. 870, 1830 (on Dipsacaceae, Geraniaceae).

E. krumbholzii U. Braun, Feddes Repert. 91, p. 440, 1980 (on *Chrysosplenium* Asia).

E. limonii Junell, 1.c., p. 225 (on Plumbaginaceae, Eur.-As.).

E. liriodendri Schw., Syn. Fung. Am. Bor., p. 269, 1834 (on *Liriodendron*, N. Am.).

E. lythri Junell, 1.c., p. 223 (on *Lythrum*, Eur.-As.).

E. parnassiae (Halst.) Jacz., Karm. opr. grib., p. 165, 1927 (on *Parnassia*, N. Am.). land).

E. pileae (Jacz.) Bunkina ex U. Braun, Feddes Repert. 92, p. 503, 1981 (on *Pilea*, Asia).

E. pisi DC. ex St.-Am., FL Agen., p. 614, 1821 (on *Fabaceae*) a. var. *pisi* (on Fabaceae).

b. var. *cruchetiana* (Blumer) U. Braun stat. nov.

Bas.: *Erysiphe cruchetiana* Blumer, Beitr. Krypt.-Fl. Schweiz 7(1), p. 193, 1933. The variety differs in the frequently branched appendages. They are branched in a coral-like manner. It is to be found on *Ononis*, on *Lathyrus*, and may be on some additional hosts.

E. polygoni DC. ex St.-Am., l.c., p. 614 (on Polygonaceae).

E. ranunculi Grey., Fl. Edin., p. 461, 1824 (on *Ranunculaceae*).

E. sambuci Ahmad, Biologia (Lahore) 6(2), p. 118, 1961 (on *Sambucus*, Asia). a. var. *sambuci*.

b. var. *crassitunicatae* Zheng & Chen, Acta Microbiol. Sinica 20(1), p. 47, 1980 (on *Sambucus*, China).

E. sedi U. Braun, 1.e., p. 502 (on *Sedum*, Asia).

E. sepulta Ell. & Everh., Bot. Gaz. 14, p. 286, 1889 (on *Bigelovia*, N. Am.).

E. thesii Junell, 1.c., p. 216 (on *Thesium*, Eur.-As.).

E. urticae Blumer, l.c., p. 224 (on *Urtica*, Eur.-As.).

E. viciae-unijugae (Homma) U. Braun, 1.c., p. 499 (on *Vicia*, Asia).

Erysiphe sect. *Erysiphe* subsect. Polysporae U. Braun subsect. nov.

Asci 6-8-, rarely 5-spored. – Asci 6-8-sporis, rare 5.

Typus: *Erysiphe ulmariae* Desm.

The following species belong to the subsection:

Erysiphe abeliae Zheng & Chen, Acta Microbiol. Sinica 20(1), p. 45, 1980 (on *Abelia*, China).

E. aggregata (Peck) Farl., Bull. Buss. Inst. 2, p. 227, 1878 (*Alnus*, N. Am.).

E. bunkiniana U. Braun, Feddes Report. 91, p. 441, 1980 (on *Plectranthus*, Far East of the Soviet Union).

E. carpophila Syd., Ann. Myc. 22, p. 294, 1924 (on *Weinmannia*, New Zealand).

E. densa Berk., in Hook., Fl. Nov. Zeal. 2, p. 208, 1855 (on *Aristotelia*, New Zealand).

E. glycines Tai, Lingnan Sci. Journ. 18(4), p. 457, 1939 (on *Glycine*, China).

Note: The appendages are up to 10 times as long as the cleistothecial diameter. The host is a member of the family *Fabaceae*. Because of the long appendages, the species seems to be close to *Microsphaera trifolii* (= *Erysiphe trifolii*). It should be transferred to *Microsphaera* (Braun 1981). Unfortunately, the type of *E. glycines* was not available for observation to Braun. The status of the species, however, can only be cleared up on the basis of the type. Therefore, Braun (1981) retained it provisory in *Erysiphe*.

E. lagerstroemiae West., Phytopathol. 23(10), p. 817, 1933 (on *Lagerstroemia*, USA).

E. mayorii Blumer, Beitr. Krypt.-Fl. Schweiz 7(1), p. 174, 1933 (on *Cirsium*, *Cicerbita*, Eur.-As.).

E. rhododendri Kapoor, Ind. Phytopathol. 18, p. 90, 1965 (on *Rhododendron*, India).

E. rubicola (Murr.) Boesewinkel, Trans. Brit. mycol. Soc. 67(1), p. 144, 1976 (on Rubus, New Zealand).
E. sikkimensis Chona, Kapoor & Gill, Ind. Phytopathol. 13, p. 72, 1960 (on *Castanopsis*, India, China).
E. symploci Kapoor, l.c., p. 91 (on *Symplocos*, India).
E. ulmariae Desm., Ann. Sci. Nat., bot., 3 ser., 6, p. 66, 1846 (on *Filipendula*, Eur.-As.).
E. vernalis Karst., Bidr. Kan. Finl. nat. o. folk 23, p. 193, 1873 (on *Alnus*, Eur.).
Note: *Uncinula paradoxa* Simonjan (Izv. akad. nauk. Armen. SSR, biol. nauk, 12(2), p. 88, 1958) seems to belong to *Erysiphe* (subsect. *Polysporae*). The species was described on *Acer assyricus* Pojark. from Armenia. The whole appendage or at least the upper half is loosely curved. *Uncinula* is characterized by a different type of appendages. The ultimate tips are uncinate. However, Braun (1981) met with loosely curved upper parts of appendages in *Erysiphe hiratae* Braun, a species on *Quercus*.

3.4.6 *Sect.* Golovinomyces

The cleistothecial states are rather uniform within the section. They hardly provide differences for a subgeneric arrangement. The conidial states, however, allow separating the section into two groups (Braun 1980a, b). *E. cichoracearum* and closely allied species possess conidiophores with cylindric foot cells and ellipsoid (barrel-shaped) conidia. The foot cells are usually shorter than 100 μm. A second group is characterized by very long foot cells, often longer than 100 μm, which are increasing from base to top. The conidia show characteristic constrictions at the ends. *E. verbasci*, *E. depressa*, and *E. echinopis* belong to this group.

1. Erysiphe sect. *Golovinomyces* subsect. *Golovinomyces*
Foot cells of the conidiophores cylindric, straight to curved, mostly shorter than 100 μm, conidia ellipsoid to barrel-shaped, without constrictions

The following species belong to the subsection:
Erysiphe artemisiae Grev., Fl. Edin., p. 459, 1824 (on *Artemisia*, Eur.-As. and ? N. Am.).
E. asperifoliorum Grev., l.c., p. 461 (on Boraginaceae).
var. *asperifoliorum* (on *Boraginaceae*, cosmopol.?).
var. *anchusae* U. Braun var. nov. (on *Anchusa*, Eur.).
Diagn.: Differt a typo ascosporis (2-) 3-4(-5).
Holotypus: on *Anchusa officinalis* L. (Klotzsch, Herb. viv. myc., cent. 10, 952, Dresden 1846, sub *Erysiphe horridula* Wallr. a. Asperifol. var. *anchusae*, Lasch) HAL.
The asci in var. *asperifoliorum* are 2-3-spored, very rarely 4-spored. The constantly high number of ascospores in the powdery mildew on *Anchusa* is very striking. Salmon (1900) places it in *Erysiphe polygoni* emend. nov.

E. biocellata Ehrh., N. Acta phys.-med. Acad. Caes. Leop.-Carol. nat. cur. 10, p. 211, 1821 (on *Lamiaceae*, cosmop.?).

Syn.: *E. labiatarum* ss. Blumer (1933).

E. cichoracearum DC. ex Merat, Nouv. fl. env. Paris, ed. 2, 1, p. 132, 1821 (on diverse hosts, cosmopol.). Note: The cleistothecia provide only few characteristics in the *E. cichoracearum* complex. Braun (1980a, b) studied on numerous hosts. They are very constant within the complex. That is why Braun (1981) considers the imperfect state as the most important feature in the *E. cichoracearum* complex. Some taxa, previously treated as species, deviate only slightly from *E. cichoracearum* s.str. A complete overlapping is to be noticed. The conidial states correspond to *E. cichoracearum* and indicate a close connection. These taxa, biologically specialized, are hardly more than a variety of *E. cichoracearum.*

var. *cichoracearum* (on diverse hosts).

Bas.: *Erysiphe fischeri* Blumer, Beitr. Krypt.-Fl. Schweiz 7(1), p. 262, 1933.

var. *plantaginis* (Link) U. Braun comb. nov. (on *Plantago*, cosmopol.?).

Bas.: *Erysiphe lamprocarpa* var. *plantaginis* Link, in L., Spec. plant., ed. 4, p. 109, 1824.

Syn.: *Erysiphe lamprocarpa* ss. Blumer (1933). *E. plantaginis* (Link) Sawada (1927) non Castagne (1845). *E. sordida* Junell (1965).

var. *magnicellulata* (U. Braun) U. Braun stat. nov. (on *Phlox*, N. Am., introduced in Europe and Asia).

Bas.: *Erysiphe magnicellulata* U. Braun, Feddes Repert. 88, p. 656, 1978.

Note: Braun (1981) studied the holotype of *Erysiphe phlogis* Schw. (ex PH). Salmon (1900) cited *E. phlogis* as a synonym of *E. cichoracearum.* However, Braun (1981) failed to find cleistothecia in the original specimen. Three powdery mildews are known on *Phlox*, *E. cichoracearum* s. lat., *Sphaerotheca fuliginea* s. lat., and *Sph. macularis* s.lat. Therefore, the identity of *E. phlogis* remains unclear. So, it was regarded as an uncertain species (nom. dub.). The original description said nothing on the number of asci and ascospores.

E. galii Blumer, l.c., p. 283 (on *Galium*, Eur., As., N. Am.).

E. macrocarpa Speer, Anz. iisterr. Akad. Wiss., math.-nat. K1., 106(1–4), p. 245, 1970 (on *Tanacetum*, Austria).

E. monardae Nagy, Phytopath. Z. 88, p. 285, 1977 (on *Monarda*, Hungaria).

E. moroczkovskii Geljuta, Ukr. Bot. Zhur. 37(2), p. 54, 1980 (on *Linosyris* and *Galatella*, USSR, Ukraine).

E. poonaensis Kamat in Chiddarwar, Curr. Sci. 24, p. 421, 1955 (on *Gonicaulon*, India).

E. riedliana Speer, 1.c., p. 244 (on *Galium*, Austria).

E. salviae (Jacz.) Blumer, l.c., p. 273 (on *Salvia*, Eur.-?).

Erysiphe sect. *Golovinomyces* subsect. Depressa U. Braun subsect. nov.

Foot cells of the conidiophores long, often longer than 100 μm, increasing from base to top, conidia with constricted ends.

Conidiophores ex hyphis sterilibus orientibus, cum cellula basali longa, saepe super 100 μm, sursum latiore; conidiis catenulatis, apicem et basim versus angustatis.

Typus: *Erysiphe depressa* (Wallr. ex) Schlecht.

The following species belong to this subsection:

Erysiphe depressa (Wallr. ex) Schlecht., Fl. Berol. 2, p. 164, 1824 (on *Arctium*, may be some additional hosts — *Onopordum*, *Centaurea montana* and *cyanus* Eur., N. Am., Central Asia).

E. echinopis U. Braun, Feddes Repert. 92, p. 504, 1981 (on *Echinops*, Eur.).

E. verbasci (Jacz.) Blumer, Beitr. Krypt.-Fl. Schweiz 7(1), p. 284, 1933 (on *Verbascum*, Eur.-Central Asia, introduced in New Zealand and N. Am

Oidium helichrysi Boesewinkel, Rev. Mycol. 41, p. 495, 1977 (on *Helichrysum*, New Zealand, introduced from Europe?).

3.4.7 Erysiphaceae: A Survey of the Genera

The family is usually divided into three subfamilies. Palla (1899) established the subfamily *Phyllactinioideae* (mycelium endophytic). Homma (1937) added the *Leveilluloideae* in order to accommodate the genus *Leveillula*. This system has been maintained by different authors, e.g. Blumer (1967). Katumoto (1973) amended the genus *Cystotheca* and proposed the new subfamily *Cystothecoideae*, owing to the intramatrical hyphae of the newly included *Cystotheca tijbodensis* (Gaum.) Katumoto (= *Lanomyces tijbodensis*). The habit of the mycelium is of submitted importance. The endo- or ectophytic appearance is strongly influenced by surrounding conditions, e.g. temperature. Yarwood (1963) and Jarvis (1964) effected internal mycelium even in *Sphaerotheca*, a genus near to *Cystotheca*. The perfect stage of *Lanomyces tijbodensis* fits nicely the cleistothecia of *Cystotheca* in all respects. Special aerial hyphae are also present. Therefore, Braun (1981) agrees with Katumoto that the species has to be transferred to *Cystotheca*. But the species is still insufficiently known. The *Oidium* stage has not been detected. It is impossible to conclude from a single collection whether it is not formed at all or merely undiscovered. The conidial stages of *Cystotheca lanestris* and *wrightii* show close affinity to *Sphaerotheca* and *Podosphaera*. The conidia are in chains, and fibrosin bodies are present. Braun (1981) could not find any reason to establish a separate subfamily for *Cystotheca*, despite the partly internal mycelium of *C. tijbodensis*. The internal mycelium of the genera *Leveillula*, *Phyllactinia*, and *Pleochaeta* is not the only difference against the genera of subfamily *Erysiphoideae*. They deviate by quite different types of conidial states, belonging to *Ovulariopsis* and *Oidiopsis*. The cleistothecia of the three genera show some affinities. They are always large with numerous asci. Asci and ascospores are also large. The differences between the conidial states of *Phyllactinia* and *Pleochaeta* (*Ovulariopsis*, incl. *Streptopodium*) and *Leveillula* (*Oidiopsis*) are small. They are not larger than between the *Euoidium* and *Pseudoidium* type in subfamily *Erysiphoideae*. Maybe, both genera should be merged as proposed by *Ciccarone* (acc. to Blumer 1967). Because of the expounded facts, Braun (1981) kept *Leveillula* in subfam. *Phyllactinioideae*.

***Erysiphaceae* Lev. (1851)**

1. Subfamily ***Erysiphoideae***

Syn.: *Cystothecoideae* Katumoto, Rept. Tottori Mycol. Inst. 10, p. 444, 1973.

Type: *Erysiphe* DC. ex Fr.

Mycelium ectophytic, very rarely partly endophytic, conidial states belonging to the genus *Oidium*, conidiophores arising from the outer mycelium.

Genera: *Arthrocladiella, Blumeria, Brasiliomyces, Bulbouncinula, Californiomyces, Cystotheca, Erysiphe, Kokkalera, Medusosphaera, Microsphaera, Podosphaera, Sawadaea, Sphaerotheca, Typhulochaeta, Uncinula, Uncinuliella.*

2. Subfamily *Phyllactinioideae* Palla, Ber. Deutsch. Bot. Ges. 17 p. 71, 1899, emend. nov.

Syn.: subfam. *Leveilluloideae* "Arn .", Homma, J. Fac Agric. Hokkaido Imp. Univ. 38, p. 422, 1937.

Type: *Phyllactinia* Lev.

Mycelium always partly endophytic, conidial stages belonging to the genera *Oidiopsis* or *Ovulariopsis*.

(incl. *Streptopodium*), conidiophores arising from the inner or outer mycelium.

Genera: *Leveillula, Phyllactinia, Pleochaeta.*

Key to the Recognized Genera of the Family

1. Mycelium ectophytic, very rarely partly endophytic, conidial stage belonging to the genus *Oidium* (subfam. *Erysiphoideae*) --------------------------- 2
1. Mycelium always partly endophytic, conidial stages belonging to the genera *Oidiopsis* or *Ovulariopsis* (subfam. *Phyllactinioideae* emend.) ---------------------------17
2. Cleistothecia with a single ascus, conidia with fibrosin bodies------------------- 3
2. Cleistothecia with two or more asci-- 6
3. Appendages of the cleistothecia absent or mycelium-like, simple or irregularly branched -- 4
3. Appendages with dichotomously branched tips------------------ *Podosphaera* (14)
4. Cleistothecial wall differentiated, composed of two layers which are easily separated from each other, special filiform or falcate hyphae present, on *Fagaceae* ---*Cystotheca* (6)
4. Cleistothecial wall simple, layers firmly connected, special aerial hyphae absent---5
5. Cleistothecial appendages absent, on *Croton*, India ----------------- *Kokkalera* (8)
5. Appendages present, usually on herbs---------------------------- *Sphaerotheca* (16)
6. True appendages absent, gelatinous cells (penicillate cells acc. to Tai 1946) attached to the upper part of the cleistothecium, cells of the wall ± radiately arranged-- *Typhulochaeta* (17)
6. Gelatinous cells absent, true appendages present or without appendages, cells usually not radiately arranged------------------------- 7
7. Appendages absent, peridium thin, one layer---------------------*Brasiliomyces* (3)
7. Appendages present ---8

8. Fruit bodies not dorsiventral, appendages mostly attached to the lower part of the cleistothecium, mycelium-like, often interwoven with the mycelium or with each other, simple or irregularly branched (tips not dichotomously branched or uncinate)--- 9
8. Fruit bodies ± dorsiventral, appendages mostly more or less equatorially or apically inserted, usually not mycelioid, not interwoven, tips simple to dichotomously branched or recurved, uncinate to helicoids -----------------------------11
9. Cleistothecia with thin peridium, semi-transparent, one conspicuous layer, on *Quercus agrifolia*, California---------------------------------- *Californiomyces* (5)
9. Cleistothecia with thick peridium, multilayered, dark, not transparent --- 10
10. Coloured secondary mycelium present, foot cells of the conidiophores with bulbous swelling, haustoria digitate, epicortex not differentiated, on *Poaceae* -- *Blumeria* (2)
10. Secondary mycelium not formed, haustoria not digitate, foot cells of the conidiophores ± cylindric, without swelling, epicortex differentiated-- *Erysiphe* (7)
11. Tips of the appendages simple to dichotomously branched-------------------- 12
11. Tips curved, uncinate to helicoid -- 14
12. Two types of cleistothecial appendages present, the equatorially inserted ones are helically twisted and dichotomously branched; simple, stiff, straight appendages are attached in a second circle, on Rosa----------------- *Medusosphaera* (10)
12. Only one type of appendages present--- 13
13. Conidia in chains, appressoria nipple-shaped, asci 2-spored, appendages dichotomously branched, first branching point near the basis or middle, on *Lycium*-- *Arthrocladiella* (1)
13. Conidia singly formed, appressoria lobed, asci (2-) 3-8-spored, appendages dichotomously branched, first branching point from the middle upward --*Microsphaera* (11)
14. Conidia in chains, with fibrosin bodies, appendages attached to the upper half of the fruit bodies, at least partly dichotomously branched, sometimes trichotomously--- *Sawadaea* (15)
14. Conidia singly formed, without conspicuous fibrosin bodies, appendages simple (very rarely few appendages branched)--------------------------------- 15
15. Only one type of appendages present, they are long and show distinctly recurved, uncinate to helicoid tips------------------------------ *Uncinula* (18)
15. Two types of appendages present, long appendages with uncinate tips and short ones which are straight to bent, simple, tapered, or capitates--16

4 U. flexuosa (Peck) U. Braun: mycelium thin, evanescent (conidia ellipsoid, oblong, hyaline, 26.2–30.6 × 10.7–15.3 gm, acc. to Parmelee 1977). Cleistothecia amphigenous, (85-)100-140(-170), μm diam., cells irregularly polygonal, 8.5–25 μm diam., long appendages ± equatorially inserted, about 25–60 in number, ± straight, hyaline, rarely brown near the base, aseptate, thin-walled, thicker

towards the base, helically twisted in the upper half (one-quarter to one-half), tips uncinate, length of the appendages 0.5–1.5 times diam. of cleistothecium, mostly equaling the diameter, 5–7.5 μm wide in the lower half, hooked and twisted part enlarged, 7.5–12.5 μm wide; a second circle of short, rough, straight to hooked, tapered appendages is present, they are mostly 15–40 × 3.5–7.5 μm; asci 5–12 per fruit body, 45–65 × 30–40 μm, without or with short stalk, 8-spored, 15–22(−25) × 9–13 μm.

At first sight, the short appendages could be taken for immature "normal" appendages. But this special type of appendages is always confined to a second circle above the long ones. Furthermore, they differ in the very rough appearance.

16. Base of the long appendages bulbous, tips uncinate, short appendages capitate, on *Koelreuteria* -----------------------------*Bulbouncinula* (4)
16. Base of the long appendages not bulbous, tips uncinate, short appendages needle-shaped, straight to bent ----------------*Uncinuliella* (19)[4]
17. Appendages mycelioid, conidial state belonging to the genus *Oidiopsis* ---*Leveillula* (9)
17. Appendages mycelioid, conidial state belonging to *Ovulariopsis* (incl. *Streptopodium*)-- 18
18. Appendages bristle-like, tapered, base swollen (bulbous).-- *Phyllactinia* (12)
18. Tips of the appendages uncinate (corresponding to *Uncinula*).------------------- ---------------------------------------*Pleochaeta* (13)

3.4.8 The Relationships Within the Family: A Hypothesis

Different authors tried to explain the phylogenetical relations between the powdery mildew genera. The proposed systems are relatively different. *Leveillula* is the most ancestral genus in the system of Arnaud (1921), whereas it represents a higher developed, derived type in Blumer (1933). The different types of conidial states are constant within larger, well-characterized, homogeneous groups of powdery mildews. But their distribution in the family is only explained in Hirata (1942, 1955). He used it in order to reveal the phylogenetical relations (Fig. 3.3). The mentioned systems tried to derive the present genera from each other. This is impossible or only to a certain extent. Agreeing with Blumer (1933), Braun (1981) considered 8-spored asci, numerous per cleistothecium, and mycelioid appendages as the most primitive features in the family. Blumer (1933) considered the *Euoidium* type of the imperfect stage as primitive and the *Pseudoidium* type as derived. But, Braun (1981) do not agree. The *Euoidium* is characteristic for *Erysiphe* sect. *Golovinomyces* and the genera with a single ascus per cleistothecium. The reduced number of asci represents a derived development. *Erysiphe* sect. *Golovinomyces* is characterized by 2-spored asci, a derived feature, and the species of the section parasitize usually *Sympetalae* (441 hosts), rarely Archichlamydeae (20 hosts). These facts respecting the host range were pointed out by Hirata (1969).

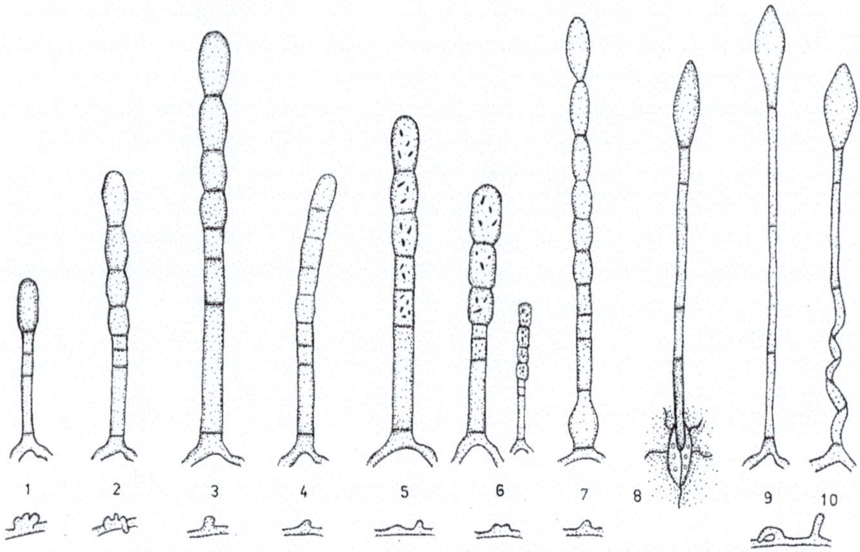

Fig. 3.3 Important types of conidial states in the *Erysiphaceae*. (1) *Pseudoidium* type of *Erysiphe* sect. *Erysiphe, Microsphaera*, and *Uncinuliella*; (2) *Euoidium* type of *Erysiphe galeopsidis*; (3) *Euoidium* type of *Erysiphe* sect. *Golovinomyces*; (4) *Euoidium* type of *Arthrocladiella*; (5) *Euoidium* type of *Sphaerotheca, Podosphaera*, and *Cystotheca* (with fibrosin bodies); (6) *Euoidium* type of *Sawadaea* (with fibrosin bodies, macro and micro conidia); (7) *Oidium* type of *Blumeria*; (8) *Oidiopsis* of *Leveillula*; (9) *Ovulariopsis* of *Phyllactinia*; (10) *Ovulariopsis* of *Pleochaeta* and *Phyllactinia dalbergiae* (foot cells twisted, = genus *Streptopodium* Zheng and Chen (1978a, b)). The conidiophores are in the first line and the appressoria in the second (Braun 1981)

Leveillula, Phyllactinia, and *Pleochaeta* are quite different from the genera with *Oidium* imperfect state. Their conidial states belong to the genera *Oidiopsis* and *Ovulariopsis*. These facts indicate a long, isolated development. The genera seem to be early derived from the ancestors of the family. The large cleistothecia, and asci, and the reduced number of ascospores in *Leveillula* show some affinity to *Phyllactinia* and *Pleochaeta*. These genera represent an isolated branch of the family. Therefore, they are accommodated in a separate subfamily. *Phyllactinia* and *Pleochaeta* are linked by corresponding conidial states. The conidiophores of *Pleochaeta* are helically twisted. The same type is known in *Phyllactinia dalbergiae* Piroz. (= Ph. *yarwoodii* Padw.). It can be presumed that the ancestors of the genera with imperfect states belonging to the *Oidium* type have been species with mycelioid appendages and numerous 8-spored asci. Such ancestral species are known in the genus *Erysiphe*. They are included by Braun (1981) in *Erysiphe* sect. *Erysiphe* subsect. *Polysporae*, e.g. *Erysiphe ulmariae, E. vernalis, E. aggregata, E. sikkimensis, E. rhododendri, E. rubicola, E. carpophila, E. symploci*, or *E. abeliae*. It should be noted that these species mostly parasitize trees and shrubs.

The genera *Sphaerotheca, Podosphaera, Cystotheca*, and *Kokkalera* are closely related. They possess a single ascus per cleistothecium and corresponding conidial states (fibrosin bodies are present). *Sawadaea* has a similar imperfect state. Conidiophores, conidia, and germ tubes match the *Euoidium* of these genera. The cleistothecia contain, however, numerous asci, and the appendages are dichotomously branched. The tips are uncinate. Nevertheless, *Sawadaea* is not very near to *Uncinula*. *Erysiphe* sect. *Golovinomyces* is linked with the former group by catenate conidia and unlobed appressoria. The conidial states of the species of *Erysiphe* sect. *Golovinomyces* subsect. *Depressa* (*E. verbasci*, *E. depressa*, *E. echinopis*) show some affinities to the genera with a single ascus per fruit body. The conidia are characterized by constricted ends. One can meet with similar conidia in *Sphaerotheca alchemillae* (Grey.) Junell. The germination pattern of the conidia in the *E. depressa*-group resembles the pannosa type in the genera *Podosphaera* and *Sphaerotheca* (pannosa type according to Hirata 1955). These facts have been pointed out by Braun (1977).

The genera *Microsphaera*, *Uncinula*, and *Erysiphe* sect. *Erysiphe* (*Typhulochaeta*, *Brasiliomyces*, *Medusosphaera*, *Bulbouncinula*, *Uncinuliella*) represent a third line, connected by the *Pseudoidium*, lobed appressoria, numerous asci, and a variable number of ascospores (2–8). *Arthrocladiella* has dichotomously branched appendages like *Microsphaera*. The conidial state, however, is near to the genera with *Euoidium* type. *Erysiphe* sect. *Erysiphe* has a central position. Especially the 8-spored species on trees and shrubs are regarded as the most primitive representatives of the family. The tips of the appendages in *Uncinula* are uncinate. But that is also the case with numerous species of *Microsphaera*. The tips of the appendages in the *M. penicillata* group are usually recurved. Accordingly, the main difference between both the genera is in the presence of dichotomous branchings. Hirata (1976) pointed out that the combination *Microsphaera*/*Uncinula* has only few common hosts. A direct derivation of *Uncinula* from *Microsphaera*, as often done, is hardly possible. Intermediate forms are not known with regard to the perfect state. It can be presumed that *Microsphaera* and *Uncinula* have been derived from *Erysiphe*-like ancestors separately. *Erysiphe abeliae* Zheng and Chen (1980) and *Uncinula paradoxa* Simonjan appear to be intermediate taxa between *Erysiphe* and *Uncinula*. *Erysiphe* sect. *Erysiphe* and *Microsphaera* are closely linked. Many transitions are known, especially in the *Trichocladia* group.

Uncinuliella, *Bulbouncinula*, and *Typhulochaeta* are considered as derivates of *Uncinula*-like ancestors. The more or less radiately arranged wall cells of *Typhulochaeta* are also to be found in *Uncinula septata* Salm. and *U. curvispora* Hara. *Medusosphaera* has to be regarded as derived from *Microsphaera*. *Brasiliomyces* and *Californiomyces* are fairly primitive genera of the family. They seem to be early derived from *Erysiphe*-like ancestors. The same is the case with *Blumeria*.

Polyphagous taxa are known in the genera with primitive morphological features, e.g. *Erysiphe*, *Sphaerotheca*, and *Leveillula*. Up to date they are able to expand their host ranges. They represent progressive lines. The further development of the family is based on those groups. Such "active" groups are also known in *Microsphaera* and *Phyllactinia*. Figure 3.4 shows the possible phylogenetic relationships within the family (Braun 1981).

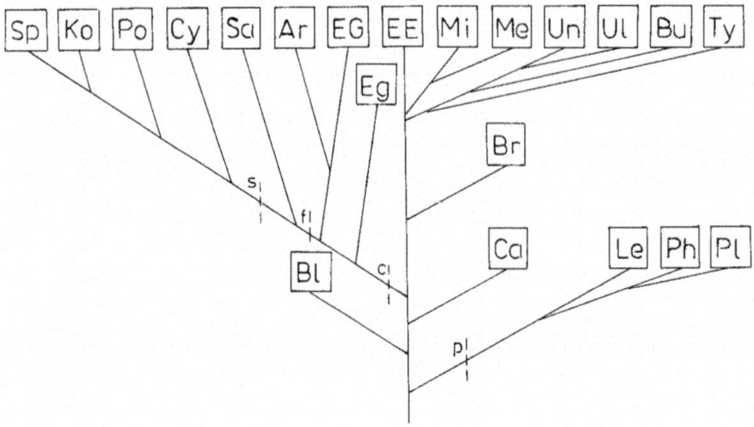

Fig. 3.4 Phylogenetical relationships within the *Erysiphaceae*. Ar *Arthrocladiella*, B1 *Blumeria*, Br *Brasiliomyces*, Bu *Bulbouncinula*, Ca *Californiomyces*, Cy *Cystotheca*, EE *Erysiphe* sect. *Erysiphe*, EG *Erysiphe* sect. *Golovinomyces*, Eg *Erysiphe* sect. *Galeopsidis* (=*E. galeopsidis*), Ko *Kokkalera*, Le *Leveillula*, Me *Medusosphaera*, Mi *Microsphaera*, Ph *Phyllactinia*, PI *Pleochaeta*, Po *Podosphaera*, Sa *Sawadaea*, Sp *Sphaerotheca*, Ty *Typhulochaeta*, Un *Uncinula*, Ul *Uncinuliella*; c conidia in chains, f fibrosin bodies present, p subfam. *Phyllactinioideae*, s a single ascus per cleistothecium (Braun 1981)

3.5 General Morphology

Morphological characters of the powdery mildew pathogens infecting different crucifers have been described at both asexual and sexual states. The pathogen is an obligate parasite. The mycelium of the fungus is amphigenous, white, dense, septate, spreading, and persistent. Conidia are hyaline, borne singly or in short chains, 25–45 × 12–16 μm in size, and cylindrical in shape (Fig. 3.5a, b). Cleistothecia are scattered, globose at first, yellowish orange becoming brown to dark brown, and black with maturity, 90–130 μm in diameter. Cells of cleistothecia are irregular, brown, and 10–25 μm in diameter (Fig. 3.5c). Appendages are numerous, mycelioid, present all over the surface of cleistothecia, narrow hyaline to faintly coloured, seldom branched, often unequal in length, and up to three times the size of the cleistothecia (Fig. 3.5d). Asci are oval to pyriform, 3–12 in number, usually 6–8 with a indistinct stipe, and 50–70 × 30–45 μm in size (Fig. 3.5e). Ascospores (Fig. 3.5f) are ovoid, 2–7 in number, and 16–22 × 11–14 μm in size (Purnell and Sivanesan 1970; Mehta et al. 2005). Conidia germinate only from one end. However, germ tube is branched having up to three branches.

Perfect stage of the fungus develops rarely, but it occurs quite abundantly under Haryana, Indian conditions (Saharan and Kaushik 1981). The taxonomy of powdery mildew on crucifers has been described by Junell (1967) who revised *Erysiphe communis* (Wallr) Link sensu Blumer and divided it into seven species including *E. cruciferarum* (Blumer 1933, 1967).

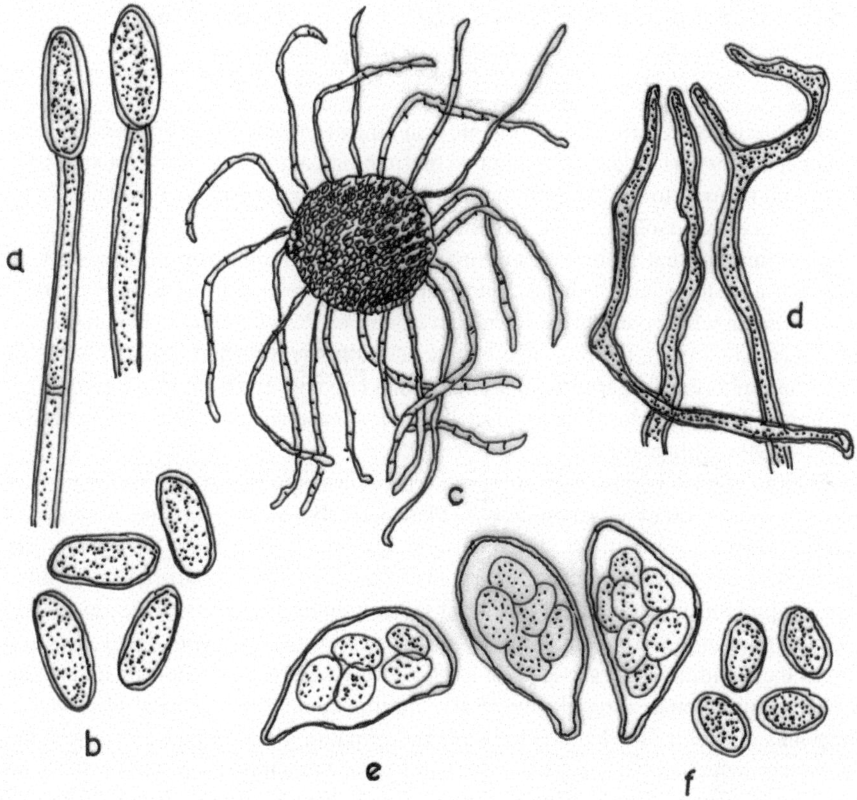

Fig. 3.5 Powdery mildew (*Erysiphe cruciferarum*); (a) conidiophores with conidia; (b) conidia; (c) cleistothecium; (d) mycelioid appendages; (e) asci; (f) ascospores (Mehta et al. 2005)

3.5.1 Morphology of the Pathogen on Brassica *Species*

The size of conidia measured on different species, and varieties of *Brassica,* varies from, 8.32–20.80 × 20.80–45.76 μm with an average range of, 12.48–14.93 × 31.03–36.94 μm. Cleistothecia/chasmothecia were pinkish brown when young, and brown to dark brown after maturity appearing black to unaided eyes. They were scattered or concentrated and globose to sub-globose with numerous hyphae like brownish septate appendages. Cleistothecia measures from 83–20 × 137.28 μm in diameter (av. 104.41–119.14 μm) on different species and varieties of *Brassica*. The number of asci produced per cleistothecium of different species and varieties varied from, 3–8 with 2–6 ascospores per asci. Asci were sub-globose to broadly ovate, stalked, light brown to yellowish in colour, and measured from 24.96–37.44 × 41.60–66.56 μm with an average range of 31.70–34.53 × 52.33–61.98 μm on different species and varieties of *Brassica*. Table 3.1 shows the size of conidia, cleistothecia, and asci, and number of asci per cleistothecium on *Brassica* species and varieties (Junell 1967; Saharan and Kaushik 1981).

3.5.2 *Morphological Characteristics of the Asexual State of Powdery Mildew Isolates on* Arabidopsis

For microscopic examination, cryogenic scanning electron microscopy was used by Adam et al. (1999) to identify distinctive morphological features of the asexual state that could be used to further differentiate *Erysiphe cruciferarum* isolate UEA1 and *E. cichoracearum* isolate UCSC1 (Plate 3.1).

Important asexual features traditionally used for classification purposes are as follows: conidium size, shape, and surface features; (ii) conidiophore type (*Euoidium*, in which conidia are formed in chains, or *Pseudoidium*, in which conidia are generated singly), the presence or absence of branching, and size and shape of the conidiophore; (iii) germination pattern; and (iv) appressorium shape (indistinct or nipple-shaped versus lobed) (Junell 1967; Boesewinkel 1980; Braun 1987; Ialongo 1992; Shin and La 1993; Zeller 1995).

Conidia of UEA1 were ellipsoid to cylindrical in shape (38–46 × 13–16 μm) with a length/width ratio of approximately 3 (Plate 3.1A, B). Small, wart-like projections and fine ridges covered the conidial surface. These extended along the appressoria and hyphae but were not present in regions of new growth. A characteristic, multi-lobed appressorium formed at one end of the conidium (Plate 3.1A, 3.1A insert, B) and was normally accompanied by the emergence of the first hypha from the same end of the conidium. The subsequent hypha developed from the opposite end (Plate 3.1B). Along hyphae at regular intervals, multilobed appressoria were observed at the points of penetration leading to haustoria formation (Plate 3.1C). Initially, at the actively growing hyphal ends, the pattern of hyphal branching tended to be alternate but later opposite branches formed to give matching pairs. Hyphal branches of UEA1 typically emerged at less than 45 degrees (Plate 3.1D). On *A. thaliana,* conidia were produced singly on short erect conidiophores (75–89 × 7.5–8.5 μm) composed of three cells rising perpendicularly from the upper hyphal surface (Plate 3.1E). On *B. napus* cv. Cobra, and on susceptible *Brassica* spp. generally, conidia were generated more rapidly and abundantly. Particularly when undisturbed, these mature conidia remained loosely attached to the conidiophore forming lengthy chains of conidia of similar shape and size (Plate 3.1E insert). Such chains were rarely observed on *A. thaliana* (Plate 3.1E). This fact together with the less prolific mycelial growth on *A. thaliana* may account for the difference in powdery mildew symptoms visible to the naked eye between susceptible.

Plate 3.1 (continued) 2 days post-inoculation (d.p.i.), lobed appressoria (ap) are present at regular intervals along the hyphae. **D,** At 3 d.p.i., secondary hyphae emerge at less than 45°, forming a network. **E,** At 7 d.p.i., conidia are produced singly from relatively short conidiophores (cp). On *Brassica napus* cv. Cobra in undisturbed conditions, conidia remain attached to conidiophore, forming chains (insert). **F** and **G,** At 12–36 h.p.i., conidia are ovoid to ellipsoid with a smooth surface. Appressorial germ tube is unlobed (club shape) and often cannot be distinguished from the first hypha emerging at the opposite end. **H,** At 2 d.p.i., hyphal appressoria are unlobed, indistinct, and apparent as small protuberances along the hyphae. **I,** At 3 d.p.i., secondary hyphae emerge at a 45– 90° angle from main hypha. **J,** At 7 d.p.i., abundant conidiation is apparent. Chains of 3–5 bulbous cells are produced, representing conidia at different differentiation stages (Adam et al. 1999)

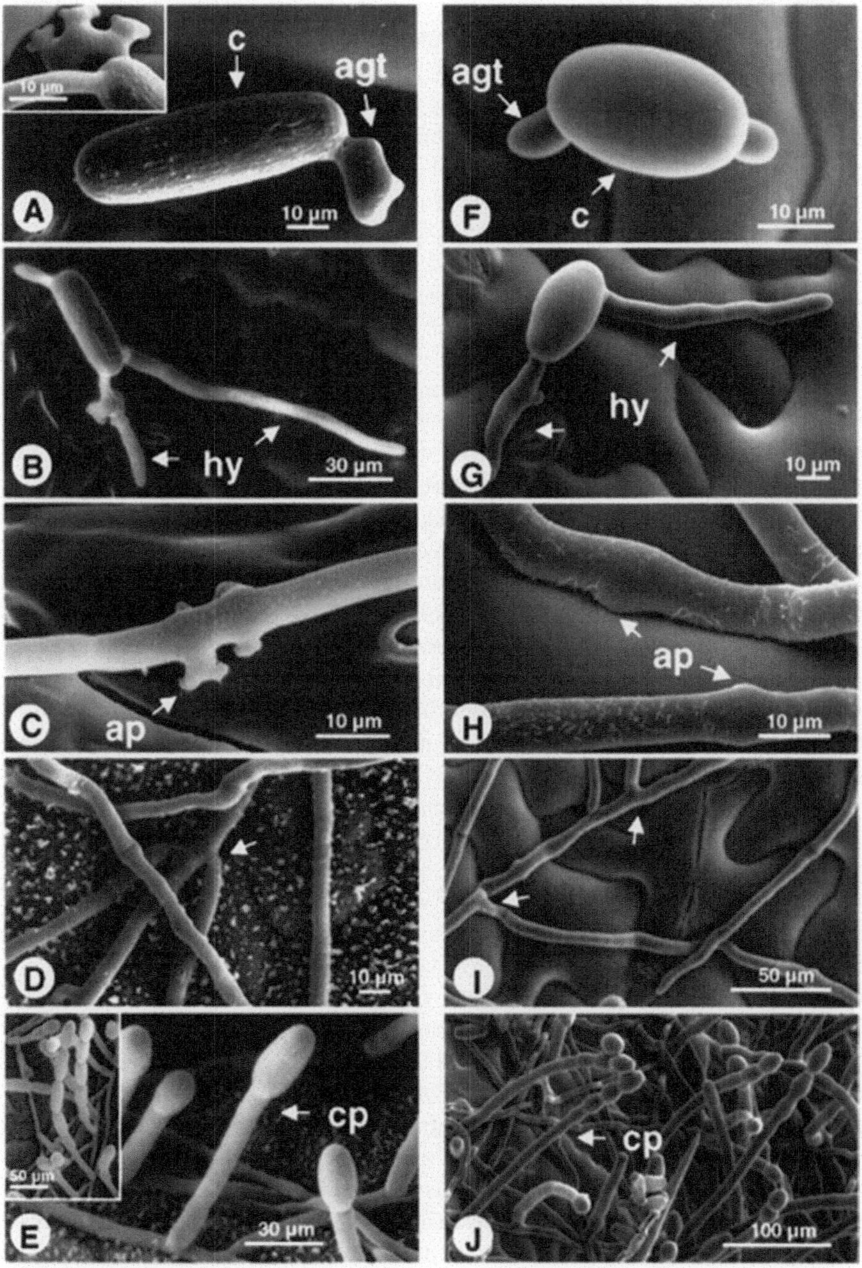

Plate 3.1 Cryogenic scanning electron micrographs illustrating morphological differences between *Erysiphe cruciferarum* UEA1 (left column) and *E. cichoracearum* UCSC1 (right column). **A,** At 12 h post-inoculation (h.p.i.), conidium (c) is ellipsoid to cylindrical with a ridged surface. Appressorial germ tube (agt) is typically lobed but can be a complex (insert) multilobed structure. **B,** At 24–36 h.p.i, first hypha (hy) has emerged from the same end as the appressorial germ tube, which has continued to elongate. Third hypha followed from the distal end. **C,** At

In *A. thaliana* and *Brassica* spp. as previously reported (Adam and Somerville 1996), UCSC1 had ovoid to ellipsoid conidia (26–33 × 16–19 μm) with a length/width ratio of less than 2 (Plate 3.1F). UCSC1 conidia were smooth surfaced with germ tubes emerging at one end bearing unlobed, club-shaped appressoria. The second germ tube usually developed from the opposite end as early as 24 h post-inoculation (h.p.i.) (Plate 3.1F, G). Hyphal appressoria were unlobed, indistinct, and often evident only as small protuberances along the hyphae (Plate 3.1H). The third and fourth hyphae arose from conidia by 3–5 d.p.i. Early in development, hyphal branches generally alternated to either side at regular intervals (Plate 3.1I) but were often found as opposing pairs later in development. These branches usually emerged at 45–90 degrees relative to the main hypha (Plate 3.1I). The most striking difference between UCSC1 and UEA1 was in the type of conidiophore. Conidiophores of UCSC1 were significantly longer, and conidia were formed in chains of 3–5 bulbous cells, representing conidia at different stages of differentiation (Plate 3.1J). The UCSC1 conidiophore stalks (88–160 × 9.5–11 μm) consisted of three elongated cells including a curved foot cell (Adam et al. 1999).

Characteristics of *E. cruciferarum* UEA1 and its virulence *on A. thaliana* accessions were revealed through light and electron microscopy by Xiao et al. (1997). The sexual stage of *E. cruciferarum* is rarely found on any of its natural hosts except *B. juncea* (Saharan and Kaushik 1981); therefore, UEA1 was identified as a member of this species on the basis of morphology of its conidia, which were 32.2–46.0 μm × 13.8–18.4 μm, and formed in short chains with the top conidium substantially larger than the lower one, its lobed appressoria (Plate 3.2a, inset; Boesewinkel 1980), and because UEA1 infected *B. napus* cv. Cobra and related species (Dixon 1981). When *B. napus* cv. Cobra plants were infected with *E. cruciferarum* UEA1, white star-shaped colonies appeared on the leaf surface within 7 days, and the colonies coalesced to cover the surface, accumulating white powdery spore masses within 2–3 weeks. A survey of the reaction of 50 accessions of *A. thaliana* to inoculation with *E. cruciferarum* UEA1 (Hall, 1994) revealed that Col-0 and La-*er* were susceptible; accessions, Bur-0, Can-0, Ct-1, LI-0, Mh-0, Nd-0, Per-l, and Su-0 were partially resistant and supported hyphal growth but not sporulation; and accessions, Ms-0, Kas-1, C-24, Sy-0, and Wa-1 were very resistant and supported no hyphal growth after germ tubes had emerged from the conidia.

Twenty-four hours after the susceptible accession La-*er* was inoculated with *E. cruciferarum* UEA1, more than 90% of the conidial inoculum had germinated and produced short germ tubes terminating in appressoria. The surface hyphae extended rapidly from secondary germ tubes and produced characteristic lobed appressoria which attached to epidermal cells that remained turgid (Plate 3.2a). By 8–10 days after inoculation, the surface mycelium had formed an extensive, branched network (Plates 3.2a and 3.3a) more than 30 mm in length (Fig. 3.6a) in colonies, 2–4 mm diameter. Conidiophores and conidia could be detected 5–6 days after inoculation (Fig. 3.7a) and became a prominent feature of colonies after 8–10 days (Plates 3.2a and 3.3a). The infected host tissues generally showed red autofluorescence from chloroplasts, at least until 10 days after inoculation, indicating that the underlying mesophyll cells were viable. No symptoms were visible to the unaided eye on Ms-0 leaves inoculated with *E. cruciferarum* UEA1. More than half of the conidial inoculum from *E. cruciferarum* UEA1 did not germinate and those that did rarely proceeded beyond formation of the first appressorium (Plates 3.2b and 3.3b). The fungal hyphae and one or two underlying host epidermal cells subsequently collapsed (Plate 3.2b). The underlying mesophyll cells normally gave red autofluorescence (Plate 3.3b), indicating they were viable. However, necrotic mesophyll cells (i.e. giving no red autofluorescence) were occasionally observed beneath germinated conidia (not shown). There was no further growth of the pathogen (Fig. 3.6a), and no conidia were produced (Fig. 3.7a).

F_1 hybrids from a cross La-*er* × Ms-0 were disease resistant, although they supported slightly more growth of the fungus than did the resistant parent, Ms-0 (Plates 3.2c and 3.3c and Fig. 3.6a). There was a collapse of several epidermal cells at the

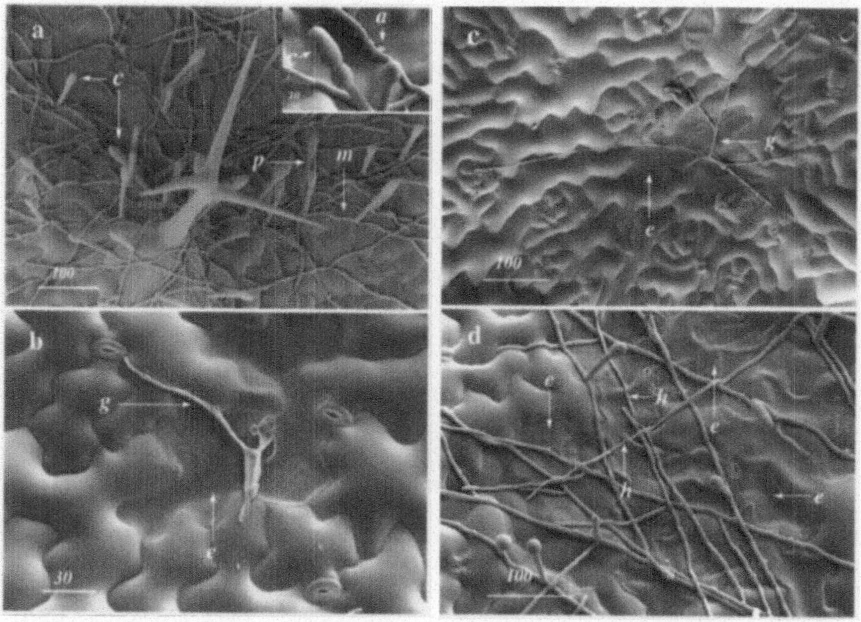

Plate 3.2 Scanning electron micrographs of *E. cruciferarum* on *A. thaliana* leaves, 8 days after inoculation. (a) Leaves of the susceptible *A. thaliana* accession, La-*er*, supported an extensive mycelial (m) network bearing conidiophores (p), and conidia (c), shown here surrounding a branched leaf trichome. Insert: Lobed appressoria (a) attached to epidermal cells of La-*er* leaves and large terminal conidium (c) were characteristic features of *E. cruciferarum*. (b) Resistance of *A. thaliana* accession Ms-0 was associated with collapse of germ tubes (g), shown here arising from a conidium. The underlying host epidermal cells (e) have also collapsed. (c) Resistance of F_1 plants from a cross La-*er* x Ms-0 was associated with partial collapse of hyphal germ tubes (g) from germinating conidium and collapse of several host epidermal cells (e) around the germinated conidium. (d) The intermediate reaction on plants from a F_3 line was characterized by a mycelial network, with conidiophores, and conidia, similar to the susceptible reaction shown in plate (a). However, in the intermediate reaction, many host epidermal cells underlying hyphae had collapsed (e). Some hyphae had also collapsed (h). Bars represent length in 1μm (Xiao et al. 1997)

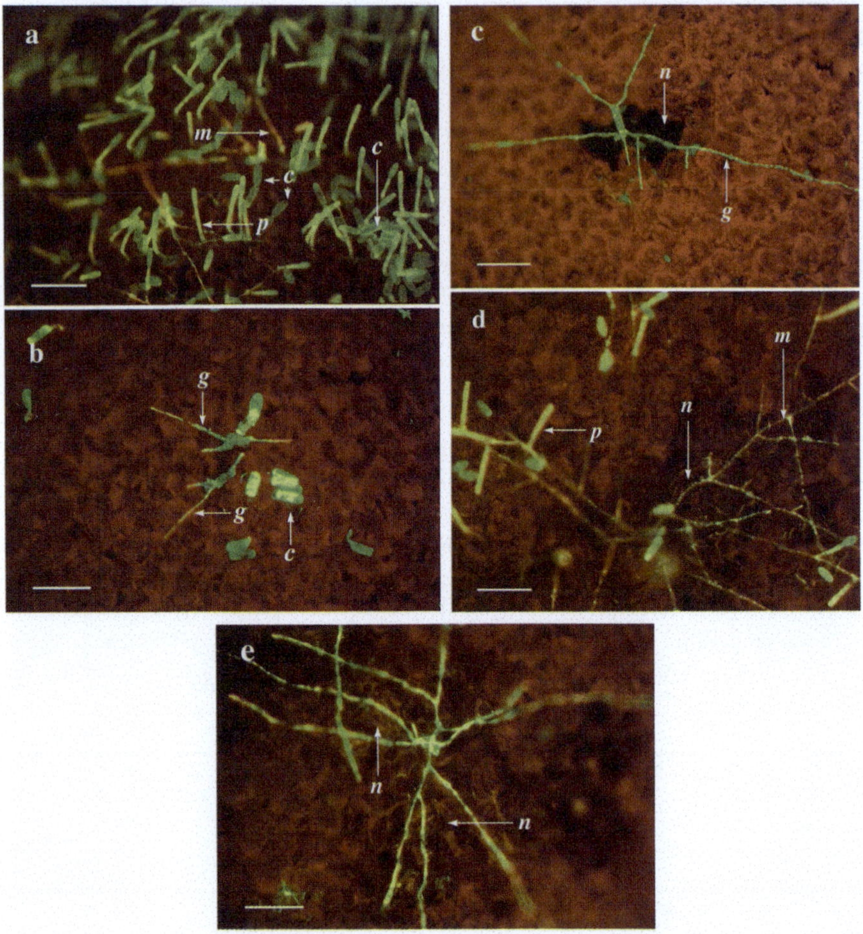

Plate 3.3 Disease reactions, viewed by epifluorescence microscopy, of *A. thaliana* leaves 10 days after inoculation with powdery mildew pathogens, fungal structures fluorescence yellow or green, whereas living host mesophyll cells are red due to chlorophyll autofluorescence, which is absent from necrotic host mesophyll cells. (a)-(d) Leaves were inoculated with *E. cruciferarum* UEA1; (e) leaves were inoculated with *E. cichoracearum* UCSCI. (a) There was an extensive mycelial network (m), partially obscured by masses of conidia (c) on susceptible *A. thaliana* accession La-*er*. Host tissues showed red autofluorescence (not visible in this plane of focus) and were apparently living. (b) Many conidia (c) in the inoculum had not germinated on inoculated leaves of resistant *A. thaliana* accession Ms-0; growth of hyphal germ tubes (g) from two germinated conidia was apparently arrested. (c) Disease resistance of F_1 plants from a cross La-*er* x Ms-0 was associated with limited growth of hyphal germ tubes (g) from conidia overlying necrotic host mesophyll cells (n). (d) The intermediate reaction seen in some F_3 lines from a cross La-*er* × Ms-0 was characterized by development of a mycelial network (m), conidiophores (p), and conidia, on host tissues which contained large patches of necrotic mesophyll cells (n). (e) Resistance f Ms-0 to *E. cichoracearum* UCSC1 was associated with limited hyphal growth from conidia over patches of necrotic host mesophyll cells (n). Bar is 100 μM (Xiao et al. 1997)

site of infection (Plate 3.2c), and the underlying mesophyll cells were necrotic (Plate 3.3c). Although growth of the fungus extended somewhat beyond the necrotic lesion (Plate 3.3c), these hyphae had collapsed (Plate 3.2c), and there was no further growth of the pathogen (Fig. 3.6a), and no conidia were produced (Fig. 3.7a).

Morphological comparison of the MGH (*Erysiphe orontii*) and UCSC (*E. cruciferarum*) isolates on *A. thaliana* leaves revealed that the MGH isolate could be easily distinguished from *E. cruciferarum* by a variety of morphological characters (Table 3.3). *Erysiphe cruciferarum* bears only one conidium per conidiophore, while the MGH isolate bears conidia in chains of up to 20 conidia per conidiophore. Conidia of the MGH isolate usually produced four to five germ tubes (Plate 3.4a), whereas conidia of UCSC isolate generally formed three germ tubes (Plate 3.4b). *Erysiphe cruciferarum* (UEA isolate) is also distinguished from the MGH isolate by its irregularly lobed appressoria and elongated pyriform to irregularly shaped haustoria. On the other hand, the MGH isolate was similar to the UCSC isolate described by Adam and Somerville (1996). Plotnikova et al. (1998) observed morphological and host response differences between the two isolates and differences in the nucleo-

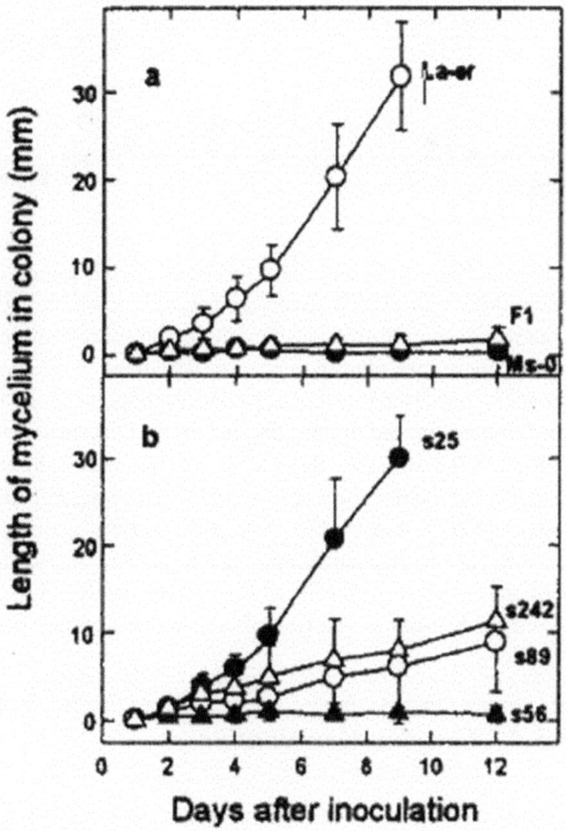

Fig. 3.6 Growth of mycelium of *E. cruciferarum* UEA1 in colonies on leaves of *A. thaliana* plants with different disease resistance genotypes. (**a**) Susceptible accession, La-*er* (O); resistant accession, Ms-0 (●); F_1 plants from a cross La-*er* × Ms-0 (Δ). (**b**) F_3 families from a cross La-*er* x Ms-O, homozygous for the indicated disease phenotype, and with indicated resistance genotype inferred from markers flanking *RPW6* and *RPW7* (in brackets), s25, susceptible, (*rpw6/rpw6, rpw7/rpw7*, ●); s56, resistant (*RPW6/RPW6, RPW7/RPW7*, ▲). s242, intermediate (*RPW6/RPW6, rpw7/rpw7*, Δ); s89, intermediate (*rpw6/rpw6, RPW7/RPW7*, O). Each point is for 20 colonies, and bars give SD (Xiao et al. 1997)

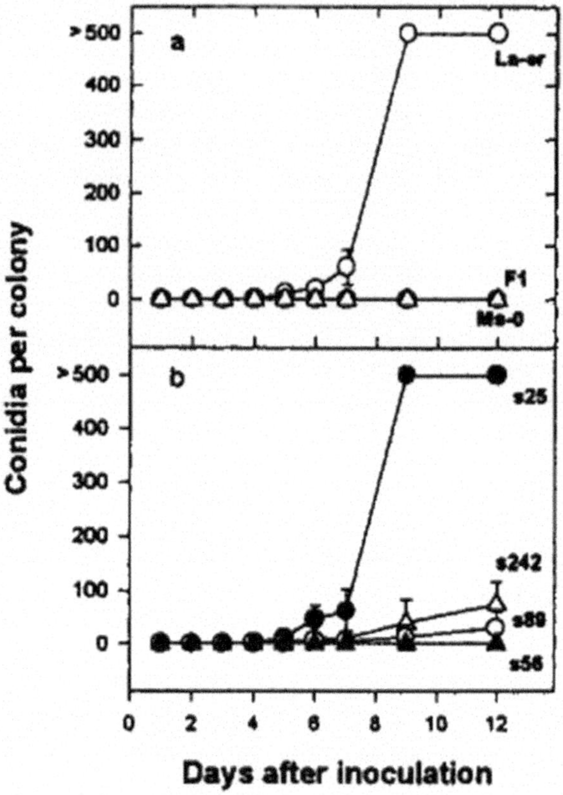

Fig. 3.7 Production of conidia in colonies of *E. cruciferarum* on leaves of *A. thaliana* plants with different disease resistance genotypes. See legend to Fig. 3.6 for explanation of plates (Xiao et al. 1997)

tide sequence of the internal transcribed spacer region of the ribosomal RNA genes of the MGH and UCSC isolates. The distinction between strains in this taxonomic complex deals in formal systematic manner information, and approaches that can be of use for future plant-parasite studies using *A. thaliana*. It is possible that other powdery mildews that are reported pathogens of Brassicaceae will be found on *Arabidopsis* in the future. Braun (1987) records several taxa on Brassicaceae. Although the status of these taxa has not been addressed, viz. *Erysiphe cichoracearum*, *E. cruciferarum*, *E. orontii*, *E. rorippae*, *E. arabidis*, *Leveillula taurica*, and *Sphaerotheca drabae*. Of these, *E. arabidis* is only questionably distinct from *E. cichoracearum* according to Braun (1987). Phylogenetic studies based on DNA sequence analysis of various *Erysiphales* species may give a new insight to the taxonomy of *Erysiphales* and possibly lead to the revision of powdery mildew taxonomy.

3.5.3 Conidial Germination

Conidial germination in relation to temperature, relative humidity, light, and substrate was investigated by Saharan and Sheoran (1988).

Table 3.3 Morphological comparison of the MGH (*Erysiphe orontii*), UCSC (*Erysiphe cruciferarum*), and *E. cichoracearum* isolates on *Arabidopsis thaliana* leaves[a] (Plotnikova et al. 1998; Adam and Somerville 1996)

Morphological characteristics	MGH isolate	UCSC isolate	*E. cichoracearum* isolate
Conidium length × diam[b, e]	29–35 × 15–19 μm	34–40 × 17–21 μm	26–33 × 6–19 μm
Number of conidia on a conidiophore[c]	Chains 7–20	Chains 3–5	Chains 3–5
Conidiophore shape[b]	Straight	Bent	Straight, foot cell curved
Conidial shape	Ovoid–ellipsoidal	Ovoid–ellipsoidal	Ovoid–ellipsoidal
Conidiophore length × diam[b,f]	70–100 × 10 μm	88–160 × 9.5–11 μm	88–160 × 9.5–11 μm
Haustorial shape[b]	Spherical	Ellipsoidal	Oval
Number of germ tubes per germinated conidium[d,g]	4–5	3–4	One
Appressoria	Nipple shaped	Lobed 2–4	Unlobed, nipple shaped
Papillae	Formed	Formed	Formed at several site
Conidial germ tube	Twisted, bent, or helical	Bent	Unlobed
Germ tube apex	Often swollen	Swollen	Swollen club shaped
Hyphae branch	At right angles 5.8–6.6 μm wide	Angular	Right angles

[a]Fungi were inoculated on to leaves of *A. thaliana* Col-0 as described in Methods and incubated at 19 °C and 4200 lux

[b]6 dpi

[c]12 dpi

[d]24 hpi

[e]The mean values for conidia of the MGH isolate were 33.8 μm (SD = 1.3) × 16.5 μm (SD = 0.9), n = 83. The mean values for conidia of the UCSC isolate were 36.9 μm (SD = 1.8) × 18.9 μm (SD = 1.1), n = 77. The means for both length and diameter were found to be significantly different in an unpaired *t* test

[f]Ranges were based on measurements of approximately ten conidiophores of each isolate

[g]The mean values were 4.4 germ tubes/germinated conidium (SD = 0.5, n = 63) for the MGH isolate and 3.3 germ tubes/germinated conidium (SD = 0.5, n = 57) for the UCSC isolate. The mean values were found to be significantly different in an unpaired *t* test

3.5.3.1 Effect of Temperature

Per cent spore germination was maximum (40%) at 20 °C followed by 25 °C (36.1%), but germ tube length remained the same (15 μm). However, there was no statistical difference in per cent spore germination at both temperatures. Below 20 °C and above 25 °C, there was significant reduction in per cent spore germina-

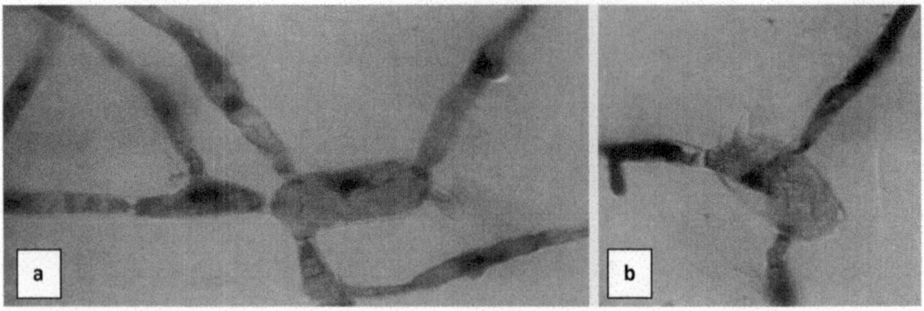

Plate 3.4 (**a**) Comparison of the MGH and UCSC isolate germination. Initial conidium of the MGH isolate with five germ tubes, LM × 1300. (**b**) Initial conidium of the UCSC isolate with three germ tubes, LM × 1300 (Plotnikova et al. 1998)

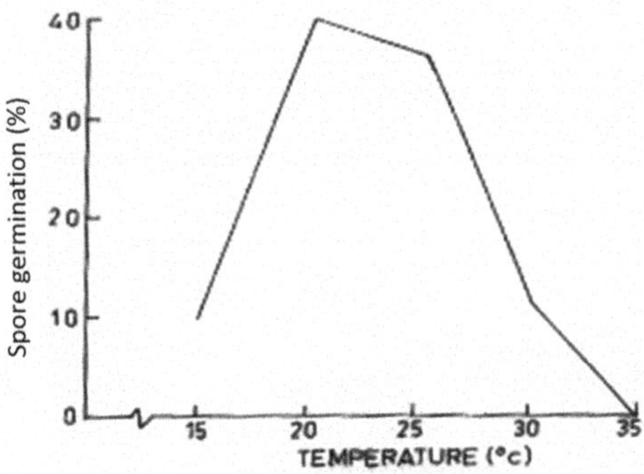

Fig. 3.8 Effect of temperature on conidial germination of *Erysiphe cruciferarum* (Singh 1984)

tion. But reduction in germ tube length at these temperatures was non-significant. Per cent appressoria formation was 100% indicating no effect of temperature regimes on appressorium formation. No spore germination was observed below 15 °C and above 30 °C. Similar observations have been made by Ashraf and Yadav (2009) who found maximum conidial germination between 20 °C and 25 °C temperatures (Fig. 3.8).

3.5.3.2 Effect of Relative Humidity

The maximum spore germination (36.7%) was recorded at 40% relative humidity followed by 50% where only 35% spores germinated. However, this difference was not statistically significant. At 30% and 60% relative humidity, there was significant decrease in per cent spore germination. Above 60% and below 30%, there was no

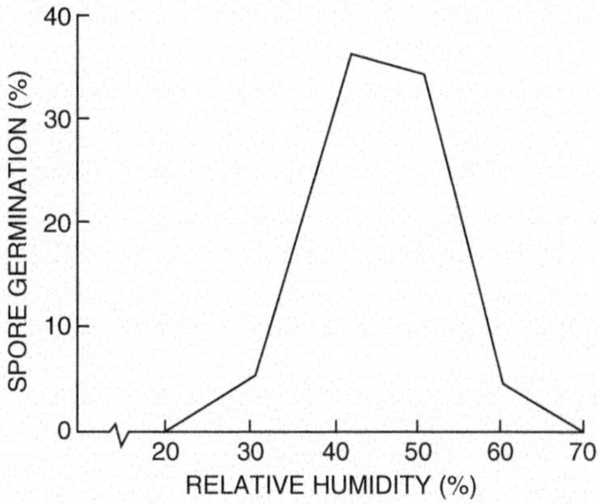

Fig. 3.9 Effect of relative humidity on conidial germination of *Erysiphe cruciferarum* (Singh 1984)

spore germination at all. Germ tube length was found 15 μm at 40% and 50% relative humidity. Although germ tube length was recorded 12.5 μm at 30% and 60% relative humidity, this difference was non-significant. There was no effect of difference in relative humidity on percentage of appressorium formation. All the germ tubes formed appressoria. Conidial germination was from one end with single germ tube per conidium (Fig. 3.9).

3.5.3.3 Effect of Light

There was slight increase in per cent germination of spores (36.7–40%) and germ tube length, when they were kept in light for 24 h as compared to dark conditions. But this increase in per cent spore germination and germ tube length was not statistically significant. Per cent appressorium formation was higher in the case of light (100%) than under darkness (87.5%). The increase in per cent appressorium formation was statistically significant. In the studies carried out earlier in *Erysiphaceae*, conidia designated as group A germinated with a short tube terminated by a lobed appressorium. The rate of germination in vitro conditions at 21 ± 1 °C at 100% RH in the dark reached <70% in 5 h.

3.5.3.4 Effect of Substrate

To record the influence of surface/substrate on powdery mildew conidial germination, the spores were incubated at room temperature on 1% agar and dry glass slides. Forty per cent relative humidity was provided. No significant differences were observed on per cent spore germination, germ tube length, and appressorial formation. It was indicated that conidial germination is not influenced by the substrate (Saharan and Sheoran 1988).

3.5.3.5 Effect of Temperature on Conidial Germination and Germ Tube Length

The conidia obtained from infected mustard leaves were ellipsoid to cylindrical in shape measuring 27.5–35 × 12.5–17.5 μm in size without fibrosin bodies (Table 3.4). The conidia germinated by the formation of straight but two types of germ tube, i.e. short but slightly lobed appressoria and long unlobed appressoria. Simple and forked tube emerged apically and basally with or without appressorium. The germination of conidia was observed at 24 ± 2 °C after intervals of 24, 48, and 72 h of germination. It was found that after 24 h, the length of germ tube ranged 10–15 μm, while after 48 h, it ranged between 23 μm and 35 μm, and after 72 h, it was 27–40 μm. The average length of germ tube was found to be 20–30 μm. It was found that at 5 °C temperature, after time interval of 12 h, there was no germination, while after 24, 48, and 72 h, there was germination in traces. The maximum germination was observed at 20 °C after 72 h of germination. The conidia germinated very well at temperature between 20–25 ± 2 °C up to 72 h of germination. The minimum germination was observed up to 24 h, and 72 h later conidia started getting deformed (Table 3.5; Ashraf and Yadav 2009).

3.5.3.6 Effect of Host Growth Stage

Powdery mildew infection, symptoms development, and pathogens morphological features are greatly influenced by host growth stages. For detailed description, see Chap. 6, at Sect. 6.5.

Table 3.4 Measurement of conidia and germ tube of *Erysiphe cruciferarum* (Ashraf and Yadav 2009)

		Conidia (μm)[a]		Length of germ tube at 24 °C at different intervals (μm)			
Sr. No.	Cultivars	Length	Width	24 h	48 h	72 h	Mean
1.	Alankar	10.0	15.0	12	28	32	24
2.	Kranti	30.0	15.0	12	38	32	24
3.	Pusa bahar	27.5	12.5	10	23	27	20
4.	RH-30	27.5	12.5	10	23	27	20
5.	Varuna	35.0	17.5	13	35	40	30

[a]Each value is an average of ten replicates

Table 3.5 Germination of conidia (%) of *Erysiphe cruciferarum* at different temperature and time intervals (Ashraf and Yadav 2009)

Temperature (°C)	12 h	24 h	48 h	72 h
5	N	T	T	T
20	22.3	32.1	40.7	47.3
25	21.8	30.2	34.6	40.3
30	22.0	26.3	D	D
35	N	D	D	D

N = No germination, *T* = Germination in traces, *D* = Conidia deformed

3.6 Phylogenetics

Phylogenetics relationship of powdery mildew of crucifers with other powdery mildew species has been determined through sequence analysis of the ITS rDNA (Choi et al. 2009) and sequence comparison of DNA encoding the 5.8S, rRNA, ITS1, and ITS2 (Adam et al. 1999).

3.6.1 Sequence Analysis of the ITS rDNA

The rDNA gene sequences covering both highly conserved ribosomal subunits and variable internal transcribed spacer (ITS) regions are widely used to evaluate the polygenic relationships among related fungal species and races within species (Saenz and Taylor 1999; Takamatsu and Kano 2001). The PCR amplification and sequencing of the ITS rDNA were performed by Choi et al. (2009) using the following primer set: ITS 1: 5′TCCGTAGGTGAACCTGCGG-3′ and ITS4: 5′-TCCTCCGCTTATTGATATGGC-3′ (White et al. 1990). The resulting sequence of the complete ITS rDNA of *Erysiphe* sp. KUS-F23994 (accession no. FJ548627) has been deposited in the GenBank of National Centre for Biotechnology Information (NCBI, http://www.ncbi.nlm.nih.gov/). The nearly complete ITS rDNA of *Erysiphe* sp. KUS-F23994 was aligned with other *Erysiphe* spp. nucleotide sequences from the NCBI GenBank database.

Molecular phylogenetic reconstructions of these ITS rDNA sequences were done using MEGA4, version 4.0 (Tamura et al. 2007), for neighbour-joining analysis using Tajima-Nei distances. To test the reproducibility of results, 1000 bootstrap replications (Felsenstein 1985) were performed by the parameters in default values. As shown in ITSbased phylogenetic tree (Fig. 3.10), the *Erysiphe* sp. KUS-F23994 formed a well-supported group (a highly boot strap value of 97%) with three sequences of powdery mildew fungus *E. cruciferarum* which is able to infect *Arabidopsis thaliana, Brassica campestris,* and *B. rapa*. However, sequence divergences between the *Erysiphe* sp. KUS-23994 and other *Erysiphe* species, except *E. cruciferarum,* were more than 3%. Taken together, the polygenetic analysis revealed that the *Erysiphe* sp. KUS-F23994 is indeed identical to *E. cruciferarum* (Choi et al. 2009).

3.6.2 Sequence Comparison of DNA Encoding the 5.8S rRNA, ITS1, and ITS2

To distinguish the powdery mildew isolates of *E. cruciferarum* (UEA1) and *E. cichoracearum* (UCSC1), sequence comparison of DNA was performed by Adam et al. (1999): The rDNA gene cluster contains both highly conserved ribosomal subunits

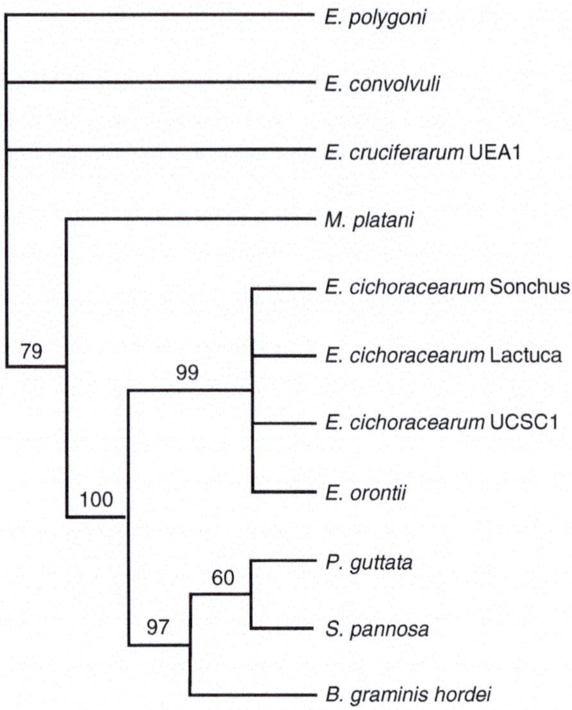

Fig. 3.10 Phylogenetic analysis of *UEA1, UCSC1*, and nine other powdery mildew fungi based on the sequences shown in Fig. 3.11 (Saenz and Taylor 1999). Strongly supported (95%) branches are shown in bold. Numbers shown are frequencies of branches in phylogenies estimated from bootstrap sampled data bases (Felsenstein 1985). Only values >60% are reported. *M. platani; Microsphaera platani*; *P. guttata; Phyllactinia guttata*; *S. pannosa; Sphaerotheca pannosa*; *B. graminis hordei; Blumeria graminis* f. sp. *hordei* (Adam et al. 1999)

and variable internal transcribed spacer (ITS) regions that can be compared to evaluate the phylogenetic relationships among related fungal species and races within species (White et al. 1990; O'Donnell 1992). The nucleotide sequences for both isolates were obtained from polymerase chain reaction (PCR) fragments defined by the primers NS7 and ITS4 (White et al. 1990). Each sequence included a portion (UCSC1, 384 bp; UEA1, 363 bp) of the 3′ end of the 18S rDNA, the complete 5.8S rDNA subunit (154 bp), and the two flanking internal transcribed spacers, ITS1 and ITS2, as well as a 57- to 60-bp portion of the nuclear large 28S rDNA. ITS1 varied in size from 223 bp to 187 bp between UEA1 and UCSC1, respectively. The ITS2 region was also smaller in UCSC1 (165 bp) than in UEA1 (184 bp). As expected, the 3′ end of 18S the 5.8S and 5′ end of 28S rDNA sequences displayed a high degree of conservation between the two *Erysiphe* isolates, with 4, 1, and 2 bp changes in the respective segments. The occurrence of base pair changes in the conserved rDNA subunit genes in addition to a number of base pair changes in ITS1 and ITS2 confirmed, at the molecular level, that both isolates were from distinct phylogenetic groups (Saenz et al. 1994; Hirata and Takamatsu 1996; Saenz and Taylor 1999). The sequences were aligned relative to each other, matching sequences of *B. graminis* f. sp. *hordei* and eight other powdery mildews for a phylogenetic reconstruction (Saenz and Taylor 1999) (Fig. 3.11). The tree is a strict consensus of the eight most parsimonious trees, which were found by a heuristic search employing the random stepwise addition option of PAUP (Swofford 1993). Branch support

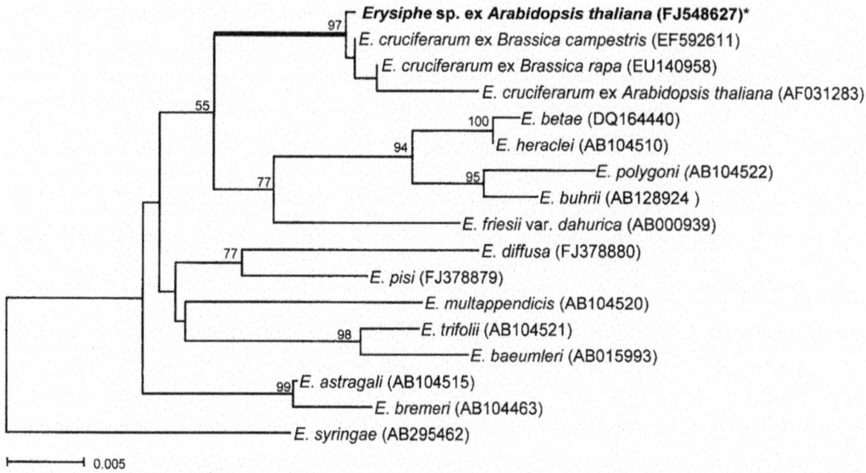

Fig. 3.11 Phylogenetic tree of *Erysiphe cruciferarum* KUS-F23994 (accession no. FJ548627) and other 16 *Erysiphe* spp. constructed by a neighbour-joining method based on the complete internal transcribed spacer (ITS) rDNA regions (ITS1, 5.8S rDNA, and ITS2). Numbers above the branches represent the bootstrap values of over 50% obtained from 1000 bootstrap replicates. Bar = Number of nucleotide substitutions per site. The GenBank accession numbers are represented in parentheses. Asterisk (∗) denotes the isolate used in this study (Choi et al. 2009)

was determined by 1000 bootstrapped data sets, which are displayed on the tree as numbers above the branches. The tree was 262 steps in length; the consistency index was 0.828, the retention index was 0.763, and the re-scaled consistency index was 0.632 (Farris 1989). This analysis showed that the UCSC1 and the UEA1 isolates are distinct *Erysiphe* spp. The UCSC1 isolate grouped with other *E. cichoracearum* isolates and *E. orontii,* while the UEA1 isolate grouped with *E. polygoni* and *E. convolvuli* (Adam et al. 1999).

3.7 Powdery Mildew Pathogen Genomics and Transcriptomes

Powdery mildew fungi have sizeable genomes, which are about four times larger than those of most other ascomycetes (average ascomycete genome size: 36.9 Mbp; Mohanta and Bae 2015). The genome of *Go*, for example, is approximately 160 Mbp in size (Spanu et al. 2010). By contrast, the number of coding genes in the powdery mildew genomes is comparatively low (average number of ascomycete coding genes: 11,129; Mohanta and Bae 2015). Only ca. 6500 genes each have been annotated in the *Bgh* and *Bgt* genomes, and ca. 7100 genes (on the basis of assembled transcript contigs) are expressed in *Go* haustoria (Spanu et al. 2010; Weßling and Panstruga 2012; Weßling et al. 2012; Wicker et al. 2013; Kusch et al. 2014).

The biological reason for the surprisingly low gene number most likely lies in the biotrophic lifestyle: due to the close association of parasite and host, the fungus acquires its nutrients from the plant. As a result, the need for the maintenance of many complex biosynthesis pathways is low, whereas the requirement to control the host cell by secreted effectors is high (Spanu et al. 2010). Associated with this unusual ratio of genome size to gene number is the presence of numerous nested retro-transposons that cover most of the powdery mildew genomes. These retro-transposons are physically closely associated with effector protein-encoding genes and are therefore thought to be involved in the rapid evolutionary adaptation of powdery mildews (Hacquard et al. 2013).

In a transcriptomic approach using a cDNA library obtained from mature *Go* haustoria extracted from heavily infected *Arabidopsis* leaves, protein-coding genes for translation and protein turnover were recognized to be most abundant (Weßling et al. 2012). This is in line with the finding that haustoria contain an abundance of cytoplasmic and endoplasmic reticulum-connected ribosomes, pointing at high levels of protein biosynthesis (Micali et al. 2011). Genes associated with mycelium development were also found to be highly represented in the *Go* haustorial transcriptome. By contrast, the transcript levels of sugar and amino acid transporters are comparatively low (Weßling et al. 2012). A substantial proportion of the transcripts are predicted to encode secreted effector proteins: 115 *Go* effector candidates (OECs) were discovered in the transcriptome of isolated haustoria (Weßling et al. 2012; Weßling et al. 2014). Among them, 84 OECs were subject of a comprehensive protein interaction study with a subset of *Arabidopsis* host proteins. In this work, identification of an interspecies effector convergence network revealed common effector target proteins (hubs) for *Arabidopsis* pathogens from three kingdoms of life, i.e. *Go* (a powdery mildew fungus), *H. arabidopsidis* (an oomycete), and *P. syringae* (a bacterium; Weßling et al. 2014). Interestingly, mutants of many of the respective host target genes show altered disease phenotypes (towards either increased resistance or higher susceptibility). This effector convergence suggests that biotrophic pathogens from different kingdoms manipulate the same host plant processes (Weßling et al. 2014). Among the common effector targets, proteins involved in cell cycle regulation/plant development are highly represented, e.g. the TFs *TCP13, TCP14*, and *TCP19* (At5g51910). Since the TF *MYB3R4* seems to be involved in powdery mildew-induced increase in polyploidy (Chandran et al. 2010), these findings may indicate that the manipulation of the host cell cycle is crucial for the *Go* infection process and that of a range of other pathogens as well (Weßling et al. 2014).

3.8 Pathogenic Variability

To determine whether a powdery mildew pathogen growing on one host species can cause infection on another host species which is known to be infected by same morphological species has been studied extensively. Neger (1902) was the first person to show that some of the morphological species of powdery mildews are broken into biologic forms, limited to definite host plants. Marchal (1902, 1903), Reed (1905, 1907, 1908, 1909, 1912), Salmon (1903a, b, 1904a, b, c, 1905a, b), Steiner (1908) and Voglino (1905) have very extensively studied regarding the occurrence of biological/physiologic specialization in different powdery mildews. The cereal powdery mildew, *Erysiphe graminis*, has been exceptionally favourable host-pathosystem for studies on biologic specialization/pathogenic variability. The biologic forms are restricted to the species of a single genus. Conidia from wheat powdery mildew will not infect barley, oats, and rye. Similarly, conidia from barley powdery mildew will not infect wheat, oats, and rye. In other words, in every case, the powdery mildew on one cereal is unable to infect the species of other cereals. It is also known as pathogenic variability or host specificity of powdery mildew pathogens. As a matter of fact, all individuals produced as result of sexual process are expected to be different from each other, and from their parents in a number of characteristics, although they retain most similarities with then and belong to the same species. It is true in the case of powdery mildews reproducing through sexual fruiting bodies (chasmothecia) and ascospores infecting crucifers. However, in asexually reproducing powdery mildew through conidia, frequency and degree of variability among the progeny are reduced greatly, but even then certain individuals among the progeny will show different characteristics. The mechanisms of variability may be hybridization, mutation, and cytoplasmic inheritance. However, so far, pathogenic variability or host specificity has not been in records in any of powdery mildew fungi-infecting crucifers. However, variations in the development of *E. cruciferarum* on two cvs. *B. napus* have been observed by Munro and Lennard (1982); genotypes of *A. thaliana* (Adam and Somerville 1996; Adam et al. 1999; Platinkova et al. 1998; Gollner et al. 2008; Kuhn et al. 2016) and genotypes of *B. juncea*, *B. napus*, and *B. carinata* have been recorded (Singh 1984; Uloth et al. 2017). Similarly, Xiao et al. (1997) observed differentiated interaction of *E. cruciferarum* isolates UEA1 on 50 accessions of *A. thaliana*. Since in India, powdery mildew pathogen of *B. juncea* completes its anamorph (asexual) and teleomorph (sexual) stages on *B. juncea* under Indian field conditions (Saharan and Kaushik 1981), there is every possibility for evolution of pathogenic variability with more virulent strains of *E. cruciferarum* (Meena et al. 2018).

References

Adam L, Somerville SC (1996) Genetic characterization of five powdery mildew disease resistance loci in *Arabidopsis thaliana*. Plant J 9:341–356

Adam L, Ellwood S, Wilson I, Saenz G, Xiao S, Oliver RP, Turner JG, Somerville S (1999) Comparison of *Erysiphe cichoracearum* and *E. cruciferarum* and a survey of 360 *Arabidopsis thaliana* accessions for resistance to these two powdery mildew pathogens. Mol Plant Microbe Interact 12:1031–1043

Arnaud G (1921) Etude sur les champignons parasites. Ann des Epiphy 7:1–115

Ashraf S, Yadav B (2009) Studies on the anamorph characters and management of powdery mildew of mustard. Trends Biosci 2(2):79–80

Bai YL, Pavan S, Zheng Z, Zappel NF, Reinstadler A, Lotti C, De Giovanni C, Ricciardi L, Lindhout P, Visser R, Theres K, Panstruga R (2008) Naturally occurring broad-spectrum powdery mildew resistance in a central American tomato accession is caused by loss of Mlo function. Mol Plant Microbe Interact 21:30–39

Blumer S (1933) Die Erysiphaceen Mitteleuropasunter besonderer Beriicksichtigung der Schweiz. Beitr Krypt-Fl Schweiz 7(1):1–483

Blumer S (1967) Echte Mehltaupilze (*Erysiphaceae*). G Fischer, Jena

Boesewinkel HJ (1976) Cleistothecia of powdery mildews in New Zealand. Trans Br Mycol Soc 67:143–146

Boesewinkel HJ (1977) Identification of Erysiphaceae by conidial characteristics. Rev Mycol 41:493–507

Boesewinkel HJ (1980) The morphology of the imperfect states of powdery mildews (Erysiphaceae). Bot Rev 46:167–224

Braun U (1977) Das Erysiphaceen-Keimungsbild als taxonomisches Merkmal und Bestimmungshilfe. Boletus 1(1):3–8

Braun U (1978) Beitrag zur Systematik und Nomenklatur der Erysiphales. Feddes Repert 88:655–665

Braun U (1980a) Morphological studies in the genus *Oidium*. Flora 170:77–90

Braun U (1980b) The genus Leveillula- a preliminary study. Nova Hedwigia 32:565–583

Braun U (1981) Taxonomic studies in the genus *Erysiphe* I. Generic delimitation and position in the system of the Erysiphaceae. Nova Hedwigia 34:679–719

Braun U (1987) A monograph of the Erysiphales (powdery mildews). Beihefte zur Nova Hedwigia 89:1–700. J Cramer Berlin, Germany

Braun U (1995) The powdery mildews (Erysiphales) of Europe. VEB G Fischer Verlag, Jena

Braun U, Cook RTA (2012) Taxonomic manual of the Erysiphales (powdery mildews), CBS biodiversity series no. 11. CBS-KNAW Fungal Diversity Centre, Utrecht

Chandran D, Inada N, Hather G, Kleindt CK, Wildermuth MC (2010) Laser microdissection of *Arabidopsis* cells at the powdery mildew infection site reveals site-specific processes and regulators. PNAS 107(1):460–465

Chiddarwar PP (1955) A new species of *Erysiphe*. Curr Sci 24:420–421

Choi HO, Choi Y, Kim DS, Hwang IS, Choi DS, Kim NH, Lee DH, Shin HD, Nam J, Hwang BK (2009) First report of powdery mildew caused by *Erysiphe cruciferarum* on *Arabidopsis thaliana* in Korea. Plant Pathol J 25(1):86–90

Chona BL, Kapoor JN, Gill HS (1960) Studies on powdery mildews from India-I. Indian Phytopathol 13:72–75

Cook RTA, Inman AJ, Billings C (1997) Identification and classification of powdery mildew anamorphs using light and scanning electron microscopy and host range data. Mycol Res 101(8):975–1002

De Candolle AP (1805) Flore Francaise 2:272–275

De Candolle AP (1815) Flore Francaise 6:104–109

De Schweinitz LD (1834) Synopsis fungorum in America. Trans Am Philos Soc 4:269–270

Dixon GR (1978) Powdery mildews of vegetable and allied crops. In: Spencer DM (ed) The powdery mildews. Academic, London, pp 495–525

Dixon G (1981) Pathogens of *Brassicas*. In: Vegetable crop diseases. Macmillan, London, pp 116–119

Farris JJ (1989) The retention index and the rescaled consistency index. Cladistics 5:417–419

Felsenstein J (1985) Confidence limits on phylogenies: an approach using the bootstrap. Evolution 39:783–791

Fries EM (1829) Systema mycologicum 3:234–247

Gollner K, Schweizer P, Bai Y, Panstruga R (2008) Natural genetic resources of *Arabidopsis thaliana* reveal a high prevalence and unexpected phenotypic plasticity of RPW8-mediated powdery mildew resistance. New Phytol 177:725–742

Golovin N (1958) Obsor rodov semeistva *Erysiphaceae. Sborn. Rabot.* Inst Prikl Zool I Fitop 5:101–139

Hacquard S, Kracher B, Maekawa T, Vernaldi S, Schulze-Lefert P, Ver Loren van Themaat E (2013) Mosaic genome structure of the barley powdery mildew pathogen and conservation of transcriptional programs in divergent hosts. PNAS 110(24):E2219–E2228

Hall D (1994) Interactions of *Arabidopsis* with fungal pathogens. Ph.D. thesis. Norwich, University of East Anglia

Hirata K (1942) On the shape of the germ tubes of Erysipheae. Bull Chiba Coll Hort 5:34–49

Hirata K (1955) On the shape of the germ tubes of Erysipheae (II). Bull Fac Agric Niigata Univ 7:24–36

Hirata K (1969) Notes on host range and geographic distribution of the powdery mildew fungi II. Trans Mycol Soc Jpn 10(2):42–72

Hirata K (1976) Notes on host range and geographic distribution of the powdery mildew fungi VI. Distribution of the hosts of powdery mildew fungi in the families of angiosperms. Trans Mycol Soc Jpn 17:35–62

Hirata K, Kojima M (1962) On the structure and sack of some powdery mildews, with some considerations on the significance of the sack. Trans Mycol Soc Jpn 3:43–46

Hirata T, Takamatsu S (1996) Nucleotide sequence diversity of rDNA internal transcribed spacers extracted from conidia and cleistothecia from several powdery mildew fungi. Mycoscience 37:283–288

Homma Y (1937) Erysiphaceae of Japan. J Fac Agric Hokkaida Univ 38:183–461

Ialongo MT (1992) Taxonomic study of some species of the genus *Erysiphe*. Mycotaxon 44:251–256

Jarvis W (1964) Thermal and translocated introduction of endophytic mycelium in two powdery mildews. Nature 203:895

Jones H, Whipps JM, Gurr SJ (2001) The tomato powdery mildew fungus *Oidium neolycopersici*. Mol Plant Pathol 2:303–309

Jorgensen JH (1988) *Erysiphe graminis*, powdery mildew of cereals and grasses. Adv Plant Pathol 6:135–157

Junell L (1965) Nomencultural remarks on some species of Erysiphaceae. Trans Br Mycol Soc 48(4):539–548

Junell L (1967) Erysiphaceae of Sweden. Symbolae Botanicae Upsalienses 19:1–117

Kapoor JW (1965) Two powdery mildews from Sikkim. Indian Phytopathol 18:90–91

Katumoto K (1973) Notes on the genera *Lanomyces* Gaum. and *Cystotheca* Berk. et Curt. Rep Tottori Mycol Inst 10:437–446

Kirk PM, Cannon PF, Minter DW, Stalpers JA (2008) Ainsworth Bisby's dictionary of the fungi, 10th edn. CAB International, Wallingford

Koch E, Slusarenko AJ (1990) Fungal pathogens of *Arabidopsis thaliana* (L.) Heyhn. Bot Helv 100:257–268

Kuhn H, Kwaaitaal M, Kusch S, Acevedo-Garcia J, Wu H, Panstruga R (2016) Biotrophy at its best: novel findings and unsolved mysteries of the *Arabidopsis*-powdery mildew pathosystem. Arabidopsis Book 14:e0184. https://doi.org/10.1199/tab.0184

Kusch S, Ahmadinejad N, Panstruga R, Kuhn H (2014) *In silico* analysis of the core signaling proteome from the barley powdery mildew pathogen (*Blumeria graminis* f.sp. *hordei*). BMC Genomics 15(1):843

Leveille J (1851) Organisation et disposition methodique des especes qui composent le genre Erysiphe. Ann Sci Nat Bot III Ser 15:109–179

Linnaeus C (1753) Species plantarum. Tomus I. Impensis Laurentii Salvii, Holmiae

Marchal E (1902) De la specialization du parasitisme chez l'*Erysiphe graminis*. Comptes Rendus 135:210–212

Marchal E (1903) De la specialization de la parasitism chez.-I: *Erysiphe graminis*. Compt Rend Acad: des Sci, Paris 135:210–212. 1067–1068; 136, 1280–1281
Meena PD, Mehta N, Rai PK, Saharan GS (2018) Geographical distribution of rapeseed-mustard powdery mildew disease in India. J Mycol Plant Pathol 48(3):284–302
Mehta N, Sangwan MS, Saharan GS (2005) Fungal diseases of rapeseed-mustard. In: Saharan GS, Mehta N, Sangwan MS (eds) Diseases of oilseed crops. Indus Publication Co, New Delhi, pp 15–86
Merat FV (1821) Nouvelle Flore des Environs de Paris, 2nd, vol I. Paris
Micali C, Gollner K, Humphry M, Consonni C, Panstruga R (2008) The powdery mildew disease of *Arabidopsis*: a paradigm for the interaction between plants and biotrophic fungi. Bio One
Micali C, Neumann U, Grunewald D, Panstruga R, O'Connell R (2011) Biogenesis of a specialized plant–fungal interface during host cell internalization of *Golovinomyces orontii* haustoria. Cell Microbiol 13:210–226
Mohanta TK, Bae H (2015) The diversity of fungal genome. Biol Proced Online 17:8
Munro JM, Lennard JH (1982) Variation in the development of *Erysiphe cruciferarum* Opiz. Ex.L Junnel on two cultivars of *Brassica napus*. Cruciferae Newslett 7:68–69
Nagy GS (1977) Erysiphe monardae sp. nov. Phytopathol Z 88:285–286
Neger FW (1901) Beitrage zur Biologie der Erysipheen. Flora 88:333–370
Neger FW (1902) Beitrage zur Biologie der Erysipheen. Flora 90:221–272
O'Donnell KO (1992) Ribosomal DNA internal transcribed spacers are highly divergent in the phytopathogenic ascomycete *Fusarium sambucinum* (*Gibberella pulicaris*). Curr Genet 22:213–220
Palla E (1899) Uber die gattung *Phyllactinia*. *Ber.* Deustch. Bot Ges 17:64–72
Parmelee JA (1977) The fungi of Ontario. II. *Erysiphaceae* (mildews). Can J Bot 55:1940–1983
Persoon CH (1796) Observations mycologicae sur descriptions tan novorum quam notibilium fungorum exhibitae. Part 1:115
Persoon CH (1801) Synopsis methodica fungorium, vol I, pp 124
Plotnikova JM, Reuber TL, Ausubel FM, Pfister DH (1998) Powdery mildew pathogenesis of *Arabidopsis thaliana*. Mycol Soc Am 90(6):1009–1016
Purnell TJA, Sivanesan (1970) *Erysiphe cruciferarum* Opiz ex Junell. CMI Descr Pathog Fungi Bact 26:251–260
Rayss T (1940) Nouvelle contribution a letude de la mycoflore de Palestine (*Deuxième partie*). Palestine J Bot Jerusalem Ser 1:313–335
Rayss T (1947) Nouvelle contribution a letude de la mycoflore de Palestine (Quartième partie). Palestine J Bot Jerusalem Ser 4(2):59–76
Rebentish JF (1804) Prodromus florae neomarchicae, pp 360–361
Reed GM (1905) Infection experiments with *Erysiphe graminis* DC. Trans Wis Acad Sci Arts Lett 15:135–162
Reed GM (1907) Infection experiments with the mildew on cucurbits, *Erysiphe cichoracearum* DC. Trans Wis Acad Sci Arts Lett 15:527–547
Reed GM (1908) Infection experiments with *Erysiphe cichoracearum* DC. Bull Univ Wisconsin Sci Ser 3:337–416
Reed GM (1909) The mildews of the cereals. Bull Torrey Bot Club 36:353–388
Reed GM (1912) Infection experiments with the powdery mildew of wheat. Phytopathology 2:81–87
Saenz GS, Taylor JW (1999) Phylogeny of the Erysiphales (powdery mildews) inferred from internal transcribed spacer ribosomal DNA sequences. Can J Bot 77:150–168
Saenz GS, Taylor JW, Gargas A (1994) 18S rRNA gene sequences and supraordinal classification of the Erysiphales. Mycologia 86:212–216
Saharan GS, Kaushik JC (1981) Occurrence and epidemiology of powdery mildew of *Brassica*. Indian Phytopathol 35:17–21
Saharan GS, Sheoran BS (1988) Conidial germination, germ tube elongation and appressorium formation of *Erysiphe cruciferarum*. Indian Phytopathol 41(1):157–159

Salmon ES (1900) A monograph of the Erysiphaceae. Mem Torrey Bot Club 9:1–292
Salmon ES (1903a) Infection powers of ascospores in Erysipheae. J Bot 41(159):204–212
Salmon ES (1903b) On specialization of parasitism in the Erysiphaceae. Beihefte *zum botanischen* Centralblatt 14:261–315
Salmon ES (1904a) On specialization of parasitism in the Erysiphaceae. New Phytol 3:109
Salmon ES (1904b) Mycological notes. J Bot 42:182–186
Salmon ES (1904c) On *Erysiphe graminis* DC. and its adaptive parasitism within the genus Bromus. Ann Mycol 2:255–267
Salmon ES (1905a) Cultural experiments with an *Oidium* on *Euonymus japonicas* Linn. f. Ann Mycol 3:1–15
Salmon ES (1905b) On the variation shown by the conidial stage of *Phyllactinia corylea* (Pers.) Karst. Ann Mycol 3:493–505
Sankhla HS, Dalela GG, Mathur RL (1967) Occurrence of perithecial stage of *E. polygoni* on *B. campestris* var. Sarson and *B. juncea*. Plant Dis Rep 51:800
Sawada K (1927) On the systematic investigation of Erysiphe in Formosa. Formosa Dept Agric Govt Res Inst Rep 24:55
Sawada K (1949) Fungi from the north-eastern region of Honshu, Japan (1). Erysiphaceae, Tohuku. Biol Res 1:2–8
Sawada K (1951) Researches on fungi in the Tohoku district of Japan (I). Erysiphaceae. Bull Govt Forest Exp Sta No. 50
Sawada K (1959) Descriptive catalogue of the Formosan Fungi XI. *Spec.* Bull Coll Agric Nat Taiwan Univ 10:16–24
Shin HD, La YJ (1993) Morphology of edge lines of chained immature conidia on conidiophores in powdery mildew fungi and their taxonomic significance. Mycotaxon 66:445–451
Singh B (1984) Epidemiology and control of rapeseed-mustard powdery mildew caused by Erysiphe cruciferarum. M.Sc thesis, CCS HAU, Hisar 61p
Spanu PD, Abbott JC, Amselem J, Burgis TA, Soanes DM, Stuber K, Ver Loren van Themaat E, Brown JK, Butcher SA, Gurr SJ, Lebrun MH, Ridout CJ, Schulze-Lefert P, Talbot NJ, Ahmadinejad N, Ametz C, Barton GR, Benjdia M, Bidzinski P, Bindschedler LV, Both M, Brewer MT, Cadle-Davidson L, Cadle-Davidson MM, Collemare J, Cramer R, Frenkel O, Godfrey HJ, Hoede C, King BC, Klages S, Kleemann J, Knoll D, Koti PS, Kreplak J, López-Ruiz FJ, Lu X, Maekawa T, Mahanil S, Micali C, Milgroom MG, Montana G, Noir S, O'Connell RJ, Oberhaensli S, Parlange F, Pedersen C, Quesneville H, Reinhardt R, Rott M, Sacristán S, Schmidt SM, Schon M, Skamnioti P, Sommer H, Stephens A, Takahara H, Thordal-Christensen H, Vigouroux M, Wessling R, Wicker T, Panstruga R (2010) Genome expansion and gene loss in powdery mildew fungi reveal tradeoffs in extreme parasitism. Science 330(6010):1543–1546
Speer EO (1975) Untersuchungen zur morphologie und systematic der Erysiphaceen I. Die Gattung *Blumeria* Golovin und ihre Typusart *Erysiphe graminis* DC. Sydowia 27:1–6
Steiner JA (1908) Die Specialisation der Alchemillenbewohnenden *Sphaerotheca humuli* (DC.) Burr. Centrallblatt fur Bakt., Parasitenkunde, und Infextious Krankheiten, Abstr II 21:677–736
Stevens FL (1925) Plant disease fungi. The Macmillan Company, New York
Swofford DL (1993) Phylogenetic analysis using parsimony (PAUP version 3.1.1). Illinois Natural History Survey, Champaign
Tai LF (1946) Further studies on the *Erysiphaceae* of China. Bull Torrey Bot Club 73(2):108–130
Takamatsu S, Kano Y (2001) PCR primers useful for nucleotide sequences of rDNA of the powdery mildew fungi. Mycoscience 42:135–139
Tamura K, Dudley J, Nei M, Kumar S (2007) MEGA4: molecular evolutionary genetics analysis (MEGA) software version 4.0. Mol Bio Evol 24:1596–1599
Uloth MB, You MP, Barbetti MJ (2017) Plant age and ambient temperature: significant drivers for powdery mildew (*Erysiphe cruciferarum*) epidemics on oilseed rape (*Brassica napus*). Plant Pathol. https://doi.org/10.1111/ppa.12740

Vogel J, Somerville S (2000) Isolation and characterization of powdery mildew-resistant *Arabidopsis* mutants. Proc Natl Acad Sci 97(4):1897–1902
Voglino P (1905) Contribuzione allo studio della *Phyllactinia corylea*. Nuovo Giornale Bot Italiano 12:313–327
Wallroth FW (1819) Naturgeschichte des Mucor *Erysiphe* L. Berl Ges Nat Freunde Verhandl I:6–45
Weltzien HC (1963) *Erysiphe betae* (Vanha) comb, nov., the powdery mildew of beets. Phytopathol Z 47(2):123–128
Wen Y, Wang W, Feng J, Luo MC, Tsuda K, Katagiri F, Bauchan G, Xiao S (2011) Identification and utilization of a sow thistle powdery mildew as a poorly adapted pathogen to dissect post-invasion non-host resistance mechanisms in *Arabidopsis*. J Exp Bot 62:2117–2129
Weßling R, Panstruga R (2012) Rapid quantification of plant-powdery mildew interactions by qPCR and conidiospore counts. Plant Methods 8:35
Weßling R, Schmidt SM, Micali CO, Knaust F, Reinhardt R, Neumann U, Ver Loren van Themaat E, Panstruga R (2012) Transcriptome analysis of enriched *Golovinomyces orontii* haustoria by deep 454 pyrosequencing. Fungal Gent Biol 49(6):470–482
Weßling R, Epple P, Altmann S, He Y, Yang L, Henz SR, McDonald N, Wiley K, Bader KC, Glaßer C, Mukhtar MS, Haigis S, Ghamsari L, Stephens AE, Ecker JR, Vidal M, Jones JDG, Mayer KFX, Ver Loren van Themaat E, Weigel D, Schulze-Lefert P, Dangl JL, Panstruga R, Braun P (2014) Convergent targeting of a common host protein-network by pathogen effectors from three kingdoms of life. Cell Host Microbe 16(3):364–375
West E (1933) Powdery mildew of crape myrtle caused by *Erysiphe lagerstroemiae* n. sp. Phytopathology 23(10):814–819
Whipps JM, Budge SP, Fenlon JS (1998) Characteristics and host range of tomato powdery mildew. Plant Pathol 47:36–48
White TJ, Bruns TD, Lee S, Taylor J (1990) Amplification and direct sequencing of fungal ribosomal genes for phylogenetics. In: Innis MA, Gelfrand DH, Sninsky JJ, White TJ (eds) PCR protocols. Academic, San Diego, pp 315–322
Wicker T, Oberhaensli S, Parlange F, Buchmann JP, Shatalina M, Roffler S, Ben-David R, Dolezel J, Simková H, Schulze-Lefert P, Spanu PD, Bruggmann R, Amselem J, Quesneville H, Ver Loren van Themaat E, Paape T, Shimizu KK, Keller B (2013) The wheat powdery mildew genome shows the unique evolution of an obligate biotroph. Nat Genet 45(9):1092–1096
Xiao S, Ellwood S, Findlay K, Oliver RP, Turner JG (1997) Characterization of three loci controlling resistance of *Arabidopsis thaliana* accession Ms-0 to two powdery mildew diseases. Plant J 12(4):757–768
Yarwood CE (1963) Predisposition to powdery mildew. Phytopathology 53:1144–1145
Yarwood CE (1978) History and taxonomy of powdery mildews. In: Spencer DM (ed) The powdery mildews. Academic, London, pp 1–37
Zeller KA (1995) Phylogenetic relatedness within the genus *Erysiphe* estimated with morphological characteristics. Mycologia 87:525–531
Zhang D, Ouyang SH, Wang LL, Cui Y, Wu QH, Liang Y, Wang ZZ, Xie JZ, Zhang DY, Wang Y, Chen YX, Liu ZY (2015) Comparative genetic mapping revealed powdery mildew resistance gene MlWE4 derived from wild emmer is located in same genomic region of Pm36 and Ml3D232 on chromosome 5BL. J Integr Agric 14:603–609
Zheng RY, Chen GQ (1978a) Taxonomic studies on the genus *Pleochaeta* of China. I. A new species on *Salicaceae: Pleochaeta salicicola* sp. Nov. Acta Microbiol Sin 18:118–121
Zheng RY, Chen GQ (1978b) Taxonomic studies on the genus *Pleochaeta* of China: II. The imperfect state of *Pleochaeta*: *Streptopodium* gen. nov. Acta Microbiol Sin 18:181–188
Zheng Y, Chen GQ (1980) Taxonomic studies on the genus *Erysiphe* of China I. New species and new varieties on *Caprifoloaceae*. Acta Microbiol Sin 20(1):45–49

Chapter 4
Infection, Pathogenesis, and Disease Cycle

4.1 Introduction

Three successive events are essential for a powdery mildew to infect and reproduce on its host. First after landing on the host surface, the pathogen must invade an epidermal tissue cell and establish a feeding structure called haustorium, through which the fungus derives all of its nutrition. Secondly, the fungus must suppress or evade host defense mechanism and if fully activated, it would arrest or kill the pathogen. Third, since powdery mildew pathogen being an obligate cannot survive on dead host tissues, it must ensure that the underlying host cells remain viable until the pathogen's life cycle is complete. The powdery mildew pathogen of *Brassica* species over winters or over summer in the form of cleistothecia or chasmothecia wherever they are formed. In the absence of sexual stage, the pathogen survives through asexual stage on the numerous crucifers. As a primary source of inoculum, ascospores or conidia are carried by wind from sources of survival to the new host and cause infection after landing on the host surface under congenial environmental conditions. Conidia survive as secondary source of inoculum and cause infection after landing on host surface through the air currents.

4.2 Infection and Pathogenesis

In crucifers depending on the sources of survival as ascospores or conidia, after landing on leaf surface, they imbibe and germinate producing a primary germ tube that differentiates at its terminal point to produce an appressorium. From the appressorium an infection peg arises which penetrates the cuticle and epidermal cell wall into the epidermal cells. The penetrating hypha enlarges immediately upon entrance into the cell lumen and forms a globose haustorium by which the pathogen obtains its nutrients in the form of sugars and amino acids. The germ tube continues to

G. S. Saharan et al., *Powdery Mildew Disease of Crucifers: Biology, Ecology and Disease Management*, https://doi.org/10.1007/978-981-13-9853-7_4

grow and branch on the surface of the plant tissue producing a network of superficial mycelium from which conidiophores arise producing conidia. Conidia disperse through air currents and become a source of secondary inoculum to cause new infections after landing on host surface.

4.2.1 Infection and Pathogenesis of Arabidopsis *Powdery Mildew*

To visualize fungal development during the infection process, fungal structures on infected leaves were stained with trypan blue at various time points after inoculation. Within 12 h after inoculation, a conidium that has landed on a leaf germinated, producing a germ tube or an appressorial germ tube that differentiates a specialized infection structure, the appressorium. A mixture of enzymatic activities, and pressure from the appressorium, and penetration peg at the site, which the grass powdery mildew *Blumeria graminis* bleaches the underlying cell wall, are known to act on the epidermal cell to pave the way for plant cell invasion (Tucker and Talbot 2001). Successful penetration of the powdery mildew fungus led to the generation of the primary fungal feeding structure, so-called the haustorium, and subsequent development of elongated secondary hyphae from the appressorium. Haustoria of biotrophic pathogens are thought to be responsible for the uptake of sugars and amino acids from the host plants to the fungal mycelium and the active delivery of effector proteins to suppress plant immune responses (Ellis et al. 2006). Formation of mature haustoria and elongated secondary hyphae supports a compatible interaction of *Erysiphe* sp. (KUS-F23994) with *Arabidopsis thaliana* (Hwang and Heitfuss 1982). Subsequent haustorial establishment and secondary hyphal proliferation occurred, exclusively on the epidermal cells of the *A. thaliana*. Along the hyphae, new appressoria formed, and penetrated epidermal cells, and then produced secondary haustoria. Completion of the asexual life cycle of *Erysiphe* sp. (KUS-F23994) was evidenced by the occurrence of sporulation (white powder) on inoculated rosette leaves of *Arabidopsis*. In general, powdery mildew fungi do not invade mesophyll cells but rather limit their growth to the epidermal layer. The *Erysiphe* sp. (KUS-F23994) produced conidia singly on the conidiophores and sporulated little on *A. thaliana* leaves, which is consistent with the previous observation of a little sporulation of *E. cruciferarum* (Micali et al. 2008; Choi et al. 2009). In contrast to the *E. cruciferarum* sporulation, *G. cichoracearum* displays visibly abundant sporulation at 7 days after inoculation, and *G. orontii* requires 10 days for full sporulation (Micali et al. 2008; Vogel and Somerville 2002). The pathogenesis of *E. orontii* causing powdery mildew of *A. thaliana* has been studied through light and scanning electron microscopy (Plotnikova et al. 1998; Reuber et al. 1998).

An isolate of *Erysiphe orontii* (isolate MGH) that is virulent on the Columbia (Col-0) accession was obtained from a spontaneous infection of *Arabidopsis* plants (Plotnikova et al. 1998), when Col-0 is inoculated with *E. orontii* conidia using a

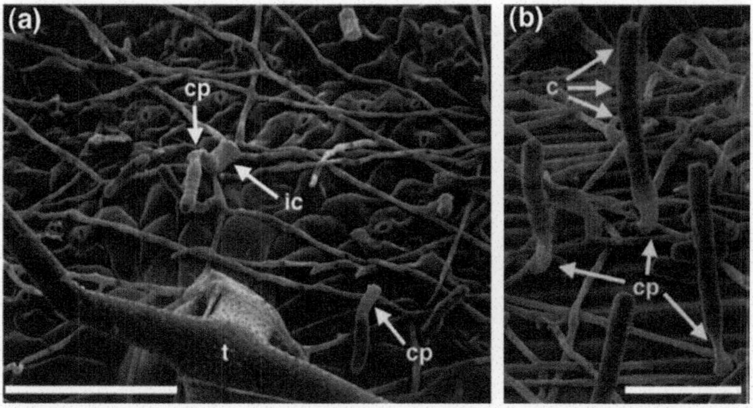

Plate 4.1 Scanning electron micrographs of *E. orontii*. (**a**) A 5-day-old colony of *E. orontii* on a Columbia leaf. i c- initial conidium; cp- conidiophore; t- trichome. Scale bar =100 µm. (**b**) Mature conidiophores on a *pad4* stem. cp- conidiophore; c- conidium. Scale = 50 µm. (Reuber et al. 1998)

settling tower, round powdery white colonies which become visible to the naked eye 5–7 days post-infection. Mature conidia (asexual spores) are produced within 5 days post-infection. Over a period of about 2 weeks, the colonies can spread to cover approximately 20% of the leaf surface. Scanning electron micrographs of *E. orontii* colonies on leaves and bolts are shown in Plate 4.1, and an *E. orontii*-infected Col-0 plant is shown in Plate 4.2a. At about 14 days post-infection, infected leaves begin to display chlorosis that eventually affects the entire leaf, which then appears to undergo premature senescence. *E. orontii* also grows on the *Landsberg erecta* (La-*er*) accession, but more poorly than on Col-0; the mycelial mat and conidiophores are sparser on La-*er* leaves than on Col-0 leaves (Plate 4.2e). Interestingly, La-*er* leaves become chlorotic more rapidly than Col-0 leaves, at about 7 days post-infection (Plate 4.2e) (Reuber et al. 1998). For details on fine structures of powdery mildew formed during pathogenesis, see Chap. 5.

4.2.2 Genetic Factors (Genes) Affecting **Arabidopsis** *Powdery Mildew Pathogenesis*

The biotrophic lifestyle of powdery mildew fungi dictates that they infect plants and establish colonies in a "quiet" manner, so as not to arouse the defense mechanisms of the host (Panstruga 2003). They likely also usurp host infrastructure for transport of nutrients and building blocks and coerce the host cell into accommodating their impressive haustorial feeding structures (Plate 4.3, Mendgen and Hahn 2002; Schulze-Lefert and Panstruga 2003). Specific host genes and/or proteins termed compatibility or susceptibility factors are believed to be essential for successful pathogenesis by a given microbe, and a lack of these factors is predicted to

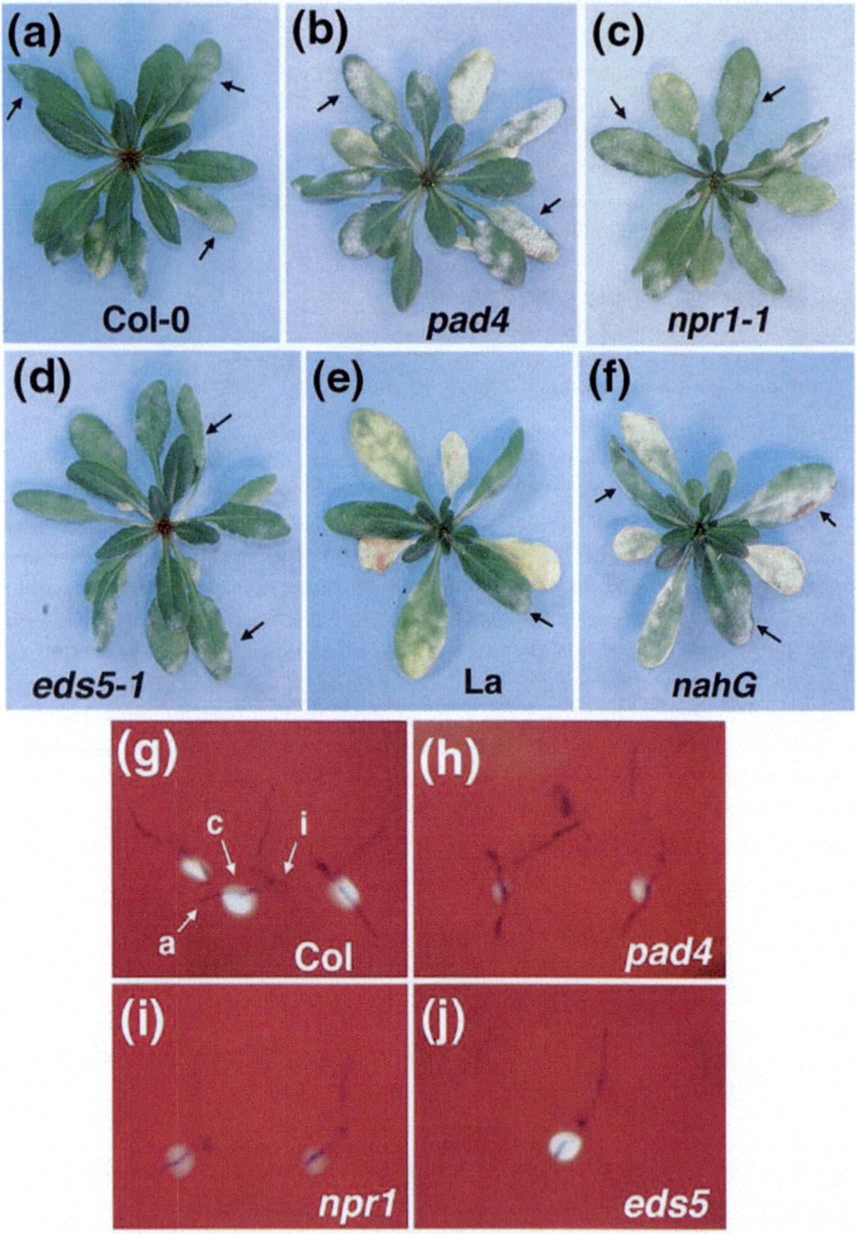

Plate 4.2 *Arabidopsis* plants infected with *E. orontii* and epifluorescence micrographs of infected *Arabidopsis* leaves stained for callose. (**a–f**) Plants were infected by settling tower. Photographs were taken 12 days postinoculation. Arrows indicate representative areas of infection. (**a**) Col-0; (**b**) *pad4–1*; (**c**) *npr1–1*; (**d**) *eds 5–1*; (**e**) La-*er*; (**f**) *nah G*. (**g–j**) Leaves stained with aniline blue to detect callose 24 h after infection with *E. orontii* (magnification = 180 x). (**g**) Col-0. *i* initial conidium, *a* appressorial germ tube, *c* cell wall apposition. (**h**) *pad4*. (**i**) *npr1–1*. (**j**) *eds5* (Reuber et al. 1998)

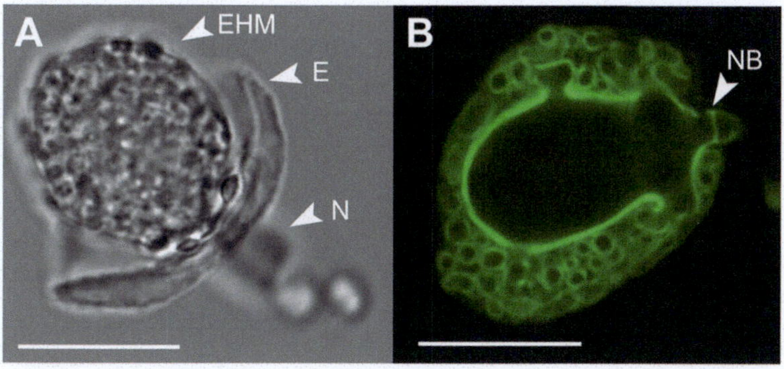

Plate 4.3 Haustorial complexes of *G. orontii*. Phase contrast (**a**) and epifluorescence (**b**) micrographs of *G. orontii* haustoria isolated from *Arabidopsis* leaves. Notice the highly convoluted and complex folding of the haustorial cell surface providing a large area for nutrient uptake from and effector delivery into the host. Haustoria were labeled with wheat germ agglutinin-FITC. *EHM* extra-haustorial membrane, *E* encasement, *N* haustorial neck, *NB* neckband. Scale bars = 20 μm (Micali et al. 2008)

result in resistance to an otherwise virulent pathogen (Vogel and Somerville 2000; Panstruga 2003). To isolate compatibility factors, two independent genetic screens were performed in the late 1990s and identified many new components that paint an ever more complex picture of *Arabidopsis* cellular mechanisms of defense. A hallmark of constitutive defense is the transcriptional activation of PR genes such as *PR1* (Loake and Grant 2007). In an attempt to identify novel elements of resistance, three powdery mildew (*G. cichoracearum*) resistant mutants that do not constitutively express *PR1* were isolated in the lab of R. Innes and named enhanced disease resistance (*edr*) 1, *edr2,* and *edr3* (Frye and Innes 1998; Frye et al. 2001; Tang and Innes 2002; Tang et al. 2005a, b, 2006). All three mutants display late onset of resistance (5–8 days postinoculation) that is characterized by accelerated cell death in mesophyll cells leading to large necrotic lesions and limited or absent conidiation of the pathogen. Resistance in all cases is SA-dependent and JA-independent. *EDR1* encodes a *CTR1*-like protein kinase and negative regulator of disease resistance (Frye et al. 2001). Presence of *EDR1* limits the transcriptional amplification of *RPW8*.1 and reduces *RPW8*-conditioned host cell death in response to pathogen attack (Xiao et al. 2005; Fig. 4.1). *EDR2* encodes a mitochondrial protein (Fig. 4.1) with a pleckstrin homology domain and a steroidogenic acute regulatory protein-related lipid transfer (START) motif that are each indicative of a role in lipid binding and signaling. Both *EDR1* and *EDR2* function in a common pathway since double mutants display resistance phenotypes identical to the respective single mutant plants (Tang et al. 2005a). In addition, both mutants display enhanced senescence in response to ethylene. In contrast, *EDR3* appears to function in an independent pathway from *EDR1* and *EDR2,* and the corresponding mutant does not display an early senescence phenotype. Like *EDR2*, *EDR3* localizes at least in part to mitochondria and encodes a dynamin-like protein with an N-terminal GTPase domain and a C-terminal GTPase effector region (Tang et al. 2006; Fig. 4.1). In

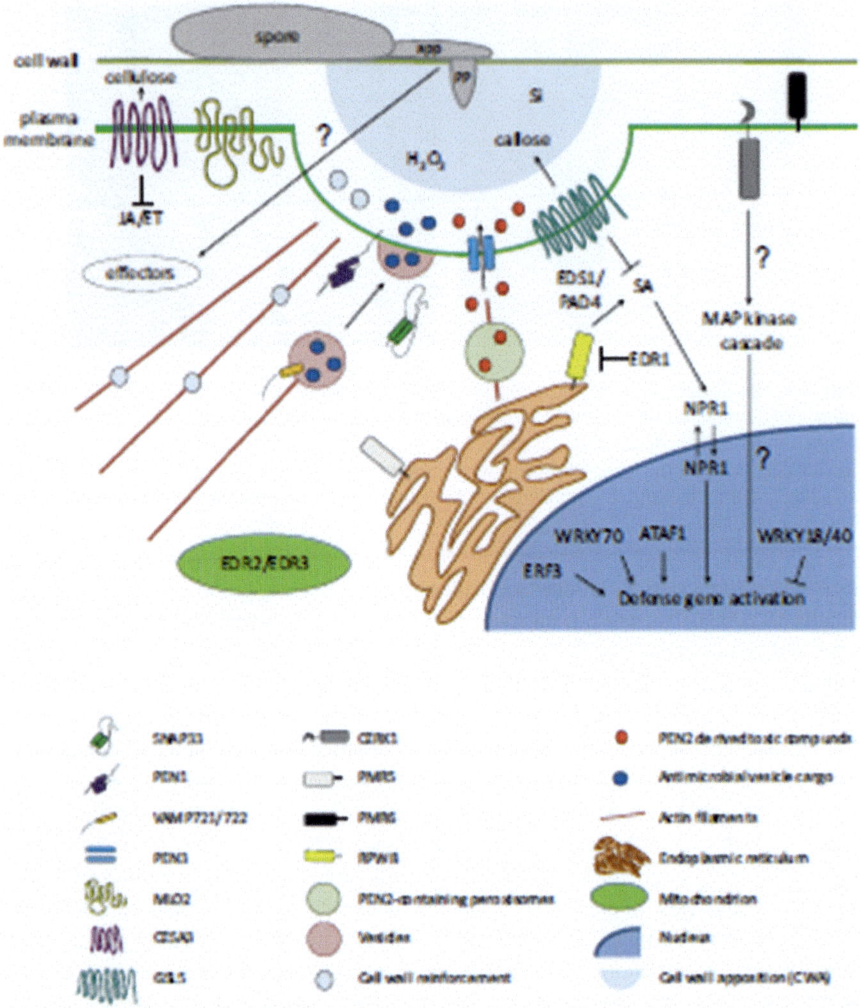

Fig. 4.1 Schematic diagram illustrating genetically anchored components in *Arabidopsis* powdery mildew susceptibility/resistance. The figure depicts a section of a host cell attacked by a powdery mildew germ tube. Components coded by shape and colour are explained in the legend below the scheme. *app* appressorium, *pp* penetration peg, *Si* silicon. Question marks indicate presumed links/activities (Micali et al. 2008)

mammals, such proteins are responsible for membrane tabulation and vesicle pinching and regulate mitochondrial dynamics associated with apoptosis (programmed cell death; Tang et al. 2006). The characterization of *edr* mutants suggests a link between SA-mediated resistance, mitochondrial function, and programmed cell death and further stresses the broad parallels that exist between animal and plant immune responses (Ausubel 2005). Indeed, a central feature of defense responses in mammalian systems is the ability to undergo apoptosis in response to pathogen (or non-self) detection (Hiscott et al. 2006). Often the apoptotic program is either initiated or amplified in mitochondria, and components of the mitochondrial apoptosome are absolutely required for completion of programmed cell death (Keeble and Gilmore 2007).

In an independent screen performed by Vogel and Somerville (2000, 2002) to recover loss of susceptibility to *G. cichoracearum* mutants in *Arabidopsis*, the powdery mildew resistant (pmr) mutants pmr1 to *pmr6* were isolated, and to date four of them have been cloned (Vogel and Somerville 2000; Vogel et al. 2002; Jacobs et al. 2003; Nishimura et al. 2003; Vogel et al. 2004; Consonni et al. 2006; Plate 4.4). *PMR2* is allelic to *MLO2, PMR4* (synonyms *CALS12* and *GSL5*) encodes a wound- and pathogen-associated callose synthase, PMR5 belongs to a large family of plant-specific genes of unknown function, and PMR6 encodes a glycosyl-phosphatidyl-inositol (GPI)-anchored pectate lyase-like protein (Vogel

Plate 4.4 Macroscopic infection phenotypes of susceptible and resistant *Arabidopsis* lines. Rosette leaves of 5–6-week-old *A. thaliana* ecotypes Col-0 (**a**), Do-0 (**b**), and Sorbo (**c**) as well as the powdery mildew resistant mutant pmr6–3 (**d**) at 13 days postinoculation with *G. orontii*. Completion of the asexual powdery mildew life cycle is evidenced by the occurrence of abundant sporulation (white powder) on inoculated rosette leaves of the susceptible accession, Col-0. Younger leaves without disease symptoms emerged after inoculation with fungal conidiospores. Note the difference in appearance of infected leaves of resistant accessions Do-0 and Sorbo. Resistance in both accessions is assumed to be governed by *RPW8* (Gollner et al. 2008)

et al. 2002; Jacobs et al. 2003; Nishimura et al. 2003; Vogel et al. 2004; Consonni et al. 2006). In pmr4, pmr5, and pmr6 mutants, resistance is entirely established post-penetration and results in reduced colony size and conidiophore formation in interactions with both *G. cichoracearum* and *G. orontii*. The *PMR* genes cloned to date are involved in different cellular activities, underscoring the diversity of plant biological processes that are required for successful powdery mildew pathogenesis (Micali et al. 2008).

PMR4 controls callose synthesis at cell wall appositions (papillae) that form beneath infection, and wound sites, and that are believed to provide a physical barrier to fungal penetration (Aist and Bushnell 1991; Plate 4.4). It therefore came as a surprise that a loss of function in a callose synthase renders a plant more resistant to powdery mildew infection (Jacobs et al. 2003; Nishimura et al. 2003). A comparative analysis of genes that are differentially expressed in wild-type and *pmr4* plants upon infection with *G. cichoracearum* revealed that resistance to powdery mildews is likely the result of constitutive activation of the SA signaling pathway that is further enhanced upon pathogen attack (Nishimura et al. 2003). Mutations in genes encoding components of the salicylic acid signaling pathway indeed abolished *pmr4*-based resistance without restoring callose deposition at papillae. Callose and/or callose synthase-based signalings were shown to activate basal defense responses that are effective against a variety of pathogens including *G. orontii* and *Hyaloperonospora parasitica* (Jacobs et al. 2003; Nishimura et al. 2003). In this sense, *PMR4* is likely less a fungal-specific compatibility molecule and more a general basal defense switch located at the cell wall that is effective against a broad range of pathogens.

Two additional pmr mutants stress the potential importance of cell wall integrity surveillance in resistance mechanisms. *PMR6* encodes a putative pectate lyase with proposed pectin-degrading activity localized at the cell wall. The *pmr6* mutant displays an increased pectin, and uronic acid content in the cell wall, and exhibits a stunted phenotype (Vogel et al. 2002). In *pmr5*, similar changes in cell wall components have been reported. *PMR5* codes for a member of a large family of plant-specific proteins of unknown function that is likely targeted to the endoplasmic reticulum/secretory pathway (Plate 4.4). Although the exact mechanism leading to powdery mildew resistance is unknown, it is independent of SA, ET, and JA signaling, since mutations in any of the known defense pathways have no effect on *pmr5*- or *pmr6*-based resistance. However, *pmr5/6* double mutants show further increased resistance compared to the respective single mutants, indicating that the two genes likely control parallel and independent defense responses. Furthermore, the fact that *pmr5* and *pmr6* are resistant to *G. cichoracearum* and *G. orontii*, but fully susceptible to unrelated pathogens such as virulent strains of either *Pseudomonas syringae* or *H. parasitica*, suggests that these two proteins, in contrast to the *PMR4* callose synthase gene, may in fact be true powdery mildew specific compatibility factors (Vogel and Somerville 2000; Vogel et al. 2002, 2004).

4.2.3 *Role of MLO Proteins (Genes) in* Arabidopsis *Powdery Mildew Pathogenesis*

Isolated in the same genetic screen as genes *PMR4*, *PMR5,* and *PMR6*, *PMR2* encoding *MLO2* (*mildew locus O*) has an essential role in the establishment of compatibility with the powdery mildew species *G. cichoracearum* and *G. orontii.* Together with the phylogenetically closely related paralogs *MLO6* and *MLO12*, which act in concert with *MLO2* in partial functional redundancy (Consonni et al. 2006), *MLO2* controls entry of powdery mildew fungi in *Arabidopsis* epidermal cells (Plate 4.5). In this respect, *MLO2*, *MLO6,* and *MLO12* and the barley homolog, and first member of this protein family to be identified – *Mlo* (Panstruga 2005) – represent the prototype for molecules that mediate compatibility between powdery mildews and their plant hosts. Loss of particular *MLO* protein (s) in barley, tomato, and *Arabidopsis* renders these plant hosts immune to powdery mildew infection (Bai et al. 2008). The conserved function of *MLO* proteins in plant defense predates the divergence of dicots and monocots and underscores the importance of this protein family in the establishment of compatible plant–powdery mildew interactions (Consonni et al. 2006). MLO proteins are integral membrane proteins with seven transmembrane domains (Plate 4.5) that have so far only been identified in plants (Devoto et al. 2003). In *Arabidopsis*, they are encoded by a family of 15 genes and appear to have diverse functions in addition to plant defense. However,

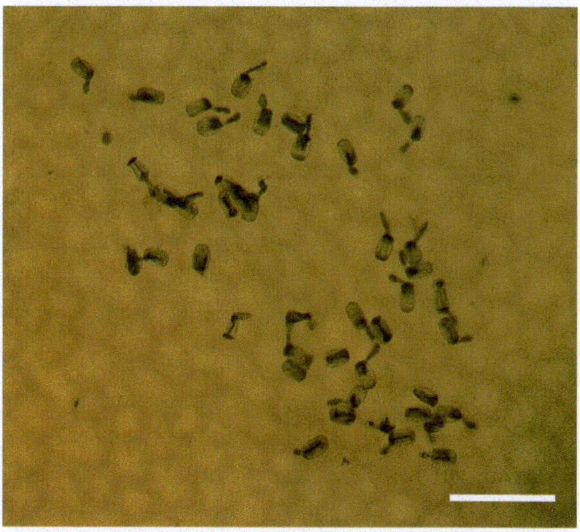

Plate 4.5 Resistance in *Arabidopsis mlo2 mlo6 mlo12* triple mutants is characterized by complete failure of successful host cell invasion. The micrograph shows attempted penetration of numerous *G. orontii* sporelings at 48 h postinoculation on a highly resistant *mlo2 mlo6 mlo12* triple mutant in the genetic background of otherwise susceptible Col-0 (Consonni et al. 2006). Note the aborted fungal entry evidenced by a lack of secondary hyphae compared to the situation in a Col-0 wild-type plant. Scale bar = 100 μm. (Micali et al. 2008)

only the three closely related members *MLO2*, *MLO6,* and *MLO12* seem to play a role in powdery mildew pathogenesis. Consistent with previous findings in barley (Jarosch et al. 1999; Kumar et al. 2001), *mlo2 mlo6* double and *mlo2 mlo6 mlo12* triple mutants show enhanced susceptibility to the necrotrophic pathogens *Alternaria alternata* and *A. brassicicola* and the hemi-biotrophic oomycete *Phytophthora infestans*, indicating that *MLO* proteins modulate the infection process of pathogens with diverse lifestyles (Consonni et al. 2006). Barley *MLO* and *MLO2* from *Arabidopsis* each interact via a conserved peptide domain in their cytoplasmic carboxyl terminus with the Ca_2+ sensor calmodulin (Kim et al. 2002; Bhat et al. 2005) and are thought to regulate defense responses against powdery mildews via PEN1 (ROR2 in barley)-dependent mechanisms at the cell periphery (Collins et al. 2003; Schulze-Lefert 2004; Panstruga 2005; Takemoto et al. 2006). Similar to non-host resistance, *mlo*-mediated resistance in *A. thaliana* does not depend on the JA/ET or SA signaling pathways (Collins et al. 2003; Lipka et al. 2005; Stein et al. 2006; Consonni et al. 2006). The physical proximity of *MLO* to the *PEN1*-dependent non-host defense machinery at the sites of fungal attack (Bhat et al. 2005), the recently observed transcriptional co-expression of these genes in *Arabidopsis* (Micali et al. 2008), and the genetic and cytological similarities between mlo-based and non-host resistance (Humphry et al. 2006) point to a tight mechanistic link between *MLO* and the basal defense apparatus. The fact that presence of *MLO* is absolutely required by powdery mildew fungi to successfully infect host plants suggests that these pathogens possibly exploit *MLO* function (s) to suppress basal defense responses (Panstruga 2005) during the process of pathogenesis.

4.2.4 Host Transcriptional Changes as an Indicator to Powdery Mildew Pathogenesis

Besides defense-related transcriptional changes induced by microbe recognition, host gene expression are potentially impacted by the action of powdery mildew effectors to promote fungal pathogenesis. Consequently, adaptation of the host metabolism to the presence of the pathogen has been reported (Wildermuth 2010). Laser microdissection-assisted site-specific profiling of transcript abundance during late *Go* infection stages (5 dpi) suggests a suppression of photosynthesis and points to a carbon source-to-sink transition in powdery mildew-infected cells (Plate 4.6; Chandran et al. 2009; Chandran et al. 2010). The powdery mildew-triggered induction of genes associated with sugar metabolism and hexose transporters associated with sink organs further reinforce this idea. The respective proteins might contribute to the availability of carbohydrates at infection sites, and consistently their elevated expression levels may reflect an increased demand for hexoses by the fungus (Fabro et al. 2008; Chandran et al. 2009, 2010). Increased transcript abundance of genes related to respiration, including glycolysis, the tricarboxylic

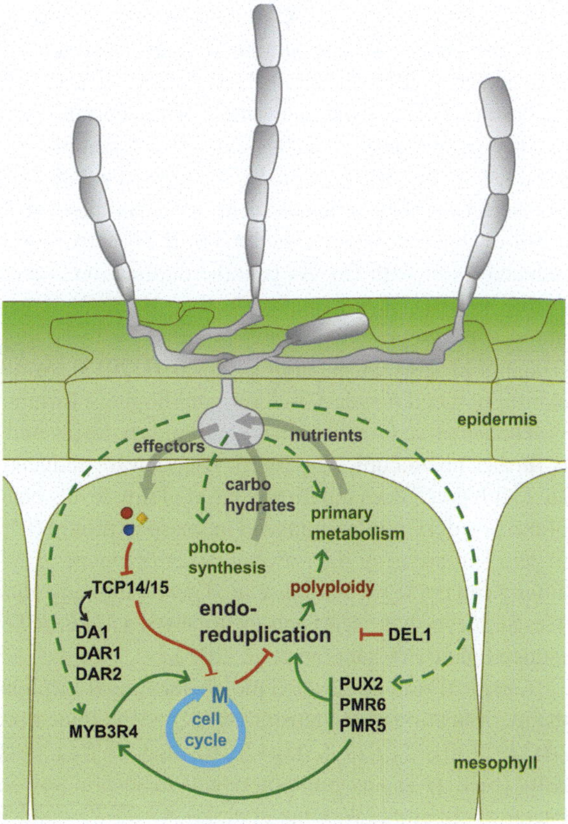

Plate 4.6 Proposed model of endo-reduplication in powdery mildew pathogenesis. The scheme depicts regulators and mechanisms involved in the control of endo-reduplication and consequences of powdery mildew-induced mesophyll polyploidy. A powdery mildew (grey)-colonized leaf epidermal cell and an underlying mesophyll cell are shown. Grey arrows indicate the proposed translocation of components between cells. Solid lines indicate proven regulatory impacts, and dashed lines indicate speculative regulatory impacts. M = mitosis (Kuhn et al. 2016)

acid cycle, and the mitochondrial electron transport chain further strengthen the notion of an adaptation to the elevated energy consumption by the infected tissue (Fabro et al. 2008; Chandran et al. 2010).

Adjustment of the plant host metabolism to support the growth of the biotrophic pathogen is consistent with an increased ploidy level of the mesophyll cells underlying infected epidermal cells at later stages of compatible interactions (Plate 4.6; Chandran et al. 2010, 2013). This correlates with the accumulation of *plant UBX domain-containing protein 2* (*PUX2*: At2g01650) transcripts after *Go* infection (Chandran et al. 2009). Strikingly, the onset of *PUX2* induction at 5 dpi overlaps with the occurrence of endo-reduplication in mesophyll cells and corresponds with fungal growth and reproduction (Chandran et al. 2013). The resulting polyploidy might compensate for the increased metabolic activity resulting from the nutritional demands of the fungus. This is supported by decreased spore formation coinciding with reduced basal ploidy in *pux2* and thus identifies endo-reduplication as a potential determinant of susceptibility to powdery mildew (Chandran et al. 2010, 2013). The presence of UBX (ubiquitin regulatory X) and PUB (peptide:

N-glycanase/UBA or UBX-containing proteins) domains in *PUX2* suggests that, like other proteins with similar domain structures, it might act as a regulatory cofactor of cell division control protein 48 (*CDC48*: At3g09840). Indeed, this AAA-ATPase (ATPase associated with diverse cellular activities) interacts with *PUX2* in vitro (Rancour et al. 2004). As *CDC48* complexes contribute to cell cycle progression, its interaction with *PUX2* might regulate cell ploidy (Rancour et al. 2004; Madsen et al. 2009; Yamanaka et al. 2012; Gallois et al. 2013). *MYB3R4* (At5g11510), a cell cycle control-associated *MYB3R TF* activating G2/M progression, is locally induced 5 dpi with *Go*. As genome duplication is a controlled process that occurs during mitosis, it is conceivable that *MYB3R4* is required for powdery mildew-induced polyploidy, which is supported by the phenotype of the *myb3r4* mutant (Haga et al. 2007; Chandran et al. 2010, 2013). Similar to *pux2, myb3r4* mutants exhibit reduced powdery mildew conidiophore formation.

The negative impact on basal cell ploidy levels and increased resistance of *pux2* is further phenocopied by *pmr6*. By contrast, reduced fungal reproduction associated with *pmr5* does not impact basal ploidy levels but correlates with a suppression of the powdery mildew-induced increase in ploidy (Chandran et al. 2013). Analysis of *pmr5* microarray data reveals an enrichment of MYB3R TF binding elements among cell cycle regulation-related genes showing altered expression in the mutant. This suggests that *PMR5* acts upstream of a *MYB3R TF* to control powdery mildew-induced ploidy (Chandran et al. 2013).

A critical role of elevated ploidy for fungal virulence is further strengthened by the identification of the teosinte branched/cycloidea/proliferating cell factor *TCP13* (At3g02150), *TCP14* (At3g47620), and *TCP15* (At1g69690) basic helix–loop–helix (bHLH) TFs as common targets of several powdery mildews, oomycetes, and bacterial effectors (Weßling et al. 2014). *TCP14* and *TCP15* repress endo-reduplication by directly regulating the expression of cell cycle genes (Peng et al. 2015). Mutation of *TCP13, TCP14*, and, to a lesser extent, *TCP15* results in increased susceptibility towards *Go* (Weßling et al. 2014). A link to ubiquitin-mediated regulation of ploidy is provided by the ubiquitin receptors *DA1* (At1g19270; "Da" is Chinese for "large"), DA-related *(DAR) 1*, and *DAR2* (At2g39830), which interact with and modulate the stability of *TCP14/15* to regulate endo-reduplication (Peng et al. 2015). Remarkably, DEL1, known to repress genes required for the onset of endo-reduplication (Vlieghe et al. 2005; Lammens et al. 2008), does not impact the *Go*-induced increase in mesophyll ploidy when mutated or overexpressed (Chandran et al. 2014). Instead, microarray analyses of *del1* plants revealed an induction of basal defense gene expression compared to wild type. The identification of effector targets involved in adaptation of the host metabolism (Weßling et al. 2014) marks one of the first steps towards elucidation of pathogen-induced reprogramming of the plant transcriptome. Further, characterization of the mechanisms by which the fungus enforces adjustment of the plant cellular program to promote its intracellular accommodation will be an important aspect of future research (Kuhn et al. 2016).

4.2.5 *Transcriptional Programming of Powdery Mildew Pathogenesis*

Transcriptional programming during powdery mildew pathogenesis was analysed by Pandey et al. (2010). The transcriptional reprogramming induced by *G. orontii* during early events of *Arabidopsis* infection, and the role of two closely related WRKY transcription factors, 18 and 40, was observed. It was observed that *WRKY18/40* negatively affects preinvasion defenses in *Arabidopsis* and deduced a subset of genes that appear to be under *WRKY18/40* control. The direct, in vivo physical interaction of *WRKY40* to W box containing promoter regions of the key regulatory genes *EDS1*, and *JAZ8*, and of the *AP2/ERF* transcription factor gene, *RRTF1* was demonstrated.

Previous transcriptomic studies mainly focused on later stages of powdery mildew infection, i.e. haustoria formation (18–24 hpi; (Fabro et al. 2008) or pathogen growth and reproduction (1–7 days postinoculation (Chandran et al. 2009). Pandey et al. (2010) found that expression of hundreds of host genes is altered within 8 hpi indicating that signaling between the pathogen and individual host cells already occurs a few hours post-contact of the spores with the leaf surface at about the stage of penetration, but certainly prior to establishment of the extra-haustorial matrix. It is important to note that *G. orontii* invades only epidermal cells (Lipka et al. 2008). Hence, local gene expression differences within such cells may actually be higher than the values observed using whole leaf samples. Interestingly, 9 out of the 24 genes most highly and differentially expressed (>threefold) in uninfected *wrky18 wrky40* plants encode transcriptional modulators. This implies that *WRKY18/40* are central negative regulators controlling distinct downstream transcriptional programs. The role of phytohormones and distinct signaling pathways during compatible interactions of *Arabidopsis* and powdery mildews is not firmly established (Chandran et al. 2009; Consonni et al. 2006; Fabro et al. 2008). Although *WRKY18/40* expression was observed to be dependent on the SA biosynthetic gene isochorismate synthases1 (*ICS1*) at later stages of *G. orontii* infection (1–7 dpi), induction of these genes at the early infection stage (6 hpi) was *ICS1* independent (Chandran et al. 2009). Compared with WT, *wrky18 wrky40* plants constitutively express 2.4–8-fold higher levels of five co-regulated JAZ family members, namely, *JAZ1, JAZ5, JAZ7, JAZ8*, and *JAZ10*, and their promoters contain several W box elements. JAZ proteins are key repressors of JA signaling (Chico et al. 2008). Pandey et al. (2010) could not detect in vivo associations of *WRKY40* at *JAZ7* and *JAZ10* promoter sites nor to *LOX2* and *AOS*, they observed specific interaction with a region of the *JAZ8* promoter containing two perfect W boxes. Currently, only one transcription factor, *MYC2*, has been identified as a direct regulator of *JAZ* promoter activity. However, expression of *JAZ7* and 8 does not require *MYC2* (Chung et al. 2009). It implies that *WRKY40* affects JA signaling by directly controlling the expression of a subset of these negative regulators. Lack of mutants for *JAZ7, –8*, and *–10* currently hampers functional testing of their role in plant–pathogen interaction. Also, the contribution of *WRKY18* in this process remains to be elucidated, although it exerts a positive regulatory function, along with *WRKY40*, in JA signaling (Wang et al. 2008).

The role of *RRTF1* in plant defense remains to be rigorously tested, but recent reports indicate its involvement in regulating redox homeostasis during stress, and expression of *RRTF1* is dependent on *COI1*, a central regulator of JA signaling (Khandelwal et al. 2008; Wang et al. 2008). Moreover, *RRTF1* is part of a core redox signaling sub-network that also includes *EDS1* and *WRKY33* (Khandelwal et al. 2008). Possibly, all three of these genes are under *WRKY18/40* control. WRKY factors form a functionally interconnected transcriptional network (Eulgem and Somssich 2007), and *WRKY18/40* may thus indirectly modulate gene expression by directly controlling other *WRKY* activities. Indeed, loss-of-*WRKY18/40* function appears to limit the pathogen-induced accumulation of *WRKY11* and *WRKY33* transcripts. PAMP-induced expression of *WRKY33* was shown to depend on W box promoter elements that are bound by *WRKY* factors in plants (Lippok et al. 2007). Compared with WT, *wrky18 wrky40* mutants displayed rapid transient induction of *CYP71A13* and an overall increased accumulation of *PAD3* transcripts. *WRKY33* binding to the *PAD3* promoter in vivo has been demonstrated, but no binding was detected within the *CYP71A13* promoter (Qiu et al. 2008). It suggests that *CYP71A13* may represent a direct target of *WRKY18/40*, while *PAD3* is an indirect target that is directly controlled by *WRKY33* (Pandey et al. 2010).

In general, expression of stress response genes is transient, and their transcription relies on a combination of distinct sets of co-activators and negative regulators (Lopez-Maury et al. 2008). It remains to be elucidated how *WRKY18/40* negatively regulate transcriptional outputs. In the case of *RRTF1* and *JAZ8*, one can envision a direct repressor function on basal gene expression. However, *CYP71A13, EDS1*, and *PAD4* regulation may be different. Expression of these genes in *wrky18 wrky40* is not significantly elevated in the absence of pathogen infection. Rather, loss-of-*WRKY18/40* function results in a pathogen-dependent but exaggerated increase in their transcript levels. A simplistic model would be that other *WRKY* factors (i.e. *WRKY60*) could substitute to form weaker but sufficient repressor complexes at these promoters. Replacement of *WRKY* factors at specific W box sites has been demonstrated for two parsley promoters (Turck et al. 2004). Activation of these genes would require an activator complex that is dependent on pathogen-triggered signaling. Elevated transcript levels observed in the mutant may be the consequence of a higher association of the activator(s) to the transcription complex due to failure of the weaker core pressers to counter this effect. This could imply that binding sites for default repressors and sites for signal-induced activators overlap, a situation often found in inducible eukaryotic promoters (Affolter et al. 2008). Naturally, alternative scenarios are conceivable that may also involve altered regulation at the post-transcriptional level.

The consequences of transcriptional reprogramming should be reflected in qualitative/quantitative altered host metabolite levels resulting in resistance of *wrky18 wrky40* plants towards *G. orontii*. However, *wrky18 wrky40* mutants accumulated high camalexin levels, and were resistant to this powdery mildew lends support to the importance of this compound in determining the outcome in this interaction.

Proteases and endo-membrane components involved in protein quality control and secretion have been associated with hormone-responsive signal transduction events and disease resistance (Beers et al. 2004; Wang et al. 2005). Several protease genes including At3g49340 were more strongly up-regulated in challenged *wrky18 wrky40* mutants than in WT plants. Genes encoding components of the endo-membrane system and endoplasmic reticulum (ER) response (e.g. *CYP71A12, CYP81F2*), wall-associated kinase (At2g34500), *BiP3* (At1g09080), and type-1 phosphodiesterase (At4g29700)) were also differentially regulated. *CYP81F2* encodes a P450 monooxygenase essential for pathogen-induced accumulation of antifungal compounds (Bednarek et al. 2009). *BiP3* has been associated with the ER response, and its up-regulation in *wrky18 wrky40* resistant plants intuitively suggests that in susceptible WT plants, *G. orontii* may manipulate part of the host ER response system by impinging on *WRKY18/40* functions. Defense activation is often preceded by mitogen-activated protein kinase (MAPK) signaling (Colcombet and Hirt 2008). Uninfected *wrky18 wrky40* plants constitutively expressed fivefold elevated transcript levels of a mitogen-activated triple kinase gene *MAPKKK19* (At5g67080).

Although *wrky18 wrky40* lines overexpressing *WRKY40* showed a high degree of susceptibility, penetration efficiency did not reach WT levels. Because *WRKY18/40/60* can form homo- and heterodimers (Xu et al. 2006), appropriate *WRKY18/40* heterodimerization may be partly required for maximal transcriptional reprogramming. Alternatively or additionally, *WRKY18* and *40* may target different subsets of genes. Due to extreme difficulties in generating plants expressing a *WRKY40* transgene under its native promoter, interpretation of results relies on ectopic expression of this gene. Despite this caveat, these lines show no obvious phenotypic differences to WT plants and do not constitutively express defense genes. Moreover, as illustrated by *ChIP*, overexpressed *WRKY40* shows binding selectivity towards some promoters and even at such sites only to a subset of the overall number of W box elements present.

During the early course of *G. orontii* infection, the JA signaling pathway is required for plant susceptibility, whereas the *EDS1* signaling pathway is needed to maintain resistance (Pandey et al. 2010). *WRKY18/40* negatively modulate the expression of positive regulators of defense such as *EDS1*, and the camalexin biosynthetic genes, but positively enhance JA signaling by partly suppressing the expression of genes encoding negative regulators of this pathway. This enables the pathogen to overcome early host defenses and ultimately leads to susceptibility. Alternatively, *WRKY18/40* regulate induced expression of genes whose products are required for pathogen growth and reproduction. However, very few such host susceptibility factors enhancing powdery mildew pathogenesis have been identified (Eichmann and Huckelhoven 2008). Moreover, despite the fact that constitutive activation of JA and SA signaling can result in enhanced resistance towards powdery mildew infection, comprehensive studies employing a range of SA, JA, and ET signaling mutants failed to reveal a clear-cut role of these pathways in *G. orontii* resistance. Thus, a model is required in which *G. orontii*, in part via the regulatory action of *WRKY18/40*, temporally alters the balance between SA and JA signaling at early stages of infections, keeping host defenses suppressed, and thereby allowing the pathogen to gain access to host nutrients. Future investigations will aim at functionally dissecting the *WRKY40/18*-dependent pathways and the role of key *WRKY40* downstream targets in modulating host plant defenses. Such studies will include extensive evaluation of phytohormone dynamics (especially of JA and SA), analysis of selected mutants, and the generation of transgenic lines modified in the expression of multiple key components identified (Pandey et al. 2010).

4.2.6 Genes Governing Powdery Mildew Pathogenesis of Host

Plant disease occurs if a virulent pathogen encounters a susceptible host plant. Since plants are considered basically resistant to most pathogenic microbes, this requires evolutionary adaptation of the pathogen to its host (Jones and Dangl 2006). Host adaptation involves pathogen effector-mediated suppression of defense mechanisms initiated by the plant, e.g. after recognition of pathogen-associated molecular patterns (PAMPs) (da Cunha et al. 2006; Koeck et al. 2011). Biotrophic pathogens often show a highly specific adaptation to their host plant. *Formae speciales* of the cereal powdery mildew fungus *Blumeria graminis* are able to colonize only one specific cereal species, while all others remain non-hosts. This may be explained by the inability of non-adapted powdery mildew fungi to suppress PAMP-triggered immunity, possibly due to an inadequate effector repertoire or timing of delivery or differences in host effector targets for susceptibility (Niks and Marcel 2009). For successful pathogenesis, powdery mildew fungi need an efficient suppression of the plant's preformed and inducible penetration resistance mechanisms as well as maintenance of the compatible interaction after host cell penetration, which may include suppression of programmed cell death (PCD; Huckelhoven and Panstruga 2011). After host cell penetration, the haustorial complex is established, which consists of the fungal haustorium, an extra-haustorial modified plant plasma membrane, and the haustorial matrix in between (Koh et al. 2005; Micali et al. 2011). The haustorial complex represents the site of the most intimate contact of the fungus with its host cell and likely functions in nutrient uptake but may also provide a platform for the delivery of fungal effectors (Micali et al. 2011). Components of defense or susceptibility to powdery mildew fungi are conserved among monocotyledonous and dicotyledonous plants. In *Arabidopsis* and barley, (non-host) penetration resistance to non-adapted powdery mildew fungi involves the functionally homologous soluble *N*-ethylmaleimide-sensitive factor attachment (SNARE) protein *PEN1/ROR2*, which form complexes with V*AMP721/722* and *SNAP33/34* (Collins et al. 2003; Douchkov et al. 2005; Kwon et al. 2008a, b). Conversely, one considers the plasma membrane-resident mildew locus O (MLO) protein as a major susceptibility factor for powdery mildew fungi in monocot and dicot plants. Lack of functional MLO protein(s) results in complete race nonspecific penetration resistance to the barley powdery mildew fungus *B. graminis* f. sp. *hordei* (*Bgh*) (Buschges et al. 1997) and abolishment of host cell penetration and conidiophore production of

Golovinomyces orontii or *G. cichoracearum* on *Arabidopsis mlo2/6/12* triple mutants (Consonni et al. 2006). The conserved cell death suppressor protein *Bax inhibitor-1* (*BI-1*) is also required for full powdery mildew susceptibility in barley. Transient or stable overexpression of *BI-1* supports penetration of *Bgh* into barley epidermal cells, while transient or stable knockdown of *BI-1* expression restricts it (Huckelhoven et al. 2003; Eichmann et al. 2006, 2010; Babaeizad et al. 2009). *BI-1* proteins are endoplasmic reticulum-resident proteins, which are suppressors of cell death responses to biotic and abiotic stress stimuli in plants and animals (Huckelhoven 2004; Watanabe and Lam 2009; Henke et al. 2011; Ishikawa et al. 2011; Robinson et al. 2011). Moreover, having six or seven predicted transmembrane domains, *BI-1* amino acid sequences harbour the '*BI-1-like protein family motif*' (NCBI conserved domain database motif cd06181) or 'inhibitor of apoptosis-promoting *Bax1-related motif*' (Inter Pro IPR006214). This motif spans the transmembrane domains of the proteins and hence represents the overall domain architecture of these proteins and is of unknown function (Huckelhoven 2004). In mammals, various transmembrane *BI-1*-like protein family motif-containing proteins have been identified, which do not exhibit obvious sequence similarity to *BI-1* proteins. Among them are the glutamate-binding protein (*GBP*, also known as glutamate-binding subunit of a *NMDA* receptor, *GRINA*), responsive to centrifugal force and shear, Golgi anti-apoptotic protein (*GAAP*), and *FAS* apoptotic inhibitory molecule 2, which is also known as lifeguard (*LFG*) (Reimers et al. 2006, 2008; Hu et al. 2009). Comprehensive structure prediction and phylogenetic analyses showed recently that all of these proteins belong to the evolutionarily conserved *LFG* protein family in eukaryotes, named after the mammalian member with the best-annotated function (Hu et al. 2009). Some members of the mammalian *LFG* protein family (e.g. human *LFG* and *GAAP*s) possess an anti-apoptotic activity, which they have in common with *BI-1* proteins (Gubser et al. 2007; Reimers et al. 2008; Hu et al. 2009). *LFG* proteins are conserved in plants (Reimers et al. 2006, 2008; Hu et al. 2009), but plant LFG protein functions are largely uncharacterized. The barley (*Hordeum vulgare* L.) and *Arabidopsis thaliana* genomes harbour at least five *LFG* genes (*HvLFGa–HvLFGe* and *AtLFG1–AtLFG5*). At least *HvLFGa*, *AtLFG1*, and *AtLFG2* are involved in susceptibility to adapted powdery mildew fungi, further supporting functional similarity between *BI-1* and *LFG* proteins.

Pathogens manipulate eukaryotic cells during invasion and target plant proteins to achieve host susceptibility. *Bax inhibitor-1* (*BI-1*) is an endoplasmic reticulum-resident cell death suppressor in plants and animals and is required for full susceptibility of barley to the barley powdery mildew fungus *Blumeria graminis* f. sp. *hordei*. Lifeguard (*LFG*) proteins resemble *BI-1* proteins in terms of predicted membrane topology, and cell-death-inhibiting function in metazoans, but display clear sequence-specific distinctions. The barley (*Hordeum vulgare* L.) and *Arabidopsis thaliana* genomes harbour five *LFG* genes, *HvLFGa–HvLFGe* and *AtLFG1–AtLFG5*, whose functions are largely uncharacterized. As observed for *HvBI-1*, single-cell overexpression of *HvLFGa* supports penetration success of *B. graminis* f. sp. *hordei* into barley epidermal cells, while transient-induced gene silencing restricts it. In penetrated barley epidermal cells, a green fluorescent protein-tagged *HvLFGa* protein accumulates at the site of fungal entry, around fungal haustoria, and in endosomal or vacuolar membranes. It suggests a role of *LFG* proteins in plant–powdery mildew interactions in both monocot and dicot plants, because stable overexpression or knockdown of *AtLFG1* or *AtLFG2* also supports or delays development of the powdery mildew fungus *E. cruciferarum* on the respective *Arabidopsis* mutants (Fig. 4.2). Weis et al. (2013) have identified new

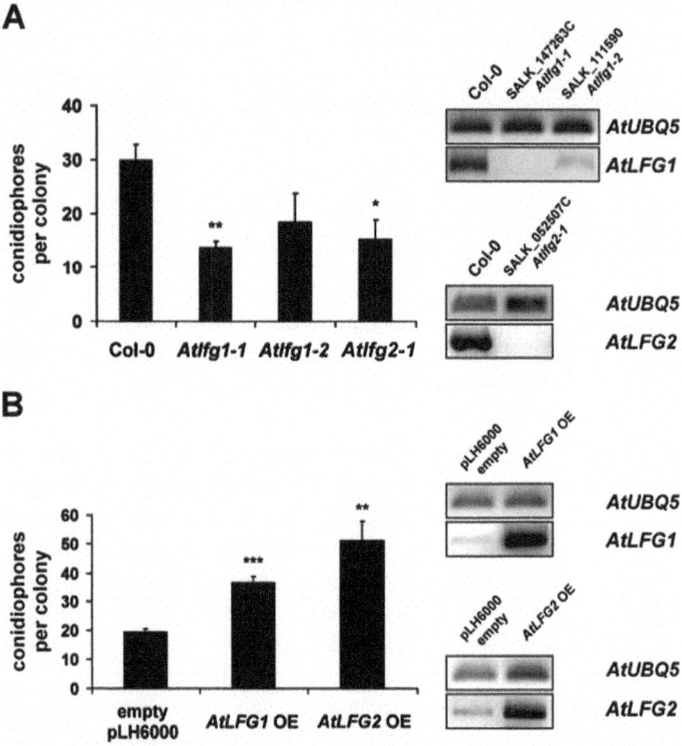

Fig. 4.2 Impact of *AtLFG1* or *AtLFG2* over- or under-expression on the outcome of the *Arabidopsis*–powdery mildew interaction. (**a**) Development of *E. cruciferarum* on 5-week-old Col-0 and *Atlfg* T-DNA insertion mutants at 5 dpi. Conidiophores per colony were counted on 5 individual plants per mutant, respectively; 50 colonies per line were evaluated. Expression of the full-length target genes was analysed by RT-PCR. Amplification of a *UBIQUITIN 5* fragment indicated similar quantity of template cDNA. The upper panel shows *AtLFG1* expression in wild-type Col-0, *Atlfg1-1*, and *Atlfg1-2* plants; the lower panel shows *AtLFG2* expression in wild-type Col-0 and in *Atlfg2-1*. (**b**) Development of *E. cruciferarum* on 5-week-old *AtLFG* overexpression mutants and empty pLH6000 vector control plants at 5 dpi. Conidiophores per colony were counted on one leaf of five individual plants per mutant, respectively. *AtLFG1* or *2* expressions in plants transformed with the empty vector pLH6000 and *AtLFG1* or *2* overexpression mutants were examined by RT-PCR. Amplification of a *UBQ5* fragment indicated similar quantity of template cDNAs. The upper panel shows *AtLFG1* expression in empty pLH6000 vector and *Atlfg1* overexpression plants; the lower panel *AtLFG2* expression in empty pLH6000 vector and in *AtLFG2* overexpression plants. Values are mean and standard errors; *, **, and *** indicate significance at $P < 0.05$, 0.01, and 0.001, respectively (Weis et al. 2013)

modulators of plant–powdery mildew interactions, which support functional similarities between *BI-1* and *LFG* proteins beyond cell death regulation.

*4.2.7 Function of Lifeguard Protein (*LGP*) in Powdery Mildew Pathogenesis*

In order to examine the function of *AtLFG* proteins in powdery mildew susceptibility, powdery mildew disease development on *Arabidopsis LFG* T-DNA insertion mutants was analysed. Two T-DNA insertion mutants were available for *AtLFG1* (SALK-147263C and SALK-111590) and one for *AtLFG2* (SALK-052507C). The integration and positions of T-DNA insertions were confirmed by PCR. Both *AtLFG1* mutants carry their T-DNA insertion at different positions in the promoter region. The T-DNA position in the *AtLFG2* mutant is in the 3′ UTR. Expression data retrieved from Genevestigator (https://www.genevestigator.com/gv/, Hruz et al. 2008) indicated a low level of expression of *AtLFG1* and a higher expression of *AtLFG2* in *Arabidopsis* leaves. To confirm suppression of *AtLFG* expression in T-DNA insertion mutants, RT-PCR was performed. *AtLFG1* expression was very low in leaves of Col-0 wild-type plants as indicated by high cycle numbers for PCR amplification. However, *AtLFG1* expression was not detectable in T-DNA insertion line SALK-147263C and only weakly detectable in T-DNA insertion line SALK-111590. The full-length transcript of *AtLFG2* could easily be amplified from Col-0 wild-type leaves but was not detectable in the *AtLFG2* T-DNA insertion line. Hereafter, *AtLFG1* lines SALK-147263C and SALK-111590 were named *Atlfg1-1* and *Atlfg1-2*, respectively, and the *AtLFG2* T-DNA insertion line SALK-052507C *Atlfg2-1*.

The inoculated *Atlfg1-1*, *Atlfg1-2,* and *Atlfg2-1* mutants were with *E. cruciferarum*, a powdery mildew fungus, which is adapted to *Arabidopsis* and analyzed the infection by macroscopic as well as microscopy. On *Atlfg* T-DNA insertion mutants, fungal development was reduced at 5 dpi, as assessed by counting the number of conidiophores per colony. *E. cruciferarum* developed 30 ± 3 conidiophores per colony on Col-0 wild-type leaves but only 14 ± 1 conidiophores per colony on *Atlfg1-1*, 18 ± 5 on *Atlfg1-2*, and 15 ± 4 conidiophores per colony on *Atlfg2-1*. This effect was statistically significant for the null mutants *Atlfg1-1* and *Atlfg2-1* but not for *Atlfg1-2*, which has residual expression of *AtLFG1*. Hence, loss of *AtLFG1* or *AtLFG2* delayed fungal development on susceptible *Arabidopsis*. As assessed by microscopy, impaired development of *E. cruciferarum* on *Atlfg* mutants did not seem to be associated with accelerated cell death responses. However, all *Atlfg* mutants showed a reproducible tendency to more cell death upon *Alternaria alternata* f. sp. *lycopersici* (AAL) toxin treatment, suggesting some potential of *AtLFG* proteins to inhibit cell death.

By means of *A. tumefaciens*-mediated transformation, stable transgenic *Arabidopsis* plants were created that express *AtLFG1* or *AtLFG2* coding sequences under the control of the CaMV *35S* promoter. Control plants were transformed with the empty vector pLH6000. Presence of the integrated T-DNA was confirmed by PCR. Transgenic plants showed elevated levels of *AtLFG1* or *AtLFG2* transcripts, respectively, when compared to empty vector control plants as analysed by RT-PCR. On *AtLFG1* as well as on *AtLFG2* overexpression mutants, *E. cruciferarum* showed an accelerated development when compared to the empty vector control plants. The number of conidiophores per colony increased from 19 ± 1 on empty vector control plants to 37 ± 2 on *AtLFG1* and to 51 ± 7 on *AtLFG2* overexpressing plants. Another two biological replications gave similar results. Together, data indicate a function of *AtLFG1* and *AtLFG2* in supporting development of *E. cruciferarum* in a compatible interaction with *Arabidopsis* (Weis et al. 2013).

4.2.8 Gene Expression Levels of Healthy Crucifers and Powdery Mildew-Infected Plants

The disease is associated with the expression of disease resistance genes in plant. To determine the ability of *B. napus* and *R. alboglabra* to resist *E. cruciferarum* infection during different time periods (1, 2, 4, 6, 8, and 10 dpi), Alkooranee et al. (2015) investigated the expression profile of defense-related genes in plants by qRT-PCR. Six primers for each genotype were used, including *PR-1, PDF1.2, PR-2, PR-3, CHI620*, and *CHI570* genes. The housekeeping gene, GAPDH, was used as reference gene. For both genotypes, the PR-1 and PR-2 genes were used as markers for the SA signal pathway, whereas the *PDF1.2* gene was used as a marker for the JA/ET signal pathway, as well as the putative marker genes *PR-3, CHI620*, and *CHI570*.

The results indicated that most of the genes were amplified (90.0%) after *E. cruciferarum* infection. The gene expression increased over time in both genotypes, which was higher in plants inoculated with the fungus pathogen compared with the healthy plants. The gene expression levels increased in *R. alboglabra* infected compared with non-infected, and with *B. napus*, it showed increased gene expression levels of *PR1* and *PR2* which were up-regulated by 218.86-fold and 43.08-fold at 1 dpi, respectively, compared with an average of twofold increased expression in non-infected (healthy) plants (Fig. 4.3a). The expression levels of the *PR-1* and *PR-2* genes increased in *B. napus* infected were up-regulated by 69.05-fold at 8 dpi and 6.14-fold at 6 dpi, respectively, compared with an average of 1.2-fold increased expression in non-infected plants (Fig. 4.3b). The *PR-3* and *PDF1.2* genes were used as markers for the JA/ET signal pathway. The expression of the *PR-3* gene increased and peaked 10.10-fold (1.04–10.50-fold) at 10 dpi in infected *B. napus* was up-regulated by 67.77-fold (from 0.117- to 7.93-fold) at 8 dpi in infected *R. alboglabra*, respectively, compared with then non-infected ones (Fig. 4.3c). The expression levels of the PDF 1.2 gene that showed significant differences between infected and non-infected plants were up-regulated by 8.151-fold (from 1.85- to 15.07-fold) at 8 dpi in *B. napus* (Fig. 4.3d). Other defense-related genes were inves-

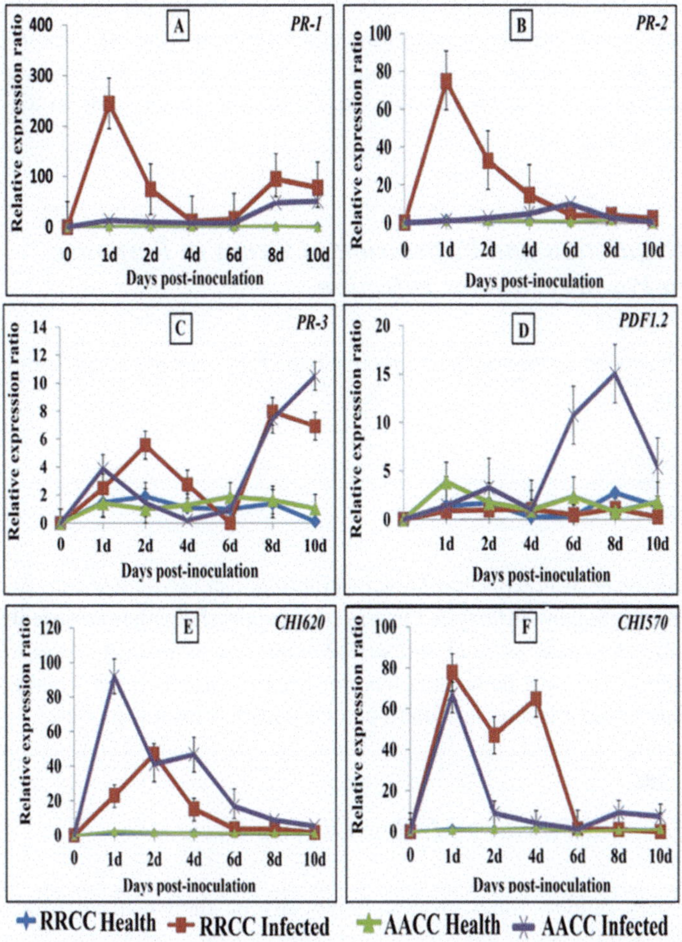

Fig. 4.3 Expression of defense-related genes in three potted for each time of *B. napus* and *R. alboglabra* genotypes of 6-week-old inoculated by pressing diseased leaves by powdery mildew onto leaves. Leaves were collected 1, 2, 4, 6, 8, and 10 days post-infection. Total RNA was extracted, and cDNA was synthesized. Expression levels of the *PR-1, PR-2, PDF1.2* (glucanase; *BGL2*), *PR-3* (basic chitinase), *CHI620*, and *CHI570* (chitinase) genes were monitored by RT q-PCR. The expression levels of genes were compared with the expression level of GAPDH. (Alkooranee et al. 2015)

tigated for the comparison between the signal transduction pathways in both genotypes. Thus, qRT-PCR was used to analyse the effects of different times and infections on the expression of the following putative marker genes, *CHI620* and *CHI570* (chitinase), which are anti-pathogenic enzymes. The expression levels of *CHI620* gene that increased clearly with time and reached its peak at 1 dpi were up-regulated by 58.62-fold (from 1.64- to 96.15-fold) in *B. napus* and reached 37.29-fold (from 1.21- to 45.13-fold) at 2 dpi in *R. alboglabra* infected compared with the non-infected (healthy) ones (Fig. 4.3e). The expression levels of the CHI570 gene increased clearly with time and reached its peak at 1 dpi in *B. napus* and in *R. alboglabra* than in non-infected (healthy), which were up-regulated by 40.86-fold (from 1.64- to 66.73-fold) and 59.87-fold (from 1.34- to 80.23-fold), respectively (Fig. 4.3e).

4.2.9 Regulation and Expression of Genes in Response to Powdery Mildew Infection

The topic has been discussed in details in Chap. 7. See detailed description at Sect. 7.6.3, Chap. 7.

4.3 Disease Cycle

The disease cycle of powdery mildew of crucifers can be divided into two phases based on the pathogens' lifestyle. On *Brassica* crops, the pathogen completes both stages of life cycle, i.e. asexual and sexual states, whereas on *Arabidopsis* sexual stage of any of the four pathogens has not been recorded so far. Therefore, only asexual state under artificial inoculation conditions has been exploited.

4.3.1 General Powdery Mildew Disease Cycle

Most powdery mildew fungi grow epiphytically on their respective host plants. Only powdery mildews of the genera *Leveillula* and *Phyllactinia* are an exception, as they infect (*L. taurica*) or form haustoria (*P. guttata*) endophytically in the leaf mesophyll tissue after entering through stomata (Boesewinkel 1980). In natural environments, powdery mildew conidiospores (mitotic, asexual spores) are mostly distributed by wind or animals. However, under laboratory or artificial inoculation field conditions, inoculations are performed by brushing, leaf-to-leaf transfer, or dusting of spores (conidia) from infected host plants onto healthy plants (Micali et al. 2008). Once dusted on a plant leaf or stem, the powdery mildew spore develops a short germ tube (Plate 4.7), and approximately 6 h postinoculation (hpi), the appressorium, a thickened infection structure, forms at the tip of this hypha. However, in the case of *Blumeria graminis* f. sp. *hordei* (*Bgh*), the appressorium builds up high pressure in order to breach the plant cuticle and cell wall (Pryce-Jones et al. 1999). Unlike in many other plant pathogenic fungi, cell wall-degrading enzymes seem to play a minor role in host cell invasion. After successful cell wall penetration, the fungus enters the host cell without disrupting the host plasma mem-

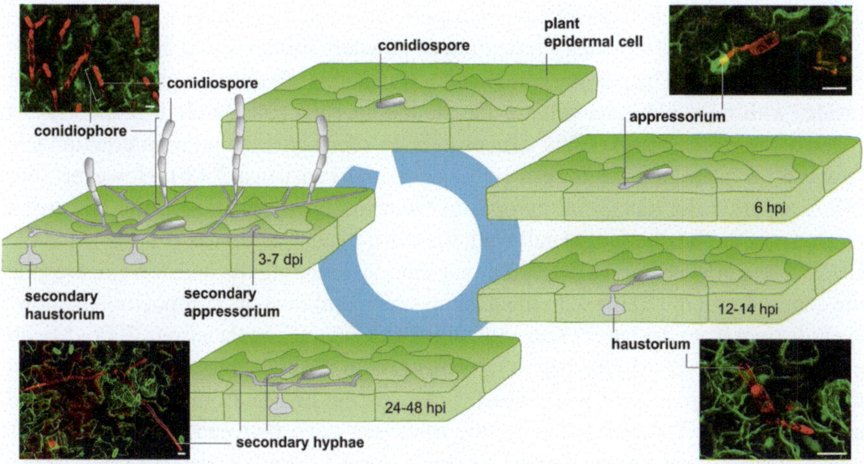

Plate 4.7 Asexual life cycle of *G. orontii* in association with *Arabidopsis*. The central part of the figure illustrates schematically the key steps of the life cycle, while the micrographs show the actual fungal infection structures. The confocal laser scanning micrographs were obtained from transgenic Col-0 plants stably expressing yellow cameleon inoculated with *Go*. Fungal infection structures were stained with FM4-64 (shown in red), while green fluorescence is representative of cytosolic yellow cameleon fluorescence. Bars = 20 μm (Kuhn et al. 2016)

brane and the haustorium, and a specialized hyphal feeding structure with protrusions for surface enlargement is formed (12–14 hpi, Plate 4.7). Haustorium development involves the formation of the extra-haustorial membrane (EHM), which separates plant and fungal structures. The haustorium represents the major interaction site between the fungus and the host plant, and it is supposed to be the hub for effector secretion (protein/genes) and nutrient uptake (O'Connell and Panstruga 2006). Once the haustorium is established, the fungus gains the nutrients necessary for its epiphytic growth. This becomes visible as secondary hyphae forming the powdery mildew colony. The secondary hyphae form new appressoria and penetrate nearby cells. The cycle concludes by the formation of conidiophores, specialized hyphae giving rise to new conidiospores at 3–7 days postinoculation (dpi; Plate 4.7). Sporulation of the pathogens becomes macroscopically visible at 7–10 dpi after 4–10 days of inoculation.

In temperate climates, powdery mildew fungi have to overwinter periods during which the host plant is either not present (annual plants) or defoliates (perennial plants). To cope with such conditions, the fungal pathogen can engage in sexual reproduction based on two compatible mating types. This process gives rise to endurable ascospores (meiospores) enclosed in asci, emerging from fruiting bodies (cleistothecia or chasmothecia). These structures are formed on infected leaves, and stems of host plants, and are visible as black–brownish pinhead bodies. The ascospores mature within the ascus enclosed in chasmothecia and are able to persist for longer periods outside the host plant. In the next season, these chasmothecia rupture to release asci which soon release ascospores to cause infection on the host tissue and serve as source of primary inoculum of powdery mildews (Saharan et al. 2005; Kuhn et al. 2016). The powdery mildew haustorium is the only fungal structure that resides within the plant, namely, inside plant epidermal cells (with the exception of *L. taurica* and *P. guttata*). This structure likely represents the main interaction site between the plant and the fungus (O'Connell and Panstruga 2006). However, four layers separate the haustorial cytoplasm from the plant cytoplasm, the haustorial plasma membrane, the fungal cell wall, the extra-haustorial matrix (EHMx), and the EHM. The EHM is a plant-derived membrane surrounding the haustorium. Despite the continuity of the EHM with the host plasma membrane, its composition is, however, distinct from the latter (Koh et al. 2005; O'Connell and Panstruga 2006; Micali et al. 2011). The EHM attaches to the haustorial neck, the contact site of the haustorium, and the plant cell wall, which separates the EHMx from the apoplast (Gil and Gay 1977). The EHMx forms the transition zone between plant and fungus and is supposed to enable both nutrient uptake and effector delivery (Bushnell 1972).

Mature *Go* haustoria are typically ca. 16-μm-wide and 10-μm-long elliptic bodies with finger-like projections coiled around the main body (Plate 4.8a–b, Micali et al. 2011). They contain a single nucleus and numerous mitochondria. In addition, the haustorial cytoplasm and the EHMx comprise a high number of vesicles, potentially due to fusion of multi-vesicular bodies (MVBs) with the plasma membrane resulting in the release of cargo vesicles into the EHMx (exosomes, Plate 4.8a). On the plant side, the endoplasmic reticulum (ER) and plant MVBs locate close to the EHM (Micali et al. 2011). Mature haustoria are often fully or partially encapsulated by encasements and cell wall appositions enclosing the EHM and EHMx, even during the compatible interaction between *Go* and *Arabidopsis* (Plate 4.8c). In fact, 20–55% of *Go* haustoria are encased to different degrees. These encapsulations depend on the age of the haustorium and contain β-1,3-polyglucans (e.g. callose), xyloglucans, rhamnogalacturonans, and arabinogalactan proteins. Deposition starts at the haustorial neck and gradually encloses the maturing haustorium (Meyer et al. 2009; Micali et al. 2011).

Compared with papillae, which represent multilayered focal cell wall reinforcements (Naumann et al. 2013), encasements seem to comprise a uniform single layer surrounding the haustoria (Micali et al. 2011). Although they typically contain callose, the formation of these encapsulations is independent from the pathogen-induced callose synthase glucan synthase-like 5/powdery mildew resistant 4 (*GSL5/PMR4*: At4g03550), suggesting that in the absence of the enzyme, other cell wall polymers replace the β-1,3-polyglucan (Meyer et al. 2009). The hypothesis that encasements indicate incomplete adaption of *Go* to *Arabidopsis* is supported by the fact that they are absent in interactions with *Gc* (Koh et al. 2005; Meyer et al. 2009). Moreover, haustoria of *Bgh* are encapsulated in leaves of the non-host plant *Arabidopsis*, but not in leaves of its host plant barley, indicating that *Bgh* effectively suppresses the encasement of haustoria in a suitable host (Meyer et al. 2009).

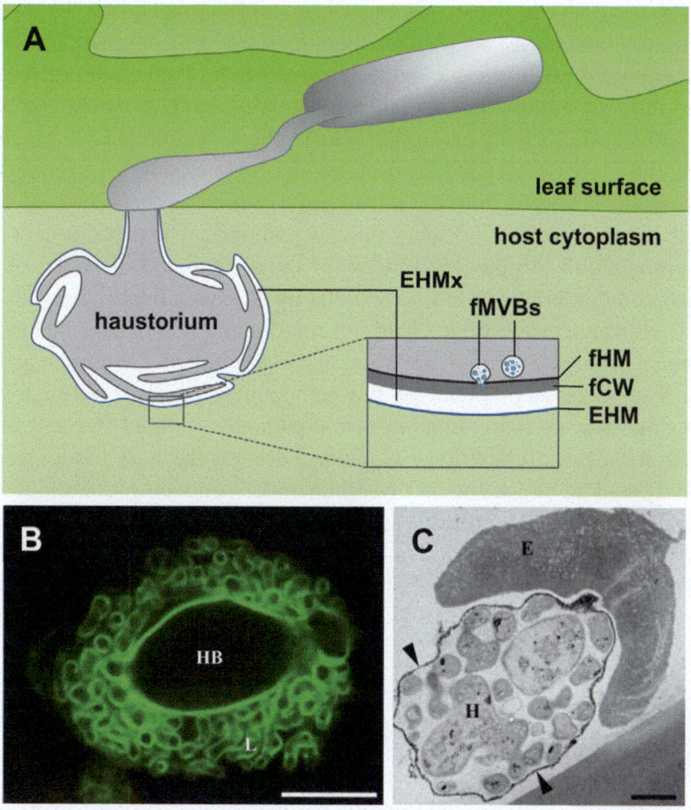

Plate 4.8 The powdery mildew haustorium. The fungal haustorium forms within cells of the leaf epidermis after penetration. (**a**). Scheme of a powdery mildew haustorium (grey) separated from the plant cytoplasm by fungal haustorial membrane (fHM), fungal cell wall (fCW), extra-haustorial matrix (EHMx), and extra-haustorial membrane (EHM). The inset depicts the proposed exocytosis of fungal multi-vesicular bodies (fMVBs) (**b**). Wheat germ agglutinin staining of chitin in an isolated haustorium of *Go*. The confocal laser scanning micrograph shows a mature haustorium body (HB) with numerous haustorial lobes (L). (**c**). Partial callose encasement of an isolated *Go* haustorium. The electron-opaque EHM (arrowheads) surrounds the haustorium (H) but not the callose-containing encasement (E). Bars = **B** 5 μm; **C** 2 μm (Micali et al. 2011)

4.3.2 Disease Cycle on Crucifers

The off-season host plants of *Brassica* species, and other weeds, may carry the fungal mycelium and conidia as source of primary inoculum. The pathogen produces abundant number of cleistothecia or chasmothecia on diseased plant of *B. juncea* leaves and siliqua tissues at the maturity stage of the crop. The possibility to serve as source of primary inoculum through cleistothecia or chasmothecia surviving on diseased plant debris in the field cannot be ruled out. However, more study is required to authenticate this theory. Cleistothecia or chasmothecia are formed on

Brassica species under Indian conditions Hisar and Haryana, during heavy sporulation on infected host leaves, stem, and siliquae favoured by alternate low and moderate temperature, low nutrition condition of host, low relative humidity, dry soils, and aging of the host including overall stress conditions for the host and pathogen (Saharan and Kaushik 1981). In the next season, these chasmothecia after absorbing water rupture to release asci and ascospores which act as source of primary inoculum. Chasmothecia remain dormant from season of production to the next season to continue the sexual cycle of powdery mildew pathogen. The predominant mode of asexual reproduction in powdery mildew of crucifers is the dispersal of asexual conidia and conidiophores, produced abundantly on infected leaf, stem, and siliqua surfaces. Conidia are produced in chains on conidiophores and disperse in nature by air currents or wind and through agitation of diseased plants by human and animal activities. The secondary spread of the pathogen takes place through airborne conidia. Long-distance dissemination of the pathogen is rapid through wind currents under low humid conditions. Conidia fallen on the host tissues germinate, grow, and spread in the form of mycelium, later producing conidiophores and conidia in the form of white mildew growth (Fig. 4.4). The pathogen is an obligate parasite and has been reported to produce cleistothecia. It is likely to carry over from season to season through cleistothecia or as mycelium on volunteer plants (Saharan et al. 2005).

4.3.3 Disease Cycle on Arabidopsis

Under artificial inoculation conditions of *Golovinomyces orontii* on *Arabidopsis* leaves, within 1 or 2 h, a conidium that has landed on a leaf will imbibe, and germinate, producing a primary germ tube that differentiates at its terminus a specialized infection structure, the appressorium (Plates 4.9, 4.10, and 4.11a). The appressorium defines the site at which the fungal pathogen attempts to breach the underlying cell wall. A mixture of enzymatic activities and pressure from the penetration peg acting on the epidermal cell pave the way for plant cell invasion. Penetration attempts are typically accompanied on the host side by the production of cell wall appositions (CWA, also termed pillae), which are thought to represent a physical and chemical blockade deployed to arrest infection. Successful penetration results in host plasma membrane invagination to accommodate the primary feeding structure of the fungus, the haustorium (Plates 4.8, 4.11b, O'Connell and Panstruga 2006). In biotrophic pathogens, haustoria are believed to be responsible for the uptake of sugars and amino acids from the plant to the fungal mycelium (Hahn et al. 1997; Voegele et al. 2001) and are thought to actively deliver proteins (effectors) to suppress plant immune responses as is the case for rust fungi (Dodds et al. 2004; O'Connell and Panstruga 2006). Interestingly, powdery mildew fungi do not invade deeper plant tissues, such as mesophyll cells, but rather limit their growth to the epidermal layer. It has been suggested that the lack of or lower photosynthetic activity of this plant cell layer makes it a powerful nutrient sink that is efficiently

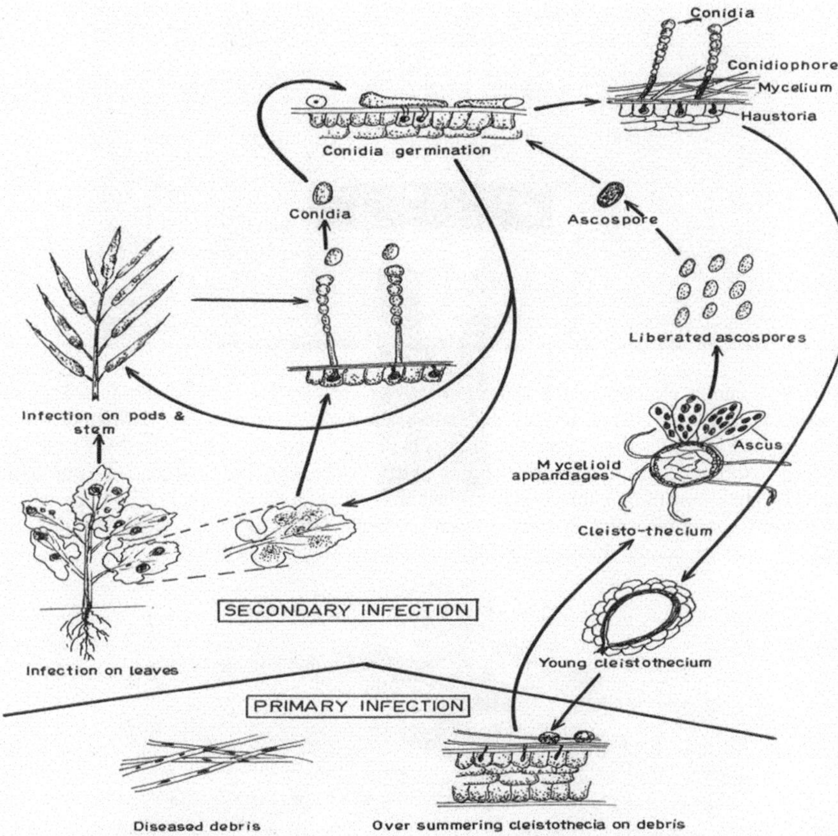

Fig. 4.4 Disease cycle of powdery mildew. (Saharan et al. 2005)

exploited by the powdery mildew pathogens (Zeyen et al. 2002). Subsequent to haustorial establishment, secondary hyphal proliferation occurs exclusively on the surface of the leaf or stem (Plate 4.11b–c). Along the hyphae, new appressoria can form, and penetrate epidermal cells, and further develop secondary haustoria (Plate 4.10). Conidiation (the formation and release of asexual spores) completes the vegetative life cycle of the pathogen or asexual cycle of powdery mildew disease on *Arabidopsis*. However, so far production of cleistothecia or chasmothecia by any of the powdery mildew fungi on *Arabidopsis* has not been observed. It indicates absence of sexual cycle of powdery mildew on *Arabidopsis*. The possibility of formation of sexual fruiting bodies may be explored under high temperature and stress conditions of plant growth.

The better-studied powdery mildew species on *Arabidopsis*, namely, *G. cichoracearum*, *G. orontii,* and *E. cruciferarum*, exhibit different infection intensities that are subject to environmental conditions such as temperature, humidity, and light intensity. *G. cichoracearum* displays abundant conidiation visible to the naked eye,

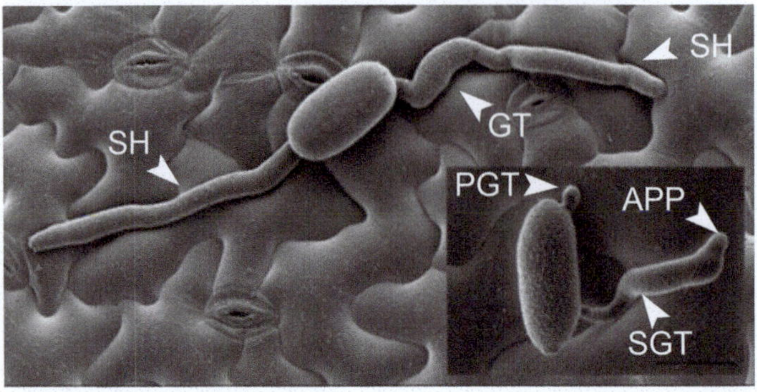

Plate 4.9 Scanning electron micrographs of *Arabidopsis* leaf surface carrying germinated spores of adapted *G. orontii* (large image) and non-adapted *B. graminis* f. sp. *hordei* at 48 h postinoculation. Note that *Bgh* forms a primary **(PGT)** and a secondary germ tube **(SGT)**, the latter of which differentiates into an appressorium **(APP)**, while *G. orontii* produces only one germ tube **(GT)**. In the case of *Bgh*, infection is arrested at this stage in approximately 95% of the cases. In contrast, *G. orontii* has already formed secondary hyphae **(SH)** indicating successful host cell penetration and haustorium formation. Scale bar = 20 μm (Micali et al. 2008)

Plate 4.10 *Oidium neolycopersici* growing on *Arabidopsis* Col-0 at 4 days after inoculation. Notice the lobate appressoria (arrowheads) that form at regular intervals along the secondary hyphae. The disease index on Col-0 is usually 2.6 (approximately 30% leaf coverage at 15 days postinoculation; Bai et al. 2008). Scale bar = 10 μm (inset), 100 μm (large picture) (Micali et al. 2008)

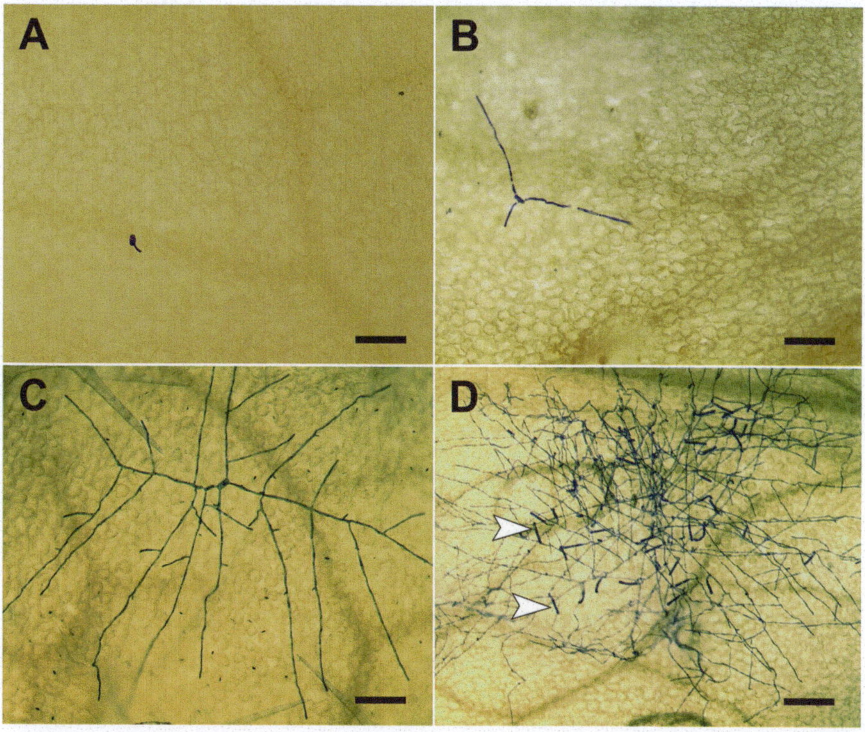

Plate 4.11 Microscopic analysis of the development of a powdery mildew colony. The micrographs show the expansion of a *G. orontii* colony on the surface of a Col-0 rosette leaf. The series of events starts with a germinated spore at 24 h postinoculation (**a**) and continues with initial hyphal elongation (following successful establishment of the first haustorium inside a host cell) at 48 h postinoculation (**b**). Subsequently, a multi-branched mycelium develops (**c**; photo taken at 63 h postinoculation), and the appearance of numerous conidiophores (arrowheads) from a fully expanded fungal colony from 5 days postinoculation onwards completes the asexual life cycle (**d**). Fungal structures were highlighted by Coomassie Blue staining of cleared leaf samples. Scale bar = 100 μm (a–c), 200 μm (**d**) (Micali et al. 2008)

starting at 7 days postinoculation (dpi), whereas *G. orontii* requires 10 days for full conidiation, and *E. cruciferarum* sporulates little on *Arabidopsis* (Vogel and Somerville 2002). The variability in infection phenotypes correlates with the phylogenetic relationship of the three species (the more virulent species *G. cichoracearum* and *G. orontii* being closely related to each other, and each being more distantly related to *E. cruciferarum*; Adam et al. 1999) and is likely indicative of the level of adaptation of each fungus to *Arabidopsis*. In this sense, the study of the three different powdery mildew species provides a large body of information on the molecular pathways that control this interaction (Micali et al. 2008).

4.4 Determinant Factors for Crucifer's Powdery Mildew Infection and Pathogenesis

There are a number of factors which govern infection, and pathogenesis of crucifer's powdery mildew. The details of influencing factors, and their effects have been given at Chap. 6, Sect. 6.7.

References

Adam L, Somerville SC (1996) Genetic characterization of five powdery mildew disease resistance loci in *Arabidopsis thaliana*. Plant J 9:341–356

Adam L, Ellwood S, Wilson I, Saenz G, Xiao S, Oliver RP, Turner JG, Somerville S (1999) Comparison of *Erysiphe cichoracearum* and *E. cruciferarum* and a survey of 360 *Arabidopsis thaliana* accessions for resistance to these two powdery mildew pathogens. Mol Plant Microbe Interact 12:1031–1043

Affolter M, Pyrowolakis G, Weiss A, Basler K (2008) Signal-induced repression: the exception or the rule in developmental signaling? Dev Cell 15:11–22

Aist JR, Bushnell WR (1991) Invasion of plants by powdery mildew fungi, and cellular mechanisms of resistance. In: Cole GT, Hoch HC (eds) The fungal spore and disease initiation in plants and animals. Plenum Press, New York, pp 321–345

Alkooranee JT, Liu S, Aledan TR, Yin Y, Li M (2015) First report of powdery mildew caused by *Erysiphe cruciferarum* on *Brassica napus* in China. Plant Dis 99(11):1651

Ausubel FM (2005) Are innate immune signaling pathways in plants and animals conserved? Nat Immun 6:973–979

Babaeizad V, Imani J, Kogel KH, Eichmann R, Huckelhoven R (2009) Over-expression of the cell death regulator BAX inhibitor-1 in barley confers reduced or enhanced susceptibility to distinct fungal pathogens. Theor Appl Genet 118:455–463

Bai YL, Pavan S, Zheng Z, Zappel NF, Reinstadler A, Lotti C, De Giovanni C, Ricciardi L, Lindhout P, Visser R, Theres K, Panstruga R (2008) Naturally occurring broad-spectrum powdery mildew resistance in a central American tomato accession is caused by loss of Mlo function. Mol Plant-Microbe Interact 21:30–39

Bednarek P, Pislewska-Bednarek M, Svatos A, Schneider B, Doubsky J, Mansurova M, Humphry M, Consonni C, Panstruga R, Sanchez-Vallet A et al (2009) A glucosinolate metabolism pathway in living plant cells mediates broad-spectrum antifungal defense. Science 323:101–106

Beers EP, Jones AM, Dickerman AW (2004) The S8 serine, C1A cysteine and A1 aspartic protease families in *Arabidopsis*. Phytochemistry 65:43–58

Bhat RA, Miklis M, Schmelzer E, Schulze-Lefert P, Panstruga R (2005) Recruitment and interaction dynamics of plant penetration resistance components in a plasma membrane microdomain. Proc Natl Acad USA 102:3135–3140

Boesewinkel HJ (1980) The morphology of the imperfect states of powdery mildews (Erysiphaceae). Bot Rev 46:167–224

Buschges R, Hollricher K, Panstruga R, Simons G, Wolter M, Frijters A, van Daelen R, van der Lee T, Diergaarde P, Groenendijk J, Topsch S, Vos P, Salamini F, Schulze-Lefert P (1997) The barley *Mlo* gene: a novel control element of plant pathogen resistance. Cell 88(5):695–705

Bushnell WR (1972) Physiology of fungal haustoria. Annu Rev Phytopathol 10:151–176

Chandran D, Tai YC, Hather G, Dewdney J, Denoux C, Burgess DG, Ausubel FM, Speed TP, Wildermuth MC (2009) Temporal global expression data reveal known and novel salicylate-impacted processes and regulators mediating powdery mildew growth and reproduction on *Arabidopsis*. Plant Physiol 49:1435–1451

Chandran D, Inada N, Hather G, Kleindt CK, Wildermuth MC (2010) Laser microdissection of *Arabidopsis* cells at the powdery mildew infection site reveals site-specific processes and regulators. PNAS 107(1):460–465

Chandran D, Rickert J, Cherk C, Dotson BR, Wildermuth MC (2013) Host cell ploidy underlying the fungal feeding site is a determinant of powdery mildew growth and reproduction. Mol Plant Microbe Interact 26:537–545

Chandran D, Rickert J, Huang Y, Steinwand MA, Marr SK, Wildermuth MC (2014) Atypical E2F transcriptional repressor DEL1 acts at the intersection of plant growth and immunity by controlling the hormone salicylic acid. Cell Host Microbe 15:506–513

Chico JM, Chini A, Fonseca S, Solano R (2008) JAZ repressors set the rhythm in jasmonate signaling. Curr Opin Plant Biol 11:486–494

Choi HO, Choi Y, Kim DS, Hwang IS, Choi DS, Kim NH, Lee DH, Shin HD, Nam HBK (2009) First report of powdery mildew caused by *Erysiphe cruciferarum* on *Arabidopsis thaliana* in Korea. Plant Path J 25(1):86–90

Chung HS, Niu Y, Browse J, Howe GA (2009) Top hits in contemporary JAZ: an update on jasmonate signaling. Phytochemistry 70:1547–1559

Colcombet J, Hirt H (2008) *Arabidopsis* MAPKs: a complex signalling network involved in multiple biological processes. Biochem J 413:217–226

Collins NC, Thordal-Christensen H, Lipka V, Bau S, Kombrink E, Qiu JL, Huckelhoven R, Stein M, Freialdenhoven A, Somerville SC, Schulze-Lefert P (2003) SNARE-protein-mediated disease resistance at the plant cell wall. Nature 425:973–977

Consonni C, Humphry ME, Hartmann HA, Livaja M, Durner J, Westphal L, Vogel J, Lipka V, Kemmerling B, Schulze-Lefert P, Somerville SC, Panstruga R (2006) Conserved requirement for a plant host cell protein in powdery mildew pathogenesis. Nat Genet 38:716–720

Da Cunha L, McFall AJ, Mackey D (2006) Innate immunity in plants: a continuum of layered defenses. Microbes Infect 8:1372–1381

Devoto A, Piffanelli P, Elliott C, Simmons C, Taramino G, Goh CS, Cohen FE, Emerson BC, Schulze-Lefert P, Panstruga R (2003) Molecular phylogeny and evolution of the plant-specific seven transmembrane MLO family. J Mol Evol 56:77–88

Dodds PN, Lawrence GJ, Catanzariti AM, Ayliffe MA, Ellis JG (2004) The *Melampsora lini* AvrL567 avirulence genes are expressed in haustoria and their products are recognized inside plant cells. Plant Cell 16:755–768

Douchkov D, Nowara D, Zierold U, Schweizer P (2005) A high-throughput gene-silencing system for the functional assessment of defense-related genes in barley epidermal cells. Mol Plant Microbe Interact 18:755–761

Eichmann R, Huckelhoven R (2008) Accommodation of powdery mildew fungi in intact plant cells. J Plant Physiol 165:5–18

Eichmann R, Dechert C, Kogel KH, Huckelhoven R (2006) Transient over-expression of barley BAX inhibitor-1 weakens oxidative defence and *MLA12*-mediated resistance to *Blumeria graminis* f. sp. *hordei*. Mol Plant Pathol 7:543–552

Eichmann R, Bischof M, Weis C, Shaw J, Lacomme C, Schweizer P, Duchkov D, Hensel G, Kumlehn J, Huckelhoven R (2010) BAX INHIBITOR-1 is required for full susceptibility of barley to powdery mildew. Mol Plant Microbe Interact 23:1217–1227

Ellis J, Catanzariti AM, Dodds P (2006) The problem of how fungal and oomycete avirulence proteins enter plant cells. Trends Plant Sci 11:61–63

Eulgem T, Somssich IE (2007) Networks of WRKY transcription factors in defense signaling. Curr Opin Plant Biol 10:366–371

Fabro G, Di Rienzo JA, Voigt CA, Sevchenko T, Dehesh K, Somerville S, Alvarez ME (2008) Genome-wide expression profiling *Arabidopsis* at the stage of *Golovinomyces cichoracearum* haustorium formation. Plant Physiol 146:1421–1439

Frye CA, Innes RW (1998) An *Arabidopsis* mutant with enhanced resistance to powdery mildew. Plant Cell 10:947–956

Frye CA, Tang DZ, Innes RW (2001) Negative regulation of defense responses in plants by a conserved MAPKK kinase. Proc Natl Acad Sci U S A 98:373–378

Gallois JL, Drouaud J, Lecureuil A, Guyon-Debast A, Bonhomme S, Guerche P (2013) Functional characterization of the plant ubiquitin regulatory X (UBX) domain-containing protein AtPUX7 in *Arabidopsis thaliana*. Gene 526:299–308

Gil F, Gay JL (1977) Ultra-structural and physiological properties of host interfacial components of haustoria of *Erysiphe pisi in vivo* and *in vitro*. Physiol Plant Pathol 10:1–12

Gollner K, Schweizer P, Bai Y, Panstruga R (2008) Natural genetic resources of *Arabidopsis thaliana* reveals a high prevalence and unexpected phenotypic plasticity of RPW8-mediated powdery mildew resistance. New Phytol 177:725–742

Gubser C, Bergamaschi D, Hollinshead M, Lu X, van Kuppeveld FJ, Smith GL (2007) A new inhibitor of apoptosis from vaccinia virus and eukaryotes. PLoS Pathog 3:e17

Haga N, Kato K, Murase M, Araki S, Kubo M, Demura T, Suzuki K, Muller I, Voß U, Jurgens G, Ito M (2007) R1R2R3-Myb proteins positively regulate cytokinesis through activation of *KNOLLE* transcription in *Arabidopsis thaliana*. Development 134:1101–1110

Hahn M, Neef U, Struck C, Gottfert M, Mendgen K (1997) A putative amino acid transporter is specifically expressed in haustoria of the rust fungus *Uromyces fabae*. Mol Plant Microbe Interact 10:438–445

Henke N, Lisak DA, Schneider L, Habicht J, Pergande M, Methner A (2011) The ancient cell death suppressor BAX inhibitor-1. Cell Calcium 50:251–260

Hiscott J, Rongtuan L, Peyman N, Paz S (2006) Master CARD: a priceless link to innate immunity. Trends Mol Med 12:53–56

Hruz T, Laule O, Szabo G, Wessendorp F, Bleuler S, Oertle L, Widmayer P, Gruissem W, Zimmermann P (2008) Genevestigator v3: a reference expression database for the meta-analysis of transcriptomes. Adv Bioinforma 420747, 1–5. https://doi.org/10.1155/2008/420747

Hu L, Smith TF, Goldberger G (2009) LFG: a candidate apoptosis regulatory gene family. Apoptosis 14:1255–1265

Huckelhoven R (2004) BAX inhibitor-1, an ancient cell death suppressor in animals and plants with prokaryotic relatives. Apoptosis 9:299–307

Huckelhoven R, Panstruga R (2011) Cell biology of the plant–powdery mildew interaction. Curr Opin Plant Biol 14:738–746

Huckelhoven R, Dechert C, Kogel KH (2003) Overexpression of barley BAX inhibitor 1 induces breakdown of *mlo*-mediated penetration resistance to *Blumeria graminis*. Proc Natl Acad Sci USA 100:5555–5560

Humphry M, Consonni C, Panstruga R (2006) Mlo-based powdery-mildew immunity: silver bullet or simply non-host resistance. Mol Plant Pathol 7:605–610

Hwang BK, Heitfuss R (1982) Characterization of adult plant resistance of spring barley to powdery mildew (*Erysiphe graminis* f. sp. *hordei*). J Phytopathol 104:168–178

Ishikawa T, Watanabe N, Nagano M, Kawai-Yamada M, Lam E (2011) Bax inhibitor-1: a highly conserved endoplasmic reticulum-resident cell death suppressor. Cell Death Differ 18:1271–1278

Jacobs AK, Lipka V, Burton RA, Panstruga R, Strizhov N, Schulze-Lefert P, Fincher GB (2003) An *Arabidopsis* callose synthase, GSL5, is required for wound and papillary callose formation. Plant Cell 15:2503–2513

Jarosch B, Kogel KH, Schaffrath U (1999) The ambivalence of the barley Mlo locus: mutations conferring resistance against powdery mildew (*Blumeria graminis* f. sp. *hordei*) enhance susceptibility to the rice blast fungus *Magnaporthe grisea*. Mol Plant Microbe Interact 12:508–514

Jones JD, Dangl JL (2006) The plant immune system. Nature 444:323–329

Jones H, Whipps JM, Gurr SJ (2001) The tomato powdery mildew fungus *Oidium neolycopersici*. Mol Plant Pathol 2:303–309

Keeble JA, Gilmore AP (2007) Apoptosis commitment-translating survival signals into decisions on mitochondria. Cell Res 17:976–984

Khandelwal A, Elvitigala T, Ghosh B, Quatrano RS (2008) *Arabidopsis* transcriptome reveals control circuits regulating redox homeostasis and the role of an AP2 transcription factor. Plant Physiol 148:2050–2058

Kim MC, Panstruga R, Elliott C, Muller J, Devoto A, Yoon HW, Park HC, Cho MJ, Schulze-Lefert P (2002) Calmodulin interacts with MLO protein to regulate defence against mildew in barley. Nature 416:447–451

Koeck M, Hardham AR, Dodds PN (2011) The role of effectors of biotrophic and hemibiotrophic fungi in infection. Cell Microbiol 13:1849–1857

Koh S, Andre A, Edwards H, Ehrhardt D, Somerville S (2005) *Arabidopsis thaliana* sub-cellular responses to compatible *Erysiphe cichoracearum* infections. Plant J 44:516–529

Kuhn H, Kwaaitaal M, Kusch S, Acevedo-Garcia J, Wu Hanstruga R (2016) Biotrophy at its best: novel findings and unsolved mysteries of the *Arabidopsis*-powdery mildew pathosystem. The *Arabidopsis* Book:0184

Kumar J, Huckelhoven R, Beckhove U, Nagarajan S, Kogel KH (2001) A compromised Mlo pathway affects the response of barley to the necrotrophic fungus *Bipolaris sorokiniana* (teleomorph: *Cochliobolus sativus*) and its toxins. Phytopathology 91:127–133

Kwon C, Neu C, Pajonk S, Yun HS, Lipka U, Humphry M, Bau S, Straus M, Kwaaitaal M, Rampelt H, El Kasmi F, Jurgens G, Parker J, Panstruga R, Lipka V, Schulze-Lefert P (2008a) Cooption of a default secretory pathway for plant immune responses. Nature 451:835–840

Kwon C, Panstruga R, Schulze-Lefert P (2008b) Les liaisons dangereuses: immunological synapse formation in animals and plants. Trends Immunol 29:159–166

Lammens T, Boudolf V, Kheibarshekan L, Zalmas LP, Gaamouche T, Maes S, Vanstraelen M, Kondorosi E, La Thangue NB, Govaerts W, Inzé D, De Veylder L (2008) Atypical E2F activity restrains APC/CCCS52A2 function obligatory for endocycle onset. Proc Natl Acad Sci U S A 105:14721–14726

Lipka V, Dittgen J, Bednarek P, Bhat R, Wiermer M, Stein M, Landtag J, Brandt W, Rosahl S, Scheel D, Llorente F, Molina A, Parker J, Somerville S, Schulze-Lefert P (2005) Pre- and post invasion defenses both contribute to non-host resistance in *Arabidopsis*. Science 310:1180–1183

Lipka U, Fuchs R, Lipka V (2008) *Arabidopsis* non-host resistance to powdery mildews. Curr Opini Plant Biol 11:404–411

Lippok B, Birkenbihl RP, Rivory G, Brummer J, Schmelzer E, Logemann E, Somssich IE (2007) Expression of AtWRKY33 encoding a pathogen-/PAMP-responsive WRKY transcription factor is regulated by a composite DNA motif containing W box elements. Mol Plant Microbe Interact 20:420–429

Loake G, Grant M (2007) Salicylic acid in plant defense: the players and protagonists. Curr Opin Plant Biol 10:466–472

Lopez-Maury L, Marguerat S, Bahler J (2008) Tuning gene expression to changing environments: from rapid responses to evolutionary adaptation. Nat Rev Genet 9:583–593

Madsen L, Seeger M, Semple CA, Hartmann-Petersen R (2009) New ATPase regulators—p97 goes to the PUB. Intl J Biochem Cell Biol 41:2380–2388

Mendgen K, Hahn M (2002) Plant infection and the establishment of fungal biotrophy. Trends Plant Sci 7:352–357

Meyer D, Pajonk S, Micali C, O'Connell R, Schulze-Lefert P (2009) Extracellular transport and integration of plant secretory proteins into pathogen-induced cell wall compartments. Plant J 57:986–999

Micali C, Gollner K, Humphry M, Consonni C, Panstruga R (2008) The powdery mildew disease of *Arabidopsis*: a paradigm for the interaction between plants and biotrophic fungi. Arabidopsis Book 6:0115

Micali CO, Neumann U, Grunewald D, Panstruga R, O'Connell R (2011) Biogenesis of a specialized plant-fungal interface during host cell internalization of *Golovinomyces orontii* haustoria. Cell Microbiol 13:210–226

Naumann M, Somerville S, Voigt C (2013) Differences in early callose deposition during adapted and non-adapted powdery mildew infection of resistant *Arabidopsis* lines. Plant Signal Behav 8:24408

Niks RE, Marcel TC (2009) Non-host and basal resistance: how to explain specificity? New Phytol 182:817–828

Nishimura MT, Stein M, Hou BH, Vogel JP, Edwards H, Somerville SC (2003) Loss of a callose synthase results in salicylic acid-dependent disease resistance. Science 301:969–972

O'Connell RJ, Panstruga R (2006) Tete a Tete inside a plant cell: establishing compatibility between plants and biotrophic fungi and oomycetes. New Phytol 171:699–718

Pandey SP, Roccaro M, Schon M, Logemann E, Somssich IE (2010) Transcriptional reprogramming regulated by WRKY18 and WRKY40 facilitates powdery mildew infection of *Arabidopsis*. Plant J 64:912–923

Panstruga R (2003) Establishing compatibility between plants and obligate biotrophic pathogens. Curr Opin Plant Biol 6:320–326

Panstruga R (2005) Serpentine plant MLO proteins as entry portals for powdery mildew fungi. Biochem Soc Trans 33:389–392

Peng Y, Chen L, Lu Y, Wu Y, Dumenil J, Zhu Z, Bevan MW, Li Y (2015) The ubiquitin receptors DA1, DAR1, and DAR2 redundantly regulate endoreduplication by modulating the stability of TCP14/15 in *Arabidopsis*. Plant Cell 27:649–662

Plotnikova JM, Reuber TL, Ausubel FM (1998) Powdery mildew pathogenesis on *Arabidopsis thaliana*. Mycologia 90:1009–1016

Pryce-Jones E, Carver TIM, Gurr SJ (1999) The roles of cellulase enzymes and mechanical force in host penetration by *Erysiphe graminis* f.sp. *hordei*. Physiol Mol Plant Pathol 55:175–182

Qiu JL, Fiil BK, Petersen K, Nielsen HB, Botanga CJ, Thorgrimsen S, Palma K, Suarez-Rodriguez MC, Sandbech-Clausen S, Lichota J, Brodersen P, Grasser KD, Mattsson O, Glazebrook J, Mundy J, Petersen M (2008) *Arabidopsis* MAP kinase4 regulates gene expression through transcription factor release in the nucleus. EMBO J 27:2214–2221

Rancour DM, Park S, Knight SD, Bednarek SY (2004) Plant UBX domain-containing protein 1, PUX1, regulates the oligomeric structure and activity of *Arabidopsis* CDC48. J Biol Chem 279:54264–54274

Reimers K, Choi CY, Mau-Thek E, Vogt PM (2006) Sequence analysis shows that lifeguard belongs to a new evolutionarily conserved cytoprotective family. Int J Mol Med 18:729–734

Reimers K, Choi CY, Bucan V, Vogt PM (2008) The Bax Inhibitor-1 (BI-1) family in apoptosis and tumorigenesis. Curr Mol Med 8:148–156

Reuber TL, Plotnikova JM, Dewdney J, Rogers EE, Wood W, Ausubel FM (1998) Correlation of defense gene induction defects with powdery mildew susceptibility in *Arabidopsis* enhanced disease susceptibility mutants. Plant J 16:473–485

Robinson KS, Clements A, Williams AC, Berger CN, Frankel G (2011) Bax inhibitor 1 in apoptosis and disease. Oncogene 30:2391–2400

Saharan GS, Kaushik JC (1981) Occurance and epidemiology of powdery mildew of *Brassica*. Indian Phytopathol 34(1):54–57

Saharan GS, Naresh M, Sangwan MS (2005) Diseases of oilseed crops. Indus Publishing Company, New Delhi, p 643

Schulze-Lefert P (2004) Knocking on heaven's wall: pathogenesis of and resistance to biotrophic fungi at the cell wall. Curr Opin Plant Biol 7:377–383

Schulze-Lefert P, Panstruga R (2003) Establishment of biotrophy by parasitic fungi and reprogramming of host cells for disease resistance. Annu Rev Phytopathol 41:641–467

Stein M, Dittgen J, Sánchez-Rodríguez C, Hou BH, Molina A, Schulze-Lefert P, Lipka V, Somerville S (2006) *Arabidopsis* PEN3/PDR8, an ATP binding cassette transporter, contributes to non-host resistance to inappropriate pathogens that enter by direct penetration. Plant Cell 18:731–746

Takemoto D, Jones D, Hardham A (2006) Re-organisation of the cytoskeleton and endoplasmic reticulum in the *Arabidopsis* pen1-1 mutant inoculated with the non-adapted powdery mildew *Blumeria graminis* f. sp. *hordei*. Mol Plant Pathol 7:553–563

Tang D, Innes RW (2002) Overexpression of a kinase-deficient form of the EDR1 gene enhances powdery mildew resistance and ethylene-induced senescence in *Arabidopsis*. Plant J 32:975–983

Tang D, Ade J, Frye CA, Innes RW (2005a) Regulation of plant defense responses in *Arabidopsis* by EDR2, a PH and START domain containing protein. Plant J 44:245–257

Tang D, Christiansen KM, Innes RW (2005b) Regulation of plant disease resistance, stress responses, cell death and ethylene signaling in *Arabidopsis* by the EDR1 protein kinase. Plant Physiol 138:1018–1026

Tang D, Ade J, Frye CA, Innes RW (2006) A mutation in the GTP hydrolysis site of *Arabidopsis* dynamin-related protein 1E confers enhanced cell death in response to powdery mildew infection. Plant J 47:75–84

Tucker SL, Talbot NJ (2001) Surface attachment and pre-penetration stage development by plant pathogenic fungi. Annu Rev Phytopathol 39:385–417

Turck F, Zhou A, Somssich IE (2004) Stimulus-dependent, promoter specific binding of transcription factor WRKY1 to its native promoter and the defense-related gene PcPR1-1 in parsley. Plant Cell 16:2573–2585

Vlieghe K, Boudolf V, Beemster GTS, Maes S, Magyar Z, Atanassova A, de Almeida EJ, De Groodt R, Inzé D, De Veylder L (2005) The DP-E2F-like gene *DEL1* controls the endocycle in *Arabidopsis thaliana*. Curr Biol 15:59–63

Voegele RT, Struck C, Hahn M, Mendgen K (2001) The role of haustoria in sugar supply during infection of broad bean by the rust fungus *Uromyces fabae*. Proc Natl Acad Sci U S A 98:8133–8138

Vogel J, Somerville S (2000) Isolation and characterization of powdery mildew resistant *Arabidopsis* mutants. Proc Natl Acad Sci U S A 97:1897–1902

Vogel J, Somerville S (2002) Powdery mildew of *Arabidopsis*: a model system for host-parasite interactions. In: Belanger RR, Bushnell WR, Dik AJ, Carver TLW (eds) The powdery mildews, a comprehensive treatise. The American Phytopathological Society Press, St Paul, pp 161–168

Vogel JP, Raab TK, Schiff C, Somerville SC (2002) PMR6, a pectate lyase-like gene required for powdery mildew susceptibility in *Arabidopsis*. Plant Cell 14:2095–2106

Vogel JP, Raab TK, Somerville CR, Somerville SC (2004) Mutations in PMR5 result in powdery mildew resistance and altered cell wall composition. Plant J 40:968–978

Wang D, Weaver ND, Kesarwani M, Dong X (2005) Induction of protein secretory pathway is required for systemic acquired resistance. Science 308:1036–1040

Wang Z, Cao G, Wang X, Miao J, Liu X, Chen Z, Qu L-J, Gu H (2008) Identification and characterization of COI1-dependent transcription factor genes involved in JA-mediated response to wounding in *Arabidopsis* plants. Plant Cell Rep 27:125–135

Watanabe N, Lam E (2009) Bax inhibitor-1, a conserved cell death suppressor, is a key molecular switch downstream from a variety of biotic and abiotic stress signals in plants. Int J Mol Sci 10:3149–3167

Weis C, Huckelhoven R, Eichmann R (2013) LIFEGUARD proteins support plant colonization by biotrophic powdery mildew fungi. J Exp Bot 64:3855–3867

Weßling R, Epple P, Altmann S, He Y, Yang L, Henz SR, McDonald N, Wiley K, Bader KC, Glaßer C, Mukhtar MS, Haigis S, Ghamsari L, Stephens AE, Ecker JR, Vidal M, Jones JDG, Mayer KF, Ver Loren van Themaat E, Weigel D, Schulze-Lefert P, Dangl JL, Panstruga R, Braun P (2014) Convergent targeting of a common host protein-network by pathogen effectors from three kingdoms of life. Cell Host Microbe 16:364–375

Whipps JM, Budge SP, Fenlon JS (1998) Characteristics and host range of tomato powdery mildew. Plant Pathol 47:36–48

Wildermuth MC (2010) Modulation of host nuclear ploidy: a common plant biotroph mechanism. Curr Opin Plant Biol 13:449–458

Xiao SY, Calis O, Patrick E, Zhang GG, Charoenwattana P, Muskett P, Parker JE, Turner JG (2005) The atypical resistance gene, RPW8, recruits components of basal defence for powdery mildew resistance in *Arabidopsis*. Plant J 42:95–110

Xu X, Chen C, Fan B, Chen Z (2006) Physical and functional interactions between pathogen-induced *Arabidopsis* WRKY18, WRKY40, and WRKY60 transcription factors. Plant Cell 18:1310–1326

Yamanaka K, Sasagawa Y, Ogura T (2012) Recent advances in p97/VCP/Cdc48 cellular functions. Biochem Biophys Act 1823:130–137

Zeyen RJ, Carver TLW, Lyngkjaer MF (2002) Epidermal cell papillae. In: Belanger RR, Bushnell WR, Dik AJ, Carver TLW (eds) The powdery mildews, a comprehensive treatise. The American Phytopathological Society Press, St Paul, pp 107–125

Chapter 5
Fine Structures and Electron Microscopy

5.1 Introduction

The pathogenesis of *Arabidopsis thaliana* by powdery mildew pathogens, viz. *E. cruciferarum, E. orontii,* and *E. cichoracearum*, during the developmental stages of powdery mildew along with fine structures of pathogens has been revealed through light and SEM under artificial inoculation conditions. The sequence of events has been described from time of inoculation to conidial germination, germ tube elongation, appressorium formation, germination of appressorium with germ tube (infection peg or hypha), haustorium formation, haustorial neck, extra-haustorial membrane, haustorial lobes, encapsulation of haustoria, formation of papilla-like structures, development of hypha into mycelium, formation of septa, shape of different structures, production of conidiophores, and conidia resulting into characteristic powdery mildew symptoms.

The *Arabidopsis thaliana* weed, has long been recognized as a suitable subject for classical plant genetic studies (Laibach 1943). The development of plant molecular biology in recent years, its small genome (70,000 kb), and low amount of repetitive DNA, have made *Arabidopsis* an attractive subject for a whole variety of molecular genetics investigations. Traditionally, although plant geneticists could observe mutations which altered phenotype, it was difficult to identify the genetic lesion unless it occurred in a well-characterized biochemical pathway or was tagged with a transposable element. With *Arabidopsis,* it is theoretically quite easy to isolate clones of the mutated gene by coupling chromosome walking with cosegregation of the mutation with known genetic or RFLP markers (Somerville 1989). Thus, genes marked by mutation can now be identified and characterized at the molecular level. In this sense, *Arabidopsis* is becoming the "*E. coli*" of the plant world. Studies on development, metabolism, and cell biology have made rapid progress, but the use of *Arabidopsis* to gain an insight into the molecular biology of plant–pathogen interactions has been hindered by the lack of well-characterized pathogens. Thus, although several fungal pathogens of *Arabidopsis* have been mentioned briefly in the literature (Brandenburger 1985), however, detailed descriptions of host–parasite

G. S. Saharan et al., *Powdery Mildew Disease of Crucifers: Biology, Ecology and Disease Management*, https://doi.org/10.1007/978-981-13-9853-7_5

relationships of *Arabidopsis* do not exist. Therefore, more knowledge about *Arabidopsis* pathogens is needed urgently. The SEM has revealed very useful sequence of events during the process of infection and pathogenesis of powdery mildew–*Arabidopsis* host–parasite interaction with very precise results under controlled conditions.

5.2 Pathogenesis of *Erysiphe cruciferarum* Observed Through Light Microscopy

Light microscopy of whole-leaf mounts showed that by 18 hours after inoculation of *Arabidopsis* strain Weiningen, powdery mildew conidia had germinated and formed large, irregularly lobed appressoria. Appressoria were produced directly from conidia (Plate 5.1a, d), or, less frequently, they developed at the apices of well-defined germ tubes (Plate 5.1c). Typically, the first haustorium was inserted into the underlying epidermal cell (Plate 5.1b). Formation of the appressorium was generally accompanied by the formation of a hypha which originated from the same (proximal) end of the conidium as the appressorium or appressorial germ tube, respectively (Plate 5.1c, d). In few cases, this hypha emerged from the opposite (distal) end of the conidium. Appressoria, appressorial germ tubes, and hyphae were always separated from the conidium by a septum (Plate 5.1a, c, e). By 3 days after inoculation, these hyphae had grown 3–4 spore lengths, were septate, and had produced 2–3 haustoria. By this time, additional hyphae had grown from the distal ends of conidia, so that often, two hyphae were present at the distal end of the conidium and one hypha and the appressorium at the proximal end. Young haustoria were globose but later became elongated-pyriform to irregular shaped (Plate 5.1f, g). In specimens stained 5 days after inoculation, haustorial bodies could be clearly seen to be surrounded by a sac-like structure which terminated in the region of the haustorial neck (Plate 5.1h, i). This appears to be the "haustorial sac" described by Bushnell and Gay (1978), composed of the extra-haustorial membrane, the vacuolar membrane of the host cell, and a thin layer of host cytoplasm which appear as one entity in the light microscope. Granular structures were observed in the extra-haustorial matrix, i.e. the space between the haustorial sac and the haustorium (Plate 5.1h, i). These structures appeared to be haustorial lobes which extended from the haustorial body into the matrix. Hirata (1967) described two types of powdery mildew haustoria, those of *Erysiphe graminis* and those of other mildews as represented by *Sphaerotheca fuliginea*. In the latter, haustorial lobes extend convolutedly over the surface of the haustorial body, giving the entire structure a compact form and granular appearance (Hirata 1967). Lobed haustoria were observed in various members of the Erysiphaceae (Bushnell and Gay 1978), including *E. polygoni* (Staveley et al. 1969).

In a few cases, the complete encapsulation of haustoria in host cells of *Arabidopsis* was observed (Plate 5.1j). Similar encasements were observed surrounding some

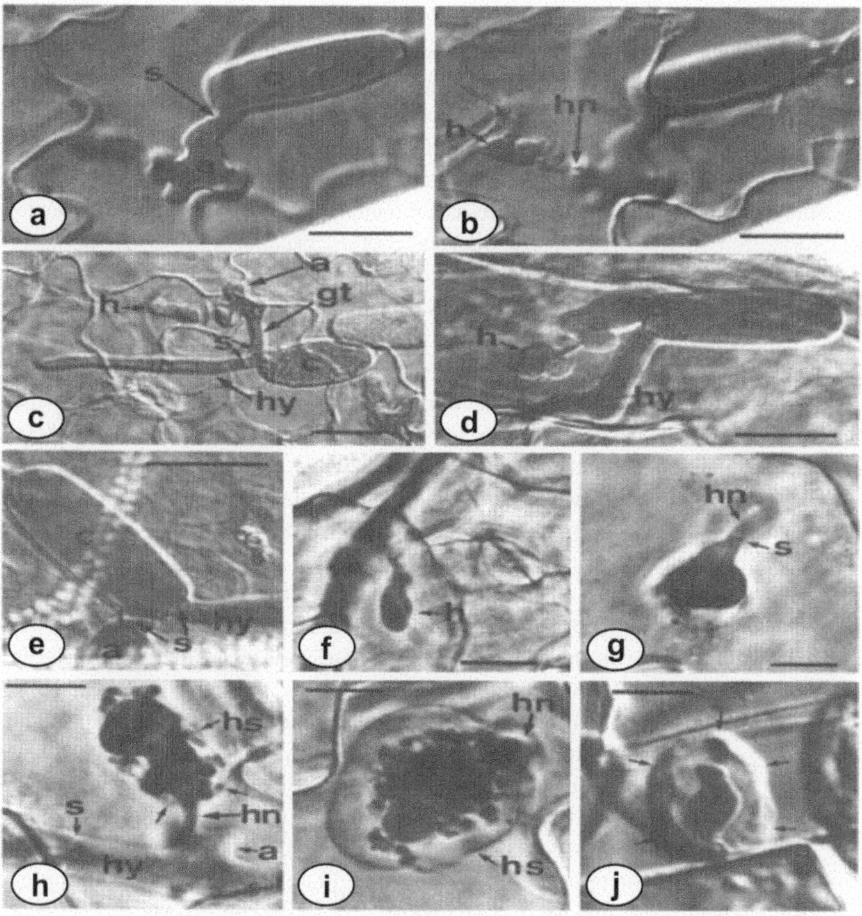

Plate 5.1 (a–j) Infection process and haustorium formation by *Erysiphe cruciferarum* on *Arabidopsis thaliana* strain Weiningen. Plate **a**. A germinated conidium with a multilobed appressorium. **b**. The same infection site as in Plate **a** but focused through to the primary haustorium in the epidermal cell. The haustorial neck extends from one of the appressorial lobes. **c**. A germinated conidium. The appressorium is positioned at the apex of a short germ tube. **d**. A germinated conidium. The appressorium developed directly from the conidium (Bar = 25 μm). **e**. Septation between conidium and appressorium and conidium and hypha. **f**. A young elongated-pyriform haustorium. (Bar = 10 μm). **g**. Haustorium showing a septum in the region of the haustorial neck. **h**. Superficial hypha with a appressorium and mature haustorium, The haustorial neck extends from one of the appressorial lobes. The haustorial sac surrounds the haustorial body and appears connected to the haustorial neck (arrows). **i**. A mature haustorium. The haustorial sac appears expanded. Numerous granular structures surround the haustorial body (Bar = 5 μm). **j**. A haustorium surrounded by a thick capsule (arrows), in an epidermal cell over a leaf vein. Bar = 10 μm (Koch and Slusarenko 1990)

haustoria formed by *Peronospora parasitica* in cells of *Arabidopsis* (Koch and Slusarenko 1990). This phenomenon could represent an attempt by the host to limit infection. Other potential resistance reactions were the deposition of granular material adjacent to haustoria, the formation of large papilla-like structures underneath the appressoria (seemingly preventing haustorium formation), and, in a few cases, necrosis of host cells invaded by haustoria.

Microscopic observation of whole-leaf mounts 5 days after inoculation revealed the formation of the asexual conidial stage of *E. cruciferarum* on *A. thaliana*. Conidiophores were erect and consisted of a few, more or less straight, cylindrical cells to which a maturing conidium was attached (Plate 5.1l). According to Boesewinkel (1980) and Braun (1987), *E. cruciferarum* belongs to those members of the Erysiphaceae in which the conidia are formed singly. Hyphal growth of *E. cruciferarum* on *Arabidopsis* was comparatively thin, and conidiophore formation remained sparse (Plate 5.2a). The leaves were symptomless to the unaided eye. Infections remained restricted to the inoculated leaves, which may reflect a rather low degree of compatibility between host and pathogen but could also be due to unfavourable environmental conditions (Koch and Slusarenko 1990).

In powdery mildews, appressoria are formed either at the end of conidial germ tubes or as lateral outgrows of superficial hyphae. They function as structures which both attach the mycelium to the host surface and form haustoria (Braun 1987). Considerable variation in the morphology of appressoria exists between different species of powdery mildews (Neger 1902; Zaracovitis 1965; Boesewinkel 1980; Braun 1987). Boesewinkel (1980) grouped *E. cruciferarum* among those Erysiphaceae with 'multilobed haustoria'. However, according to Koch and Slusarenko (1990), appressoria on hyphae of *E. cruciferarum* frequently had 2–4 lobes (Plates 5.1h, 5.2c, d). Haustoria did not extend from the centre of appressoria but from one of the appressorial lobes (Plate 5.1b, h). Appressoria were formed singly (Plate 5.1h) or on opposite sides of hyphae (Plate 5.2d). The shape of appressoria on hyphae corresponded closely with the hyphal appressoria of *E. cruciferarum* depicted by Boesewinkel (1980) and Braun (1987), but the shape of appressoria formed on germ tubes differed from those shown by both authors. Apparently, the appressoria on germ tubes of *E. cruciferarum* are variable, because the shape of these structures also differs considerably between the reports of Boesewinkel (1980) and Braun (1987).

5.3 Pathogenesis of *Erysiphe orontii* Observed Through Light and Scanning Electron Microscopy

The pathogenesis of *E. orontii* causing powdery mildew of *A. thaliana* has been studied through light and SEM by Plotnikova et al. (1998). The development of Massachusetts General Hospital (MGH) isolate of powdery mildew started with imbibitions of conidia and germination within 1–2 h (Plate 5.3a, b). The germ tubes

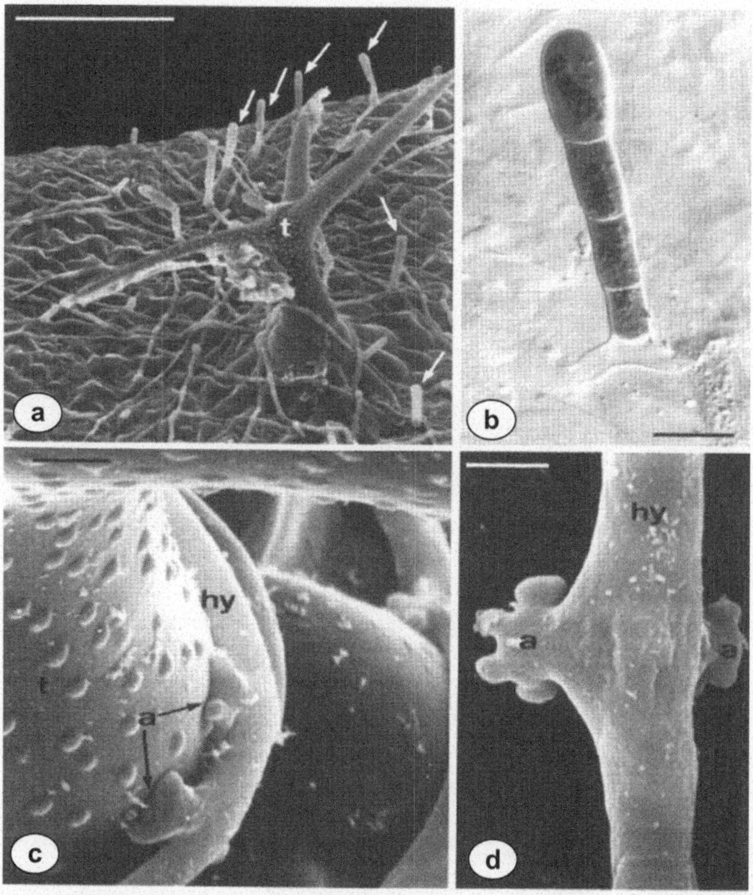

Plate 5.2 (**a–d**) Formation of conidiophores and lobed appressoria by *Erysiphe cruciferarum* on *Arabidopsis thaliana*. Plate **a**. Hyphae and conidiophores (arrows) on the leaf surface. Bar = 100 μm. **b**. A conidiophore of *E. cruciferarum*. Note the three cylindrical cells bearing a single, maturing conidium. Bar = 25 μm. **c**. Lobed appressoria (arrows) on hypha growing along the base of a trichome. Bar = 5 μm. **d**. Hypha with four-lobed appressorium. A second appressorium is present on the opposite side of the hypha. Bar = 2.5 μm (Koch and Slusarenko 1990)

were somewhat twisted, though sometimes straight, bent, or helicoids and rarely branched. The apex of the germ tube was often swollen. The first appressoria were formed by 5 h from the swollen apex of the germ tube and were separated by septa (Plate 5.3b, c). Germ tubes grew radically over the epidermal surface and gave rise to superficial hyphae by 38 h (Plate 5.3c). The mycelium of the MGH isolate was amphigenous, evanescent, superficial hyphae slightly flexuous, branched at right angles, and ca 5.8–6.6 μm wide. At 3–5 dpi, conidiophores which were straight or sometimes curved began to develop perpendicular to the leaf surface (Plate 5.3d). Conidia were produced in chains.

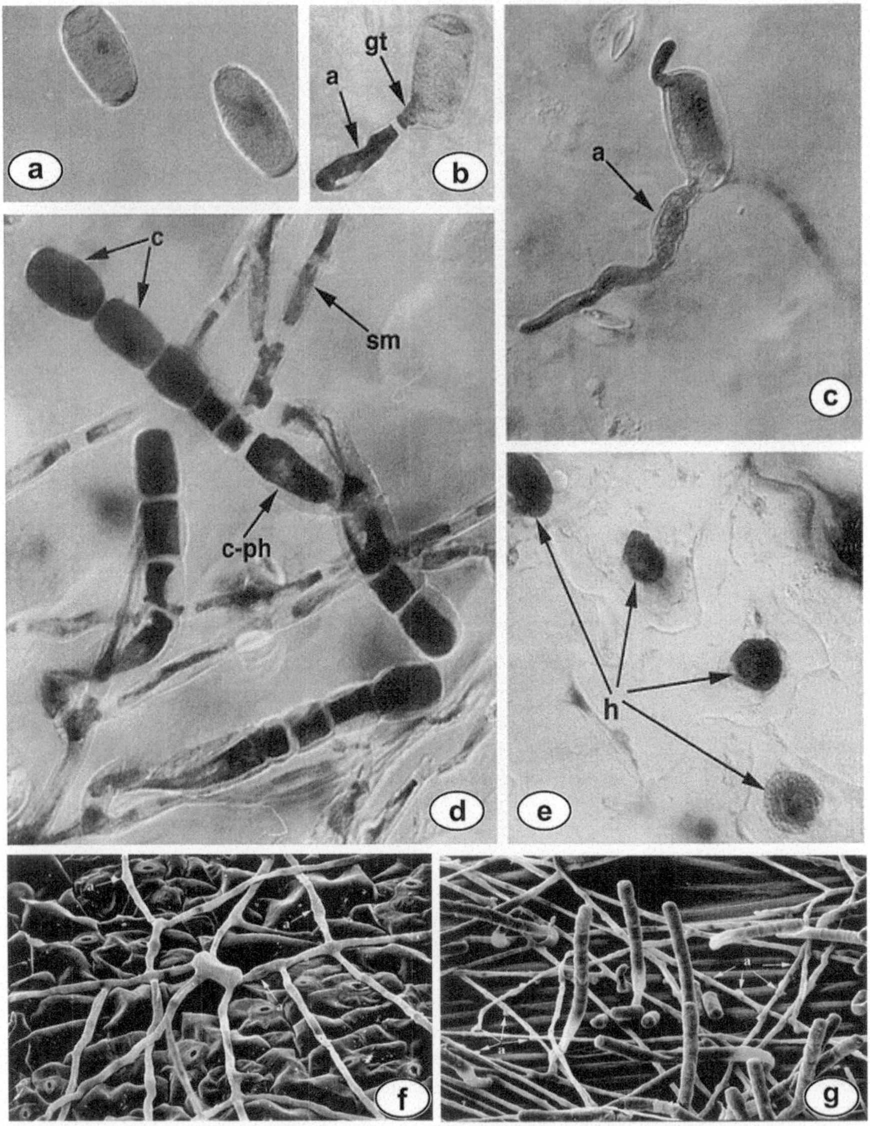

Plate 5.3 (a–e). Light microscopy (LM) of powdery mildew pathogen (MGH isolate). Plate **a**. Ungerminated conidia LM X 870. **b**. MGH isolate conidium with a short germ tube (gt) and a nipple-shaped appressorium (a), 24 hpi, LM X 900. **c**. A conidium with three germ tubes (80 hpi); an appressorium (a) formed on the left side contains two nuclei (arrow), LM X 1000. **d**. Superficial mycelium (sm) of the MGH isolate with conidiophores (c-ph) and conidial (c) chains, 6 dpi, LM X 680. **e**. Spherical intracellular haustoria (h) of the MGH isolate, 6 dpi, LM X 900. **(f–g)** Scanning electron microscopy (SEM) of young and mature colonies of powdery mildew (MGH isolate). **f**. Young 5-day-old colony of the MGH isolate on an *A. thaliana* Col-0 leaf surface. Superficial mycelium with appressoria (arrows) growing along the epidermal cell surface. SEM X 850. **g**. Powdery mildew (MGH isolate) development on an *A. thaliana* stem surface (2-week-old colony). Straight conidiophores with conidial chains and numerous appressoria (a) with nipple-shaped protuberances are seen (arrows). Mature conidia detaching from the chain apex drop onto the stem surface. SEM X 470 (Plotnikova et al. 1998)

At 4 dpi the conidiophores bore only one or two conidia on susceptible plants, but with time, the number of conidia per conidiophore increased to 7–20 or more at high humidity. New appressoria were formed along the superficial hyphae, and all the appressoria had nipple-shaped protuberances on one side. A thin penetration peg emerged from below the protuberance and pierced the wall of an epidermal cell. Some pegs were blocked by host papillae, whereas others became swollen at the apex and formed intracellular haustoria. Young haustoria of the MGH isolate were ellipsoidal; mature haustoria became spherical (Plate 5.3e). Scanning electron microscopy of *A. thaliana* Col-0 leaves infected with the MGH isolate showed that the fungus attached to the host surface firmly, growing over epidermal cells (Plate 5.3f, g). Plate 5.3f shows a 5-day-old colony of the MGH isolate on Col-0 leaf. A conidium with 5 germ tubes, superficial hyphae branched at right angles, and new appressoria can be seen. A 2-week-old infection is seen in Plate 5.3g. Superficial mycelium of the MGH isolate bears many conidiophores with conidia on an *A. thaliana* (mutant *pad4*) stem surface. Conidial chains were usually longer on *Arabidopsis* stems than on leaf surfaces. Many appressoria with nipple-shaped protuberances which give rise to infection pegs and intracellular haustoria are seen in Plate 5.3g.

Fungal growth was apparent without magnification at 5 dpi on *A. thaliana* Col-0. In the following days, hyphae spread rapidly, and adjacent colonies coalesced. At 10 dpi a white mycelial mat covered about 20% of the leaf surface of Col-0 plants. In older infections (−14 dpi), Col-0 began to display chlorosis of heavily infected leaves. Some spreading of the infection via conidia to previously uninfected leaves occurred. Isolated patches of powdery mildew were found on cauline leaves and bolts (flowering stems).

5.4 Pathogenesis of *E. cichoracearum* Observed Through Light and Cryogenic Scanning Electron Microscopy

The process of infection and pathogenesis of *Arabidopsis* by *E. cichoracearum* were observed using light and cryogenic SEM by Adam and Somerville (1996). The disease reaction phenotype in compatible interaction was characterized by the development of white, snowlike colonies, which first become apparent on the surface of the infected leaves at about 5 days postinoculation (dpi). Mycelial development and conidiation were profuse and could be observed reliably at 7 dpi (Plate 5.4a). Infection occurred on both stems and leaves in irregular patches or more commonly covering the entire surface of inoculated leaves. In the compatible interaction, fungal growth was not usually associated with any visible host reaction (i.e. chlorotic or necrotic flecks occurred at inoculation sites infrequently) within 10–12 dpi. With longer incubation periods, premature senescence of leaves, which was characterized by spreading chlorosis, dehydration, and necrosis, generally occurred.

Plate 5.4 Disease reaction phenotypes of six powdery mildew-resistant *Arabidopsis* accessions 7 days postinoculation with *E. cichoracearum*. Compatible interaction on Col-5 (**a**). Incompatible interactions with accessions Wa-1 (**b**), Kas-1(**d**), Stw-0 (**f**), Si-0(**h**), Te-0 (**j**), and Su-0 (**l**). The disease reaction phenotypes of the F_1 (Col-5 x resistant accessions) are shown in the lower half of each panel for We-1(**c**), Kas-1(**e**), Stw-0(**g**), Si-0 (**i**), Te-0 (**k**), and Su-0(**m**). (Adam and Somerville 1996)

To provide more detailed information on the morphology of the UCSC isolate and on the infection sequence, whole leaf samples, prepared over an interval of 1–7 dpi, were observed by light and cryogenic SEM. Conidia were ovoid–ellipsoid in shape (about 26–33 x 16–19 μm), with a length/width ratio of approximately 2. An unlobed germ tube arose from the end of a conidium less than 1–2 h postinoculation (hpi) (Plate 5.5a). Often, a second germ tube emerged from the opposite end of conidium which 24 hpi (Plate 5.5b–c). By 24 hpi, most conidia (more than 90%) had germinated. The first germ tube could be distinguished from the second by its slightly enlarged, swollen, or club-shaped appearance (Plate 5.5b, d). Because of its distinct morphology, the first germ tube was referred to as the appressorial germ tube. The appressorial germ tube produced a penetration peg, which often successfully penetrated the cell wall and established a haustorium. This observation is supported by the development of papillae underneath germ tubes, adjacent to conidia (Plate 5.5a, b), as well as the development of haustoria by 9 hpi (Plate 5.5e, f). The haustorium of *E. cichoracearum* had a simple, unlobed oval shape, which is retained. At 24 hpi, no plant cell death was observed in the compatible interaction. At 2 dpi, both hyphae had elongated to 60 μm or more (Plate 5.5d). The second hypha attempted to penetrate the host epidermis but at some distance from the conidium (Plate 5.5c, d). At 3–4 dpi, (Plate 5.5e, f) the hyphae had elongated and branched more or less at right angles to form a network on the leaf surface. In SEM micrographs, the appressorium appeared as unlobed, nipple-shaped protuberances along the hyphae (arrowheads) (Plate 5.5f). Consistent with this, papillae formed at several sites along hyphae (Plate 5.6d) indicating that several additional attempts to penetrate the plant cell surface were made by the pathogen. As they elongated, hyphae followed the uneven surface of the leaves except at points of interaction between two hyphae (Plate 5.5f). At 4 dpi, conidiophores began to develop perpendicular to the leaf surface. The conidiophore foot cell was often curved (Plate 5.5g). Conidia were produced in chains from conidiophores (88–160 x 9.5–11 μm), which consisted of three cylindrical cells followed by a number of bulging cells of variable size (Plate 5.5h). Immature conidia could be distinguished from mature, well-defined conidia, which occurred at the ends of the conidiophores (Plate 5.5h). At 4 dpi, the conidiophores consisted of only one or two conidia (Plate 5.5g), but with time, the number of conidia per conidiophores increased (n = 3–5) (Plate 5.5h, i). Fungal growth was apparent to the naked eye at 5 dpi. In the following days, fungal mycelia spread rapidly, and adjacent patches of mycelia coalesced.

In places where hyphae were removed from the leaf surface mechanically during cryofixation, a flocculent material, which appeared to be derived from the hyphae (Plate 5.5c), was observed on the leaf surface (Plate 5.5 j, k). The material resembles the amorphous extracellular material observed under primary and appressorial germ tubes of *E. graminis* f. sp. *avenae* (Carver et al. 1995). Penetration holes occurred at regular intervals along the track left by a hypha on the leaf surface (Plate 5.5k). At transverse section of a fragmented epidermal cell, a penetration peg was observed extending across the cell wall and connecting to a haustorium in the epidermal cell (Plates 5.5l and 5.6f).

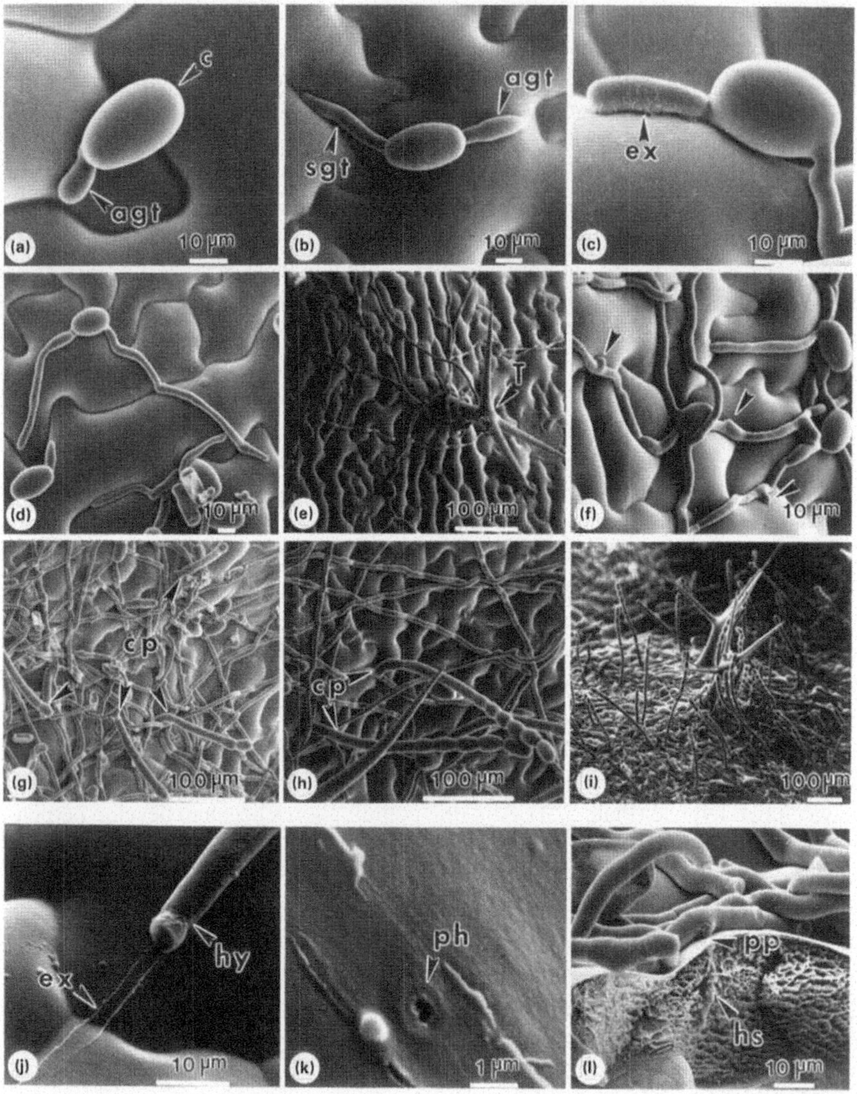

Plate 5.5 Cryogenic scanning electron micrographs illustrating the infection sequence of *E. cichoracearum* on the susceptible Col-1. (**a–c**). One day postinoculation. The appressorial germ tube (agt) has emerged from the conidium (c). A second functional germ tube (sgt) emerged at the opposite end of the conidium from the appressorial germ tube. The appressorial germ tube can be distinguished from the second germ tube by the swollen appearance of the former. In (**c**), a flocculent material (ex) can be observed on the germ tube. (**d**) Two days postinoculation. Hyphae from both germ tubes have elongated. The hyphae follow the leaf surface closely. (**e** and **f**). Three days postinoculation. The hyphae have formed a network on the leaf surface. A trichome (T) can be seen in (**e**). In (**f**) appressoria (arrowheads) which have a simple unlobed shape in *E. cichoracearum* can be observed along a hypha. Occasionally, a hypha will leave the leaf surface in cross another hypha as illustrated in (**f**). (**g**) Four days postinoculation. Conidiophores (cp) start to form

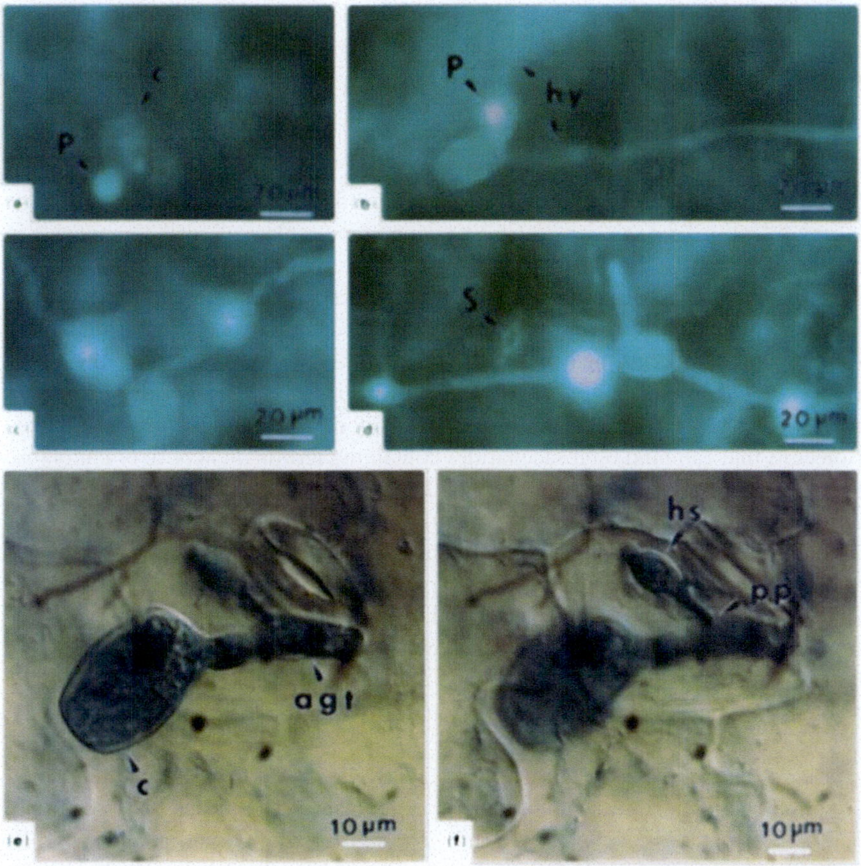

Plate 5.6 Light micrographs of Col-5 infected with *E. cichoracearum*. (**a–d**) Papilla formation by Col-5 at 12 h (**a–b**) or 3 days (**c–d**) after inoculation with *E. cichoracearum*. Papillae (P) are displayed as callose-rich deposits stained with aniline blue and visualized by epifluorescent microscopy. Conidia (**c**), hyphae (hy), and stomatal surface of guard cells (s) are faintly autofluorescent and can also be observed. (**e–f**) Haustorium formation at 12 hpi. In (**e**), the fungal structures on the surface of leaf, the conidium (**c**), and appressorial germ tube (agt) are in focus. In (**f**), the haustorium (ha), which is located below the leaf surface in the epidermal cell, is in focus. The penetration peg (pp) joining the appressorial germ tube to the haustorium is visible. This sample was stained with trypan blue. (Adam and Somerville 1996)

Plate 5.5 (continued) (**h**) Five days postinoculation. Conidiophores formation is profuse, and well-developed chains of conidia are apparent on the conidiophores (cp). Powdery mildew growth first becomes apparent by eye. (**i**) Seven days postinoculation. Dense mycelium and abundant conidiation are apparent. This time point was used to access the disease reaction phenotype. (**j**) The deposition of flocculent material or exudates (ex) underneath a hypha (hy) is shown. A section of hypha was mechanically removed prior to cryofixation of the sample. (**k**) A penetration hole (ph) was exposed by mechanical removal of a section of hypha prior to cryofixation of the sample. (**l**) A cross section showing both a penetration peg (pp) extending an epidermal cell and a haustorium (hs) in the cell interior. (Adam and Somerville 1996)

5.5 Location and Amount of Chitin in Powdery Mildew Fungus Through TEM

Chitin is present at the tips of growing hyphae and in conidia of *E. cichoracearum* during infection of *Arabidopsis.* Chitin is one of the major cell wall components in many fungi. Fungal cell walls change during growth and development, and the deposition and subsequent removal of chitin is particularly important during septation (Klis et al. 2001; Adams 2004). However, the distribution of chitin in *E. cichoracearum* structures formed during infections of *Arabidopsis* was unknown. Cell wall components and chitin were localized in growing *E. cichoracearum* using labeled wheat germ agglutinin (WGA). WGA contains a group of closely related isolectins that can bind oligosaccharides containing terminal GlcNAc or chitobiose and membrane glycoproteins (Peters and Latka 1986). WGA has been used in several studies on chitin distribution in fungal walls (Galun et al. 1976; Tronchin et al. 1981). Laser scanning confocal microscopy visualized WGA-488 Alexa conjugates at the growing tips of appressoria and hyphae (Plate 7.4 A, B). WGA staining was also observed in the cell walls of conidia developing on conidiophores (Plate 7.4 C). As the conidia matured and detached from the conidiophore, chitin was localized to both ends of conidia (Plate 7.4 D). When conidia germinated, the region where the appressorial germ tube developed accumulated chitin. Furthermore, although mature conidia have chitin at their ends (Plate 7.4 D), germinated conidia (Plate 7.4 A) lack detectable chitin in this position, suggesting temporal and spatial regulation of chitin localization. Transmission electron microscopy confirmed that WGA-gold colloidal conjugates were present at the growing tip of appressorium (Plate 7.4 E) and also accumulated in the cell wall of the fungal haustorium, a feeding structure that forms in epidermal cells (Plate 7.4F). Fungal structures directly in contact with the plant and likely to be exposed to plant chitinases, such as appressoria and haustoria, show relatively high content of chitin. Therefore, it is likely that chitooligomers will be generated during the course of powdery mildew infections (Ramonell et al. 2002). For detailed description, see Chap. 7 at Sect. 7.4.7.1.

References

Adams DJ (2004) Fungal cell wall chitinases and glucanases. Microbiology 150(7):2029–2035

Adam L, Somerville SC (1996) Genetic characterization of five powdery mildew disease resistance loci in *Arabidopsis thaliana*. Plant J 9:341–356

Boesewinkel HJ (1980) The morphology of the imperfect states of powdery mildews (Erysiphaceae). Bot Rev 46:167–224

Brandenburger W (1985) Parasitische Pilze an Gefaßpflanzen in Europe. Gustav Fischer, Stuttgart, New York, 1248 S

Braun U (1987) A monograph of the Erysiphales (Powdery mildews). Beihefte zur Nova Hedwigia 89: 1–700, J Cramer Berlin, Germany

Bushnell WR, Gay J (1978) Accumulation of solutes in relation to the structure and function of haustoria in powdery mildews. In: Spencer DM (ed) The powdery mildews. Academic, London, pp 183–235. 565

Carver TLW, Thomas BJ, Ingerson-Morris SM (1995) The surface of *Erysiphe graminis* and the production of extracellular material at the fungus-host interface during germling and colony development. Can J Bot 73:272–287

Galun M, Braun A, Frensdorff A, Galun E (1976) Hyphal walls of isolated lichen fungi. Arch Microbiol 108(1):9–16

Hirata K (1967) Notes on haustoria, hyphae and conidia of the powdery mildew fungus of barley, *Erysiphe graminis* f. sp. *hordei*. Mem Fac Agri Niigata University 6:205–259

Klis FM, Smits GJ, van den Ende H (2001) Differential regulation of cell wall biogenesis during growth and development in yeast. Microbiology 147(4):781–794

Koch E, Slusarenko AJ (1990) Fungal pathogens of *Arabidopsis thaliana* (L.) Heyhn. Bot Helv 100:257–268

Laibach F (1943) Arabidopsis thaliana (L.) Hehyn. als Objekt fur genetische und entwicklungsphy siologische Untersuchungen. Bot Archiv 44:439–455

Neger FW (1902) Beitrage zur Biologie der Erysipheen. Flora 90:221–272

Peters W, Latka I (1986) Electron microscopic localization of chitin using colloidal gold labelled with wheat germ agglutinin. Histochemistry 84(2):155–160

Plotnikova JM, Reuber TL, Ausubel FM, Pfister DH (1998) Powdery mildew pathogenesis of *Arabidopsis thaliana*. Mycol Soc Am 90(6):1009–1016

Ramonell KM, Zhang B, Ewing RM, Chen Y, Xu D, Stacey G, Somerville S (2002) Microarray analysis of chitin elicitation in *Arabidopsis thaliana*. Mol Plant Pathol 3(5):301–311

Somerville C (1989) *Arabidopsis* Blooms. Plant Cell 1:1131–1135

Staveley JR, Pillai A, Hanson EW (1969) Electron microscopy of the development of *Erysiphe polygoni* in resistant and susceptible *Trifolium pratense*. Phytopathology 59:1688–1693

Tronchin G, Poulain D, Herbaut J, Biguet J (1981) Localization of chitin in the cell wall of *Candida albicans* by means of wheat germ agglutinin. Fluorescence and ultra-structural studies. Eur J Cell Biol 26:121–128

Zaracovitis C (1965) Attempts to identify powdery mildew fungi by conidial characters. Trans Brit Mycol Soc 48:553–558

Chapter 6
Epidemiology and Disease Forecasting

6.1 Introduction

If a plant disease like powdery mildew is to become severe under field conditions, particularly if it is to spread over a large area and develop into an epidemic, the right combinations of environmental factors must occur and spread either constantly or repeatedly and at frequent intervals over a large area. A plant disease epidemic implies the development and rapid spread of a disease on a particular kind of crop plant cultivated over a large area. The first component of plant disease epidemic is a large area planted with one, more or less genetically uniform crop plant and the field being close together. The second component of an epidemic is the presence or appearance of a virulent pathogen near the cultivated host plants. The third component is the combinations of the congenial environmental factors (temperature, relative humidity, wind velocity) coinciding with the susceptible stage/age of the host and with the pathogen inoculum production, spread, inoculation, penetration, infection, and reproduction phages. The crucifer's powdery mildew is a polycyclic disease. The incubation and latent periods of the pathogen are very short on all the cruciferae host plants. Therefore, the disease may get epidemic form within a very short period of time after infection, if congenial conditions prevail during the growth stages of the host crop. Crucifer's powdery mildew epidemic development is favoured by age of the host, age of the leaf tissues, tolerance/resistance level of different crucifers genotypes coinciding with congenial environmental conditions for conidial germination, infection (temperature and relative humidity), and spread of pathogen (wind velocity). Date of planting has great bearing on disease development under field conditions. A disease forecasting system of powdery mildew of crucifers has been developed based on crop age, temperature, maximum severity of disease, crop age at first appearance of disease, crop age at peak severity of the disease, and cultivar grown at different locations.

G. S. Saharan et al., *Powdery Mildew Disease of Crucifers: Biology, Ecology and Disease Management*, https://doi.org/10.1007/978-981-13-9853-7_6

6.2 Disease Development in Relation to Environmental Conditions

In general, the information on the epidemiology of powdery mildew disease is very meagre (Price 1970; Saharan 1970; Yarwood 1957). At Hisar, Haryana (India), powdery mildew on *Brassica* appeared in mild form during 1978–1979. The disease probably could not progress due to low mean temperature (13.7 and 14.6 °C) in the month of February during these 2 years along with washing effects of rainfall, i.e. 33.3 and 66.3 mm, respectively. During 1980, 16.1 °C mean temperature in the month of February proved highly favourable along with negligible rainfall of 0.6 mm for the onset of powdery mildew epidemic. Contrary to preceding 2 years in February 1980, maximum temperature (25.1 °C) and minimum temperature (7.1 °C) remained quite high along with low humidity (65%) to favour the start of powdery mildew epidemic. In March, during the year 1980 also, temperature variable coupled with relative humidity variables proved more favourable than the previous 2 years (Saharan 1980). However, there was rainfall of 21.1 mm, but its washing effects were nullified due to rise in temperature and low relative humidity variables. After analysing weather variables prevalent during 3 years, it appeared that for onset and epidemic development of powdery mildew of *Brassica*, moderate temperature and low humidity were favourable along with dry season in the months of February and March. Rainfall during these months suppressed the disease due to its washing effects on the conidial form, thus reducing the inoculum potential of the pathogen (Saharan and Kaushik 1981). Yarwood (1936, 1939, 1957) used water spray for the control of powdery mildews and indicated dry weather favourable and rainfall injurious for disease development. The optimum temperature for the germination of conidia, germ tube growth, and appressorium formation is 20–25 °C (Fig. 3.8, Chap. 3). There is no conidial germination below 15 °C and above 30 °C (Table 6.1) (Singh 1984; Saharan and Sheoran 1988; Uloth et al. 2017). In North Saurashtra, India, 27.2–28.9 °C temperature and 27–42% relative humidity favoured the powdery mildew development of mustard (Kanzaria et al. 2013). Powdery mildew appears on mustard in Rajasthan, India, in the first week of February (Shivpuri et al. 1997) and in Jharkhand, Ranchi, India, in December (Saharan 2016 personal observation), and in North Saurashtra, it appears at 66–74 days after sowing and gradually increases until harvest (Kanzaria et al. 2013).

Table 6.1 Effect of temperature on conidial germination of *Erysiphe cruciferarum* (Singh 1984)

Temperature °C	Spore germination (%)	Germ tube length (μm)	Appressorium formation (%)
15	10.0	12.5	100
20	40.0	15.0	100
25	36.1	15.0	100
30	12.0	12.5	100
35	0.0	0.0	0
C.D.	5.52	7.03	–

These occurrences were correlated with increased temperature and ageing or maturity of the crop. It is likely that maturity of the crop influences both the onset and development of powdery mildew epidemics as it is observed with other species of powdery mildews (Cole 1966; Asalf et al. 2014). In Australia, the development of powdery mildew on *Brassica napus* was slower, and final severity reduced at a day/night temperature 14/10 °C compared with 22/17 °C (Figs. 6.1and 6.2). In vitro maximum growth of the germ tubes from conidia of *E. cruciferarum* was at 15–20 °C, and survival of conidia reduced by temperature >30 °C (Table 6.1). The epidemics are most severe in the two warmer cropping regions, viz. the northern agricultural region of Western Australia and New South Wales (Uloth et al. 2017).

6.2.1 *Influence of Weather Variables on Cleistothecial Formation*

It is essential to have infected plants under constant observation for detecting formation of cleistothecial stage. Cleistothecia start appearing in the form of pinkish bodies on maturing leaves, stem, and pods of *Brassica* in the last week of March under Hisar, Indian, conditions. In the first week of April, abundant number of cleistothecia appeared on all infected plant parts in the form of dark brown to black scattered or concentrated bodies. *Brassica nigra* var. *matopobka* showed numerous cleistothecia on both sides of succulent green young leaves in the first week of April since other varieties matured earlier than this one. The moderate temperature variables (temperature maximum 27.5, temperature minimum 11.2, and mean temperature 19.2 °C) in the month of March and early April 1980 favoured cleistothecial formation than in preceding years. However, in general the perfect stage of powdery mildew occurs much less frequently in tropical countries (Jhooty 1967) but, under favourable temperature conditions, may appear in abundance which requires

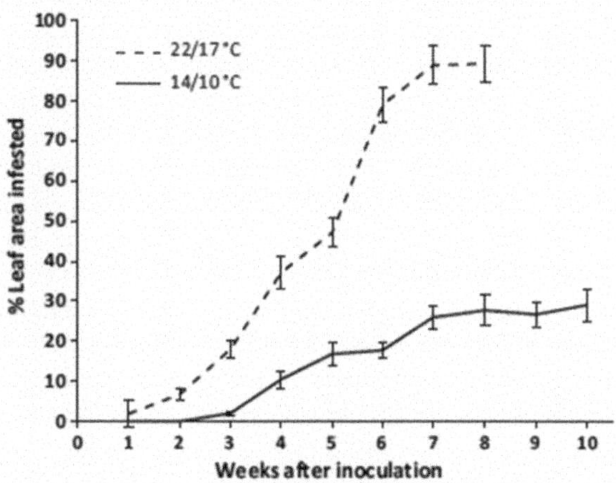

Fig. 6.1 Progress of leaf infection by powdery mildew (*Erysiphe cruciferarum*) of *Brassica napus* plants under two temperature regimes (14/10 °C, 22/17 °C). Error bars show the standard error of the mean (Uloth et al. 2017)

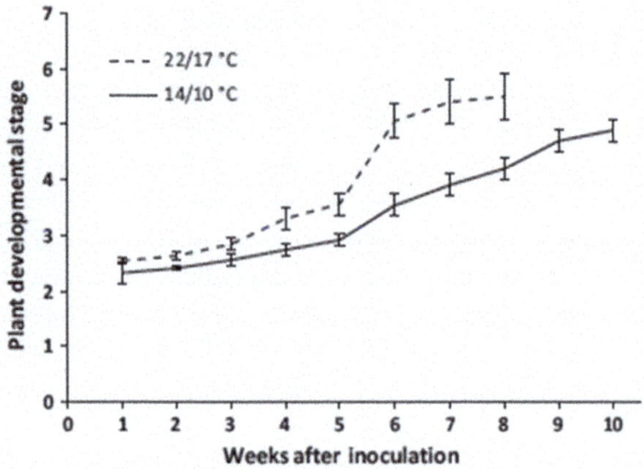

Fig. 6.2 Progress of *Brassica napus* developmental stage (Sylvester-Bradley and Makepeace 1984) under two temperature regimes (14/10 °C, 22/17 °C). Error bars show the standard error of the mean (Uloth et al. 2017)

constant observation. Since the temperature conditions normally become favourable near maturity of the crop for appearance of perfect stage, scientists lose interest thinking that season is over for this crop. It appears that cleistothecial formation in *Brassica* had been favoured due to alternating low and moderate temperature, heavy sporulation, low nutrition conditions of the host, low relative humidity, dry soil, and ageing of the host as in the case of other powdery mildews (Yarwood 1957; Saharan and Kaushik 1981).

6.3 Disease Development in Relation to Host Age and Temperature

The extent of powdery mildew infection on oilseed rape is determined by both the plant age at which infection occurs and the ambient temperature. The infection by *E. cruciferarum* increases as the host plant ages or matures (Saharan and Kaushik 1981; Rudgard and Wheeler 1985; Parry 1990; Kennedy 2010; Uloth et al. 2017). Findings implicate existence of an effective resistance mechanism operating in the juvenile leaf tissue of *B. napus*. Plant age at the time of infection dramatically influenced the speed with which powdery mildew developed on the leaves of *B. napus*. Disease symptoms appeared very quickly (9 dai) in plants that were already flowering when exposed to *E. cruciferarum* but took up to 5 weeks longer to appear on younger plants. It was notable that infection was never seen on young plant tissue, as powdery mildew was never observed on the leaves of plants younger than 37 days old nor initially on any fresh new growth (Uloth et al. 2017) that appeared on more mature infected plants (Figs. 6.3, 6.4, and 6.5; Tables 6.2 and 6.3). The studies of Alkooranee et al. (2015a, b) showed similar findings for powdery mildew affecting *B. napus* in China. It was described that *E. cruciferarum* affecting *B. napus* seedlings indicated that these 'seedlings' were in fact 40–45 days old when powdery mildew was first observed and 30 days old at the time of inoculation (Alkooranee et al. 2015 a, b).

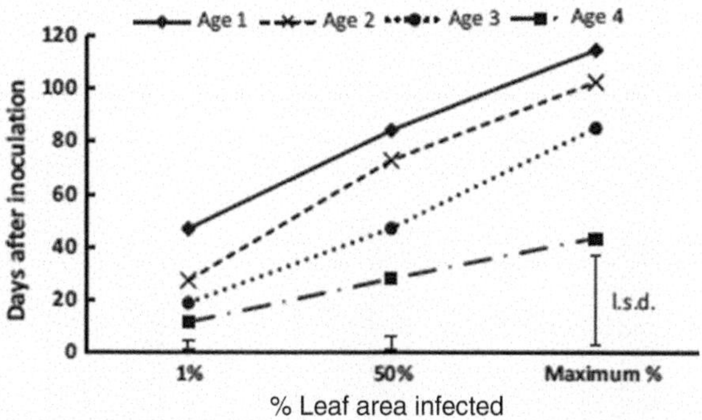

Fig. 6.3 Mean time taken after exposure to powdery mildew (*Erysiphe cruciferarum*) inoculum for the disease in *Brassica napus* plants of four different ages (0, 14, 28, and 42 days after seeding for ages 1, 2, 3, and 4, respectively) to reach 1%, 50%, and the maximum percentage of leaf area infected. Bars represent the least significant difference (l.s.d at $P = 0.05$) (Uloth et al. 2017)

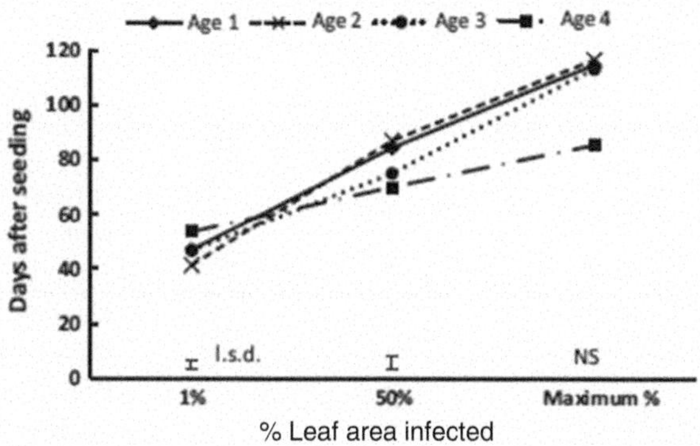

Fig. 6.4 Mean age of *Brassica napus* plants (days after seeding) at which powdery mildew (*Erysiphe cruciferarum*) disease reached 1%, 50%, and the maximum percentage of leaf area infected in plants of four different ages (0, 14, 28, and 42, days after seeding for ages 1, 2, 3, and 4, respectively). Bars represent the least significant difference (l.s.d. at $P = 0.05$) (Uloth et al. 2017)

Although powdery mildew developed more slowly on all tissues of the younger plants, final levels of pod disease did not differ between the age cohorts. This may be because plants younger when inoculated grew for longer (128 days for the youngest age treatment and 58 days in the oldest age treatment), allowing disease levels to 'catch up'. Hence, it is notable that maximum disease was reached on all plants with same length of time after seeding. In the field, final levels of disease on the pods would probably be more affected by the time of infection as there is a finite growing period available to rain-fed crops. Very rapid development of powdery mildew on more mature plants of oilseed rape will allow powdery mildew to severely affect a crop even if it is exposed late in the growing season, provided that environmental conditions are favourable to the disease. It has been observed relatively in the warmer northern agricultural regions of Western Australia and New South Wales (Uloth et al. 2017).

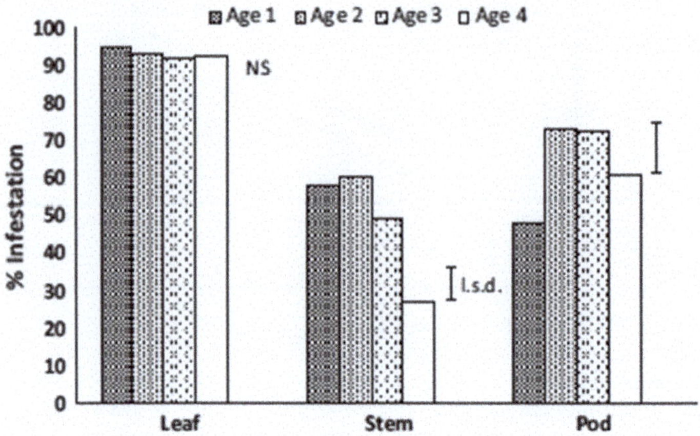

Fig. 6.5 Average maximum percentage powdery mildew (*Erysiphe cruciferarum*) infection of three different *Brassica napus* tissues (leaf, stem, pod). Bars represent the least significant difference (P < 0.05). Ages 1, 2, 3, and 4 equate to 0, 14, 28, and 42 days after seeding, respectively. Bars represent the least significant difference (l.s.d. at $p = 0.05$) (Uloth et al. 2017)

Table 6.2 Simplified Sylvester-Bradley stages of development in *Brassica napus* (Modified from Sylvester-Bradley and Makepeace 1984) (Uloth et al. 2017)

Sylvester-Bradley stage	Description
0	Dry seed
1	Leaf production
2	Stem extension
3	Flowering
4	Flower bud development
5	Pod development
6	Seed development

Table 6.3 Percent conidial germination and length of germ tubes (average) of *Erysiphe cruciferarum* from *Brassica napus* stems after 24 h of incubation on glass slides at four different temperatures (Uloth et al. 2017)

Temperature °C	Mean length of germ tubes (μm)	Germination (%)
5	33.6	0.1
15	58.8	5.8
25	181.2	6.8
35	63.6	0.8
P-value	< 0.001	<0.001
l.s.d. at *P* < 0.05	39.2	2.8

Initial infection occurred at varying developmental stages, ranging from stem elongation to early flowering, depending on the age at which plants were first exposed to powdery mildew. However, plants from all treatments reached the maximum level of infection at a similar age, regardless of the age at which they were first infected. This implies that for *B. napus*, the age of the plant is critical in determining when the powdery mildew disease epidemic peaks. It is widely accepted that mechanisms of defense against pathogens are intertwined with plant development (Whalen 2005). Resistance (or susceptibility) to diseases may develop gradually as the plant matures and can also be significantly affected by major transitions in the plant, such as the onset of flowering (Poethig 2003; Baurle and Dean 2006). This appears to be particularly evident in the powdery mildews, although the effect is not the same for all species.

This intriguing diversity within the powdery mildews suggests that elucidating the mechanisms governing the interaction of host age, developmental stage, and environmental factors in the success of *E. cruciferarum* and other powdery mildews would be of great value in improving the overall understanding of host resistance to powdery mildew. Powdery mildew infection on the stem was also affected by the age at which exposure occurred, but in contrast to leaf infection, the plants that were oldest when exposed to the disease had the lowest final level of stem infection. The youngest plants were exposed to powdery mildew for much longer overall, so it appears that the degree of stem infection is also determined by the length of the time exposure as well as changes in the maturity level of the plant. The observation that infection was always visible on the peduncle of pods before appearing on the pod itself is of interest and may occur because the peduncle is exposed to the pathogen for longer than the pod, as the pod is surrounded by the petals of the flower in the early stages of its development. If that had been the case, then the level of infection on the peduncles of immature inflorescences will be a useful predictor of likely final levels of infection in the field.

Mild temperatures favoured the development of *E. cruciferarum*. The maximum growth of *E. cruciferarum* germ tubes was reached at 15–20 °C, and survival of conidia was reduced by temperatures greater than 30 °C (Fig. 6.6; Table 6.3). Ambient temperature is one of the factors that dramatically affect the success of powdery mildew infections, and mild temperatures generally favour the disease. Final disease severity was significantly reduced at the lower temperature, and this was associated with greatly reduced growth of the germ tubes of *E. cruciferarum* and a reduction in spore viability. Further, it may be that the slow growth of powdery mildew also gives the plant more time to institute successful resistance mechanisms (Uloth et al. 2017).

6.4 Disease Development in Relation to Date of Sowing and Cultivars

Severity of powdery mildew on Indian mustard was higher in later-sown crops as depicted in Fig. 6.7. Initiation of the powdery mildew disease on Indian mustard occurred during 50–120 das with higher frequencies recorded in later part of crop growth. Disease intensity is reported to increase with plant age (Saharan and

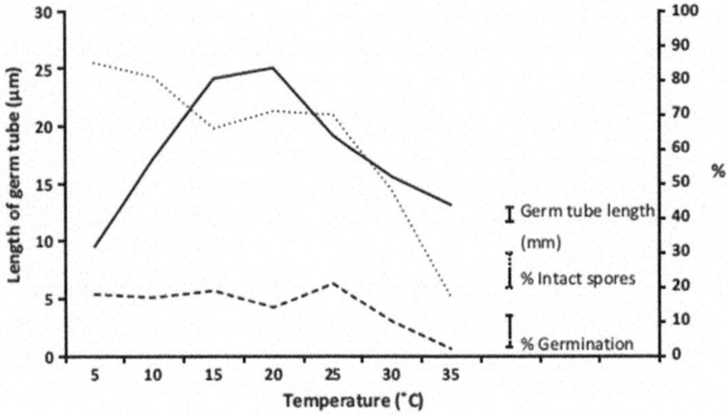

Fig. 6.6 Average percentage germination, percentage of conidia taken from infected leaves of *Brassica napus* with intact appearance, and average length of germ tube for *Erysiphe cruciferarum* conidia after 24 h of moist incubation on glass microscope slides at temperatures of 5, 10, 15, 20, 25, 30, and 35 °C. Bars represent the least significant difference (p = 0.05) for each parameter assessed (Uloth et al. 2017)

Kaushik 1981), which could be the reason for higher frequencies of disease initiation observed in older plants (Desai et al. 2004). The powdery mildew severity on the plants was favoured by >5 days of ≥9.1 h of sunshine (R^2 = 0.9), > 2 days of morning (maximum) RH of <90% (R^2 = 0.7), afternoon (minimum) RH 24–50% (R^2 = 0.92), minimum temperature > 5 °C (R^2 = 0.87), and a maximum temperature of 24–30 °C (R^2 = 0.83). Earlier, Crowton and Kennedy (1999) estimated percentage infection being negatively correlated to the period of exposure to wetness and long periods of exposure to free water responsible for significant reduction in the ability of conidia to germinate. The information on conditions favouring severity of powdery mildew on Indian mustard is useful for developing prediction models.

A delayed sowing results in coincidence of the vulnerable growth stage of plants as indicated earlier with warm (maximum temperature 24–30 °C; minimum temperature >5 °C; >9.1 h sunshine) and lower (morning <90%; afternoon 24–50%) RH conditions. The sustenance of such favourable conditions decides the longevity of the period of mildew attack and further build-up on the crop, which consequently affects yield. Thus, the damage caused to a crop by powdery mildew is likely to be related to sowing date, i.e. late sowing results in higher mildew severity (Fig. 6.7; Saharan and Kaushik 1981; Desai et al. 2004). Thus, it would be appropriate to sow the crop at the time to enable escape or non-coincidence of the vulnerable disease development stage with favourable weather factors leading to higher mildew severity on the crop. The later-sown crop matures quicker than the timely sown one due to rising of temperature towards end of crop season, which leads to faster maturity of the former that results in lower yield. However, it may also be noted that early-sown crops have encountered high disease severity on the crop due to coincidence of favourable weather factors with vulnerable crop stage (Kolte 1985). Hence, it would also be unwise to consider a thumb rule of escaping the disease in early-sown

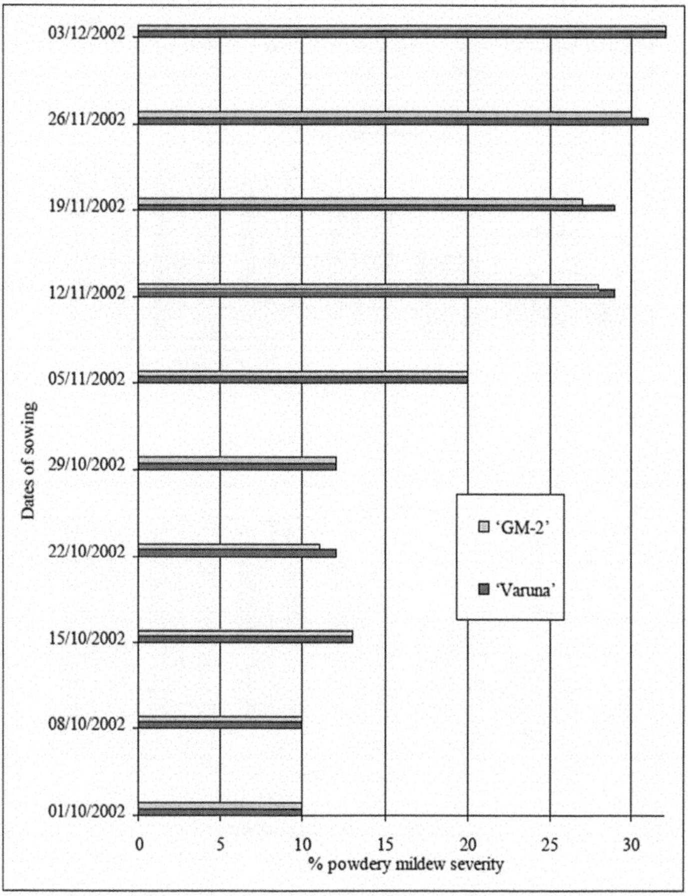

Fig. 6.7 Effect of dates of sowing on powdery mildew severity of mustard cvs. (Desai et al. 2004)

crop, whereby importance of forecasting needs to be underlined in this prominent disease of the crop in India (Desai et al. 2004).

Development and progression of powdery mildew on four different cultivars of mustard as influenced by variation and interaction of different environmental factors under field conditions is depicted in Fig. 6.8. Average maximum speck size 5.99 mm was observed when crop was sown on second fortnight of Nov. (Nov. 22) on all the four varieties followed by crop sown on first fortnight of Nov. (Nov. 15) and first fortnight of Dec (Dec. 2), respectively. Disease progression was maximum during mid of March on all the four varieties in all the three staggering dates of sowing, when Temp. Max. 32.5 °C, Temp. Min. 12.7 °C, RHM 94.5%, RHE 38.5%, Avp. M 12.30 mm, Avp. E 14.30 mm, wind speed 2 km/hr., and sunshine 8.80 h prevailed (Figs. 6.9, 6.10, and 6.11). Disease intensity increased from 48 to 74% with delay in date of sowing. An apparent infection rate varied between 0.10 to 0.37 on all the three dates of sowing. Maximum apparent infection rate was observed

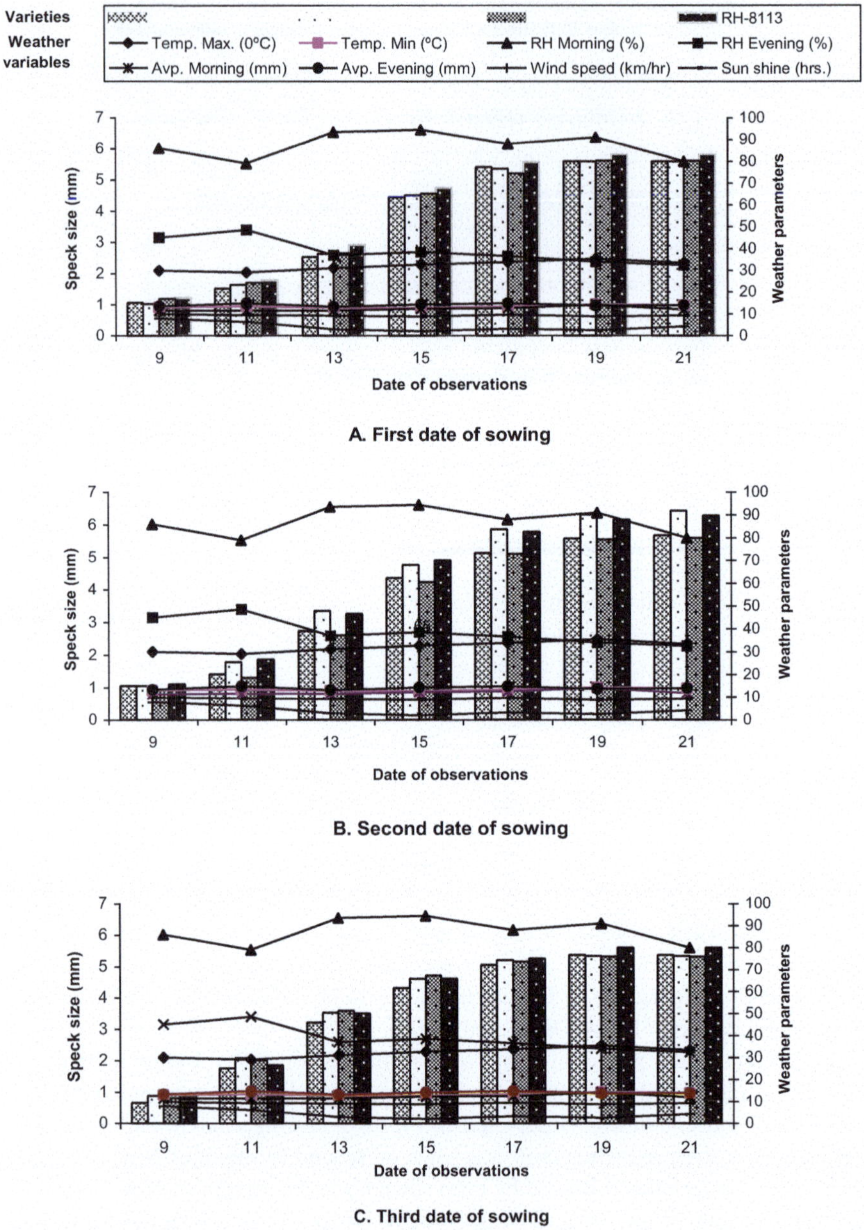

Fig. 6.8 Progression of powdery mildew on different varieties of mustard planted on different dates of sowing in relation to weather variables (Singh et al. 2008)

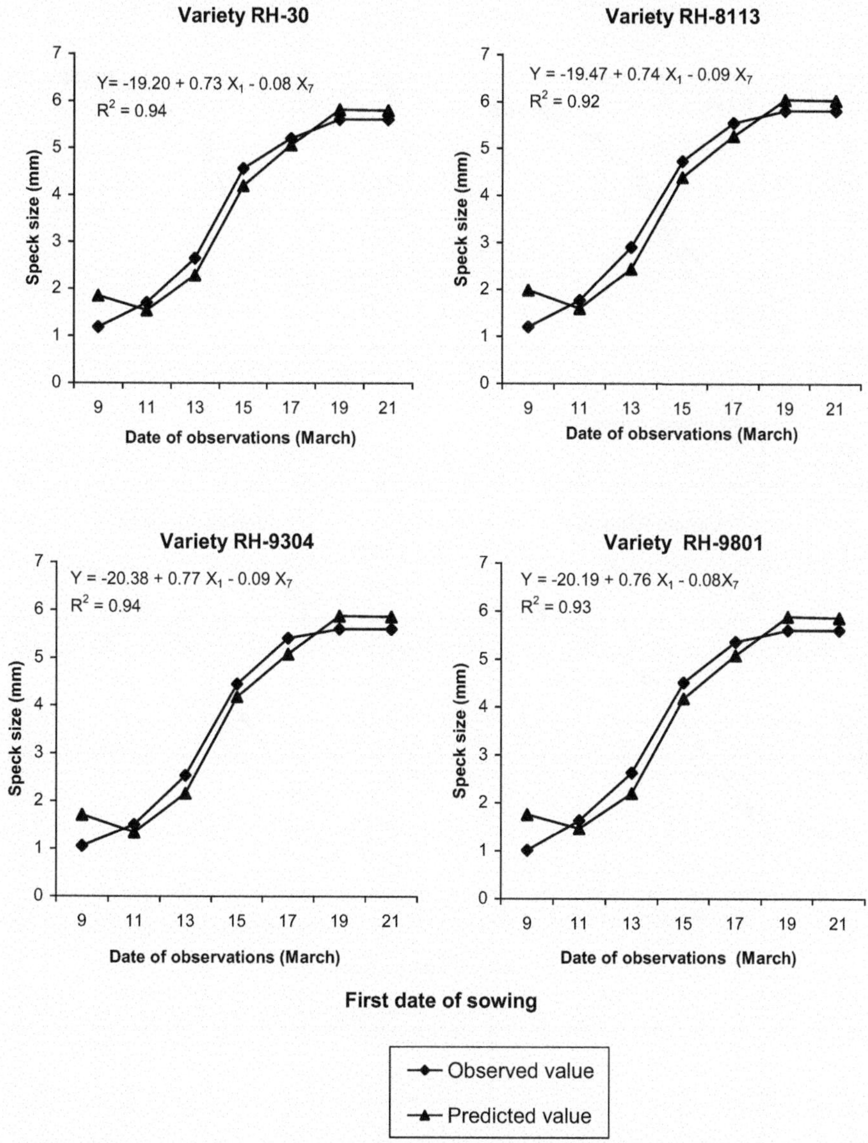

Fig. 6.9 Effect of environmental variables on the progression of powdery mildew (predicted/observed) on different varieties of mustard crop sown on Nov. 15 (Singh 2004)

during mid of March on all the varieties during first two dates of sowing except third date of sowing where it was delayed by 3 days, i.e. on 18 March (Table 6.4).

Similarly, AUDPC also increased with the delay in date of sowing from 326 to 440 (Fig. 6.12). The stepwise multiple regression analysis of data in relation to

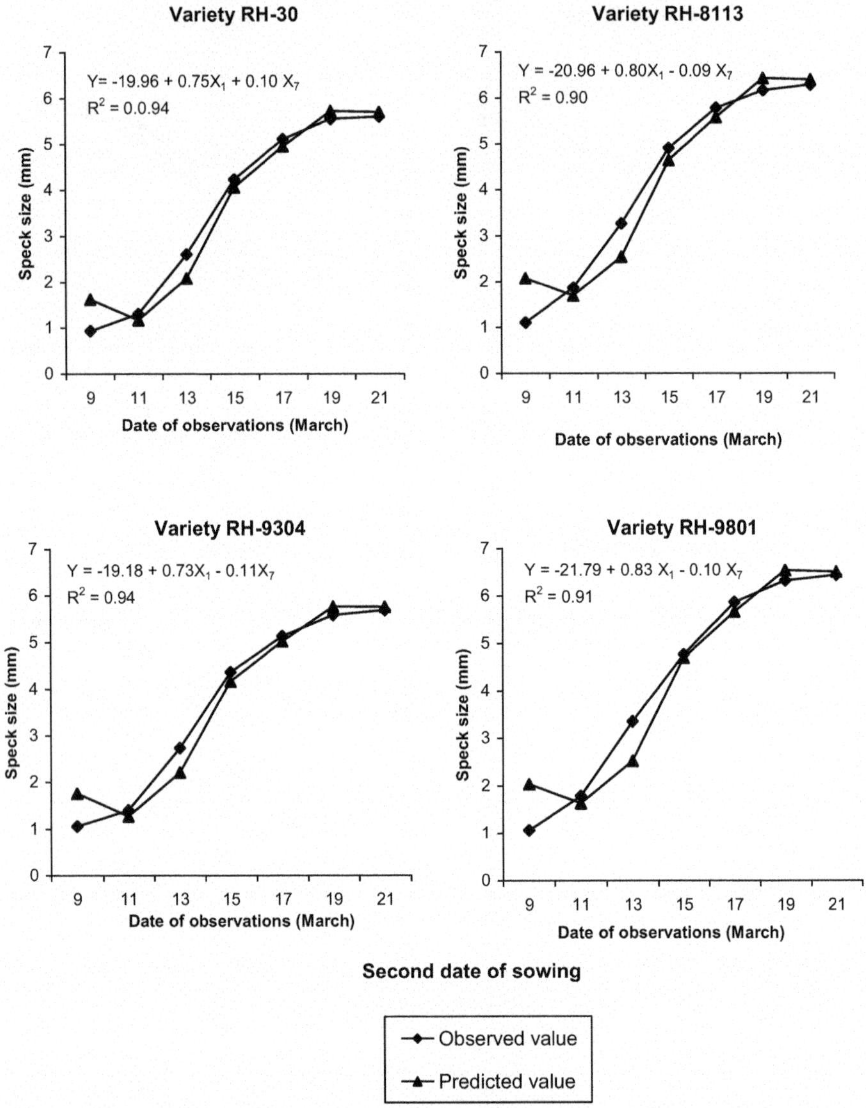

Fig. 6.10 Effect of environmental variables on the progression of powdery mildew (predicted/observed) on different varieties of mustard crop sown on Nov. 22 (Singh 2004)

weather variables revealed more R^2 value (>0.90) in all the dates of sowing and on all the varieties. R^2 values lie between 0.90 and 0.94. The regression equations (Table 6.5) developed also revealed that Temp. Max. (X_1) played major role in the development of powdery mildew irrespective of all the varieties and staggering dates of sowing. However, some role of wind speed (X_7) during first two dates of

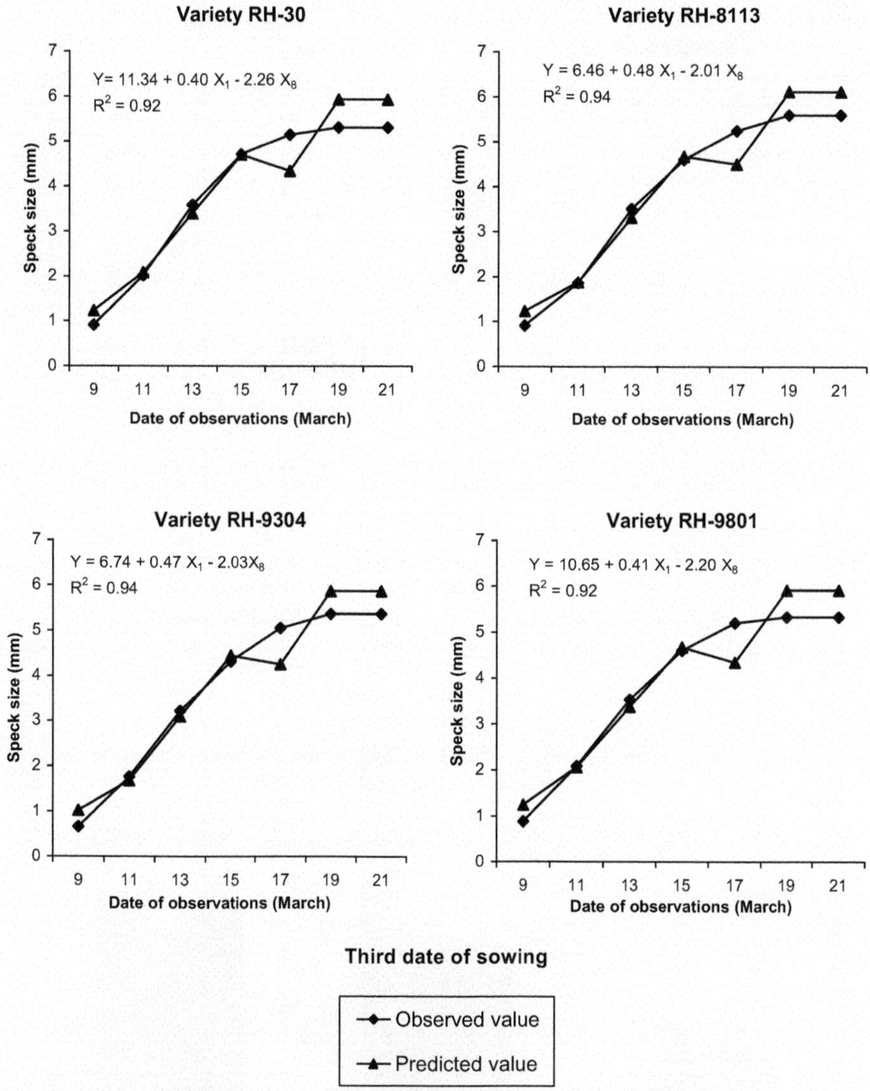

Fig. 6.11 Effect of environmental variables on the progression of powdery mildew (predicted/observed) on different varieties of mustard crop sown on Dec. 02 (Singh 2004)

sowing and sunshine (X_8) during third date of sowing has also been observed in addition to some other unknown factors. Correlation matrix for the progression of powdery mildew in relation to weather variables also revealed that Temp. Max. and Avp. M have positive and significant correlation in disease development on all the varieties and during all the three dates of sowing, whereas RHE has negative and

Table 6.4 Regression equations for the progression of powdery mildew on different varieties of mustard in relation to weather parameters under different dates of sowing (Singh 2004)

Varieties	Regression equations	R^2
	Ist DOS*	
RH-30	$Y = -19.20 + 0.73X_1 - 0.08X_7$	0.94
RH-8113	$Y = -19.47 + 0.74X_1 - 0.09X_7$	0.92
RH-9304	$Y = -20.38 + 0.77X_1 - 0.09X_7$	0.94
RH-9801	$Y = -20.19 + 0.76X_1 - 0.08X_7$	0.93
	IInd DOS	
RH-30	$Y = -19.90 + 0.75X_1 - 0.10X_7$	0.94
RH-8113	$Y = -20.96 + 0.80X_1 - 0.09X_7$	0.90
RH-9304	$Y = -19.18 + 0.73X_1 - 0.11X_7$	0.94
RH-9801	$Y = -21.79 + 0.83X_1 - 0.10X_7$	0.91
	IIIrd DOS	
RH-30	$Y = 11.34 + 0.40X_1 - 2.26X_8$	0.92
RH-8113	$Y = 6.46 + 0.48X_1 - 2.01X_8$	0.94
RH-9304	$Y = 6.74 + 0.47X_1 - 2.03X_8$	0.94
RH-9801	$Y = 10.65 + 0.41X_1 - 2.20X_8$	0.92

DOS* – Date of sowing
X_1 = Temperature (maximum)
X_2 = Temperature (minimum)
X_3 = Relative humidity (morning)
X_4 = Relative humidity (evening)
X_5 = Average evaporation (morning)
X_6 = Average evaporation (evening)
X_7 = Wind speed
X_8 = Sunshine

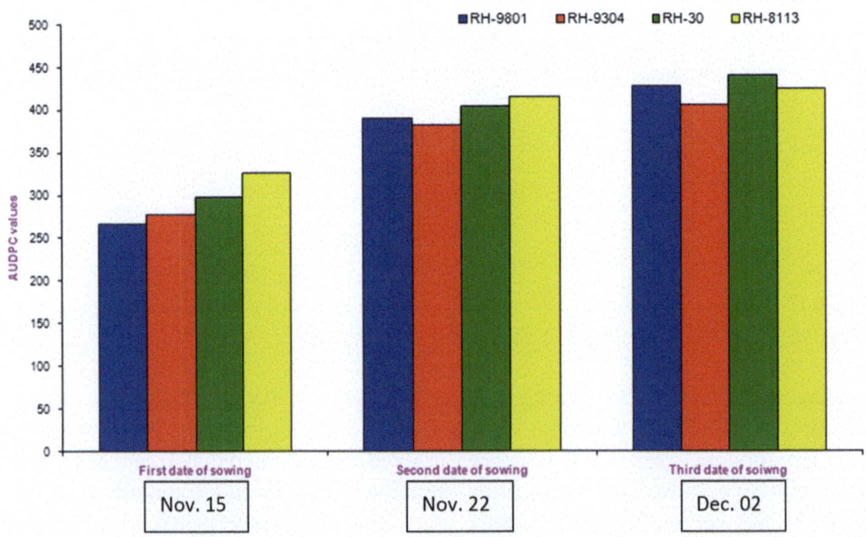

Fig. 6.12 Area under disease progress curve of powdery mildew on mustard cultivars on different dates of sowing (Singh 2004)

Table 6.5 Regression equations for the progression of powdery mildew on various mustard cultivars/varieties in relation to weather parameters (Singh 2004; Meena et al. 2018)

Cultivars/varieties	Regression equations	R^2
RH-30	$Y = -19.38 + 0.74X_1 - 0.10X_7$	0.91
RH-8812	$Y = -11.71 + 0.48X_1 - 0.08X_7$	0.83
RH-9304	$Y = -14.12 + 0.57X_1 - 0.10X_7$	0.81
RH-9801	$Y = -12.74 + 0.52X_1 - 0.11X_7$	0.83
RH-9901	$Y = -15.14 + 0.60X_1 - 0.08X_7$	0.86
RC-781	$Y = -11.34 + 0.47X_1 - 0.10X_7$	0.81
Purple mutant	$Y = -9.67 + 0.40X_1 - 0.07X_7$	0.86
GSL-1	$Y = -0.80 + 0.08X_1 - 0.01X_7$	0.47
HC-9603	Disease did not appear	–

X_1 = Temperature (maximum)
X_2 = Temperature (minimum)
X_3 = Relative humidity (morning)
X_4 = Relative humidity (evening)
X_5 = Average evaporation (morning)
X_6 = Average evaporation (evening)
X_7 = Wind speed
X_8 = Sunshine

Table 6.6 Time taken for powdery mildew (*Erysiphe cruciferarum*) infection to appear (days after inoculation, dai), for plants to reach pod fill (dai), mean age of plants at pod fill (days after seeding, das), and mean area under the disease progress curve (AUDPC) at 42, 55, and 83 dai for six *Brassica napus* cultivars (Uloth et al. 2017)

Cultivars	Mean time to 1% leaf infection (dai)	Mean time to pod fill (dai)	Mean age at pod fill (das)	AUDPC		
				42 dai	55 dai	83 dai
Banjo	33.3	67.4	94.5	80.0	123.3	163.2
Karoo	23.7	66.6	93.7	122.2	161.3	200.8
Lantern	24.4	75.6	102.8	112.0	168.4	223.4
Thunder-TT	29.4	75.0	102.1	102.1	159.4	206.3
Tribune	24.8	75.8	103.0	136.4	188.0	238.7
Trilogy	22.5	48.9	76.1	183.6	225.3	259.1
P-value	<0.001	<0.001	<0.001	<0.001	<0.001	<0.001
L.s.d. at $P < 0.05$	2.1	5.4	5.4	23.1	23.2	30.2
P-value for interaction between plant age category and cultivars	0.003	<0.001	<0.001	<0.001	<0.001	n.s.

significant correlation in all the cases for powdery mildew progression (Singh et al. 2008; Meena et al. 2018).

In Australia, powdery mildew development on *B. napus* in relation to leaf, stem, and pod infection of six cultivars was observed by Uloth et al. (2017) under artificial inoculation conditions. Leaf susceptibility to powdery mildew differed between cultivars (Table 6.6; Fig. 6.13). Leaves of Trilogy were the most susceptible with the highest AUDPC of 259.1 at 83 dai, followed by Tribune, Karoo, Lantern, Thunder-TT, and Banjo ($P < 0.001$), and this order was largely followed for other

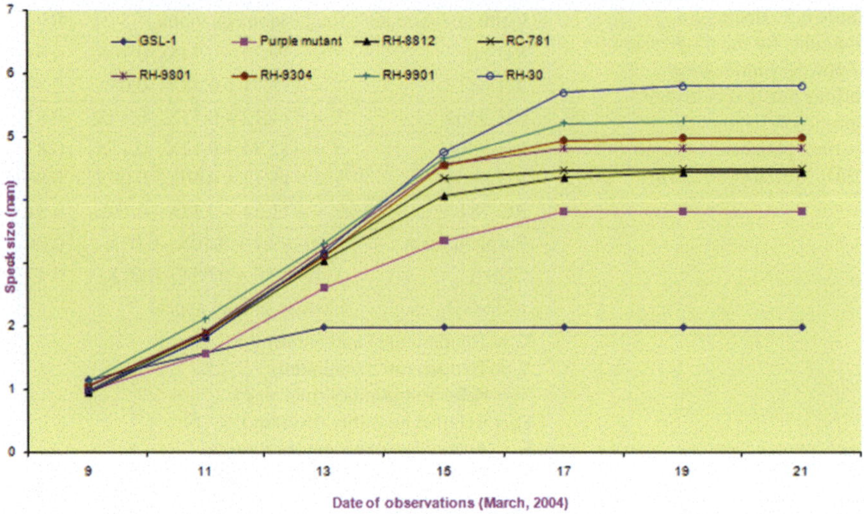

Fig. 6.13 Progression of powdery mildew on eight varieties of mustard (Singh 2004; Meena et al. 2018)

Fig. 6.14 Average maximum percentage powdery mildew (*Erysiphe cruciferarum*) infection of leaf, stem, and pod for six *Brassica napus* cultivars Banjo, Karoo, Lantern, Thunder-TT, Tribune, and Trilogy. Bars represent the least significant difference (l.s.d. at $p = 0.05$) (Uloth et al. 2017)

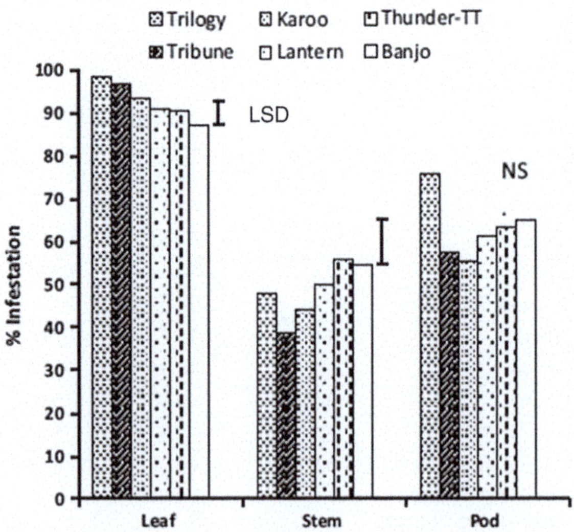

measures of leaf susceptibility, namely, AUDPC at 55 dai, AUDPC at 42 dai (Table 6.6), and maximum per cent dai (Fig. 6.13). These cultivars also differed in the amount of powdery mildew infection visible on stems ($P < 0.001$; Fig. 6.13), but order of susceptibility of the cvs. was different from that of leaf infection with Thunder-TT having the most stem infection with a maximum of 56% of the stem covered with powdery mildew, followed by Banjo, Lantern, Trilogy, Karoo, and Tribune. The amount of pod infection did not differ between cultivars ($P < 0.001$). AUDPC at 83 dai was negatively correlated with the time at which the first symptoms of powdery mildew appeared on a particular cv. (Fig. 6.14; $P < 0.005$, $R^2 = 0.61$, $n = 6$).

6.5 Disease Development in Relation to Crop Growth Stage

Powdery mildew infection and symptoms development is correlated with the growth stages of the *B. napus* crop. After inoculation at young age 1 (Table 6.2), it takes long period of 44 dai (days after inoculation) to appear the symptoms on leaves. Plants from the youngest age group took on an average of 115 days to reach maximum infection levels, but it took only 43 days on plants from the oldest age group (Fig. 6.3). Plants of growth stages 1 and 2 showed leaf symptoms during the stem elongation phase. However, powdery mildew reached its maximum level on plants at the same age, regardless of the time at which they were first exposed to the pathogen (Fig. 6.4). Powdery mildew was not observed on the leaves of very young seedlings. The plants which initiated new side shoots at pod set phase and having infection showed no symptom of powdery mildew on these new growths up to 10 days after its appearance (Plate 6.1). The final severity of leaf infection was independent of the age at which plants were inoculated. Plants that were older when infected had a lower maximum level of stem infection than younger plants, while the opposite was true for pod infection (Fig. 6.5). Stem infection was first observed at 3 weeks after inoculation (wai) with mycelium evident on the lower part of the stem and on some leaf petioles (Plate 6.1b). Once pods started to form, infection on the lower part of the stem gradually disappeared or became non-significant, leaving small brownish black areas at the site of pervious infection. Subsequently, high levels of stem infection were recorded with powdery mildew growth on stem and inflorescence/raceme as the flowering stalk became infected. The luxuriant white sporulation around stem areas in the vicinity of pods usually covered approximately one-third of the whole stem (Plate 6.1c). On severely infected plants, many pods were aborted from flowering racemes. The infection on a pod was always preceded by infection of the pod peduncle (Plate 6.1d, e) with mycelium often visible on the peduncle even when petioles were still present. After visible powdery mildew infection coverage of 90–100% on any part of the plant, the fungus showed its ageing, becoming grey, and the extent and coverage of obvious mycelium and/or conidia on the tissue declined (Plate 6.1f, Uloth et al. 2017).

6.6 Disease Development in Relation to Host Resistance

To assess the nature of powdery mildew resistance in *Brassica* crops, seven cvs. of *Brassica juncea* and one each of *Brassica napus* and *Brassica carinata* were selected for evaluation of slow mildewing components by Singh (2004). Various components of slow mildewing, viz. incubation period, latent period, no. of colony/speck per leaf, no. of conidia per colony/speck, progression of the disease, and disease intensity, were recorded under field conditions during 2003– 2004 crop season. The incubation period of test cvs. ranged from 3 to 4 days. However, powdery mildew was not observed on variety HC-9603 even under artificial inoculation conditions. The maximum incubation period of 4 days was recorded on the varieties RH-9304 and

Plate 6.1 Infection patterns on *Brassica napus* plants by powdery mildew (*Erysiphe cruciferarum*). (**a**) Unaffected regrowth of side shoots on heavily infected plants where terminal shoot and inflorescence of cv. Banjo have collapsed and died from powdery mildew (arrows indicate leaves and stem previously killed by *E. cruciferarum*); (**b**) early infection of stem; (**c**) powdery mildew covering both pods and stem in heavily infected inflorescence of cv. Thunder-TT; (**d, e**) peduncle infected by powdery mildew before any powdery mildew growth was visible on the pod (arrows indicate visible powdery mildew infection); (**f**) 'aged' infection on cv. Thunder-TT showing greying of mycelium and darker damaged areas on pod and stem (Uloth et al. 2017)

RH-9801, whereas in rest of the varieties, it was 3 days. However, the non-significant differences in incubation period in the case of *Brassica* cvs. (Table 6.7) were recorded. The latent period of test cultivars ranged between 1 and 3 days. The variety HC-9603 did not contract powdery mildew. The latent period was 1–2 days in varieties belonging to the *B. juncea*. However, it was 3 days in the case of variety GSL-1belonging to the *B. napus* (Table 6.7). The results also revealed that there was no significant difference in the latent period in the varieties belonging to *B. juncea*. However, it differed in the case of *B. napus* where it was slightly higher. The numbers of powdery mildew specks/leaf were also recorded on all the nine varieties as a test of slow mildewing components. The variety GSL-1 showed minimum number of specks/leaf (5.52), whereas the variety RH-9801 contracted maximum number of the specks/leaf (39.40). It was followed by the variety RH-9304 (33.97). On the other varieties, viz. RH-8812, RH-9901, RC-781, and Purple Mutant, the number of specks ranged between 18–27 per leaf, which is moderate. The variety GSL-1 contracted less number of specks per leaf which significantly differed from the other varieties (less than ten specks/leaf) which may be considered as resistant, whereas other varieties such as RH-9801, RH-9304, and RH-30 had higher number of specks/leaf (more than 30) and may be termed as susceptible. The other varieties such as RH-9901, RH-8812, RC-781, and Purple Mutant contracted the powdery mildew

Table 6.7 Components of resistance to powdery mildew on mustard cultivars/varieties (Singh 2004)

Cultivars/ varieties	Disease intensity (%)	No. of specks/leaf	No. of conidia/ microscopic field*	Incubation period (days)	Latent period (days)
RH-30	46.20 (42.82)	31.16	83.75	3	2
RH-8812	33.60 (35.43)	18.33	50.08	3	1
RH-9304	47.80 (43.74)	33.97	75.08	4	2
RH-9801	49.50 (44.71)	39.40	81.66	4	2
RH-9901	38.90 (38.59)	21.08	72.75	3	1
RC-781	43.40 (41.21)	27.30	41.58	3	1
Purple Mutant	36.70 (37.29)	21.67	62.58	3	1
GSL-1	4.10 (11.68)	5.52	10.00	3	3
HC-9603	–	–	–	–	–
C.D. at 5%	**8.45**	**7.85**	**17.77**		

Figures in parentheses are angular transformed values
*Five speck suspended in 1 ml of water
– Disease did not appear

specks ranging between 10 and 30 specks/leaf and considered as moderately susceptible. The disease intensity was recorded on all the varieties except HC-9603 on which disease did not appear till the end of the crop season to test the behaviour of varieties against powdery mildew. The maximum disease (49.50%) was recorded on the variety RH-9801 followed by the variety RH-9304 (47.80%) and RH-30 (46.20%) though statistically at par. The minimum disease (4.10%) was observed on the variety GSL-1, whereas in other varieties such as RH-8812, RH-9901, Purple Mutant, and RC-781, the disease intensity ranged between 33.60 and 43.40%. There were no significant differences in all the varieties in relation to disease intensity except GSL-1 and HC-9603 which appeared as resistant to powdery mildew disease and others as susceptible (Table 6.7).

The number of conidia produced in each speck by *E. cruciferarum* on different cultivars/varieties of mustard was examined under the compound microscope (10 × 10x). The minimum number of conidia/speck was recorded on the variety GSL-1 (10 conidia/speck). It was followed by the variety RC-781 where it was 41.58 conidia/speck. In other varieties such as RH-30, RH-9801, RH-9304, RH-8812, RH-9901, and Purple Mutant, the conidial production ranged between 50.08 and 83.75 conidia per speck being maximum on the variety RH-30 (83.75) and minimum on RH-8812 (50.08). It was revealed that the considerable amount of conidia per speck was produced in all the susceptible varieties belonging to *B. juncea* except GSL-1 and HC-9603 which belongs to *B. napus* and *B. carinata,* respectively (Table 6.7).

The progression of the powdery mildew on all the nine varieties was recorded from the appearance of the disease till the maturity of the leaves on ten randomly selected marked tagged leaves from each replication. The disease appeared on all the varieties in the first week of March except on HC-9603 (disease did not appear). The minimum speck size was recorded in the variety GSL-1 (1.98 mm) whereas maximum in the variety RH-30 (5.80 mm). It was followed by the variety RH-9901 (5.25 mm) and RH-9304 (4.98 mm), whereas on other varieties, the speck size ranged between 3.81 and 4.81 mm which is moderate (Table 6.8). The progression of powdery mildew on different varieties/cultivars of mustard presented in Fig. 6.15 revealed that the progression of powdery mildew was maximum up to mid of March; after that the disease was slowed down. The minimum progression was recorded in the variety GSL-1 where it was almost static after initiation of the disease. Similarly, on Purple Mutant variety also, the disease progression was slow as compared to the other varieties (Meena et al. 2018).

The progression of the powdery mildew in relation to weather variables was evaluated which revealed the maximum R^2 value, i.e. 0.91 was on variety RH-30 followed by RH-9901 (0.86), Purple Mutant (0.86), RH-9801 (0.83), and RH-8812 (0.83). The minimum value (R^2 = 0.47) was recorded in variety GSL-1 which indicated that in addition to weather variables included here, other factors such as varietal resistance and some unknown factors have significant role in the disease development (Table 6.5). The varieties GSL-1 and HC-9603 appeared as resistant to powdery mildew with the expression of slow mildewing components, whereas other

Table 6.8 Progression of powdery mildew (speck size, mm) on different cultivars/varieties of mustard in relation to slow mildewing assessment (Singh 2004; Meena et al. 2018)

Date of observations	Speck size on various cultivars/varieties of mustard								
	RH-30	RH-8812	RH-9304	RH-9801	RH-9901	RC-781	Purple Mutant	GSL-1	HC-9603
9–3-04	0.98∗	0.94	1.06	1.03	1.11	0.98	0.98	1.15	–
11–3-04	1.81 (0.83)	1.83 (0.89)	1.88 (0.82)	1.90 (0.87)	2.11 (1.0)	1.90 (0.92)	1.56 (0.58)	1.58 (0.43)	–
13–3-04	3.16 (1.35)	3.03 (1.20)	3.11 (1.23)	3.23 (1.33)	3.31 (1.20)	3.13 (1.23)	2.60 (1.04)	1.98 (0.40)	–
15–3-04	4.75 (1.59)	4.06 (1.03)	4.55 (1.44)	4.55 (1.32)	4.65 (1.34)	4.33 (1.20)	3.36 (0.36)	1.98 (0.0)	–
17–3-04	5.70 (0.95)	4.35 (0.29)	4.93 (0.38)	4.81 (0.26)	5.21 (0.56)	4.46 (0.13)	3.81 (0.45)	1.98 (0.0)	–
19–3-04	5.80 (0.10)	4.43 (0.08)	4.98 (0.05)	4.81 (0.0)	5.25 (0.04)	4.48 (0.02)	3.81 (0.0)	1.98 (0.0)	–
21–3-04	5.80 (0.0)	4.43 (0.0)	4.98 (0.0)	4.81 (0.0)	5.25 (0.0)	4.48 (0.0)	3.81 (0.0)	1.98 (0.0)	–

Figures in parentheses are the periodical progression
∗Size of specks (mm)
–Disease did not appear

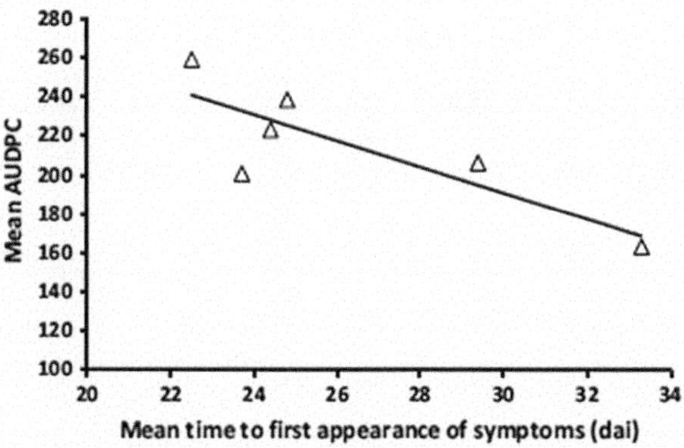

Fig. 6.15 Relationship between the time the first visible symptoms of powdery mildew (*Erysiphe cruciferarum*) infection appeared (days after inoculation, dai) and area under the disease progress curve (AUDPC) at 83 dai (average for six *Brassica napus* cultivars Banjo, Karoo, Lantern, Thunder-TT, Tribune, and Trilogy) (Uloth et al. 2017)

varieties belonging to *B. juncea* group appeared as susceptible to the disease showing faster powdery mildew development under field conditions.

The correlation matrix for progression of the powdery mildew in relation to weather variables on all the test varieties/cultivars was analysed. It was observed that Temp. Max. (X_1) had significant and positive role in the progression of powdery mildew on all the varieties/cultivars except GSL-1 where it was positive but non-significant. Similarly, Avp. M (X_5) also had significant and positive role in the disease progression on all the varieties/cultivars except HC-9603 (disease did not appear). The RHE (X_4) has negative and significant correlation in the disease development on all the varieties/cultivars except GSL-1 where it was negative but non-significant. Similarly, Sunshine (X_8) has negative and significant correlation in all the varieties/cultivars. Other weather variables such as Temp. Min. (X_2), RHM (X_3), and Avp. E (X_6) have positive but non-significant correlation in disease progression on all the varieties/cultivars (Table 6.9).

The Progression of powdery mildew on different varieties/cultivars of mustard presented in Fig. 6.13 revealed that it was maximum up to mid of March; after that the progression was slowed down. The minimum progression was recorded in the variety GSL-1 where it was about static after initial progression. Similarly, in Purple Mutant variety also, the progression was slow as compared to the other varieties. Prediction and observed values on the progression of powdery mildew on different varieties/cultivars of mustard are very close to each other (Fig. 6.16a, b). It may be a very good indicator to develop powdery mildew forecasting models.

Table 6.9 Correlation matrix between powdery mildew (speck size, mm) and weather parameters in mustard cultivars/varieties (Singh 2004; Meena et al. 2018)

	Correlation matrix of powdery mildew and weather variables on various cultivars/varieties								
Weather variables	RH-30	RH-8812	RH-9304	RH-9801	RH-9901	RC-781	Purple Mutant	GSL-1	HC-9603
Temperature maximum (X_1)	0.95*	0.90*	0.92*	0.90*	0.92*	0.88*	0.91*	0.68	—
Temperature minimum (X_2)	0.41	0.46	0.42	0.43	0.45	0.45	0.43	0.62	—
Relative humidity morning (X_3)	0.56	0.59	0.59	0.61	0.57	0.61	0.60	0.54	—
Relative humidity evening (X_4)	−0.85*	−0.83*	−0.84*	−0.83*	−0.83*	−0.82*	−0.84*	−0.68	—
Average evaporation morning (X_5)	0.82*	0.84*	0.82*	0.82*	0.84*	0.83*	0.83*	0.88*	—
Average evaporation evening (X_6)	0.28	0.25	0.26	0.26	0.28	0.25	0.27	0.20	—
Wind speed (X_7)	−0.63	−0.63	−0.64	−0.64	−0.62	−0.63	−0.64	−0.48	—
Sunshine (X_8)	−0.72	−0.79*	−0.78*	−0.79*	−0.77*	−0.81*	−0.75*	−0.80*	—

*Significant at 5% ($P = 0.05$)

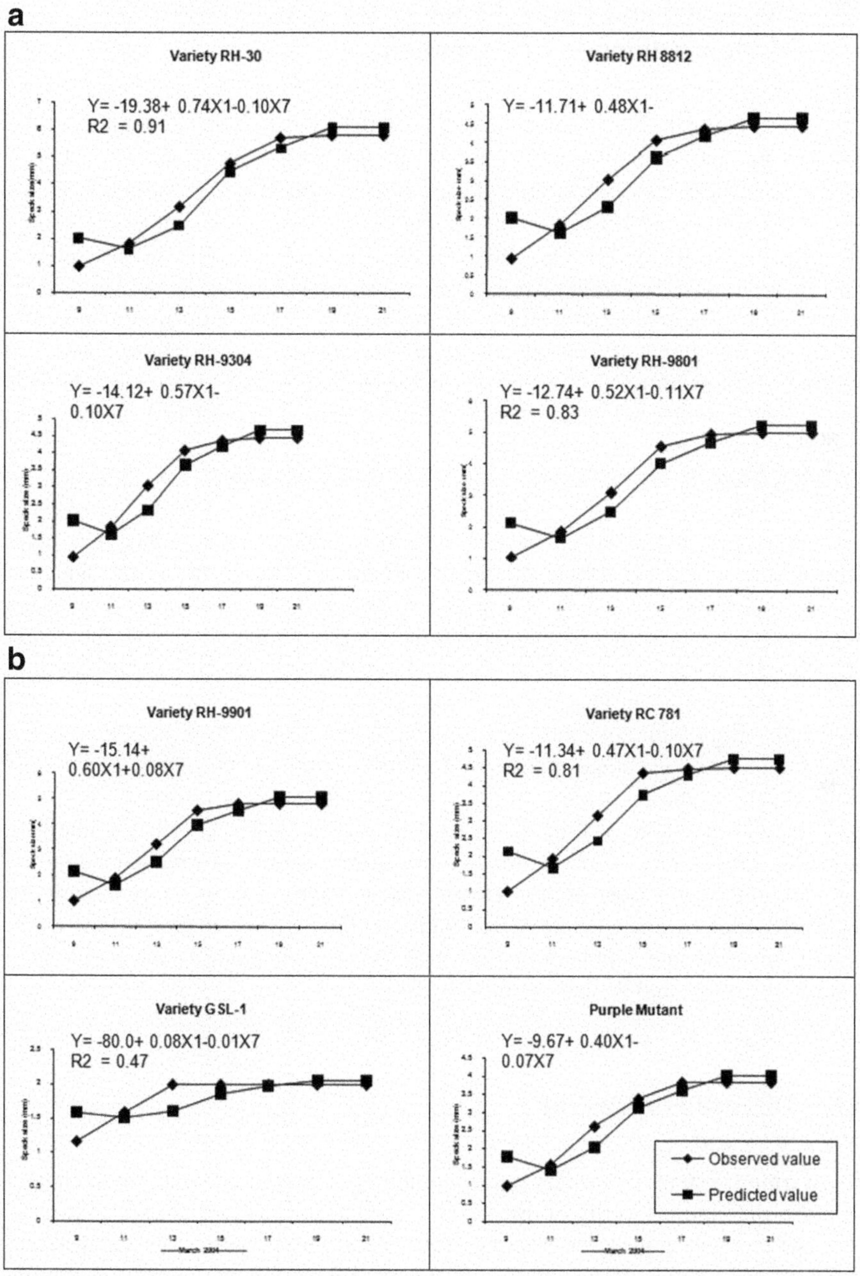

Fig. 6.16 (**a**) Effect of weather variables on the progression of powdery mildew (predicted/observed) on different slow mildewing cultivars/varieties of mustard (Singh 2004). (**b**) Effect of weather variables on the progression of powdery mildew (predicted/observed) on different slow mildewing cultivars/varieties of mustard (Singh 2004)

6.7 Congenial and Critical Factors for Crucifers Powdery Mildew Epidemic Development

There are a number of environmental factors which are very crucial to influence the powdery mildew development of crucifers into epidemic form after host–pathogen interaction. These factors determine the progress of powdery mildew on host plants with their influence and effects on interacting partners, host, and pathogen. It causes the infection in susceptible host after landing of pathogen conidia on host surface, their germination and formation of appressoria are maximum between 15 and 20 °C temperatures. Their germination is greatly reduced at >30 °C temperature. Infection and disease development is faster with the influence of mean temperature (16–22 °C), minimum temperature (>7 °C), maximum temperature (25–28 °C), relative humidity (27–65%), sunshine hours (>9 h/day), wind velocity (2 km/h), and ageing of the host plants. Infection rate is positively favoured by host age and ambient temperature. Infection rate increases with ageing host tissues. There is no infection on younger than 37 days host and freshly emerging new leaves. Disease develops at fast rate if host and pathogen interact coinciding with favourable host age, plant growth stages, and environmental factors. Stem infection is maximum with the increase in length of time they are exposed to the pathogen and maturity level of the host. Symptoms are visible at anamorph state or asexual stage with the development of pathogens mycelium, conidiophores, and conidia on host surface. Date of crop planting has significant bearing on disease epidemiology under late-sown conditions coinciding with congenial and critical factors at 40–120 days after sowing. Teleomorph or sexual stage appears in the form of dark brown spherical bodies of cleistothecia or chasmothecia embedded in powdery mass of host leaf, stem, and pods at maturity stage of crop when temperature is 11–27 °C (19 °C), alternate moderate temperature, heavy sporulation, low host nutrition, low relative humidity, dry soil, and ageing host tissues. Host resistance and progression of disease are measured using parameters like AUDPC, disease intensity, incubation period, latent period, infection rate, number of colony/leaf, number of conidia/ microscopic field (sporulation rate), and R^2 values (Table 6.10).

6.8 Disease Forecasting

Forewarning system of powdery mildew in mustard crop based on different characters, viz. maximum severity of disease (Y_1), crop age at first appearance of disease (Y_2), and crop age at peak severity of disease (Y_3), for two varieties sown at different locations has been developed by Kumar et al. (2013), using neural network models with MLP and RBF architectures. Weather data from 1999–2000 to 2005–2006 were considered for models development. The data for subsequent years were used for testing the models. The developed weather indices were used as input variables, and various characters were used as output in the neural network models. The avail-

Table 6.10 Congenial and critical factors for crucifers powdery mildew epidemics*

Sr. No.	Determinant characters	Influencing factors/effects
1.	Conidial germination and appressoria formation	Maximum between 15 and 30 °C temperature
2.	Conidial survival	Reduced at >30 °C temperature
3.	Infection and disease development	Faster at mean temp. 16–22 °C, minimum temp. >7 °C, maximum temp. 25–28 °C, relative humidity 27–65%,sunshine >9 h/day, wind velocity 2 km/h, ageing host plant
4.	Infection rate	Favoured by host age, ambient temperature, no infection on younger than 37 days host and fresh new leaves. Infection rate increases with ageing host tissues
5.	Interaction	Host age/growth stage/environmental factors – faster disease development
6.	Stem infection	Maximum with length of time exposure and maturity level of host
7.	Anamorph state	Mycelium, conidiophores, and conidia show symptoms on host
8.	Date of planting	Maximum disease on late-sown crop coinciding with congenial and critical factors, 40–120 das
9.	Cleistothecial/ teleomorph state	Temp. 11–27 °C (19 °C), alternate low and moderate temp., heavy sporulation, low host nutrition, low RH, dry soil, host ageing, visible as dark spherical small bodies
10.	AUDPC	Measures of disease progression and host resistance
11.	Disease intensity	-do-
12.	Incubation period	-do
13.	Latent period	-do
14.	Infection rate	-do
15.	Number of specks or colony/leaf	-do
16.	Number of conidia/ microscopic field	-do
17.	R^2 values	-do

*Based on information provided by Saharan and Sheoran (1988); Saharan and Kaushik (1981); Singh (2004); Desai et al. (2004); Singh et al. (2008); Uloth et al. (2017)

able data set was divided into three sets, namely, training, validation, and testing set. The details of data sets are given in Table 6.11.

6.8.1 *Weather Indices-Based Regression Models*

The models developed indices were used in developing forecast models through regression approach. The form of the model was developed as per formula of Agrawal and Mehta (2007) as mentioned below:

Table 6.11 Weather-based forecasting models for powdery mildew of mustard (Kumar et al. 2013)

Character	Variety	Training set	Validation set	Testing set
Maximum severity (*Y*1)	Varuna	45	10	10
Age at first app (*Y*2)		45	10	10
Age at peak severity (*Y*3)		45	10	10
Maximum severity (*Y*1)	GM-2	45	10	10
Age at first app (*Y*2)		45	10	10
Age at peak severity (*Y*3)		45	10	10

Number of data points in different sets for various characters in two varieties of mustard at S.K. Nagar

$$Y = a_0 + \sum_{i=1}^{P} \sum_{j=0}^{1} \text{aij Zif} + \sum_{i \neq 1}^{P} \sum_{j=0}^{1} \text{bii j Zii j} + \varepsilon$$

where *Y* is variable to forecast; a_0, a_{ij}, $b_{ii\,j}$ are constants; and ε is error term, and other symbols have same meaning as explained earlier. Stepwise regression technique was used for selecting important variables to be included in the model.

6.8.2 *Multilayer Perception (MLP) and Radial Basis Function (RBF) Architecture-Based Neural Network Models (NNM)*

Neural network models using MLP architecture with different hidden layers and different number of neurons in a hidden layer with hyperbolic function as an activation function with varying learning rates and RBF architecture were obtained, and best architecture was selected having lowest mean absolute percentage error (MAPE).

6.8.3 *Performance Measure*

The forecasting performance of various artificial neural network (ANN) models and regression models was judged by mean absolute percentage error (MAPE):

$$\text{MAPE} = {}^{1}\!/_{\text{n}} \Sigma \frac{Yt - Ft}{\text{Yt}} \times 100$$

where *Yt* is actual observation, Ft is the forecast from model, and *n* is the total number of test data point.

Table 6.12 Models to forecast different characters of powdery mildew in mustard crop along with coefficient of determination in two varieties of mustard (Kumar et al. 2013)

Variety	Character	Model	R^2
Varuna	*Y1*	$Y = 133.40 + 0.11\ Z_{120} + 12.71\ Z_{21}$	0.84
	Y2	$Y = 59.52 + 0.02\ Z_{120} – 0.01\ Z_{241} + 0.06\ Z_{351}$	0.84
	Y3	$Y = 126.35 + 1.17\ Z_{41} – 0.04\ Z_{141}$	0.65
GM-2	*Y1*	$Y = -52.46 + 0.06\ Z_{121} – 0.08\ Z_{131} + 1.23\ Z_{31}$	0.56
	Y2	$Y = 53.12 + 0.16\ Z_{251} – 0.01\ Z_{241} + 0.003\ Z_{130}$	0.85
	Y3	$Y = 106.27 + 0.22\ Z_{11} + 0.005\ Z_{341}$	0.69

Weather indices (WI)-based regression models were developed for various characters, and models have been validated using data on subsequent years not included in developing the models. The analysis has been done by using SAS (Statistical Analysis System) Version 9.2 software package available at Indian Agricultural Statistics Research Institute, New Delhi. The models are given in Table 6.12. Neural network models using MLP architecture with different hidden layers (one and two) and different number of neurons (4, 5, and 6) in a hidden layer with hyperbolic function as an activation function with varying learning rates (from 0.3 to 0.8) and RBF architecture were obtained, and best architecture was selected having lowest mean absolute percentage error (MAPE). The analysis has been done by using Statistical Neural Networks Version 6.1 available at Indian Agricultural Statistics Research Institute, New Delhi. The mean absolute percentage error (MAPE) for different characters of powdery mildew in mustard crop in two varieties for various developed models is presented in Table 6.13. This table reveals that the neural network models using MLP have lowest MAPE as compared to other developed models in most of the cases. The ANN model has non-linear pattern recognition capability which is valuable for modeling and forecasting complex non-linear problems in practice. Kumar et al. (2013) found that neural network model using multilayer perception (MLP) architecture is better than RBF and weather indices-based regression models in terms of MAPE. Therefore, reliable forewarning system indicates for maximum severity of disease, at crop age at first appearance of disease; crop age at peak severity of disease on two different varieties of mustard crop for powdery mildew is possible well in advance (Kumar et al. 2013; Table 6.13).

Powdery mildew prediction model based on crop age and weather variables have been devised by Desai et al. (2004). Crop age at first appearance of the mildew on the crop (Y_x), crop age at highest severity of the mildew (Y_y), and peak disease severity (Y_z) were related with weather variables in different weeks including presowing week, and the interactions were found significant. The regional and cultivar specific models devised using data of initial 4 years thereby could predict the crop age at which powdery mildew first appears on the crop (Table 6.14), crop age at highest mildew severity (Table 6.15), and the peak disease severity (Table 6.16). The predictions were possible at least 3 weeks ahead of first appearance of the disease on the crop, thus allowing growers to undertake timely fungicidal sprays. The disease was never found to appear before 50 das or 8th week after sowing, while the

Table 6.13 Mean absolute percentage error (MAPE) of various models for powdery mildew of mustard (Kumar et al. 2013)

Character	Variety	MLP	RBF	WI
Maximum severity (*Y*1)	Varuna	26.5	56.2	35.1
Age at First app (*Y*2)		12.2	15.1	20.2
Age at Peak Severity (*Y*3)		9.5	13.5	12.8
Maximum severity (*Y*1)	GM-2	6.7	50.2	64.7
Age at First app (*Y*2)		21.4	15.3	21.7
Age at Peak Severity (*Y*3)		12.8	14.1	13.8

MLP multilayer perception, *RBF* radial basis function

Table 6.14 Models to forecast crop age (Yx) at first appearance of powdery mildew on Indian mustard (Desai et al. 2004)

Location	Cultivar	Crop age (week) of prediction	Model	R^2
S. K. Nagar	Varuna	5	Yx = −80.99 + 0.93 Z max-temp +2.03 Z ssh	0.99
S.K. Nagar	GM-2	3	Yx = 4.11 + 1.67 Z min-temp	0.98

max-temp maximum temperature, *min-temp* minimum temperature, *ssh* sunshine hours

Table 6.15 Models to forecast crop age (Y_y) at highest severity of powdery mildew on Indian mustard (Desai et al. 2004)

Location	Cultivar	Crop age (week) of prediction	Model	R^2
S. K. Nagar	Varuna	4	Yy = 123.11 + 0.029 Z aft-r. h. × ssh + 1.33Zssh	0.92
S.K. Nagar	GM-2	3	Yy = −124.38 + 0.04 Z min-temp × aft-r. h. + 0.53Zmorn-r. h.	0.91
Bharatpur	Varuna	4	Yy = 2.34 + 0.96 Z min-temp	0.99
Bharatpur	PCR-7	2	Yy = −10.46 + 4.88 Z min-temp + 0.11 Z max-temp × min-temp	0.98

max-temp maximum temperature, *min-temp* minimum temperature, *morn* morning, *aft* afternoon, *r. h.* relative humidity, *ssh* sunshine hours

Table 6.16 Models to forecast highest severity (Yz) of powdery mildew on Indian mustard (Desai et al. 2004)

Location	Cultivar	Crop age (week) of prediction	Model	R^2
S.K. Nagar	Varuna	5	Yz = 12.58 + 0.01 Z max-temp x ssh	0.96
S.K. Nagar	GM-2	3	Yz = 4.72 + 0.01 Z min-temp x ssh +0.02 Z morn-r. h.	0.98

max-temp = maximum temperature, *min-temp* = minimum temperature, *morn-r. h.* = relative humidity, *ssh* = sunshine hours prediction for crop age at first appearance of mildew was possible for both the locations in the beginning of 5th week (Desai et al. 2004).

6.8.4 The Forms of Two Indices

$$\text{Ziij} = \sum_{w=n1}^{n2} \text{iiwXiwXiw}^{\text{rj}}$$

X_{iw} is value of ith weather variables in wth week.
R_{iw} is correlation coefficient between variable to forecast and ith weather variable in wth week.
R_{iiw} is correlation coefficient between variable to forecast and product of X_i and X_i in wth week.
p is number of weather variable.
n_1 is the initial week for which weather data are included in the model.
n_2 is the final week for which weather data are included in the model.

Using these weather indices as independent variables and variable to forecast as dependent variable, two types of neural network architecture, namely, multilayer perception (MLP) and radial basis function (RBF) were attempted and compared with weather indices-based regression model.

To develop the prediction model of powdery mildew development on different cvs. of oilseed *Brassica*, Singh (2004) compared the observed values of powdery mildew development on nine varieties/cultivars with predicted values. The varieties of *B. juncea* such as RH-9901, RH-9801, RH-9304, RH-8812, RH-30, and RC-781 showed faster disease progression, whereas the variety GSL-1 showed slow disease progression after the initiation of the disease. Slightly slow disease progression was also observed in the variety Purple Mutant. Almost similar trends were observed in the case of observed values and predicted values of powdery mildew progression on all the varieties/ cultivars of mustard (Fig. 6.16a, b).

In Australia, powdery mildew of oilseed rape (*B. napus*) is of sporadic nature in most southern areas, whereas epidemics are most severe in the two warmer cropping regions, viz. the northern agricultural region of Western Australia and New South Wales. With the rise in winter temperatures during future climate change, earlier and more severe powdery mildew outbreaks in Australia are predicted (Uloth et al. 2017). The extent of powdery mildew infection on oilseed rape is determined by plant age at which infection occurs and the ambient temperature. Infection by powdery mildew pathogen increases as the host plant ages or matures ((Fig. 6.3); Saharan and Kaushik 1981; Uloth et al. 2017). In the case of *B. napus*, the age of the plant is critical in determining when the powdery mildew disease epidemic peaks. Powdery mildew infection on the stem is also affected by the age at which it gets exposure to the pathogen, but in contrast to leaf infection, the oldest plants' stems have the lowest level of infection. Final disease severity is significantly reduced at the lower temperature, and it is associated with reduced growth of the germ tube of the pathogen along with reduction in spore viability. In oilseed *Brassicas*, increase in temperature and plant age both contribute towards increase in disease intensity late in the growing period (Uloth et al. 2017).

Generally, young plant tissues of *Arabidopsis* under artificial inoculation have shown more susceptibility than ageing tissues to *E. cruciferarum*. However, sporulation on *Arabidopsis* is sparse, and macroscopically visible disease symptoms are absent. From a qualitative point of view, the inoculated plants are susceptible to the powdery mildew pathogen (Koch and Slusarenko 1990). Conidial state of this fungus is often thin and rather inconspicuous on *Arabidopsis*, whereas it covers the other hosts completely and conidia are formed abundantly (Junell 1967).

References

Agrawal R, Mehta SC (2007) Weather based forecasting of crop yields, pests and diseases – IASRI models. J Indian Soc Agril Stat 61:255–263

Alkooranee JT, Yin Y, Aledan TR, Jiang Y, Lu G, Wu J, Li M (2015a) Systemic Resistance to Powdery Mildew in *Brassica napus* (AACC) and *Raphanus alboglabra* (RRCC) by *Trichoderma harzianum* TH12. PLoS One 10(11):0142177

Alkooranee JT, Liu S, Aledan TR, Yin Y, Li M (2015b) First report of powdery mildew caused by *Erysiphe cruciferarum* on *Brassica napus* in China. Plant Dis 99(11):1651

Asalf B, Gadoury DM, Tronsmo AM, Seem RC, Dobson A, Peres NA, Stensvand A (2014) Ontogenic resistance of leaves and fruit, and how leaf folding influences the distribution of powdery mildew on strawberry plants colonized by *Podosphaera aphanis*. Phytopathology 104:954–963

Baurle I, Dean C (2006) The timing of developmental transitions in plants. Cell 125:655–664

Cole JS (1966) Powdery mildew of tobacco (*Erysiphe cichoracearum* DC): susceptibility of proximal and distal parts of leaves from different stalk positions on intact and topped field plants in relation to free amino nitrogen and carbohydrate content. Ann Appl Biol 58:61–69

Crowton OWB, Kennedy R (1999) Effects of humidity and wetness duration on the germination and infection of *Erysiphe cruciferarum.* In: Proceedings of the 1st Intel powdery mildew conference, 29 August–3 September, 1999, Abstr, Avignon, France, 1999

Desai A, Chattopadhyay C, Agrawal R, Kumar A, Meena RL, Meena PD, Sharma KC, Srinivasa Rao M, Prasad YG, Ramakrishna YS (2004) *Brassica juncea* powdery mildew epidemiology and weather-based forecasting models for India – a case study. J Plant Dis Prot 111:429–438

Jhooty JS (1967) Identity of powdery mildew of cucurbits in India. Plant Dis Reptr 51:1079–1080

Junell L (1967) Erysiphaceae of Sweden. Symbolae Botanicae Upsalienses 19:1–117

Kanzaria KK, Dhruj IU, Sahu DD (2013) Influence of weather parameters on powdery mildew disease of mustard under North Saurashtra agroclimatic zone. J Agromet 15(1):86–88

Kennedy R (2010) Brassicas: forecasting light leaf spot and powdery mildew in vegetable Brassica crops based on 'in field' detection of airborne spores. In: Annual Report, Agriculture and Horticulture Development Board, UK. FV333. University of Worcester, Worcester

Koch E, Slusarenko AJ (1990) Fungal pathogens of *Arabidopsis thaliana* (L.) Heyhn. Bot Helv 100:257–268

Kolte SJ (1985) Diseases of annual edible oilseed crops, Vol. II Rapeseed-mustard and sesame diseases. CRC Press Inc, Boca Raton, p 135

Kumar A, Agrawal R, Chattopadhyay C (2013) Weather based forecast models for diseases in mustard crop. Mausam 64(4):663–670

Meena PD, Mehta N, Rai PK, Saharan GS (2018) Geographical distribution of rapeseed-mustard powdery mildew disease in India. J Mycol Pl Pathol 48(3):284–302

Parry R (1990) The occurrence of powdery mildew on rutabagas in southern Ontario. Canad Plant Dis Surv 70:15–16

Poethig RS (2003) Phase change and the regulation of developmental timing in plants. Science 301:334–336

Price TV (1970) Epidemiology and control of powdery mildew (*Sphaerotheca pannosa*) on roses. Ann Appl Biol 65 (2:231–248

Rudgard SA, Wheeler BEJ (1985) Relationships between establishment of *Erysiphe cruciferarum* and soluble amino-acids in leaves of Brussels sprouts. Plant Pathol 34:609–615

Saharan GS (1970) Powdery mildews and their epidemiology at Jobner (Rajasthan). Pesticides 4(10):16–25

Saharan GS (1980) Epidemiology of powdery mildew of rape and mustard. Indian J Mycol Plant Pathol (Abstr) 10:XLVI

Saharan GS (2016) ICAR AICRPRM QRT visit of Birsa Agricultural University Ranchi, Jharkhand during 2016 crop season

Saharan GS, Kaushik JC (1981) Occurrence and epidemiology of powdery mildew of *Brassica*. Indian Phytopathol 35:17–21

Saharan GS, Sheoran BS (1988) Conidial germination, germ tube elongation and appressorium formation of *Erysiphe cruciferarum*. Indian Phytopathol 41(1):157–159

Shivpuri A, Chipa HP, Gupta RBL, Sharma KN (1997) Field evaluation of white rust, powdery mildew and stem rot. Ann Arid Zone 36:387–389

Singh RS (1984) Introduction to principles of plant pathology, 3rd edn. Oxford and IBH Publications, New Delhi, p 608

Singh K (2004) Studies on the ecofriendly management of powdery mildew (*Erysiphe cruciferarum* Opiz ex. Junell) of mustard [*Brassica juncea* (Linn.) Czern & Coss]. M. Sc Thesis, CCS HAU, Hisar 108 + xix

Singh K, Mehta N, Sangwan MS (2008) Influence of weather parameters on the progression of powdery mildew on four varieties of rapeseed-mustard in Haryana. Plant Dis Res 23(2):39–45

Sylvester-Bradley R, Makepeace RJ (1984) A code for stages of development in oilseed rape (*Brassica napus* L.). Aspects Appl Biol 6:399–419

Uloth MB, You MP, Barbetti MJ (2017) Plant age and ambient temperature: significant drivers for powdery mildew (*Erysiphe cruciferarum*) epidemics on oilseed rape (*Brassica napus*). Plant Pathol. https://doi.org/10.1111/ppa12740

Whalen MC (2005) Host defence in a developmental context. Mol Plant Pathol 6:347–360

Yarwood CE (1936) The tolerance of *Erysiphe polygoni* and certain other powdery mildews to low humidity. Phytopathology 26:845–849

Yarwood CE (1939) Control of powdery mildews with a water spray. Phytopathology 29:288–290

Yarwood CE (1957) Powdery mildews. Bot Rev. XXIII:235–301

Chapter 7
Host Resistance

7.1 Introduction

On host surface, the first line of defense of plants against pathogen is their surface which the pathogen has to penetrate to cause infection. Some structural defense barriers are present on the host surface even before the pathogens come in contact with the host plants. Such structures include the amount, structure, and quality of wax and cuticle that covers the epidermal cells; the structure of the epidermal cell walls; the size, location, and shape of stomata and lenticel number; the length and density of hairs on leaf surface; and the presence on the plant of tissues made up of thick-walled cells that hinder the entry and advance of the pathogens. The second line of defense or resistance after penetration of the pathogen involves the formation of histological defense structures, cellular defense structures, and cytological defense reactions in the form of hypersensitive response (HR). To protect themselves against pathogens, plants have evolved intricate immune responses that include accumulation of reactive oxygen species (ROS), deposition of callose, formation of papilla, phenolic compounds, enhanced expression of defense-related genes, and biosynthesis of phytoalexins. In crucifers, there are two pathways through which defense genes are activated. One pathway requires salicylic acid (SA) as a signaling intermediary, and the other requires the simultaneous perception of ethylene and jasmonic acid (JA). Numerous mutations that affect signaling through these pathways have been identified, and some of the corresponding genes have been cloned. In *Brassica* crops, resistance to powdery mildew is controlled by a single dominant gene with modifiers. In *Arabidopsis*, resistance to powdery mildew is either polygenic, based on the atypical resistance gene, RPW8, or in combinations thereof, with RPW8 representing the major quantitative trait locus. In general, more accessions are resistant to *E. cruciferarum* (76%) than to *G. cichoracearum* (63%) and *G. orontii* (26%). However, as in the case of most R genes, RPW8-mediated resistance is associated with the expression of pathogenesis-related (PR) genes and triggers a hypersensitive response (HR) evidenced by whole-cell callose deposition, H_2O_2 accumulation, and

G. S. Saharan et al., *Powdery Mildew Disease of Crucifers: Biology, Ecology and Disease Management*, https://doi.org/10.1007/978-981-13-9853-7_7

cell death in response to pathogen attack. Since 1990, the *Arabidopsis*–powdery mildew pathosystem has become a popular model to study molecular processes governing powdery mildew infection process to answer how biotrophic pathogens overcome plant defense system. The mechanisms of host defense against powdery mildew of crucifers operate at both pre- and post-penetration stages. Recently, molecular and genetical mechanisms of host resistance in crucifers to powdery mildew have been dissected using adapted and non-adapted powdery mildew pathogens through *Arabidopsis*–powdery mildew host pathosystem. Loss-of-function mutant alleles of mildew resistance locus (MLO) encoding seven-trans-membrane, calmodulin-binding protein confer a broad-spectrum resistance towards adapted powdery mildew of *Arabidopsis*. Mechanisms of non-host resistance in crucifers to powdery mildew have also been unravelled. Non-host resistance is strong, and durable, and bears the potential for targeted employment of this valuable trait to manage the host pathogens. However, because non-host resistance operates at the plant–species level, it often has been beyond the reach of breeding programmes that are based on allelic combinations within a given crop species.

7.2 Structural and Functional Components of Host Resistance

In *Arabidopsis* and other crucifer species, infected cells react to penetration attempts through the formation of cell wall apposition (CWAs) (Plate 4.1, Chap. 4), which (despite sparse experimental evidence supporting this view) are commonly thought to provide both physical cell wall reinforcements and a chemical antimicrobial blockade against the invading pathogen (Hardham et al. 2007; Huckelhoven 2007). Papillae are formed within a few hours after fungal penetration attempts in response to adapted and non-adapted pathogens. Remarkably, physical damage (needle puncture) can also induce the formation of papilla-like structures, which suggests that it is the pathogen-triggered wounding of plant cells that prompts the formation of CWAs (Aist 1976). However, detailed analysis of the structures formed in response to non-biotic, mechanical wounds revealed that they are distinct in their composition and likely in their function from fungus-induced CWAs (Russo and Bushnell 1989). In addition, no papilla formation was detected in response to mechanical (needle) stimulation of parsley suspension culture cells, despite detection of other responses associated with pathogen recognition such as reactive oxygen intermediate (ROI) production and induction of several elicitor-response genes (Gus-Meyer et al. 1998). Although there is no absolute association between papilla formation and resistance to fungal penetration, delayed formation of CWAs is correlated with enhanced fungal infection (Assaad et al. 2004). CWAs are composed of an apparently amorphous mixture of cellulose, pectin, callose, lignins, phenolics, silicon, H_2O_2, and derivatives as well as dedicated de novo-synthesized antimicrobial metabolites (phytoalexins) and fungal enzyme inhibitors (Plate 4.1,

Chap. 4). Some of these compounds are delivered within secretory vesicles, while others are synthesized on the spot by cell wall-resident enzymes (Huckelhoven 2007). Each component of the CWA likely plays a role in defense. Structural components such as pectin, callose, and cellulose are thought to provide physical reinforcement to the plant cell wall at the site of attack (Zeyen et al. 2002). In addition, released moieties of these carbohydrate polymers as well as enzymes involved in their synthesis have potent signaling capabilities and are thought to be involved in the modulation of downstream defense responses. Callose synthase *GSL5/PMR4* protein (and/or callose itself) appears to be a negative regulator of SA-dependent defense responses, and mutations in this gene render *Arabidopsis* resistant to *G. orontii* and *G. cichoracearum* (Jacobs et al. 2003; Nishimura et al. 2003). Silicon (Si) is another component found in papillae (Plate 4.1, Chap. 4) that was long thought to provide structural reinforcement to the cell wall (Zeyen et al. 2002). More recently, a comparative transcriptome analysis of *Arabidopsis* plants either supplied with or lacking Si revealed that silicon modulates gene expression in a pathogen-dependent manner (Fauteux et al. 2006). Specifically, it appears to dampen general stress responses associated with pathogen infection and partially restores expression of genes encoding components of primary metabolism to normal conditions without affecting the expression of defense-related genes. Reduction of biotic stress responses results in reduced powdery mildew density and conidiation on leaves, effects that had been historically reported for a variety of other plant species treated with Si (Ghanmi et al. 2004). To complete defense at the cell wall, pathogen-induced antimicrobial compounds (phytoalexins) are actively delivered to the site of fungal contact. Components of the basal defense machinery such as the PEN2 glycosyl hydrolase (associated with peroxisomes; Plate 4.1, Chap. 4) and the plasma membrane-resident *PEN3* ABC transporter (Plate 4.1, Chap. 4) are thought to cooperate and generate toxic antimicrobial compounds that are pumped at the site of attack (Lipka et al. 2005; Stein et al. 2006).

The delivery of cell wall components and antimicrobial compounds at CWAs is thought to occur at least in part in membrane-bound vesicles that move along tracks of actin microfilaments (Plate 4.1, Chap. 4). This view is supported by the observation of massive cytoskeletal reorganization of plant cells upon pathogen infection. Reminiscent of findings in other plant–powdery mildew interactions (Kobayashi et al. 1997a; Opalski et al. 2005), fluorescence microscopy of GFP-tagged *Arabidopsis* cytoskeletal proteins revealed a rapid concentration of actin arrays towards the site of attack by the non-adapted fungus *B. graminis* f. sp. *hordei*. This rearrangement was accompanied by a mobilization of the nucleus, and endoplasmic reticulum to the site of the emergent penetration peg, and later towards the developing haustorium (Takemoto et al. 2006). Golgi-derived bodies circulate and make frequent stops below the incipient penetration peg, consistent with the hypothesis that the plant protein synthesis and secretory machinery are actively recruited to the emerging infection structures (Hardham et al. 2007). Mutations in PEN1, originally identified as a component of powdery mildew non-host resistance, did not affect actin reorganization upon challenge with the non-adapted barley powdery mildew fungus (Takemoto et al. 2006), suggesting that this protein acts downstream of actin reorganization. However, in barley cells, chemical and genetic interference with actin polymerization significantly affects basal defense and blocks at least partially non-host resistance against inappropriate powdery mildews (Kobayashi et al. 1997a, b; Miklis et al. 2007). The scenario is similar in *Arabidopsis* (Yun et al. 2003) and suggests an important functional as well as structural role for cytoskeletal components in plant defense responses (Schmidt and Panstruga 2007).

It is interesting to note that plant cytoskeleton and organelle responses documented for powdery mildew infections are strikingly similar to those observed during infection of several plant species with beneficial arbuscular mycorrhizal (AM) fungi (Strack et al. 2003). In this case, the plant cell guides the formation of intracellular arbuscules (analogous in structure and function to haustoria) through the formation of microtubule tracks and by reorganization of the actin network. In tomato, inhibition of microfilament formation interferes with the proper development of the fungus on root tissue (Timonen et al. 2006). In pathogenic powdery mildew–plant interactions, components of both fungal and plant origin are implicated in the formation of the haustorium. During post-penetration, the plant plasma membrane invaginates, and the haustorium starts to develop, initially as a bud and then as a relatively smooth globular body, and finally matures into an intricate, highly convoluted, and branched structure (Koh et al. 2005; Plate 4.4, Chap. 4). The haustorium is a unicellular structure shielded from the plant environment by its own plasma membrane, cell wall, extra-haustorial matrix (an amorphous environment separating plant and fungal structures), and extra-haustorial membrane (EHM), a unique interface between the host and the pathogen, thought to be composed of a modified plant plasma membrane (Bracker 1968).

In the studies of *G. cichoracearum* haustoria on *Arabidopsis*, fluorescently labeled host plasma membrane markers spread as far as the two haustorial neckbands that separate fungal from plant structures. Below the neckbands, the EHM has a distinct architecture, and function from the plant, and fungal membranes (Koh et al. 2005). ATPase and other enzymatic activities normally found at the plant plasma membrane are lacking in the EHM (Spencer-Phillips and Gay 1981). The EHM enveloping haustoria of the pea powdery mildew, *Erysiphe pisi*, is twice as thick as normal host plasma membrane, is highly convoluted in healthy haustorial complexes, and is resistant to detergent treatments (Gil and Gay 1977). The biogenesis, and molecular composition, of EHMs covering the haustoria of biotrophic pathogens represents indeed one of the major mysteries of plant–powdery mildew interactions. Under optimal conditions, the life-span and transport capacity of a single haustorium appear to be sufficient to allow the fungus to complete its life cycle (Shirasu et al. 1999). However, defense responses deployed by the host can interfere with optimal haustorium development and function. *Arabidopsis* cells infected with *G. orontii*, mature haustoria are frequently encased (partially or fully) in what appears to be an extension of the papillary structure, characterized by presence of callose (Micali et al. 2008). These encasements, when complete, appear to crush haustorial bodies and may restrict passage of nutrients to the fungus and pathogen effectors to the host. As such they may represent one of the several layers of resistance to powdery mildew infection (Micali et al. 2008).

7.3 Pre-penetration Resistance Mechanisms

Most of the information on crucifer host resistance to powdery mildew has been generated using *Arabidopsis*–powdery mildew host pathosystem. The first barriers of powdery mildew pathogen encounter during infection are the cuticle and epicuticular waxes overlying the plant cell wall (Malinovsky et al. 2014). Powdery mildew presumably employs mainly hydrostatic pressure to penetrate this preformed perimeter of epidermal cells. Accordingly, the plant can sense the pathogen in several ways. Firstly, the pressure exerted on the plant cell might activate plant mechano-sensors (Bhat et al. 2005; Ellinger and Voigt 2014a). Secondly, damage-associated molecular patterns (DAMPs) released by the breakdown of the plant cell wall, or pathogen-associated molecular patterns (PAMPs) released by the fungus, can be detected by pattern recognition receptors (PRRs) and activate immune signaling (Boller and Felix 2009).

The carbohydrate polymer chitin is a major constituent of fungal cell walls and when exogenously applied to *Arabidopsis* activates PAMP-triggered immune responses. Chitin is perceived by the membrane-localized PRRs CHITIN ELICITOR RECEPTOR KINASE 1 (*CERK1*: At3g21630) (Miya et al. 2007) and the LYSIN MOTIF RECEPTOR-LIKE KINASEs 4/5 (*LYK4/5*: At2g23770/At2g33580) (Cao et al. 2014). Application of PAMPs, including chitin, leads to the accumulation of defense-related proteins and the deposition of callose at seemingly random locations in treated tissues (Gomez-Gomez et al. 1999; Luna et al. 2011; Underwood and Somerville 2013). This phenomenon is similar to the localized formation of papillae and suggests that PAMP-induced PRR activation alone can trigger the establishment of papilla-like structures. The adapted powdery mildew *Golovinomyces cichoracearum (GC)* shows increased sporulation on *cerk1* mutants in comparison to wild-type plants, which suggests that signaling through *CERK1* contributes to basal resistance to the powdery mildew disease (Wan et al. 2008). It is presently unknown whether *lyk4* or *lyk5* mutants are more susceptible to powdery mildew as well. Presumably, plants can perceive further powdery mildew-derived PAMPs. However, the only other known molecule from a powdery mildew pathogen that activates defense gene expression and decreases fungal growth in various cereals after application is a soluble carbohydrate elicitor isolated from conidia of the wheat powdery mildew pathogen, *Blumeria graminis* f. sp. *tritici* (*Bgt*) (Schweizer et al. 2000).

7.3.1 *Role of Papilla Formation*

Plant cell responses to the early detection events include polarization of cellular organelles and rearrangement of cytoskeletal elements (microtubules and actin filaments) below the attack site (Schmelzer 2002; Huckelhoven and Panstruga 2011). Underneath the attempted penetration site, defense-related proteins focally accumulate (Assaad et al. 2004; Bhat et al. 2005; Kwon et al. 2008b; Meyer et al. 2009; Kwaaitaal et al. 2010). Furthermore, the plasma membrane is altered locally and gains lipid raft-like properties (Bhat et al. 2005). Both virulent and non-virulent powdery mildew fungi induce the formation of a small, dome-like structure called papilla below the incipient fungal appressorium (Collins et al. 2003; Assaad et al. 2004; Koh et al. 2005). Among other components such as membranous vesicles, the papilla contains callose, silicon, reactive oxygen species (ROS), and phenolic compounds. The resulting structure is believed to reinforce the cell wall to prevent fungal invasion (Zeyen et al. 2002). This hypothesis is supported by a correlation between the timing of papilla formation and powdery mildew resistance. Mutation of the target membrane-soluble N-ethylmaleimide-sensitive factor attachment protein receptor (t-SNARE) penetration *(PEN) 1*/syntaxin of plants (SYP) 121 (At3g11820) and the vesicle-associated SNAREs vesicle-associated membrane proteins (VAMPs) 721/722 (At1g04750/At2g33120) delays papilla formation and increases *Bgh* penetration success (Assaad et al. 2004; Kwon et al. 2008b; Bohlenius et al. 2010). The ubiquitin ligase *Arabidopsis* Tóxicos en Levadura (*ATL31*: *At5g27420*), which is a regulator of responses to changes in the cellular carbon/nitrogen ratio, interacts with *PEN1* in co-immunoprecipitation experiments. Transient expression assays in *Nicotiana benthamiana* leaves inoculated with *Blumeria graminis* f. sp. *hordei* (*Bgh*) show accumulation of ubiquitination activity-deficient *ATL31C143S*-GFP around papillae and in vesicle-like structures near papillae, while fluorescence is undetectable for *ATL31*-GFP. Furthermore, overexpression of *ATL31* in *Arabidopsis pen1*-1 mutants results in enhanced penetration resistance to *Bgh* and faster formation of papillae (Maekawa et al. 2014). The same observation was made for *RAB GTPASE homolog 4c* (*RABA4c*: *At5g47960*) overexpression lines (Ellinger et al. 2014), suggesting that only the timely formation of papillae can restrict invasion by powdery mildew fungi.

7.3.2 Role of Callose Deposition

Accumulation of the β-1,3-polyglucan callose, as the major constituent of papillae, is a generic response to pathogen challenge (Ellinger and Voigt 2014b). Together with the β-1,4-polyglucan cellulose, callose generates a three-dimensional network of ~250 nm fibrils, which can provide protection against cell wall hydrolysis by fungal enzymes (Plate 7.1; Eggert et al. 2014). Callose could not be unambiguously linked to resistance for a long time, and work on a mutant of the *GSL5/PMR4* gene initially even suggested that loss of papillary callose reduces sporulation of adapted powdery mildews (Jacobs et al. 2003; Nishimura et al. 2003). However, the inhibition of post-invasive fungal growth in *pmr4* mutants relies on hyper-induced salicylic acid (SA) responses upon powdery mildew attack. When the *pmr4* knockout mutation is combined with a further mutation that leads to a loss of SA biosynthesis or signaling, the increased resistance is compromised (Nishimura et al. 2003).

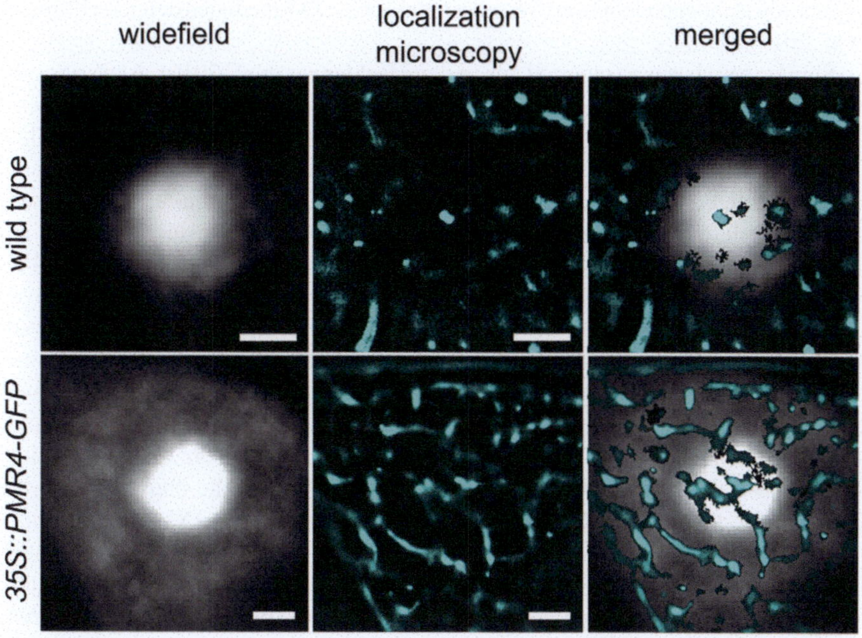

Plate 7.1 Nanoscale resolution of callose polymer fibrils in pathogen-induced cell wall papillae. Three-week-old *Arabidopsis* wild-type and pathogen-resistant *PMR4-GFP* overexpressing lines (*P35S::PMR4-GFP*) were inoculated with the adapted powdery mildew *Gc*. Localization microscopy (dSTORM: direct stochastical optical reconstruction microscopy) of aniline blue-stained callose polymer fibrils in pathogen-induced papillae at sites of attempted fungal penetration at 12 hpi in rosette leaves. Scale bars = 2 μm (Kuhn et al. 2016)

Except for the loss of callose, papillae of *pmr4* mutant plants have a similar appearance as papillae of wild-type plants (Nishimura et al. 2003). While loss of callose in the *pmr4* mutant has only limited impact on penetration resistance (Jacobs et al. 2003; Ellinger et al. 2013), increased callose deposition after powdery mildew attack caused by *PMR4* overexpression results in full penetration resistance to both *Gc* and *Bgh*. The latter effect seems to correlate with structural differences of papillae in *PMR4* overexpression lines compared to the wild type, as the transgenic lines show larger cores of callose-dense deposits, whereas wild-type papillae display a more diffuse structure (Naumann et al. 2013). Together these findings indicate that additional papillary components support the contribution of callose to prevent fungal penetration (Ellinger et al. 2013). The *PMR4-GFP* fusion protein focally accumulates at the powdery mildew attack site, and its presence coincides with the occurrence of callose deposits. The callose accumulations in the *PMR4-GFP* overexpression line are not only enlarged but also deposited in a layer facing the fungus on top of the cellulose microfibrillar network (Eggert et al. 2014). The increase in the proportion of callose presumably protects the cellulose component of papillae from enzymatic digestion (Eggert et al. 2014). In contrast to the increased post-penetration resistance in *pmr4* mutants, the increase in resistance caused by *PMR4-GFP* overexpression is independent from SA or jasmonic acid (JA)-mediated defense (Ellinger et al. 2013).

Similar to what was reported for barley (Bohlenius et al. 2010), ADP ribosylation factor-GTP exchange factor (ARF-GEF)-mediated vesicle trafficking is essential for callose accumulation in papillae in *Arabidopsis* (Nielsen et al. 2012). This ARF-GEF dependence indicates that either *PMR4* accumulation at fungal attack sites, the delivery of callose precursors, and/or the callose deposition process itself involves vesicle-mediated transport processes. *PMR4* interacts with and acts as an effector of the small GTPase of the Ras (rat sarcoma) superfamily, *RABA4c* (Ellinger et al. 2014). *RABA4c* expression is transiently up-regulated prior to callose deposition in response to biotic stress. Knockouts of *RABA4c* exhibit a delayed increase of callose synthase activity, slightly reduced numbers of callose deposits, and slightly increased *Gc* penetration rates. By contrast, overexpression of *RABA4c* results in full penetration resistance to *Gc,* and hastens, and increases callose deposition. Both effects depend on the presence of *PMR4* and *RABA4c* GTPase activity. The *RABA4c* localization to membranes is independent on the prenylation of its C-terminal CaaX motif ('C' cysteine, 'a' aliphatic amino acid, 'X' variable amino acid). A C-terminal *RABA4c*-mCitrine fusion lacking this lipid modification still localizes to membranes, though solely when PMR4 is present, which supports the finding that both proteins physically interact in planta (Ellinger et al. 2014). *Rab* (Ras-related in brain) GTPases play major roles in virtually all vesicle trafficking processes in eukaryotic cells. To what extent *PMR4* localization to the plasma membrane or to focal accumulation sites depends on *RABA4c*-dependent vesicle trafficking pathways is unknown (Kuhn et al. 2016).

7.3.3 Role of Extracellular Deposition of Proteins into Papillae

The discovery that components of a SNARE protein complex are involved in penetration resistance suggests that these proteins directly control vesicle fusion at the powdery mildew attack site (Collins et al. 2003; Kwon et al. 2008a, b). After vesicle fusion, and cargo release, SNARE proteins are usually recycled and stay on the cytosolic side of the plasma membrane (Kwon et al. 2008a). Surprisingly, in case of the focal accumulation of SNARE proteins at attempted fungal entry sites, this is not the case. Instead, fluorescent fusions of *PEN1*, soluble N-ethylmaleimide-sensitive factor adaptor protein *(SNAP) 33* (At5g61210; a t-SNARE), and the ATP-binding cassette (ABC) transporter *PEN3* (At1g59870) accumulate within papillae, and haustorial encasements, and therefore end up in the extracellular (apoplastic) space. Within cell wall appositions, GFP-PEN1 co-localize with the lipophilic fluorescent tracer of endosomes, FM4-64, indicating that membrane material co-accumulates with these proteins in papillae and haustorial encasements (Meyer et al. 2009; Nielsen et al. 2012). As demonstrated by electron microscopy and co-localization with the Rab-like GTPase MVB marker *ARA6/RABF1*-GFP (At3g54840), MVBs focally accumulate at pathogen attack sites. It is therefore conceivable that MVBs contribute to extracellular deposition of otherwise intracellularly localized proteins (An et al. 2006; Meyer et al. 2009; Nielsen et al. 2012). According to this hypothesis, vesicles containing *PEN1, SNAP33*, and *PEN3* may sort and incorporate into the lumen of MVBs after endocytosis. These MVBs might subsequently fuse with the plasma membrane, which could explain extracellular protein delivery (Meyer et al. 2009). The extent and significance of the extracellular deposition of otherwise intracellular proteins for powdery mildew resistance are currently unknown (Kuhn et al. 2016).

7.3.4 Role of Silicon-Mediated Resistance

Silicon (Si) contributes to powdery mildew resistance of cereals and various other plant species (Fauteux et al. 2005). Accordantly, watering *Arabidopsis* plants with silicon results in a lower powdery mildew disease incidence although *Arabidopsis* lacks dedicated Si transporters (Ghanmi et al. 2004). Accumulation of insoluble Si at powdery mildew attack sites led to the hypothesis of Si acting as a simple physical barrier (Belanger et al. 2002). However, not in every case presence of insoluble Si correlates with increased resistance to fungal penetration. Consequently, a physiological or biochemical role in mediating cellular resistance has been postulated (Belanger et al. 2003). While Si fertilization alone has a minor effect on transcript abundance, *Gc* inoculation of Si-fertilized plants versus *Gc* inoculation of non-Si-supplemented plants attenuated the magnitude of powdery mildew-induced down-regulation of genes by more than 25% (Fauteux et al. 2006). As many of these powdery mildew-repressed genes are related to primary metabolism, the Si-mediated reduced down-regulation might indicate stress alleviation. Consequently, Si feeding potentially facilitates a more efficient response to powdery mildew infection (Fauteux et al. 2006). This hypothesis is further corroborated by transgenic *Arabidopsis* plants stably expressing the wheat Si transporter *Ta*Lsi1. These plants have increased Si levels and concomitantly further enhanced powdery mildew resistance in the presence of Si compared to wild-type plants (Vivancos et al. 2015).

7.4 Post-penetration Resistance Mechanisms

It is considered as a second line of defense and *RPW8*-mediated broad-spectrum resistance which operates at sub-cuticular level of host–pathogen interaction. In many plant species that can be colonized by powdery mildew fungi, dedicated dominantly or semi-dominantly inherited resistance (*R*) genes provide isolate-specific protection as a second line of defense against the disease (Chelkowski et al. 2003; Bai et al. 2005; Marone et al. 2013). These types of genes typically encode canonical nucleotide binding site-leucine-rich repeat (*NB-LRR/NLR*) proteins (Takken and Goverse 2012). *R* genes occur typically in multiple allelic forms within plant populations. These polymorphic variants are effective against particular pathogen isolates encoding effectors that are recognized by the respective R proteins ('gene-for-gene relationship'). It is thought that plant R proteins either directly or indirectly associate with cognate effector proteins to trigger a boosted defense output that often culminates in a hypersensitive response (HR) associated with local host cell death, thereby restricting pathogen proliferation (Dangl and Jones 2001). However, till date, *R* genes that are effective against powdery mildew fungi have not been found in *Arabidopsis*, but in *Brassica* spp., powdery mildew resistance governed by dominant genes has been reported (Abraham 1993; Singh et al. 1997; Saharan and Krishnia 2001). Instead, a polymorphic genetic locus, resistance to powdery mildews *8* (*RPW8*), harbouring two unconventional non-*NB-LRR*-type powdery mildew resistance genes, is a major source of resistance in *Arabidopsis*. The *RPW8* locus has a complex arrangement that differs between *Arabidopsis* accessions. In the resistant ecotype Ms-0, the *RPW8* locus harbours five gene copies that encode small sequence-related basic proteins with a predicted N-terminal trans-membrane domain and one or two C-terminal coiled-coil domains. Of these, two paralogs (tandemly arranged *RPW8.1* and *RPW8.2*) contribute to effective resistance against powdery mildew, while other paralogs (*HR1*, *HR2*, and *HR3*) of the *RPW8* locus are inactive in this respect (Xiao et al. 2001). Phylogenetic analysis on the basis of syntenic loci in the *Arabidopsis* relatives *Arabidopsis lyrata*, *Brassica rapa,* and *Brassica oleracea* suggests that *RPW8.1* and *RPW8.2* likely evolved from a *HR3*-like ancestor gene through a series of gene duplication events and subsequent diversification by positive selection (Xiao et al. 2004).

The powdery mildew resistance-conferring *RPW8* locus, which was originally described in the accession Ms-0 (Xiao et al. 1997), shows a widespread distribution in *Arabidopsis* populations. Most powdery mildew-resistant accessions contain a 'functional' version of *RPW8.1* and/or *RPW8.2*. The locus is thus a major source of natural powdery mildew resistance in *Arabidopsis* (Orgil et al. 2007; Gollner et al. 2008). Notably, the Col-0 reference accession lacks functional copies of *RPW8.1*/*RPW8.2* and is thus susceptible to all known powdery mildew species that are capable to colonize *Arabidopsis* plants (Xiao et al. 2001). Resistance mediated by *RPW8* occurs after the establishment of haustoria and is typically associated with the accumulation of hydrogen peroxide and localized host cell death, although these responses exhibit some degree of plasticity in different ecotypes (Xiao et al. 2001; Gollner et al. 2008). *RPW8.1* and *RPW8.2* operate through a SA-dependent positive feedback loop, which also promotes transcript accumulation of the two genes (Plate 7.2; Xiao et al. 2003). Consistently, *RPW8*-mediated powdery mildew resistance requires components of SA signaling (*Enhanced Disease Susceptibility (EDS) 1* (At3g48090) and *EDS5* (At4g39030), *Phytoalexin-Deficient (PAD) 4* (At3g52430), *Arabidopsis* non-expresser of PR genes *(NPR) 1:* At1g64280*1*)) that also play a role in basal defense (Xiao et al. 2005). Overexpression or ectopic expression of *RPW8* proteins leads to enhanced resistance against diverse biotrophic pathogens (cauliflower mosaic virus and the oomycete *Hyaloperonospora arabidopsidis*) but more pronounced susceptibility to necrotrophic pathogens (*Alternaria* and *Botrytis* ssp.; Wang et al. 2007; Ma et al. 2014).

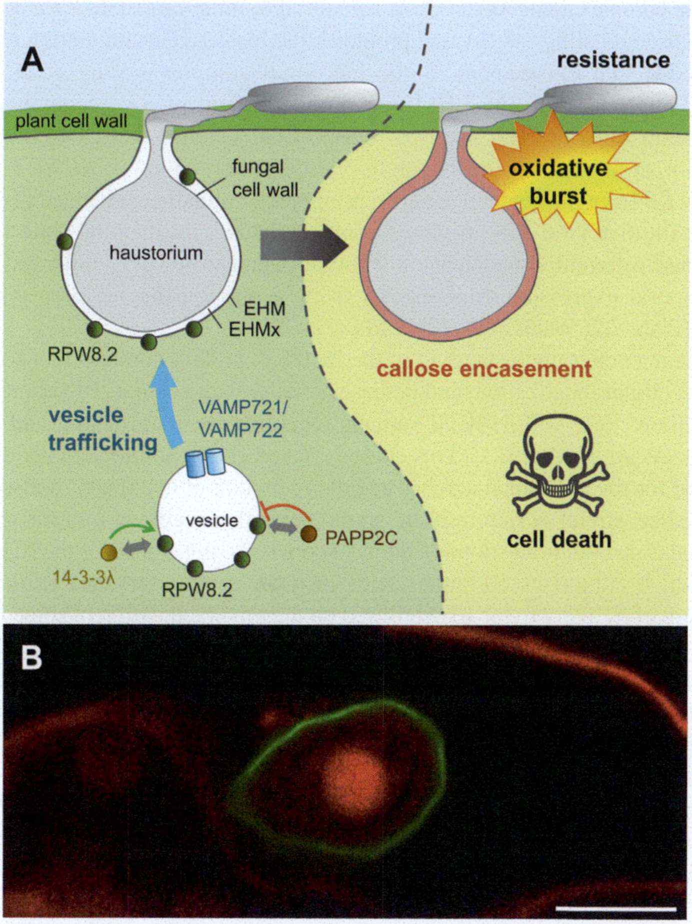

Plate 7.2 *RPW8* localizes at the EHM and contributes to cell death upon powdery mildew infection. (**a**) Scheme depicting *RPW8.2* function. Left: *RPW8.2* interactors and *RPW8.2* deposition at the EHM. Right: *RPW8.2*-triggered oxidative burst, and callose encasement of haustoria correlates with subsequent host cell death. (**b**) Confocal laser scanning micrograph of GFP-labeled *RPW8.2* (green) in the EHM. Red, propidium iodide-stained plant, and fungal structures. Bar = 10 μm (Kuhn et al. 2016)

Although *RPW8* function in the context of powdery mildew infection is rather evident, information on the protein and its in planta activity is limited. Yeast two-hybrid assays identified 14-3-3λ (AT5g10450) and the phytochrome-associated protein phosphatase type 2C (*PAPP2C*: At1g22280) as potential *RPW8.2* interactors (Yang et al. 2009; Wang et al. 2012). While genetic evidence suggests that the 14-3-3 protein is a positive regulator of *RPW8* function, *PAPP2C* seems to be a negative regulator of cell death and powdery mildew resistance. Notably, following powdery mildew attack, *RPW8.2* accumulates at the EHM (Plate 7.2; Wang et al. 2009). In fact, *RPW8.2* was the first protein described to localize to this specialized membrane compartment. At the extra-haustorial membrane (EHM), *RPW8.2* activates defense signaling via SA and promotes the localized accumulation of hydrogen peroxide and encasement of the haustorial complex (Wang et al., 2009). Targeting of *RPW8.2* to the EHM occurs independent of SA accumulation, but requires actin function, and involves transport on secretory vesicles (Wang et al. 2009; Kim et al. 2014). Interestingly, ectopic expression of *RPW8.1-YFP* or *RPW8.2-YFP* from the respective native promoters, mutually exchanged promoters, or the constitutive viral 35S promoter results in distinct localization patterns of the proteins and differential resistance phenotypes against powdery mildews. Precise spatiotemporal expression thus appears to be a prerequisite for proper *RPW8.2* function (Plate 7.2; Wang et al. 2010; Ma et al. 2014).

To obtain a better understanding of the *RPW8.2*, protein domains that contribute to its sub-cellular localization, and defense activity, Wang et al. (2013) functionally analysed more than 100 *RPW8.2* variants regarding their trafficking and defense properties (Wang et al. 2013). This study revealed single amino acid residues that are critical for the antifungal activity and the induction of cell death. It also uncovered two short stretches rich in basic amino acids that together with the predicted N-terminal transmembrane domain define a core targeting signal for the EHM. This region, which comprises 60 amino acids in total, is necessary and sufficient for localization of *RPW8.2* to the EHM. Based on the mis-localization of some *RPW8.2* mutant variants to the nucleus, and/or plastidic stromules, Kuhn et al. (2016) proposed the existence of a dedicated membrane trafficking pathway towards the EHM (Wang et al. 2013). Notably, the two short basic stretches that contribute to EHM localization apparently also play a role in nucleocytoplasmic trafficking of *RPW8.2*, suggesting that a portion of the *RPW8.2* pool might have a function in the nucleus (Huang et al., 2014). Overexpression of non-functional yet EHM-targeted *RPW8.2* versions can exert a dominant-negative effect on functional *RPW8.2*, thereby compromising *RPW8.2*-mediated powdery mildew resistance. Such dominant-negative *RPW8.2* variants also affect basal defense against powdery mildew and result in an enhanced disease susceptibility (*eds*) phenotype, suggesting the existence of further EHM-localized factors that contribute to basal levels of post-penetration resistance in *Arabidopsis* (Zhang et al. 2015b).

Widespread presence of a locus that confers broad-spectrum powdery mildew resistance (*RPW8*) might explain why no canonical cytoplasmic NB-LRR-type R proteins against this disease evolved in *Arabidopsis*. Although such genes are seemingly lacking in natural *Arabidopsis* populations, a heterologously expressed R protein from monocotyledonous barley can confer isolate-specific powdery mildew resistance in *Arabidopsis*. Transgenic expression of the barley mildew resistance locus A1 (*MLA1*) coiled-coil NB-LRR-type resistance protein in a partially immune compromised mutant background (*pen2 pad4 senescence-associated gene 101* (*sag101*: At5g14930)) results in isolate-specific resistance against the matching barley powdery mildew (*Bgh*) (Maekawa et al. 2012). This remarkable finding suggests that the signaling machinery acting downstream of *MLA1* activation is conserved between monocotyledonous and dicotyledonous plant species, two lineages that diverged ca. 200 million years ago. *MLA1* function in resistance responses towards powdery mildew in *Arabidopsis* does not require SA, JA, or ethylene (ET). As in barley, in *Arabidopsis MLA1* exhibits nucleocytoplasmic partitioning, and its activation upon powdery mildew inoculation results in pronounced and sustained transcriptional re-programming (Maekawa et al. 2012).

*7.4.1 Role of Enhanced Disease Resistance (*EDR*) Genes*

Transcriptional activation of *pathogenesis-related* (*PR*) genes is one hallmark of induced defense (Loake and Grant 2007). During a genetic screen aimed to identify novel elements of plant defense, three mutants with enhanced disease resistance (*edr1* (At1g08720), *edr2* (At4g19040), and *edr3* (At3g60190)) that do not express *PR1* (At2g14610) upon inoculation with *Gc* were isolated (Frye and Innes 1998; Frye et al. 2001; Tang and Innes 2002; Tang et al. 2005a, b, 2006). Interestingly, all three mutants show characteristics of 'late-acting' resistance (i.e. at 5 to 8 dpi), which is associated with accelerated mesophyll cell death leading to macroscopic patches of lesions and drastically either reduced or absent sporulation. Genetic epistasis analysis revealed that *edr*-mediated resistance is SA-dependent and JA-independent (Frye and Innes 1998; Tang et al. 2005b, 2006).

EDR1 encodes a mitogen-activated protein kinase kinase kinase (MAPKKK) that negatively regulates plant disease resistance (Frye et al. 2001). The *edr1* mutant displays enhanced cell death during infection with the adapted powdery mildew pathogen *Gc* and in response to drought stress (Frye et al. 2001; Tang et al. 2005a, b). Cell death associated with *edr1* resistance requires the E3 ubiquitin ligases *ATL1* (At1g04360) and KEEP ON GOING (KEG: At5g13530). Both E3 proteins are inhibited by interaction with *EDR1*, and the cell death phenotypes associated with *edr1* are suppressed upon their depletion, indicating that *EDR1* acts as a negative regulator of programmed cell death (Serrano et al. 2014). KEG possibly recruits *EDR1* to the *trans*-Golgi network (TGN), and in turn *EDR1* regulates E3 ligase activity of KEG to further suppress cell death (Gu and Innes 2011; Liu and Stone 2013). Overexpression of *ATL1* causes extensive cell death, which depends on its E3 ligase activity. Strikingly, knockdown of *ATL1* expression does not only interfere with *edr1*-mediated cell death but causes hypersusceptibility to powdery mildew infection, demonstrating that *ATL1* is a positive regulator of pathogen-induced cell death (Serrano et al. 2014). A further link of *EDR1* to suppression of cell death is provided by its inhibitory interaction with mitogen-activated protein kinase kinase *(MKK) 4* (At1g51660) and MKK5 (At3g21220) that are part of the MAPK cascade fine-tuning plant immunity (Zhao et al. 2014).

EDR2 encodes a mitochondrial protein with a pleckstrin homology domain and a steroidogenic acute regulatory protein-related lipid transfer (START) motif. Both *EDR1* and *EDR2* function in a common genetic pathway as evidenced by the *edr1 edr2* double mutant, showing resistance phenotypes that are indistinguishable from the respective single mutants (Tang et al. 2005b). In addition, *edr1* and *edr2* both display enhanced senescence in response to ET. Interestingly, mutations in the amino-transferase *AGD2-LIKE* Defense Response Protein1 (*ALD1*: At2g13810) suppress *edr2*-mediated phenotypes including powdery mildew resistance, programmed cell death, and ET-induced senescence, but not the *edr1 edr2* double mutant phenotype (Nie et al. 2011). This raises the question how *EDR1* and *EDR2* activities are coordinated during the regulation of defense, cell death, and ET-induced senescence. Different from *EDR1* and *EDR2*, *EDR3* seems to function in a separate pathway, since *edr3* does not display an early senescence phenotype. *EDR3* encodes a dynamin-like protein localized partially to mitochondria. Despite the absence of a constitutive cell death phenotype in *Arabidopsis*, the mammalian counterpart of *EDR3* plays a role in regulating mitochondrial dynamics associated with programmed cell death (Tang et al. 2006). Recently, a fourth *EDR* gene, *EDR4* (At5g05190), with unknown protein function, and preferential localization of the gene product at the plasma membrane, and endosomal compartments, has been isolated. Like previously identified *EDRs*, *EDR4* is involved in negative regulation of SA-dependent powdery mildew resistance (Wu et al. 2012). *EDR4* functions in the same pathway as *EDR1* and *EDR2* and interacts with *EDR1*, recruiting it to fungal penetration sites. The shared phenotypic features of *edr* mutants suggest a general link between SA-mediated resistance, mitochondrial function, and programmed cell death (Ausubel 2005).

*7.4.2 Role of Powdery Mildew-Resistant Mutant (*PMR*) Genes*

In a genetic screen with the aim to identify susceptibility factors involved in interactions between *Arabidopsis* and the powdery mildew pathogen *Gc*, six powdery mildew-resistant mutants, *pmr1* to *pmr6*, were isolated. Four of the corresponding genes, namely, *PMR2* (At1g11310), *PMR4/GSL5*, *PMR5* (At5g58600), and *PMR6* (At3g54920), have been cloned and to some extent functionally characterized (Vogel and Somerville 2000; Vogel et al. 2002; Jacobs et al. 2003; Nishimura et al. 2003; Vogel et al. 2004; Consonni et al. 2006). The *pmr2* mutant is defective in Mildew Resistance Locus *O* (*MLO*) *2* (At1g11310), which encodes an integral membrane protein of unknown function. *PMR5* belongs to a large plant-specific gene family of unknown function, and *PMR6* encodes a glycosylphosphatidyl-inositol (GPI)-anchored pectate lyase-like protein (Vogel et al. 2002, 2004; Jacobs et al. 2003; Nishimura et al. 2003; Consonni et al. 2006). The latter *pmr* mutants, *pmr5* and *pmr6*, are believed to impact cell wall integrity, further stressing the contribution of the cell wall to powdery mildew resistance. The *Arabidopsis pmr5* mutant exhibits resistance to the adapted powdery mildew fungi *Gc* and *Go*, and enrichment of pectin as well as reduced pectin modification occurs in the cell walls of *pmr5* plants (Vogel et al. 2004). In addition, *PMR5* contributes to *PEN2*-mediated pre-invasion resistance to the non-adapted fungus *Magnaporthe oryzae*. The *pen2 pmr5* double mutant shows enhanced penetration success of *M. oryzae* (Maeda et al. 2009), indicating that *PMR5* is involved in host and non-host resistance and emphasizing the importance of cell wall integrity for both types of resistance. *PMR6* localizes at the plant cell wall, where it might degrade pectin. In line with this assumption, the *pmr6* mutant displays increased pectin and uronic acid contents. Like *pmr5*, the *pmr6* mutant is resistant to *Gc* and *Go,* which is in both cases independent of SA, ET, and JA signaling (Vogel et al. 2002). The *pmr5 pmr6* double mutant shows increased resistance compared to the respective single mutants, suggesting that the two genes may function separately during plant defense. Furthermore, *PMR5* and *PMR6* are involved in the regulation of ploidy in mesophyll cells underlying the fungal feeding sites (Chandran et al. 2013).

7.4.3 Role of Powdery Mildew-Resistant Genes

Arabidopsis MLO susceptibility genes were isolated and characterized based on their sequence similarity to barley *MLO* (Consonni et al. 2006) and identified as the *pmr2* mutant in the above-mentioned forward genetic screen (Vogel and Somerville 2000). According to phylogenetic analyses, there are 15 *MLO* genes distributed into 5 clades in *Arabidopsis*, of which *MLO2*, *MLO6* (At1g61560), and *MLO12* (At2g39200) belong to the same clade (Devoto et al. 2003; Acevedo-Garcia et al. 2014). *mlo2* mutants display reduced penetration success and less sporulation after infection with the adapted powdery mildew fungus *Go* (Consonni et al. 2006). Interestingly, *MLO2* controls penetration success of powdery mildew fungi together with *MLO6* and *MLO12*. While the *mlo6* and *mlo12* single and double mutants do not show any resistance phenotype, they gradually increase resistance of *mlo2* if combined in double and triple mutant combinations, with the *mlo2 mlo6 mlo12* triple mutant being fully resistant (Plate 7.3; Consonni et al. 2006). *MLO* genes encode evolutionary ancient integral membrane proteins with seven transmembrane domains and unknown biochemical activity (Devoto et al. 2003; Kusch et al. 2016). Besides *Arabidopsis,* and barley, mutations of closely related *MLO*

Plate 7.3 Macroscopic infection phenotypes of Col-0 and the *mlo2 mlo6 mlo12* mutant. Five-week-old wild-type (Col-0) and *mlo2 mlo6 mlo12* plants (in Col-0 genetic background) were inoculated with *Go* and photographs 1 week after inoculation (Kuhn et al. 2016)

genes in tomato, pea, and further plants render these hosts species resistant to powdery mildew infection, indicating a similar function of the respective proteins (Bai et al. 2005; Humphry et al. 2011). Similar to NHR, *mlo2*-mediated powdery mildew resistance does not depend on major phytohormone signaling pathways such as those relying on JA, ET, or SA (Consonni et al. 2006). By contrast, all three *PEN* genes are required for *mlo2*-mediated resistance to powdery mildew (Consonni et al. 2006). These findings suggest that *mlo*-mediated resistance and NHR may share overlapping pathways in plant defense (Humphry et al. 2006). Besides the *PEN* proteins, *CYP79B2* (At4g39950) and *CYP79B3* (At2g22330), two cytochrome monooxygenases that catalyse the entry step towards the production of diverse indolic metabolites, including the *Arabidopsis*-specific phytoalexin camalexin, and indole glucosinolates are required for *mlo2*-mediated resistance. In contrast to *CYP79B2* and *CYP79B3*, another cytochrome P450 monooxygenase, *PAD3* (At3g26830), which catalyses the final step in camalexin biosynthesis, only plays a minor role in *mlo2*-mediated resistance (Consonni et al. 2010).

7.4.4 Role of R Genes in Pre- and Post-pathogenesis Resistance

The pathogen-triggered rearrangement of cellular components correlates with the formation of papillae, and a major radial reorganization of actin filaments underneath attempted powdery mildew entry sites (Kobayashi et al. 1997a, b; Takemoto et al. 2006). Pharmacological treatment of leaves with inhibitors of actin filament polymerization (cytochalasins and latrunculin B) and myosin (BDM (2,3-butanedione monoxime) and NEM (*N*-ethyl maleimide)) results in reduced recruitment of organelles and vesicles towards the site of fungal attack and decreased powdery mildew penetration resistance (Kobayashi et al. 1997a, b; Yun et al. 2003; Yang et al. 2014). Conversely, silencing of genes coding for subclass I actin depolymerization factors (ADFs) increases resistance against *Go* and results in enhanced filament bundling during early *Go* infection (Inada et al., 2016). Together these findings suggest that intact actin microfilaments and myosin motors are required for successful defense. In fact, single mutants of *MYOSIN XI* genes (*xi-1-1* (At1g17580), *xi-2-1*, *xi-2-2* (At5g43900), *xi-i-1*, *xi-i-2* (At4g33200), *xi-k-1*, *xi-k-2* (At5g20490)), one triple mutant (*xi-1-1*, *xi-2-1*, *xi-k-2*), and one quadruple mutant (*xi-1-1*, *xi-2-1*, *xi-i-1*, *xi-k-2*) exhibit higher penetration frequencies compared to Col-0 wild type upon *Bgh* inoculation. Furthermore, upon challenge with *Gc*, the quadruple mutant shows increased fungal growth, and hyphal branches at 3 dpi, and more conidiophores at 7 dpi compared to Col-0 wild type (Yang et al. 2014). Collectively, these findings indicate that transport activities along the actin cytoskeleton might be crucial for pre- and possibly post-invasive defense against powdery mildews.

SNARE proteins mediate fusion events between vesicular and target membranes. Based on the presence of a critical arginine or glutamine residue in the centre of the

SNARE domain, this family is divided into R- or Q-SNARE proteins, respectively, where the latter can be further subdivided into Qa-, Qb-, or Qc-SNAREs (Collins et al. 2003; Lipka et al. 2007). *PEN1* (Qa-SNARE), *SNAP33* (Qb + Qc-SNARE), and *VAMP721/722* (R-SNARE) form a ternary SNARE complex that focally accumulates at fungal penetration sites. This complex is required for the timely assembly of papillae and most likely for the release of pathogen-induced vesicle cargo (Assaad et al. 2004; Kwon et al. 2008b; Kwaaitaal et al. 2010). In addition to these SNARE proteins, the TGN-localized Qa-SNAREs of the *SYP4* family, which are plant orthologs of the syntaxin 16 in animals, and yeast *Tlg2* (t-SNARE affecting a late Golgi compartment), seem to be required in powdery mildew disease resistance responses. Double mutants *syp42* (At4g02195) and *syp43* (At3g05710) plants show increased secondary hyphae formation compared to the Col-0 wild type after inoculation with the non-adapted powdery mildew *E. pisi*, while *Go* infection is unaltered (Uemura et al. 2012). Interestingly, mRFP-*VAMP722* partially co-localizes with GFP-tagged *SYP43*, but not with Venus-*SYP61* (At1g28490), another TGN marker. In addition, GFP-*SYP43* localizes between the TGN cisternae (labeled with Venus-*SYP61*) and compartments labeled with mRFP-*VAMP722* (Uemura et al. 2012). Furthermore, the TGN-localized KEG ubiquitin ligase, which interacts with *EDR1* and regulates transport of membrane-associated proteins to the vacuole, is degraded following the maturation of *Gc* haustoria (Gu and Innes 2012). These observations suggest that KEG might be a plausible virulence target of the powdery mildew fungus. Together, these findings highlight the importance of the TGN during powdery mildew infection.

The ARF-GEF inhibitor brefeldin A (BFA) has been widely used to study the impact of membrane trafficking in powdery mildew interactions. The treatment with BFA hampers penetration resistance to *Bgh* in Col-0 leaves (Nielsen et al. 2012). Additionally, BFA-treated leaves of a *pen1* transgenic line expressing *GFP-PEN1* show reduced accumulation of the fusion protein and callose at the sites of attempted fungal penetration. As strong mutants of the well-studied BFA-sensitive ARF-GEF GNOM (At1g13980) are dwarfed, and therefore not suitable for detailed analysis, Nielsen et al. (2012) generated trans-heterozygote plants, carrying two different mutated alleles of GNOM (gnom $^{B409/emb30\text{-}1}$). These partially complement the respective non-functional domains of the ARF-GEF dimer. *Bgh* infection of genome $^{B409/emb30\text{-}1}$ plants reveal an increase in fungal penetration, and a delay in callose deposition, and papillary *GFP-PEN1* accumulation, thus mimicking BFA treatment (Nielsen et al. 2012). Together these findings suggest that BFA-sensitive GNOM regulates sorting of material to be transported to the papilla, including *PEN1* (Nielsen et al. 2012). Notably, BFA treatment of the above-mentioned myosin quadruple knockout mutant (*xi-1-1 xi-2-1 xi-1 xi-k-2*) results in retention of *GFP-PEN1* at the plasma membrane, which contrasts its accumulation in BFA bodies in Col-0 epidermal cells. Additionally, accumulation of *GFP-PEN1*, callose, and autofluorescent material at attempted penetration sites is reduced in the myosin quadruple mutant upon *Bgh* infection (Yang et al. 2014). This experimental outcome implies that members of the myosin XI family are involved in sub-cellular infection pathways that modulate penetration resistance to powdery mildew.

R protein *RPW8*.2 localizes to the EHM in cells attacked by the adapted powdery mildew pathogens *Gc* and *Go* (Wang et al. 2009; Micali et al. 2011). Localization studies using *RPW8*.2-YFP under the control of its native promoter in transgenic *Go*-infected Col-0 plants revealed that accumulation of *RPW8.2* occurs around mature haustoria that have been partially or completely encased (Plate 4.1, Chap. 4; Micali et al. 2011). Immunogold labeling of *RPW8.2*-YFP in plants infected with *Gc* supports localization at the EHM, which is reduced after treatment with the actin polymerization inhibitor cytochalasin E (Wang et al. 2009). Overexpression of *ADF6* (At2g31200) in Col-0 plants causes the same response, indicating that intact actin microfilaments are required for successful recruitment of *RPW8.2* to the EHM. By contrast, treatment with oryzalin, a microtubule polymerization inhibitor, does not affect the localization of the resistance protein (Wang et al. 2009). Furthermore, immunogold labeling experiments showed the presence of *RPW8.2* in vesicle-like endo-membrane compartments on the cytoplasmic side of the callose encasement of the haustorial complex (Wang et al. 2009). A recent study revealed that the same *RPW8.2*-containing vesicles co-localize with the R-SNARE proteins *VAMP721* and *VAMP722*. While in the absence of VAMP721 trafficking of *RPW8.2* to the EHM is delayed, lack of *VAMP722* has a less drastic impact. Reduced EHM targeting efficiency of *RPW8.2*-YFP in the tested mutants correlates with enhanced *Go* sporulation (Kim et al. 2014). Moreover, delivery of *RPW8.2* to the EHM is independent of SA signaling and *PEN1* function, implying that *VAMP721/722* vesicles are required for pre-invasive and post-invasive vesicle trafficking pathways in defense against powdery mildews (Wang et al. 2009; Kim et al. 2014).

Host membrane penetration plays a central role during defense against powdery mildew fungi and in other plant–microbe interactions (Dormann et al. 2014; Inada and Ueda 2014; Leborgne-Castel and Bouhidel 2014; Teh and Hofius 2014). Therefore, it is not surprising that pathogens including powdery mildews may attempt to interfere with this pathway. Consistent with this notion, the *Bgh* effecter candidate *BEC4* interacts with a member of the ARF-GTPase-activating protein (ARF-GAP) family in barley (Schmidt et al. 2014). The *Arabidopsis* ortholog of this protein is *AGD5* (At5g54310). Interestingly, *agd5* mutant alleles show considerably elevated *E. pisi*, but unaltered *Go* entry rates. Whether more powdery mildew effectors target the host trafficking machinery will be an object of further investigations (Kuhn et al. 2016).

7.4.5 Role of Powdery Mildew R Genes Through Altered Cell Wall Composition of Hosts

As in the *pmr6* mutants, three lines of evidence indicate that the resistance mechanism operating in the *pmr5* mutant does not require the activation of either the SA or JA/ethylene defense pathways. First, *pmr5* plants did not constitutively express high levels of either *PR1* or *PDF1.2* mRNA indicating that resistance is not mediated by constitutive activation of the SA or JA/ethylene pathways. Secondly, neither *PR1* nor *PDF1.2* was induced to high levels after inoculation indicating that the resistance is not mediated by a hyper-activation of either of these signal transduction pathways. Thirdly, and most convincingly, mutants or transgenes that block signaling through the SA or JA/ethylene pathways did not abolish powdery mildew resistance in *pmr5* plants. Thus, *pmr5*-mediated resistance is independent of the activation of known defense pathways. Therefore, either *PMR5* is required for fungal growth, or the *pmr5* mutation activates a novel defense pathway. Unlike resistance attributed to the majority of resistance genes, and disease-resistant mutants, the attenuation of powdery mildew growth on *pmr5* and *pmr6* did not require cell death as shown by the lack of cell death below fungal colonies. In the course of looking for resistance-associated cell death, it was noticed that both *pmr5* and *pmr6* had microlesions along veins on a subset of the oldest leaves. This phenotype was not correlated with resistance because only a small subset of the oldest leaves had lesions, yet all leaves were highly resistant. This phenotype can be phenocopied by heat treatment of wild-type plants (Vogel et al. 2002). Importantly, heat-treated plants were still susceptible to powdery mildew. It was the microlesions as a pleiotropic effect of the *pmr5* and *pmr6* mutations unrelated to disease resistance. These microlesions may be responsible for the slightly elevated basal level of *PR1* observed in *pmr5* and *pmr6*. That both *pmr5* and *pmr6* have these microlesions underscores the similarity of these mutants.

As both *pmr5* and *pmr6* were fully susceptible to *P. syringae* pv. *tomato* and *H. parasitica*, the resistance is not due to the activation of a broad-spectrum defense pathway, like systemic acquired resistance. Both *pmr5* and *pmr6* were resistant to *E. orontii* indicating that the resistance is effective against isolates from two powdery mildew species. Thus, *pmr5* and *pmr6* resistance is qualitatively different than resistance conferred by either gene-for-gene resistance genes or previously described disease-resistant mutants. The FTIR spectra from *pmr5* epidermal cell walls were similar to the spectra from *pmr6* suggesting that both mutants have increased pectin and the pectin had lower methyl esterification or O-acetylation relative to wild type. Moreover, like *pmr6*, the major FTIR spectral features associated with cellulose and xyloglucan shifted in energy in *pmr5* cell walls, suggesting an alteration in the hydrogen-bonding environment. Interestingly, the *pmr5 pmr6*-3 double mutant had higher levels of uronic acid than either *pmr5* or *prm6* indicating that these two mutants interact synergistically to increase uronic acid content. The synergistic effect on uronic acid content along with the similarity of *pmr5* and *pmr6* phenotypes (e.g. powdery mildew resistance, morphology, microlesions, cell wall composition) suggests that these two mutations affect parallel pathways that regulate some aspects of pectin biosynthesis either directly or indirectly. Furthermore, *PMR5* is predicted to be associated with the endoplasmic reticulum by a hydrophobic N-terminal signal sequence, and *PMR6* is predicted to locate to the exterior side of the plasma membrane via a glycosylphosphatidylinositol anchor. Thus, it is unlikely that these two proteins interact directly.

To address the possibility that the increase in uronic acid observed in the *pmr5 pmr6*-3 double mutant was due to an indirect effect of cell size, Vogel et al. (2004)

determined the relationship between cell size and pectin content in three dwarf mutants that were not directly related to disease resistance or pectin metabolism. Results indicated that while the wild type, the three dwarfs, and *pmr5* did show a weak correlation between cell size and pectin content, the large increase in uronic acid observed for the *pmr5 pmr6*-3 double mutant could not be attributed solely to decreased cell size. Thus, it is possible that the increase in pectin in *pmr5 pmr6*-3 cell walls restricts cell expansion and this in turn limits cell size.

The cell wall is very dynamic and responds to physiological stresses and altered substrate availability with compensating changes in organization (Gillmor et al. 2002). To assess whether the changes in pectin content inferred from the FTIR spectra were associated with any compensating changes in other components, the cell wall neutral sugar content was measured. Aside from the approximately 50% reduction in fucose in the double mutant, all other statistically significant changes in neutral sugars were modest. As approximately two-thirds of the fucose in the *Arabidopsis* leaf cell wall is found in xyloglucan, the decreased fucose in the double mutant may suggest decreased xyloglucan fucosylation (Perrin et al. 2003; Zablackis et al. 1995). The presence of relatively normal amounts of xylose in the double mutant suggests that the amount of xyloglucan is not strongly altered. *pmr5 pmr6*-3 cell walls had small but significant increases in arabinose and galactose suggesting increased abundance of the galactose- and arabinose-containing side chains of the pectin, rhamnogalacturonan I. The characterization of *pmr5* revealed that *pmr5*-mediated resistance does not require the activation of the SA or JA/ethylene defense pathways, does not require cell death, and is not broad-spectrum. In addition, the phenotype of *pmr5* plants is very similar to *pmr6* plants. Taken together, these data suggest that *pmr5* and *pmr6* employ similar mechanisms to limit fungal growth and that this mechanism is unrelated to known defense signaling pathways. There are several possible explanations for the disease resistance of the *pmr5* mutant. This mutant may be a less hospitable host for powdery mildews. It is evident that the *pmr5* extra-haustorial matrix may have altered composition, especially of modified pectins, decreasing nutrient transport to the fungus or the powdery mildew pathogen may have limited ability to digest the *pmr5* outer epidermal cell wall. Alternatively, the *pmr5* cell wall may carry latent signaling molecules that are released upon powdery mildew infections to activate novel defenses (Vorwerk et al. 2007). Whatever the basis for disease resistance, the *pmr5* and *pmr6* mutants highlight the importance of cell wall composition in plant–pathogen interactions (Vogel et al. 2004).

7.4.6 *Roles of Salicylate,* NPR1, PAD4, *and* EDS5 *in Powdery Mildew Resistance to* Arabidopsis

The powdery mildew fungus, *E. orontii*, infection of *Arabidopsis* elicits the strong accumulation of *PR1*, *BGL2*, and *PR5* mRNAs (Fig. 7.1). In several plants, salicylate has been shown to act as a signal molecule in the activation of PR gene

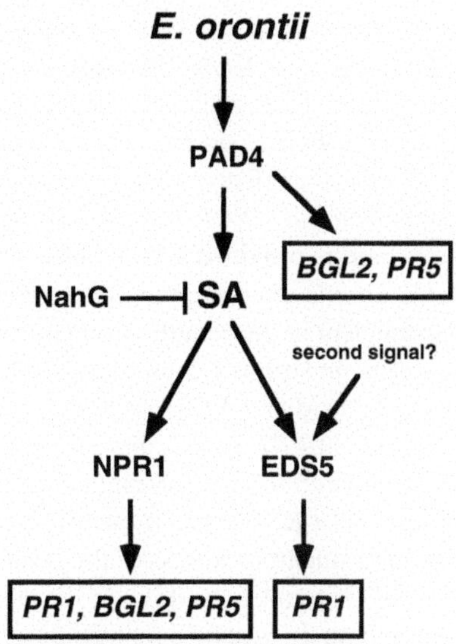

Fig. 7.1 Model of gene induction in *Arabidopsis* by *E. orontii* (Reuber et al. 1998)

expression (Yang et al. 1997), suggesting that a salicylate-dependent signaling pathway is important for limiting *E. orontii* growth in *Arabidopsis*. It is consistent with observations that *npr1* and *pad4* mutants as well as transgenic *nahG* plants exhibit enhanced susceptibility to *E. orontii*. *Arabidopsis npr1/nim1/sai1* mutants fail to activate *PR1*, *BGL2*, and *PR5* gene expression in response to exogenous salicylate, demonstrating that the *NPR1* protein acts in a signal transduction cascade which responds to a salicylate signal (Cao et al. 1994; Delaney et al. 1995). The gene *npr1* mutant accumulate essentially wild-type levels of *BGL2*, and *PR5* mRNAs, and about 10% of the wild-type levels of *PR1* mRNA in response to *P. syringae* (Glazebrook et al. 1996). Similarly, Reuber et al. (1998) observed only a modest reduction in the levels of *BGL2* and *PR5* mRNAs and about 20% of the wild-type levels of *PR1* mRNA following *E. orontii* infection of *npr1–1*. Assuming that the *npr1* mutants are not leaky, the activation of PR gene expression in the *npr1* mutants by *P. syringae* and *E. orontii* suggests that there are *NPR1*-independent pathogen-activated pathways that can lead to PR gene expression. There is a possibility that the *npr1* mutants could retain some activity, since a transgenic line that is suppressed for *NPR1* expression exhibits very low *PR1* activity after *P. syringae* infection (Cao et al. 1998). It has been observed that *nahG* plants exhibit very low levels of *PR1* induction following *E. orontii* infection suggests that salicylate-dependent pathways are required for almost all of the *PR1* expression in response to *E. orontii*. These results are consistent with those of Zhao and Last (1996), who found that *PR1* expression in response to virulent *P. syringae* is completely abolished in *nahG* plants. Moreover, the fact that *nahG* plants are more susceptible to *E. orontii* than *npr1* plants and exhibit lower levels of *PR1* induction following *E. orontii* infection suggests the existence of an *NPR1*-independent but salicylate-dependent defense gene activation pathway. There is precedent for such pathways as camalexin induction in response to *P. syringae* is salicylate-dependent but *NPR1*-independent (Zhao and Last 1996). Because the *nahG* plants retain significant *BGL2* and *PR5* expression, there must also be pathogen-activated salicylate-independent pathways leading to *BGL2* and *PR5* expression.

Further, evidence that both salicylate-independent pathways and salicylate-dependent but *NPR1*-independent pathways are involved in limiting the growth of *E. orontii* is provided by the *pad4* mutant. The *pad4* mutant is the most susceptible to *E. orontii* of all the mutants tested. The *pad4–1* mutant was recently reported to be deficient in the accumulation of salicylic acid in response to infection by virulent *P. syringae* strains (Zhou et al. 1998), indicating that *PAD4* functions upstream of salicylic acid and *NPR1*. However, Reuber et al. (1998) found less than 10% accumulation of *PR1*, *BGL2*, and *PR5* mRNAs following infection of *pad4* with *E. orontii*, compared to 20% accumulation of *PR1*, and greater than 50% accumulation of *BGL2* and *PR5* in *npr1* mutants. The small amount of PR gene expression in *pad4* could be due to leakiness of the *pad4* allele that have been used in the production of a small amount of salicylate through a *PAD4*-independent pathway or induction through an *E. orontii*-dependent but salicylate-independent pathway. Evidence for the existence of salicylate-independent defense gene activation pathways is provided by the observation that *pad4* is significantly more deficient in *BGL2* and *PR5* expression than the nahG transgenics. If this interpretation is correct, *PAD4* must clearly play a pivotal role in the presumptive salicylate-independent pathway(s).

The *eds5–1* also showed reduced *PR1* accumulation in response to *E. orontii*. Almost all of the *PR1* expression in response to *E. orontii* seems to be salicylate dependent. Therefore, it is most likely that *EDS5* also functions in a salicylate-dependent pathway. The additive effects of the *npr1–1* and *eds5–1* mutations on *PR1* expression in the *eds5–1 npr1–1* double mutant suggest that the *EDS5* and *NPR1* gene products may act in parallel signal transduction pathways, both of which are necessary for full expression of *PR1* in response to pathogen infection. If this is the case, the *EDS5*-dependent pathway must also require a second, salicylate-independent pathogen-generated signal to induce PR gene expression, since the *npr1–1* mutation completely blocks expression of PR genes in response to exogenously applied salicylate. However, since it is not clear whether the *npr1–1* mutation is a complete loss-of- function mutation, and because the nature of the *eds5–1* mutation is unknown, it is also possible that neither mutation results in a null phenotype and that the *EDS5* and *NPR1* gene products function in the same pathway. The fact that the double *npr1–1 eds5–1* mutant is more susceptible than either the *eds5–1* or *npr1–1* mutant that *EDS5* plays a significant role in limiting the extent of *E. orontii* infection. There is no accumulation of salicylate in uninfected leaves since there is no PR gene induction. This is expected for a non-necrotizing pathogen such as *E. orontii* which does not typically induce systemic acquired resistance (Reuber et al. 1998).

7.4.7 Role of Chitin Gene to Powdery Mildew Resistance

Chitin is a major component of fungal walls and insect exoskeletons. Plants produce chitinases upon pathogen attack, and chito-oligomers induce defense responses in plants, though the exact mechanism behind this response is unknown. Using the *ATH1* Affymetrix microarrays consisting of about 23,000 genes, Ramonell et al. (2002) examined the response of *Arabidopsis* (*Arabidopsis thaliana*) seedlings to chito-octamers and hydrolysed chitin after 30 min of treatment. The expression patterns elicited by the chito-octamer and hydrolysed chitin were similar. Microarray expression profiles for several genes were verified via northern analysis or quantitative reverse transcription-PCR. T-DNA insertion mutants for nine chito-oligomer-responsive genes have been characteriszed. Three of the mutants were more susceptible to the fungal pathogen, powdery mildew, than wild type as measured by conidiophore production. These three mutants included mutants of genes for two disease resistance-like proteins and a putative E3 ligase. The isolation of loss-of-function mutants with enhanced disease susceptibility provides direct evidence that the chito-octamer is an important oligosaccharide elicitor of plant defenses. Also, this study demonstrates the value of microarray data for identifying new components of uncharacterized signaling pathways. The location of chitin present in different structures of powdery mildew pathogen (*Ec*) has been revealed through TEM.

7.4.7.1 Location of Chitin in Powdery Mildew Fungus Through TEM

Chitin is present at the tips of growing hyphae and in conidia of *E. cichoracearum* during infection of *Arabidopsis.* Chitin is one of the major cell wall components in many fungi. Fungal cell walls change during growth and development, and the deposition and subsequent removal of chitin are particularly important during septation (Smits et al. 2001; Adams 2004). However, the distribution of chitin in *E. cichoracearum* structures formed during infections of *Arabidopsis* was unknown. Cell wall components and chitin were localized in growing *E. cichoracearum* using labeled wheat germ agglutinin (WGA). WGA contains a group of closely related isolectins that can bind oligosaccharides containing terminal GlcNAc or chitobiose and membrane glycoproteins (Peters and Latka 1986). WGA has been used in several studies on chitin distribution in fungal walls (Galun et al. 1976; Tronchin et al. 1981). Laser scanning confocal microscopy visualized WGA-488 Alexa conjugates at the growing tips of appressoria and hyphae (Plate 7.4a, b). WGA staining was also observed in the cell walls of conidia developing on conidiophores (Plate 7.4c). As the conidia matured and detached from the conidiophore, chitin was localized to both ends of conidia (Plate 7.4d). When conidia germinated, the region where the appressorial germ tube developed accumulated chitin. Furthermore, although mature conidia have chitin at their ends (Plate 7.4d), germinated conidia (Plate 7.4a) lack detectable

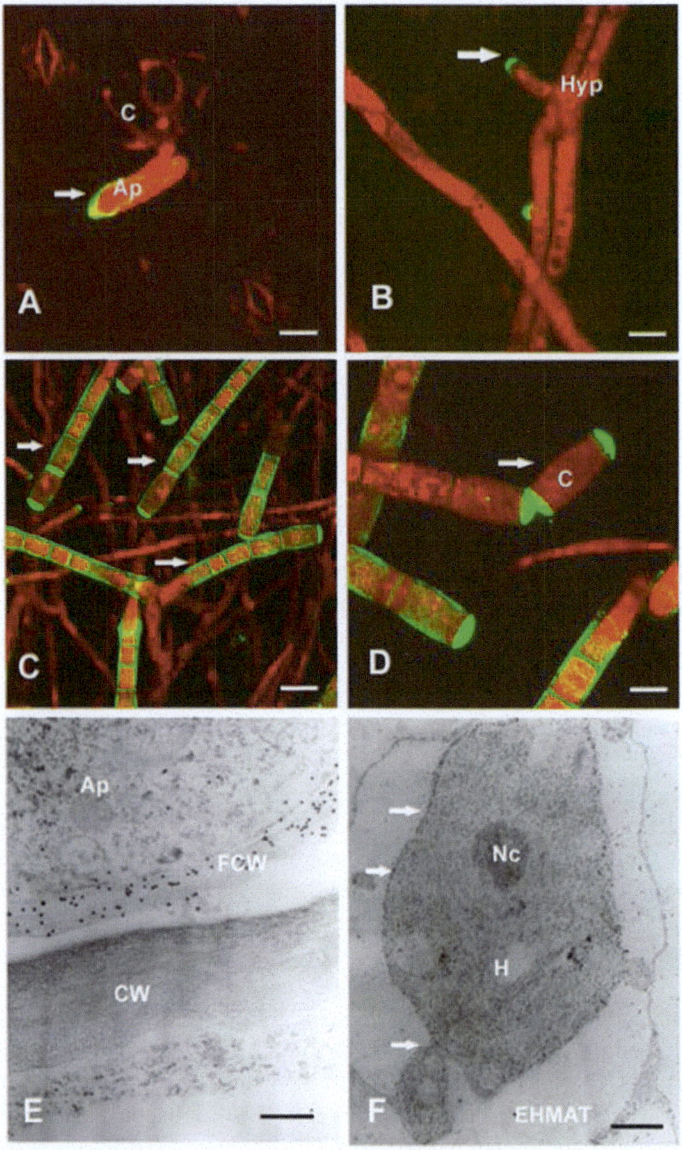

Plate 7.4 Chitin localization in the powdery mildew, *E. cichoracearum*. Confocal images are presented in **a** to **d**, and transmission electron micrographs in **e** and **f**. In **a** to **d**, samples were stained with PI (red channel) to highlight fungal structures, although plant structures can be stained with this nonspecific stain, and chitin was localized with the lectin, WGA-Alexa Fluor 488 (green channel). In **e** and **f**, WGA colloidal gold conjugates were used to localize chitin. (**a**) A merged confocal micrograph showing chitin localized to the tip of the appressorium (arrow; 1 dpi). Plant guard cells are also partially stained with PI in this image. Bar = 12 µm. *C* conidium, *Ap* appressorium. (**b**) Chitin localization at the growing tip of hypha (arrow) but not on the elongated hyphal region (3 dpi). *Hyp* hyphae. Bar = 11 µm. (**c**) Chitin labeling occurs in the developing conidia still

chitin in this position, suggesting temporal and spatial regulation of chitin localization. Transmission electron microscopy confirmed that WGA-gold colloidal conjugates were present at the growing tip of appressorium (Plate 7.4e) and also accumulated in the cell wall of the fungal haustorium, a feeding structure that forms in epidermal cells (Plate 7.4f). Fungal structures directly in contact with the plant and likely to be exposed to plant chitinases, such as appressoria and haustoria, show relatively high content of chitin. Therefore, it is likely that chito-oligomers will be generated during the course of powdery mildew infections (Ramonell et al. 2005).

7.4.8 Role of **Arabidopsis** *Powdery Mildew-Resistant Genes in Hypersensitivity*

Arabidopsis plants with multiple copies of genomic fragments containing *RPW8.1* and *RPW8.2*, transcriptionally regulated by their native promoters, developed apparently SHL that were associated with enhanced expression of *RPW8.1*, *RPW8.2*, and *PR* genes. In the transgenic line S24, SHL appeared as isolated necrotic spots that enlarged to form necroses that resembled the HR induced by powdery mildew pathogens on plants containing *RPW8.1* and *RPW8.2*, such as resistant accession Ms-0 and Col-0 transgenic line S5. SHL developed in S24 and other lines in which the transgenes *RPW8.1* and *RPW8.2* were under the control of their native promoters but not in lines in which these transgenes were under the control of the 35S promoter. However, the level of the transcripts for *RPW8.1* and *RPW8.2* was not greatly different between these different lines. To investigate this discrepancy, Xiao et al. (2003) used a reporter for the *RPW8.1* promoter and observed that in progeny of line S24, this reporter was activated only in cells at the margin of spreading lesions, which represented <10% of the leaf tissue. Apparently, there would have been at least ten times more transcripts of *RPW8.1* and *RPW8.2* in these S24 cells destined to become necrotic than in cells of lines in which the genes were expressed from the 35S promoter. These results indicate that the native promoters of *RPW8.1* and *RPW8.2* probably are required for SHL. It is possible that any lines with similarly high expression of the transgenes under the control of the 35S promoter would not survive. By contrast, overexpression of the *R* gene *Pto* controlled by the 35S promoter in tomato results in the non-propagative spontaneous death of isolated cells (Tang et al. 1999). However, these lesions were barely visible to the naked eye and much smaller than the HR lesions induced by avirulent strains of the pathogen.

Plate 7.4 (continued) attached to conidiophores (arrows; 7 dpi). Bar = 31 μm. (**d**) A mature conidium (arrow) detaching from a conidiophore shows strong chitin localization at the both ends. Bar = 16 μm. (**e**) Chitin occurs in the fungal appressorial cell wall but not on the plant cell wall. CW, plant outer epidermal cell wall; FCW, fungal cell wall. Bar = 2.15 μm. (**f**) Chitin is found in the haustorial cell wall (arrows). *EHMAT* extra-haustorial matrix, *H* haustorium, *Nc* haustorial nucleus. Bar = 9.21 μm (Ramonell et al. 2002)

There was a general positive correlation between SHL development and level of *RPW8.1* and *RPW8.2*. It indicates that transcripts *RPW8.1* and *RPW8.2* were induced by *E. cichoracearum* pathogen (Xiao et al. 2001), which led to speculate that *RPW8*-mediated HR and SHL involve a self-amplification mechanism. By introducing a *GUS* reporter for the putative *RPW8.1* promoter into a background containing multiple copies of *RPW8.1* and *RPW8.2*, Xiao et al. (2003) clearly demonstrated that transcription of *RPW8.1* was further enhanced in the localized area where SHL developed, most likely as a consequence of the expression of *RPW8.1* and *RPW8.2* above a threshold level that has yet to be determined. Thus, the promoters of *RPW8.1* and probably *RPW8.2* seem to be critical for the transcriptional self-amplification. Analysis of the 1 kb sequence upstream of the *RPW8.1* and *RPW8.2* translational starts revealed three W-box elements in the promoter region of *RPW8.1* (TTGACC at −162 bp, TTGACT at −282 bp, and AGTCAA at −135 bp in front of the translational start) and two in the promoter region of *RPW8.2* (TTGACT at −281 and −526 bp). W-boxes are *cis*-acting elements often found in promoters of many SA and pathogen-responsive genes, such as *NPR1* and *PR1* (Lebel et al. 1998; Yu et al. 2001). They are binding sites for the WRKY family of transcription factors for the transcriptional regulation of defense-related genes (Rushton and Somssich 1998; Eulgem et al. 2000). Therefore, it was anticipated that *RPW8.1* and *RPW8.2* would be induced by SA.

SA is required for HR and for the expression of disease resistance in many plants (Malamy and Klessig 1992; Dempsey et al. 1999; Shirasu et al. 1999; Alvarez 2000). In *Arabidopsis*, SA also may function in a feedback amplification circuit involving the defense signaling genes *EDS1*, *PAD4*, and *EDS5* that amplifies the resistance response (Falk et al. 1999; Jirage et al. 1999; Feys et al. 2001; Rusterucci et al. 2001). These responses have similarity to the cell death in the *lsd5* mutant, which is SA dependent and SA induced (Weymann et al. 1995). SA also is essential to *RPW8.1*- and *RPW8.2*-dependent HR and plants containing the *nahG* transgene which did not express the HR (Xiao et al. 2001). SA is critical for *RPW8*-mediated SHL, because the *nahG* transgene suppressed SHL in line S24. Moreover, S24 plants in which SHL was initiated also had high levels of endogenous SA, and this finding correlated positively with the enhanced expression of *RPW8.1* and *RPW8.2*. Significantly, environmental conditions that suppressed SHL also suppressed both the accumulation of SA and the enhanced transcription of *RPW8.1*, and *RPW8.2*. These data suggest that the expression of *RPW8.1* and *RPW8.2* above a threshold level leads to SA accumulation, because SA did not accumulate, nor did SHL develop, in the single-copy line S5. Paradoxically, perhaps, by SA application to S24 plants that SA regulates *RPW8.1* and *RPW8.2* expression directly enhancing SHL in otherwise suppressive conditions has been shown by Xiao et al. (2003). Significantly, Col-0 plants containing the *GUS* reporter for the putative *RPW8.1* promoter did not develop lesions when treated with SA, but GUS activity increased after SA application.

Xiao et al. (2003) studies support a model for the *RPW8*-dependent SHL, HR, and resistance in which SA mediates the transcriptional self-amplification of *RPW8* (Fig. 7.2). This feedback circuit amplifies the defense response stimulated by *RPW8*

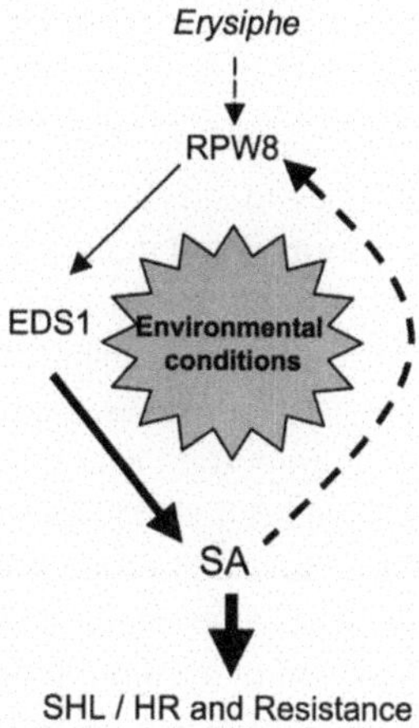

Fig. 7.2 Model for *RPW8*-mediated SHL, HR, and resistance. In this model, the recognition of *Erysiphe* pathogens by *RPW8* triggers defense responses via a SA-dependent pathway (Xiao et al. 2001), leading to SA accumulation. Increased SA further enhances the expression of *RPW8.1* and *RPW8.2* via a feedback amplification circuit, leading to HR and resistance or SHL. Environmental conditions that suppress this amplification circuit suppress HR and disease resistance or SHL (Xiao et al. 2003)

independent of the pathogen. Xiao et al. (2003) proposed that unknown local stimuli trigger this amplification circuit in line S24, leading to local necroses. *RPW8.1* promoter–*GUS* reporter experiments provide compelling independent evidence that SA enhances *RPW8.1* transcription through the promoter element, even in lines lacking the *RPW8* genes. Further, evidence for this model showed that the *RPW8.1* promoter is activated only in incipient lesions and at the leading edges of developing lesions. This finding indicates that *RPW8.1* expression and *RPW8.2* expression as well, is programmed to occur just ahead of the advancing lesion but not throughout the leaf. A corollary to this is that local signaling molecules, which include SA, must define the advance of lesions into tissue in which the transcription of *RPW8* genes is activated. This model also could explain why SHL did not develop in *35S::RPW8* lines: this promoter is not activated by SA. SA-dependent transcription of *RPW8.1* and *RPW8.2* forms part of an amplification circuit that leads to the accumulation of SA and SHL or HR. Presumably, therefore, SA formed during the expression of resistance to different pathogens also would induce the accumulation of transcripts of *RPW8.1* and *RPW8.2*. This raises the possibility that *RPW8.1* and *RPW8.2*, and possibly other members of this gene family as well (Xiao et al. 2001), contribute to the expression of HR against other pathogens. However, *RPW8.1* and *RPW8.2* couple the specific recognition of powdery mildew pathogens to the induction of this defense response, and a future challenge is to understand the basis of this specificity (Xiao et al. 2003).

7.4.9 *Role of* NPR1 *Gene in Resistance to Powdery Mildew*

Plants are very often exposed to a variety of biotic stresses and thus have evolved multidimensional defense approaches to survive or retain their fitness (Roux et al. 2014). The plants display both preformed and inducible defense mechanisms to overcome pathogen challenges. However, much stronger and long-lasting is inducible defense response such as systemic acquired resistance (SAR). Most of the PR proteins such as glucanases, chitinases, thaumatins, and defensins possess antifungal activities and are known to play an important role in disease resistance. Exogenous application of SA or its analogs have been also revealed to activate SAR pathway in plants (Durrant and Dong 2004; Makandar et al. 2006). Conversely, *Arabidopsis thaliana* plants expressing *NahG* transgene which codes for salicylate hydroxylase (SA-degrading enzyme) were deficient in accumulating SA and hence failed to activate SAR (Delaney et al. 1995). In addition to SA, a group of heterogeneous proteins are crucial for the activation of SAR. Among them are the *NPR1* protein, a key regulator in the SA-mediated SAR signal transduction pathway. The quest to discover the SA receptor led to the discovery of a regulatory or transcription co-factor protein *NPR1* (Cao et al. 1994). However, many studies have revealed that *NPR1* is linked to SA signaling; however, its role as SA receptor remains largely unknown. In this context, Wu et al. (2012) have recently reported that *NPR1* is the receptor for SA pathway in *Arabidopsis*. In addition, two *NPR1* paralogs, namely, *NPR3* and *NPR4*, bind SA and control the proteasome-mediated degradation of *NPR1* protein through their interaction with *NPR1* (Fu et al. 2012). After pathogen infection, plants produce a variety of phytohormones; their composition, quantity, and timing significantly vary among plant species and depend mainly on the pathogens' lifestyle and their mode of infection (De-Vos et al. 2005). SA pathway generally provides resistance to biotrophic pathogens, whereas jasmonic acid/ethylene (JA/ET) pathways are commonly associated with resistance to necrotrophic pathogens and to herbivorous pests (Glazebrook 2005; Bari and Jones 2009). Generally, SA and JA signaling pathways operate antagonistically, and thus, elevated resistance against biotrophs is often related with increased susceptibility to necrotrophs and vice versa (Grant and Lamb 2006). Many regulatory components involved in SA/JA crosstalk have been identified; among them is *NPR1* which plays a crucial role in regulating SA-mediated suppression of the JA pathway (Spoel et al. 2003; Pieterse et al. 2012; Thaler et al. 2012; Van der Does et al. 2013). The SA/JA antagonism is commonly found in many plant species under various taxonomic groups; therefore, it seems to be evolutionarily conserved (Thaler et al. 2012).

As first discovered in *Arabidopsis*, various *AtNPR1* homologs have been isolated thereafter in many agriculturally important crops (Chen et al. 2013; Zhong et al. 2015). *NPR1* is a multigene family in *Arabidopsis* with multifaceted functions. The genes *AtNPR1* and *AtNPR2* are notably considered as a key regulator of SAR (Cao et al. 1997, 1998; Zhang et al. 2003), while *AtNPR3* and *AtNPR4* are known as negative regulator of SAR (Fu et al. 2012). Moreover, another group of *AtNPR1* homologs are *AtBOP1* and *AtBOP2*, which are related with lateral organ development (Hepworth et al. 2005). However, most of the studies were carried out on *Arabidopsis NPR1* (*AtNPR1*). Structurally, *AtNPR1* and its homologs contain an ankyrin repeat,

N-terminal BTB/POZ broad-complex, Tramtrack, Bric a brac/poxvirus, and zinc finger domains, respectively (Cao et al. 1997; Aravind and Koonin 1999). In non-induced state, *NPR1* exists as an inactive oligomer form into cytosolic region. However, after SA accumulation, the redox status of the cell changes which leads to dissociation of the inactive oligomer *NPR1* to active monomers and their translocation to the nucleus where they bind to TGA factors, thereby inducing PR genes (Mou et al. 2003; Tada et al. 2008). Transcriptional studies have shown that *NPR1* is expressed at low levels in mock plants but is induced significantly after microbial attack or treatment with SA or its biologically active analogs. Many studies have revealed *NPR1* mutant plants are more prone to diseases and also show altered expression of defense marker PR genes compared to *NPR1*-expressing plants (Glazebrook et al. 1996; Cao et al. 1997). Furthermore, *NPR1* also plays a role in crosstalk of SA/JA signaling pathways and in antagonistic effect of SA on JA signaling (Spoel et al. 2003). Hence, *NPR1* is considered as the positive regulator of SA-mediated plant immune responses. To explore the defense role of *NPR1* against bacterial, viral, and fungal pathogens, various overexpression studies have been carried out in both model and crop plant systems. *NPR1* mediates the SA-induced expression of pathogenesis-related (PR) genes and SAR. Overexpression of *NPR1* results in the increase in transcript levels of antifungal genes like *PR1, PR2* (glucanase), and *PR5* (thaumatin), which are universally known to have antifungal activity. Many studies have revealed the potential antifungal activity of these PR genes against a wide range of fungal pathogens. PR gene activity is regulated at the level of redox-dependent nuclear transport of *NPR1*. For example, overexpression of *NPR1* in *Arabidopsis* plants confers enhanced disease resistance to bacterial and fungal infections (Cao et al. 1998; Friedrich et al. 2001). Transgenic carrot plants overexpressing *AtNPR1* exhibit high disease resistance not only to biotrophs (*Erysiphe heraclei*) but also to necrotrophic fungal pathogen (*Botrytis cinerea, Sclerotinia sclerotiorum*, and *Alternaria radicina*), respectively (Wally et al. 2009). Parkhi et al. (2010) also reported that cotton transgenic plants expressing *AtNPR1* exhibited broad spectrum of disease resistance not only to fungal pathogens but also to nematodes. Additionally, tobacco plants overexpressing *Malus hupehensis* NPR1 confer resistance to *Botrytis cinerea* as well as induces battery of pathogen-related genes. Furthermore, studies have revealed that rice and wheat plants overexpressing *NPR1* gene confer broad spectrum of disease resistance against most disastrous pathogens *Magnaporthe oryzae*, *Fusarium verticillioides*, and *Fusarium oxysporum*, respectively (Makandar et al. 2006; Quilis et al. 2008). The gene *NPR1* overexpression in *A. thaliana* was reported to result in an increase in the transcript levels of PR genes, hence proves that *NPR1*-dependent PR gene-mediated disease resistance (Cao et al. 1997; Friedrich et al. 2001). Most recently, overexpression of *NPR1* was also revealed to confer disease resistance against a broad range of pathogens in different crops (Dutt et al. 2015; Sundaresha et al. 2016). These results revealed that *NPR1* is a potential candidate gene for developing disease-resistant transgenic crops against multiple pathogens. Therefore, genetic transformation of defense regulatory genes (like *NPR1*) that controls the function of multiple defense genes is a potential source for developing broad-spectrum and long-lasting disease resistance against pathogens in *B. juncea*. *NPR1* (non-expressor of pathogen-related gene 1) is a bonafide receptor of salicylic acid (SA) which modulates multiple immune responses in plants especially activation of induced and systemic acquired resistance (SAR). Ali et al. (2017) have isolated and characterized a new *NPR1* homolog (*BjNPR1*) from *B. juncea*. The phylogenetic tree constructed based on the deduced sequence of *BjNPR1* with homologs from other species revealed that *BjNPR1* grouped together with other known *NPR1* proteins of Cruciferae family was nearest to *B. napus* (Fig. 7.3). Furthermore, expression analysis showed that *BjNPR1* was up-regulated after SA treatment and fungal infection but not by jasmonic acid or abscisic acid. To understand the defensive role of this gene, Ali et al. (2017) generated *B. juncea* transgenic lines overexpressing *BjNPR1* and further confirmed by PCR and Southern blotting. The transgenic lines showed no phenotypic abnormalities, and constitutive expression of *BjNPR1* activates defense signaling pathways by priming the expression of antifungal PR genes. Moreover, *BjNPR1* transgenic lines

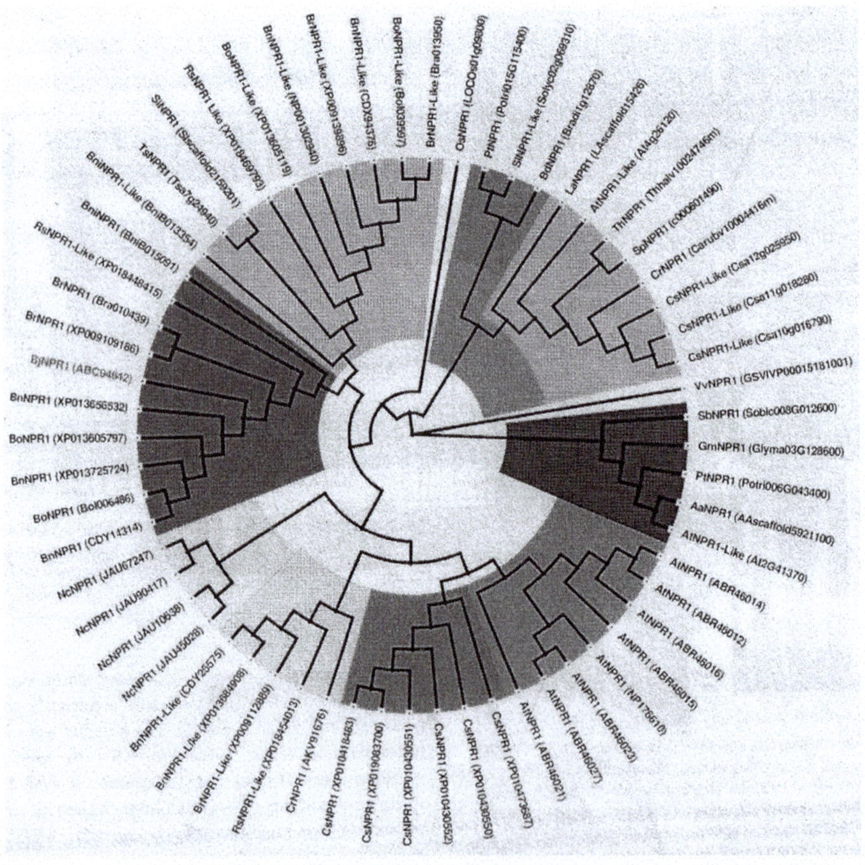

Fig. 7.3 Phylogenetic analysis of *BjNPR1* with other *NPR1* proteins from different plant species. The deduced amino acid sequences of *BjNPR1* were retrieved from NCBI GenBank and were further aligned with ClustalW using *MEGA7.1* bioinformatic tool. The tree was generated using Maximum-Likelihood (ML) method with 1000 bootstrap replicates. GenBank IDs of each *NPR1* protein sequence are given in the brackets behind the species names (Ali et al. 2017)

showed enhanced resistance to *Alternaria brassicae* and *Erysiphe cruciferarum* as there was delay in symptoms and reduced disease severity than non-transgenic plants. In addition, the rate of disease spreading to uninfected or distal parts was also delayed in transgenic plants, thus suggesting the activation of SAR. It is suggested that *BjNPR1* is involved in broad spectrum of disease resistance against fungal pathogens including powdery mildew of *B. juncea* (Ali et al. 2017).

BjNPR1 plants showed improved resistance against powdery mildew. To assess the resistance level of *BjNPR1* transgenic lines against powdery mildew, plants were infected, and disease scoring was done for 1–3 weeks. In non-transgenic plants, a higher number of *E. cruciferarum* colonies were observed than transgenic lines on the 7th, 12th, and 17th days after infection (Plate 7.5a). Based on colony count, there was approximately 50% reduction in newly formed colonies between

Plate 7.5 Screening of *BjNPR1* transgenic lines for powdery mildew disease resistance. Forty-day-old wild-type plants and *BjNPR1* transgenic plants were infected with *E. cruciferarum*, and disease scoring was done at different time intervals. (**a**, **b**) *BjNPR1* transgenic lines (L2 and L5) showed reduced number of *E. cruciferarum* colonies than wild-type plants at the 7th, 12th, and 18th dpi. (**c**) *E. cruciferarum* disease severity in *BjNPR1* transgenic lines and wild-type plants. Bar = 35 μm. The asterisks indicate statistically significant differences between the *BjNPR1* transgenic and control (non-transgenic) plants after powdery mildew infection ($P < 0.05$; $P < 0.01$) (Ali et al. 2017)

transgenic lines and non-transgenic plants (Plate 7.5b). At the 17th day of infection, transgenic plants showed powdery mildew infection with a disease scale of 3–4 (30–40%), while non-transformed leaves (wild type) revealed 7–8 (70–80%) of disease incidence, respectively (Plate 7.5c). In addition, *E. cruciferarum*-mediated cell death was examined in *BjNPR1* transgenic and wild-type plants at different time points using trypan blue staining and light microscopy. Based on microscopic observations, more cell death was observed in control than that of transgenic plants (Plate 7.6a). To further investigate the role of *BjNPR1* in improving powdery mildew disease resistance, the growth or fungal biomass of *E. cruciferarum* in *BjNPR1* transgenic lines with wild-type plants was compared using light microscopy. As shown in Plate 7.6b, overexpression lines (L2 and L5) showed

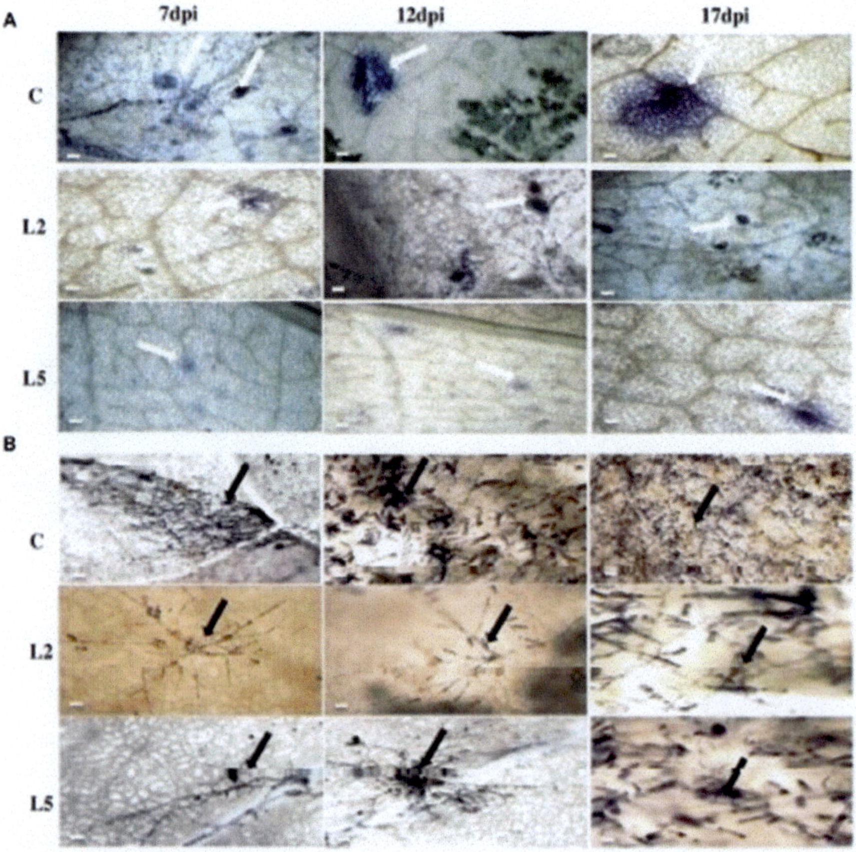

Plate 7.6 Microscopic examination of cell death and fungal biomass in *BjNPR1* transgenic and wild-type plants using trypan blue staining. (**a**) Microscopic examination of *E. cruciferarum*-mediated cell death in *BjNPR1* lines and wild-type plants are shown with bold white arrows. (**b**) *E. cruciferarum* spore load or biomass in *BjNPR1* transgenic lines and wild-type plants at various dpi after trypan blue staining are highlighted with bold black arrows. Bar = 30 mm (Ali et al. 2017)

significant reduction of fungal biomass of *E. cruciferarum* than wild-type plants at the 7th, 12th, and 17th dpi. However, overexpression of *BjNPR1* could not inhibit the growth of *E. cruciferarum* completely, thereby providing only partial resistance to powdery mildew disease. Also, more number of leaves and pods were infected in non-transgenic plants as compared to transgenic lines. These results indicate that *BjNPR1* transgenic plants exhibited partial resistance to powdery mildew infection, which was sufficient to delay the spread of infection in non-infected leaves or other parts (Ali et al. 2017).

7.4.10 Role of MAP65-3 *Gene in Powdery Mildew Resistance*

The microtubule (MT) cytoskeleton is a highly flexible and dynamic polar structure of the plant cell, assembled from tubulin heterodimers. It is involved in nuclear and cell division, in cell morphogenesis and expansion, and in intracellular transport (Wasteneys and Galway 2003; Wasteneys 2004; Hamada 2014). MTs also play a role in plant in responses to biotic and biotic stress exposure, and their rearrangements accompany both defense and successful infection by symbiotic and pathogenic microbes (Schmidt and Panstruga 2007; Hardham 2013). MT rearrangements occur during arbuscular mycorrhizal (Genre et al. 2005) and rhizobial symbiosis (Vassileva et al. 2005), during the formation of plant parasitic nematode feeding sites (Caillaud et al. 2008a; de Almeida Engler and Favery 2011), following virus attack (Martiniere et al. 2009), or following infection by filamentous oomycetes or fungi (Kobayashi et al. 1994; Baluska et al. 1995; Cahill et al. 2002; Takemoto et al. 2003; Hardham et al. 2008; Hoefle et al. 2011). Microtubule-associated proteins (MAPs) and their regulatory kinases and phosphatases are instrumental for microtubule dynamics (Wasteneys 2004; Gardiner 2013; Hamada 2014). They, and the small Rho of Plants (ROP) GTPases that regulate the MT cytoskeleton (Mucha et al. 2011), have been shown to determine plant susceptibility to viruses and fungi (Kragler et al. 2003; Ouko et al. 2010; Hoefle et al. 2011; Poraty-Gavra et al. 2013). The *Arabidopsis thaliana MAP65-3* (*AtMAP65-3*) is a critical module giant cell ontogenesis, and for successful pathogen development (Caillaud et al. 2008b), plant *MAP65*s are involved in the spatially and temporally regulated binding and bundling of MTs (Chan et al. 1999; Hamada 2014).

In *A. thaliana*, nine members of this family were identified (Hussey et al. 2002), and individual members have particular functions with respect to different MT arrays. *AtMAP65-3* is only associated with mitotic MT arrays (Muller et al. 2004; Caillaud et al. 2008b; Ho et al. 2011). The protein organizes both spindle morphogenesis and phragmoplast expansion (Muller et al. 2004; Caillaud et al. 2008b; Ho et al. 2011). Consequently, *AtMAP65-3* loss-of-function mutants are dwarf, with both shoots and roots being stunted, and polynucleate, hypertrophied cells with aberrant cell wall stubs occur frequently (Muller et al. 2004; Caillaud et al. 2008b; Ho et al. 2011). Plants protect themselves against pathogenic microorganisms by combining constitutive and induced defense mechanisms. The induction of plant defenses involves the recognition of compounds derived from the pathogen, called pathogen-associated molecular patterns (PAMPs). Pattern-triggered immunity (PTI) results from PAMP perception, which leads to the activation of signaling cascades and the subsequent induction of defense-related genes (Zipfel et al. 2004; Jones and Dangl 2006).

Pathogens are able to suppress these defenses by secreting effector proteins that manipulate host cell functions. In turn, plants evolved resistance proteins, which allow recognition of these effectors or their activities. This leads to effector-triggered immunity (ETI) and activation of the hypersensitive response (HR). The HR involves local programmed cell death that prevents pathogen spreading within the plant (Zipfel et al. 2004; Jones and Dangl 2006). Both PTI and ETI/HR involve mitogen-activated protein kinase (MAPK) cascades, the production of reactive oxygen species (ROS), and the transcriptional activation of genes, which, among others, encode antimicrobial pathogenesis-related (PR) proteins. The signaling pathways of PTI or ETI are fine-tuned by plant signaling molecules such as salicylic acid (SA), jasmonic acid (JA), and ethylene (ET) (Glazebrook 2005; Pieterse et al. 2012). The hormone SA plays a major role in plant resistance to (hemi-)biotrophic pathogens (Pieterse et al. 2012). In *A. thaliana*, SA synthesis occurs in plastids via *isochorismate synthase 1* (*ICS1* or *SID2*) and is triggered by pathogens. SA can be exported to the cytosol by the transporter enhanced disease susceptibility 5 (*EDS5*). SA accumulated in the cytoplasm can be converted to SA glucoside (SAG), which is stored in the vacuole and hydrolysed back to SA when needed. Elevated levels of total SA (free SA plus SAG) have been correlated with the induction of defense gene expression and enhanced plant resistance (Pieterse et al. 2012). In addition, SA is a key regulator of plant immunity through its antagonistic interaction with ET and JA pathways (Glazebrook 2005). Different from SA, JA and ET accumulate mainly in response to necrotrophic pathogens. An increasing number of mutants with reduced susceptibility to plant pathogens are described, and breeding for loss of susceptibility becomes a new strategy to achieve disease resistance (de Almeida-Engler et al. 2005; Dangl et al. 2013; Huckelhoven et al. 2013). Loss of disease susceptibility is frequently caused by a deregulation of SA- or JA-dependent plant defense signaling, by the impairment of cellular rearrangements, or by the limitation of nutrient supply for the pathogen (Dangl et al. 2013; Huckelhoven et al. 2013; Lapin and Van den Ackerveken 2013; Van Schie and Takken 2014). The expression of the gene encoding *AtMAP65-3* is strongly induced in *A. thaliana* upon infection by two biotrophic filamentous pathogens, the oomycete *Hyaloperonospora arabidopsidis* (*Hpa*) and the powdery mildew fungus *Erysiphe cruciferarum* (*Ec*). Both pathogens develop haustoria inside host cells that constitute the feeding structures for nutrient supply (O'Connell and Panstruga 2006). The plants mutated in *AtMAP65-3* are impaired in their susceptibility to both filamentous pathogens. Mutants accumulate increased levels of SA and constitutively express genes encoding PR proteins in the leaves. Increased SA accumulation is not responsible for the mutant dwarfism, indicating that *AtMAP65-3* exerts a dual role in positively regulating plant growth, and development, and in negatively regulating plant defense responses. After infection, powdery mildew pathogen induces the formation of intracellular bulbous structures called haustoria, which are required for the biotrophic lifestyle. The microtubule-associated protein *AtMAP65-3* plays a critical role in organizing cytoskeleton microtubule arrays during mitosis and cytokinesis. This renders the protein essential for the development of giant cells, which are the feeding sites induced by root knot nematodes. The *At-MAP65-3* expression is also induced in leaves upon infection by the downy mildew oomycete and the powdery mildew fungus. Loss of *AtMAP65-3* function in the *map65-3* mutant dramatically reduced infection by both pathogens, predominantly at the stages of leaf penetration. Whole transcriptome analysis showed an over-represented, constitutive activation of genes involved in SA biosynthesis, signaling, and defense execution in *map65-3*, whereas jasmonic acid (JA)-mediated signaling was down-regulated. Preventing SA synthesis and accumulation in *map65-3* rescued plant susceptibility to pathogens, but not the developmental phenotype caused by cytoskeleton defaults. *AtMAP65-3* thus has a dual role. It positively regulates cytokinesis, thus plant growth and development, and negatively interferes with plant defense against filamentous biotrophs. Quentin et al. (2016) suggested that downy mildew and powdery mildew stimulate *AtMAP65-3* expression to down-regulate SA signaling for infection.

7.4.11 Role of Receptor-Like Cytoplasmic Kinases in Powdery Mildew Resistance

Receptor-like cytoplasmic kinases (RLCKs) belong to the receptor-like kinase (RLK) superfamily that is involved in a variety of biological processes like plant growth, development, and immune responses (Afzal et al. 2008). RLCKs share a conserved serine/threonine (Ser/Thr) kinase domain (Afzal et al. 2008; Gish and Clark 2011) together with the large and diverse transmembrane RLK protein family. Together over 600 RLKs and RLCKs exist in *Arabidopsis,* and over 1000 in rice (Shiu et al. 2004). In contrast to RLKs, RLCKs do not possess an extracellular and transmembrane domain resulting in their cytoplasmic localization. However, some RLCKs are anchored to the plasma membrane through myristoylation motifs (Murase et al. 2004; Tang et al. 2008; Veronese et al. 2006). As protein kinases, RLCKs transmit intracellular signals through phosphorylation of target proteins or through RLK complex-mediated transphosphorylation events. Although little is known about their precise biological functions, several RLCKs were reported to be, either alone or in concert with RLKs, involved in plant development and immunity (Lin et al. 2013). In this regard, RLCKs play a role in the embryonic patterning process (Bayer et al. 2009), self-incompatibility (Murase et al. 2004), organ separation (Burr et al. 2011), ethylene (Laluk et al. 2011), and brassinosteroid signaling (Sreeramulu et al. 2013; Tang et al. 2008). In plant immunity, RLCKs are involved in both pathogen-associated molecular pattern (PAMP)-triggered immunity (PTI), and effector-triggered immunity (ETI) (Lu et al. 2010; Swiderski and Innes 2001;

Zhang et al. 2010). RLCKs are divided into 13 sub-families (RLCK I–XIII) (Shiu et al. 2004). *Arabidopsis* RLCKs of the sub-family VI A interact with the plant-specific *Rho* family of small monomeric G proteins called 'Rho of plants' (RAC/ROPs) (Jurca et al. 2008; Molendijk et al. 2008). A function of the VI sub-family of RLCKs as RAC/ROP downstream signaling effectors in plants was supported by RAC/ROP GTPase-dependent activation of RLCKs in *Arabidopsis, Medicago truncatula* (*M. truncatula*), and barley (*Hordeum vulgare* L) (Dorjgotov et al. 2009; Huesmann et al. 2012). RAC/ROP proteins regulate processes like cell development, hormone signaling, cytoskeleton rearrangement, and plant disease resistance or susceptibility via various downstream effector proteins including RLCKs (Berken 2006; Nibau et al. 2006). RAC/ROPs act as molecular switches, transducing extracellular signals into intracellular responses by shuttling between an inactive GDP-bound and an activated GTP-bound state. Several regulatory molecules including guanine nucleotide exchange factors (ROPGEFs), GTPase-activating proteins (ROPGAPs), and guanine nucleotide dissociation inhibitors (GDIs) adjust the balance between these two forms (Nibau et al. 2006). Furthermore, RAC/ROPs are divided into two phylogenetic sub-groups (type I and type II) depending on their post-translational lipid modifications, which anchor the active proteins in the plasma membrane (Winge et al. 2000). Besides the eleven characterized RAC/ROP proteins in *Arabidopsis* (Li et al. 2001), six and seven RAC/ROPs are described in barley (Schultheiss et al. 2003), and rice (Chen et al. 2010) respectively. In *Arabidopsis*, *AtROP4* and *AtROP6* are involved in the auxin binding protein 1 (ABP1)-mediated formation of lobed pavement cells through organization of the actin and microtubule cytoskeleton as well as in pathogen response (Fu et al. 2005; Fu et al. 2009; Poraty-Gavra et al. 2013; Xu et al. 2010). In addition, *AtROP4* and *AtROP6* are described as activators of *AtRLCK*s. Nevertheless, the functional knowledge about RAC/ROP-regulated RLCK signaling in plants is limited. Recently, the barley RLCK ROP binding kinase1 (*HvRBK1*), which is closely related to *AtRLCK VI A3*, was shown to serve as RAC/ROP effector in the barley–barley powdery mildew interaction (Huesmann et al. 2012). *HvRBK1* interacted with the susceptibility factor *HvRACB*, which is required for successful invasion of intact barley epidermal cells by the biotrophic fungus *Blumeria graminis* f. sp. *hordei*, the causal agent of powdery mildew disease (Hoefle et al. 2011; Schultheiss et al. 2002). All RLCKs described as RAC/ROP interactors are members of the RLCK VI sub-family that is divided into groups A and B based on their domain structure (Jurca et al. 2008; Molendijk et al. 2008). In *Arabidopsis*, the RLCK VI sub-family shows up-regulation of gene expression under abiotic stress or hormone treatments as well as in response to the pathogens *Botrytis cinerea* and *Phytophthora infestans* (Jurca et al. 2008; Molendijk et al. 2008). Reiner et al. (2014) have described the *HvRBK1*-related *Arabidopsis AtRLCK VI A3* as direct molecular interactor of *Arabidopsis* RAC/ROPs. *AtRLCK VI A3* interacts with *AtROP*s in yeast and shows increased kinase activity in the presence of constitutively activated (CA) *AtROP6* in vitro. Furthermore, *AtRLCK VIA3* mutant lines show a reduced growth phenotype and an increased number in trichome branching. Finally, a slight increase in susceptibility towards the powdery mildew fungus *Erysiphe* *cruciferarum* was observed in *AtRLCK VIA3* mutant lines suggesting a function of *AtRLCK VIA3* in the *Arabidopsis*–powdery mildew pathosystem. RLCKs of the VIA sub-family are possible downstream effectors of the small monomeric G proteins of the plant-specific *Rho* family, called 'Rho of plants' (RAC/ROPs). *Arabidopsis thaliana AtRLCK VI A3* acts as a molecular interactor of *AtROPs*. In *Arabidopsis* epidermal cells, transient co-expression of plasma membrane located constitutively activated (CA) AtROP4 or CA AtROP6 resulted in the recruitment of green fluorescent protein (GFP)-tagged *AtRLCK VI A3* to the cell periphery. Intrinsic kinase activity of *AtRLCK VI A3* was enhanced in the presence of *CA AtROP6* in vitro and further suggested a functional interaction between the proteins. In the interaction of the biotrophic powdery mildew fungus *Erysiphe cruciferarum* and its host plant *Arabidopsis*, *Atrlck VI A3* mutant lines supported enhanced fungal reproduction. Furthermore, *Atrlck VI A3* mutant lines showed slightly reduced size and an increase in trichome branch number compared to wild-type plants. There is a role of the *AtROP*-regulated *AtRLCK VI A3* in basal resistance to *E. cruciferarum* as well as in plant growth and cellular differentiation during trichome morphogenesis (Reiner et al. 2014).

*7.4.12 Arabidopsis Triple Mutants (***mlo2, mlo6, and mlo12***) Mechanism of Resistance to Powdery Mildew*

The major naturally occurring source of resistance effective against powdery mildews in *Arabidopsis* is a gene identified as *resistance to powdery mildew8* (*RPW8*) locus (Xiao et al. 2001; Gollner et al. 2008). This complex locus shows extensive intra-specific genetic variation and confers dominantly inherited resistance against multiple powdery mildew species. The respective genes encode non-canonical resistance proteins that lead to arrest of fungal pathogenesis after host cell penetration (post-penetration resistance). Effective resistance correlates with the encasement of the fungal feeding structures (haustorial complexes) in a callose-containing cell wall matrix (Wang et al. 2009). A different type of powdery mildew resistance is conferred by recessively inherited loss-of-function mutations in specific *mildew resistance locus O* (*MLO*) genes. These genes, which encode integral membrane proteins of unknown biochemical activity, comprise a family of 15 members in *Arabidopsis* (Devoto et al. 2003). Loss-of-function mutations in *MLO2* (At1g11310) result in incomplete resistance against powdery mildew attack that is characterized by a reduction in host cell entry rates by 50%. This coincides with an arrest of hyphal growth prior to the formation of conidiophores in *mlo2* plants, resulting in almost entirely abolished sporulation (Vogel and Somerville 2000; Consonni et al. 2006). Mutations in *MLO6* (At1g61560) and *MLO12* (At2g39200) do not affect powdery mildew interactions on their own. However, it was cooperatively enhanced *mlo2*-conditioned resistance and in combination with a mutation in *MLO2* causes a complete lack of host cell penetration by fungal sporelings, leading to complete immunity (pre-penetration resistance). This type of powdery mildew resistance is best known from barley, where natural and induced *mlo* mutants have been discovered more than 70 years ago and have been successfully employed in agriculture for over 35 years (Jorgensen 1992; Lyngkjaer et al. 2000). More recently, *mlo*-based resistance has been described in several other monocotyledonous and dicotyledonous plant species such as pea, tomato, and wheat. Hence, *mlo*-based resistance is a seemingly universal phenomenon within angiosperm plant species that are hosts to powdery mildew fungi. At the phenotypical and molecular level, *mlo* resistance resembles the highly effective defense against non-adapted powdery mildews (non-host resistance). The *mlo2 mlo6 mlo12* mutant shows a spectacular level of resistance against different powdery mildews. Conversely, it slightly enhanced disease symptoms and in part also pathogen proliferation interactions with some hemibiotrophic/necrotrophic pathogens such as *Alternaria alternata*, *A. brassicicola,* and *Phytophthora infestans*. These phenotypes might be the indirect consequence of deregulated mesophyll cell death in leaves of the *mlo2 mlo6 mlo12* mutant. Similar to the barley *mlo* mutant, leaves of the triple mutant are subject to spontaneous deposition of callose-containing cell wall appositions and ultimately premature senescence (Consonni et al. 2006, 2010). The role of *mlo* mutants in providing resistance to powdery mildew fungi is well-documented (Kusch and Panstruga 2017).

In order to establish a comprehensive interaction profile of the *Arabidopsis mlo2 mlo6 mlo12* triple mutant, Acevedo-Garcia et al. (2017) challenged individuals of two independent triple mutant lines with a broad panel of microorganisms along with *Golovinomyces orontii* causing powdery mildew of *Arabidopsis* that are known to be virulent on the *Arabidopsis* Col-0 accession. These two mutant lines represent two entirely independent allele combinations in the genetic background of Col-0. The resulting infection phenotypes were compared with that of Col-0 wild type, which served as control in all experiments. A novel *mlo2 mlo6 mlo12* triple mutant line with near-complete powdery mildew resistance is fully resistant to the adapted powdery mildew pathogen, *G. orontii* (Consonni et al. 2006). Acevedo-Garcia et al. (2017) generated a second triple mutant line, *mlo2*-6 *mlo6*-4 *mlo12*-8, which is based on different T-DNA insertions in the three *AtMLO* genes. As revealed by reverse transcriptase-polymerase chain reaction (RT-PCR) analysis, the T-DNA insertions in these lines result in a lack of full-length *MLO2*, *MLO6*, and *MLO12* transcripts. The two triple mutant lines grow similarly as Col-0 wild-type plants, but suffer from early leaf senescence.

This phenotype is evident by the slightly chlorotic rosette leaves of the two *mlo2 mlo6 mlo12* lines at the age of approximately 6 weeks (Plate 7.7c). The triple mutants upon challenge with *G. orontii* showed that line *mlo2*-6 *mlo6*-4 *mlo12*-8 fully resembles line *mlo2*-5 *mlo6*-2 *mlo12*-1 with respect to the macroscopic and microscopic infection phenotypes (Plate 7.7c–e). Unlike the Col-0 control plantsz which showed abundant fungal sporulation at 8 days postinoculation (dpi), individuals of both lines lacked visible powdery mildew symptoms (Plate 7.7c). This finding was consistent with analysis at the microscopic level at 48 hours postinoculation (hpi), which revealed an early abortion of fungal pathogenesis at the level of host cell entry in the two triple mutants, while Col-0 control plants showed

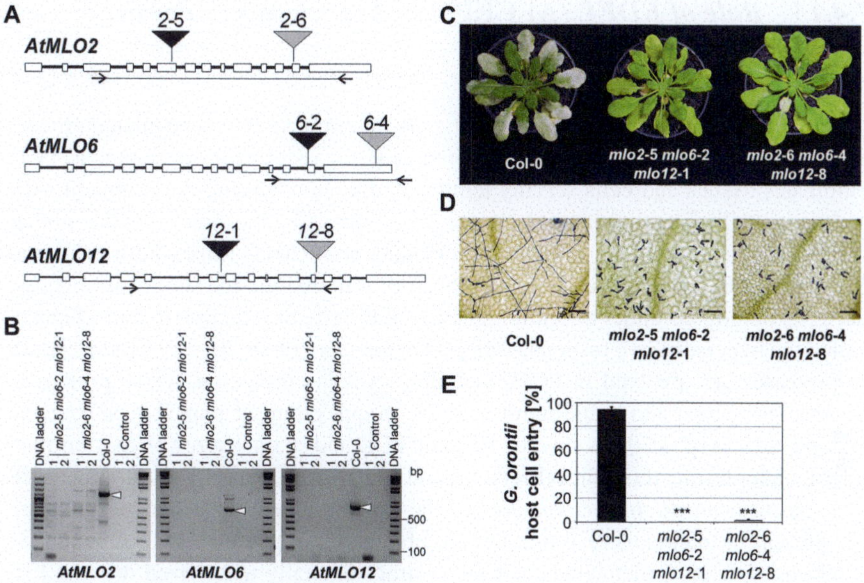

Plate 7.7 The *G. orontii* resistance phenotype of the *mlo2*-6 *mlo6*-4 *mlo12*-8 triple mutant is indistinguishable from the *mlo2*-5 *mlo6*-2 *mlo12*-1 triple mutant. Six-week-old *Arabidopsis* plants were touch-inoculated with *G. orontii* conidiospores. (**a**) Scheme depicting the T-DNA insertion sites in *MLO2*, *MLO6*, and *MLO12*. Rectangles represent exons and black lines introns. Triangles symbolize the T-DNA insertion sites of the various *mlo* alleles. Lines flanked by inverted arrows (primer binding sites) below the gene models indicate the RT-PCR amplicons used to test for *MLO* transcript accumulation in the mutant lines. (**b**) RT-PCR analysis of *MLO2*, *MLO6*, and *MLO12* transcript accumulation. Primer pairs covering the regions indicated in panel A were used to amplify the respective transcript amplicons from cDNA of lines *mlo2*-5 *mlo6*-2 *mlo12*-1 and *mlo2*-6 *mlo6*-4 *mlo12*-8 (two individuals each) as well as Col-0 wild-type plants (positive control). RT-PCR reactions without reverse transcription (control 1) and amplification without template (control 2) served as negative controls. White arrowheads indicate RT-PCR products of the expected size in case of Col-0 wild-type plants. (**c**) Representative macroscopic infection phenotypes at 8 dpi. (**d**) Light micrographs visualizing fungal pathogenesis at 48 hpi. Leaf samples were cleared in destaining solution, and fungal infection structures subsequently stained with Coomassie Brilliant Blue. Bars = 100 μm. (**e**) Quantitative assessment of host cell entry. Data show the mean ± standard error of the mean (SEM) from three experiments. In each experiment, at least 100 interaction sites from 1 to 3 leaves of 5 independent plants per genotype were assessed (total of >500 interaction sites per genotype and experiment). ***indicates a statistically significant difference from Col-0 ($P < 0.001$) according to a GLM test (binomial distribution) (Acevedo-Garcia et al. 2017)

extensive mycelial growth (Plate 7.7d). The line *mlo2*-5 *mlo6*-2 *mlo12*-1 was entirely resistant, lacking any recognizable host cell penetration (0% entry rate as judged by the absence of secondary hyphae and discernible haustoria), while line *mlo2*-6 *mlo6*-4 *mlo12*-8 allowed the occasional formation of fungal micro-colonies (Plate 7.7e). The two *mlo2 mlo6 mlo12* triple mutants are essentially equivalent with regard to the level of resistance against the obligate biotrophic powdery mildew pathogen, *G. orontii* (Acevedo-Garcia et al. 2017).

7.4.13 *Role of KDEL (*At CEP1*) in Arabidopsis to Powdery Mildew Resistance*

Programmed cell death (PCD) is a genetically determined, highly regulated process in all multicellular organisms and a prerequisite for successful development. PCD eliminates tissues and cells serving temporary functions during development such as tapetum cells in anthers and suspensor cells connecting the embryo to the mother plant or nucleus cells of a mature ovule (Zhang et al. 2014; Lopex-Fernandez and Maldonado 2015; Zhou et al. 2016). Plants furthermore limit the spread of fungal or bacterial pathogens under execution of PCD at the site of infection in a mechanism called the hypersensitive response (HR) (Dickman and Fluhr 2013). Diverse classes of proteases are involved in PCD, including cysteine proteases, serine proteases, aspartic proteases, and metalloproteases. A unique group of papain-type cysteine endopeptidases (CysEPs) is specific for plant PCD and characterized by a C-terminal KDEL endoplasmic reticulum (ER) retention signal (KDEL CysEPs) with RcCysEP from castor bean (*Ricinus communis*) as the founding member. KDEL CysEPs are not present in mammals or fungi, but are ubiquitous in plants. KDEL CysEPs are synthesized as pre-proenzymes and are co-translationally transferred into the ER, where the pre-sequence signal peptide is removed. The enzyme KDEL CysEPs can be stored as enzymatically inactive proenzymes in ER-derived compartments+ upon acidification, the KDEL CysEPs are released+ and the pro-sequence and the C-terminal KDEL endoplasmic reticulum retention signal are removed for activation of the enzyme. The mature, enzymatically active KDEL CysEPs exhibit unusual broad substrate specificity. KDEL CysEPs are unique in being able to digest not only cytoplasmic components in tissues that collapse during final stages of PCD; they furthermore are able to digest extensins that form the basic scaffold for cell wall formation. The broad substrate specificity is due to the active site cleft of the KDEL CysEPs that accepts a wide variety of amino acids including proline and glycosylated hydroxyproline of the hydroxyproline-rich glycoproteins of the cell wall. The respective amino acids, which are decisive for this generally more open appearance of the active site cleft, together with the amino acids defining the catalytic pocket are highly conserved among all known KDEL CysEPs (Schmid et al. 1998; Beers et al. 2004; Schaller 2004; Than et al. 2004; Helm et al. 2008; Hierl et al. 2012; Howing et al. 2014; Nakano et al. 2014).

In *Arabidopsis*, three KDEL *CysEPs CEP1* (At5g50260), *CEP2* (At3g48340), and *CEP3* (At3g48350) have been identified that are expressed in tissues undergoing developmental PCD. Furthermore, *CEP1* was found to be expressed in late response to biotic stress stimuli in the leaf. Two *CEP1* T-DNA insertion lines (SAIL 158 B06 and SALK 01306, both carrying the T-DNA insertion within the third exon) showed enhanced susceptibility to powdery mildew caused by *Erysiphe cruciferarum*. A translational fusion protein of *CEP1* with a threefold hemaglutinin tag and the green fluorescent protein under control of the endogenous *CEP1* promoter (*PCEP1*: pre-pro-3xHA-EGFP-AtCEP1-KDEL) rescued the pathogenesis phenotype demonstrating the function of *CEP1* in restriction of powdery mildew disease.

atcep1 knockout plants transformed with the non-functional reporter including EGFP without the mature *CEP1* subunit (*PCEP1*: pre-pro-3xHA-EGFP-KDEL) retained susceptibility to *E. cruciferarum*. The spatiotemporal *CEP1*-reporter expression during fungal infection together with microscopic inspection of the interaction phenotype suggested a function of *CEP1* in controlling late stages of the compatible interaction including late epidermal cell death (Helm et al. 2008; Hierl et al. 2014; Howing et al. 2014).

Defense responses in *Arabidopsis* are regulated by multiple signal transduction pathways in which salicylic acid (SA), jasmonic acid (JA), and ethylene (ET) function as key signaling molecules. Mutants such as *cpr5* constitutively activate these defense pathways. The *CPR5* gene (*constitutive expression of PR genes 5*, At5g64930) is a regulator of expression of pathogenesis-related genes and participates in signal transduction pathways involved in plant defense and development such as leaf senescence or flowering. It codes for nuclear envelope membrane protein. Loss of *CPR5* leads to spontaneous expression of chlorotic lesions and reduced trichome development. The *cpr5* plants were found to be constitutively resistant to virulent pathogens such as the bacterial pathogen *Pseudomonas syringae* and the oomycete *Hyaloperonospora arabidopsidis*. Howing et al. (2017) found in public expression data that *CEP1* (At5g50260, Affymetrix *ATH1* probe set ID 248545 at; GEO accession GSE5745) is constitutively up-regulated in *cpr5* mutants (www.genevestigator.com). They used the *cpr5-2* mutant allele that has a point mutation in the fourth exon leading to a premature stop codon (Trp477stop) in order to analyse a possible contribution of the *CEP1* up-regulation to chlorotic leaf lesions in *cpr5*. An increase in *CEP1* expression in *cpr5-2* mutant was measured, which coincided with the appearance of leaf lesions. The expression of *CEP1* was particularly evidenced in leaf cells that surround the chlorotic lesions and presumably underwent cell death. A strong resistance of *cpr5-2* against infection of *E. cruciferarum* and pathogenesis with cell death phenotypes in *cep1 cpr5-2* double mutants as compared to the single mutants was observed. This suggested a contribution of *CEP1* to *CPR5*-controlled cell death (Bowling et al. 1997; Boch et al. 1998; Clarke et al. 2000; Kirik et al. 2001; Zimmermann et al. 2004; Brininstool et al. 2008; Howing et al. 2017).

Programmed cell death (PCD) is a prerequisite for successful development, and it limits the spread of biotrophic pathogens in a rapid hypersensitive response at the site of infection. KDEL-tailed cysteine endopeptidases (KDEL CysEP) are a subgroup of papain-type cysteine endopeptidases expressed in tissues undergoing PCD. In *Arabidopsis*, three KDEL CysEPs (*AtCEP1*, *AtCEP2*, and *AtCEP3*) are expressed. Howing et al. (2014) have shown that AtCEP1 is a factor of basal resistance to powdery mildew caused by the biotrophic ascomycete *E. cruciferarum* and is expressed in spatiotemporal association with the late fungal development on *Arabidopsis* leaves. The endoplasmic reticulum-localized proenzyme of AtCEP1 was further visualized at the haustorial complex encased with callose. The *AtCPR5* gene (*constitutive expression of PR genes 5*) is a regulator of expression of pathogenesis-related genes. Loss of *AtCPR5* leads to spontaneous expression of chlorotic lesions which was associated with enhanced expression of *AtCEP1*. Howing et al. (2017) used the *atcpr5-2* mutant plants and the *atcep1 atcpr5-2* double mutants harbouring a non-functional reporter (*PCEP1: pre-pro-3xHA-EGFP-KDEL*) for visualization of *AtCEP1* promoter activity. It has been observed that the specific up-regulation of *AtCEP1* in direct neighbourhood of spreading leaf lesions thus likely represents cells undergoing PCD. A strong resistance of *atcpr5* mutant plants against infection with *E. cruciferarum* is observed. Loss of *AtCEP1* had no obvious influence on the strong resistance of *atcpr5-2* mutant plants against infection with *E. cruciferarum*. However, the area of necrotic leaf lesions associated with *E. cruciferarum* colonies was significantly larger in *atcpr5-2* as compared to *atcep1 atcpr5-2* double mutant plants. The presence of *AtCEP1* thus contributes to *AtCPR5*-controlled PCD at the sites of powdery mildew infection (Howing et al. 2017).

Papain-type CysEPs have diverse functions in plant defense to pathogens. The papain-type KDEL *CysEP CEP1* fulfills its function in plant defense during late development of *E. cruciferarum* in close spatial association with the fungal haustorium and haustorial callose encasements. Additionally, *CPR5* controls resistance to powdery mildew and PCD in response to infection by *E. cruciferarum*. In *cpr5* mutants, *CEP1* is overexpressed in spatiotemporal association with spontaneous cell death. *CEP1* contributes to *CPR5*-controlled PCD triggered by *E. cruciferarum* but is not required to express high-level powdery mildew resistance of *cpr5* mutants (Koh et al. 2005; Misas-Villamil et al. 2016; Howing et al. 2017).

7.4.14 Expression of Genes for Camalexin Synthesis for Powdery Mildew Resistance

The *cyp83a1-3* mutant displays enhanced resistance to powdery mildews, including *G. cichoracearum*. The levels of SA accumulation and *PR* gene expression in *cyp83a1-3* mutants were similar to the wild type, and double mutant analysis showed that resistance in *cyp83a1-3* was dependent on *PAD4*, *EDS1*, and *NPR1* but independent of *SID2* and *EDS5*. Mutations in *SID2* or *EDS5* suppressed SA accumulation induced by powdery mildew but did not suppress the disease resistance, indicating that resistance in *cyp83a1-3* is not caused by enhanced SA signaling. These observations indicate that *cyp83a1-3*-mediated resistance differs from that in the *edr1* and *edr2* mutants (Frye et al. 2001; Tang et al. 2005b). Moreover, the *cyp83a1-3* mutant accumulates more camalexin upon powdery mildew infection, and mutations in *PAD3* or *WRKY33* suppressed both the disease resistance and the high accumulation of camalexin, indicating a link between camalexin levels and responses to powdery mildew (Plate 7.8a). Consistent with a role of camalexin in powdery mildew resistance, the *PAD3*-overexpressing plants accumulated more camalexin and mimicked the resistance phenotype of the *cyp83a1-3* mutant. Taken together, these findings indicate that the higher level of camalexin contributes to the enhanced resistance to *G. cichoracearum* observed in the *cyp83a1-3* mutant (Plate 7.8b). Loss-of-function mutations in two transcription factors, *WRKY18* and

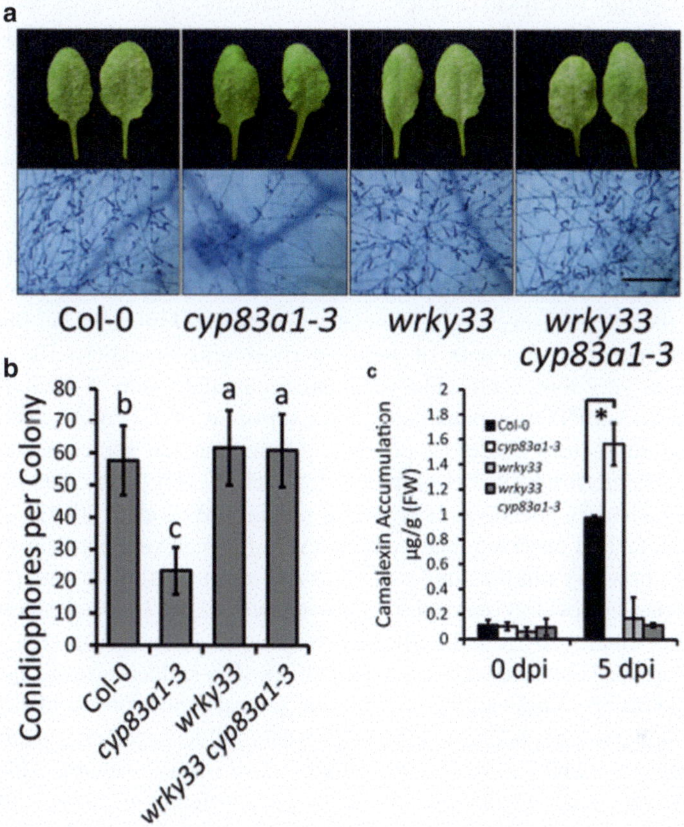

Plate 7.8 The resistance phenotype and high levels of camalexin in *cyp83a1-3* are suppressed by mutation of *WRKY33*. (**a**) Four-week-old wild-type, *cyp83a1-3*, *wrky33*, and *wrky33 cyp83a1-3* double mutant plants were infected with *G. cichoracearum*. Representative leaves were removed and stained with trypan blue at 8 dpi, bar = 200 μm. (**b**) Quantification of fungal growth of the plants in (**a**) at 5 dpi by counting the number of conidiophores per colony. Results represent the mean and standard deviation in three independent experiments ($n = 30$; $P < 0.01$, nested ANOVA). (**c**) Camalexin accumulation of the plants in (**a**) was determined at 0 and 5 dpi. Asterisk represents statistically significant difference from wild type ($P < 0.01$, nested ANOVA) (Liu et al. 2016)

WRKY40, result in the accumulation of higher levels of camalexin as well as both pre-invasive and post-invasive resistance against the powdery mildew fungus *G. orontii* (Pandey et al. 2010; Schon et al. 2013). The *PAD3* (a key enzyme in camalexin biosynthesis) is required for the pre-invasive resistance (but surprisingly not for the post-invasive resistance) against *G. orontii* in a *wrky18 wrky40* background (Schon et al. 2013). Moreover, Schon et al. (2013) also report that *wrky18 wrky40* plants do not show increased resistance against two other powdery mildews, *G. cichoracearum* and *E. cruciferarum*. It may be related to differences in camalexin levels in the two mutants or to differences in susceptibility of different

G. cichoracearum isolates to camalexin. The function of *CYP83A1* has been studied previously. An earlier study showed that overexpression of *CYP83A1* could rescue the auxin-excess phenotype of *cyp83b1/rnt1/sur2* mutants (Bak and Feyereisen 2001). Further studies showed that levels of many phenylpropanoid pathway-derived products were reduced in the *cyp83a1-2/ref2-1* mutant, indicating crosstalk between the pathways producing aliphatic glucosinolates and indole glucosinolates (Hemm et al. 2003; Naur et al. 2003). Recent work showed that *CYP83A1* interacts with *BAX INHIBITOR-1*, a cell death suppressor in plants and animals. The loss-of-function mutants *cyp83a1-1* and *cyp83a1-2* displayed enhanced resistance to the powdery mildew fungus *E. cruciferarum* (Weis et al. 2013). A more recent study measured the levels of several glucosinolates in *cyp83a1-1* mutants, but found only marginally increased amounts of indole-derived glucosinolates. The *cyp83a1* mutants lack very-long-chain aldehydes and accumulate more 5-methylthiopentanaldoxime (5-MPTO), a potentially toxic substrate of *CYP83A1* (Weis et al. 2014). Very-long-chain aldehydes promote germination and appressorium formation of *E. cruciferarum*; it was proposed that lack of very-long-chain aldehydes causes the resistance phenotypes in *cyp83a1* mutants (Weis et al. 2014).

Liu et al. (2016) observed that the high level of camalexin contributes to resistance to the powdery mildew fungus *G. cichoracearum*. It is worth noting that different species of powdery mildew, *G. cichoracearum* and *E. cruciferarum*, were used by Liu et al. (2016) and earlier by Weis et al. (2014). Although *cyp83a1* mutants displayed enhanced resistance to both powdery mildew strains, the mechanisms could differ. Consistent with this notion, several studies observed differences in infection phenotypes between different powdery mildew species in *Arabidopsis*. *G. cichoracearum* and *G. orontii* have different host ranges/responses (Plotnikova et al. 1998), and many *Arabidopsis* accessions have different responses to the powdery mildew species *E. cruciferarum UEA1* and *G. cichoracearum UCSC1* (Adam et al. 1999). Liu et al. (2016) showed that *pad3* and *wrky33* suppressed the accumulation of camalexin and the enhanced resistance in *cyp83a1-3* mutants, indicating a role of camalexin in *cyp83a1-3*-mediated resistance (Plate 7.9a). It would be interesting to examine the responses of *pad3 cyp83a1-3* and *wrky33 cyp83a1-3* mutants to *E. cruciferarum* and to measure the levels of very-long-chain aldehydes and 5-MPTO in those mutants. It is also possible that both very-long-chain aldehydes and camalexin contribute to resistance against *G. cichoracearum* and *E. cruciferarum*.

CYP83A1 functions in the biosynthesis of aliphatic glucosinolates, and it is not clear that how the mutation of the glucosinolate synthetase gene *CYP83A1* affects the accumulation of camalexin. The possibility may be that it may cause crosstalk within the complicated metabolic network. In this scenario, the biosynthetic pathway of aliphatic glucosinolates, which involves *CYP83A1*, and indole glucosinolates, which share the IAOx intermediate with camalexin (Nafisi et al. 2007; Schuhegger et al. 2007a; Bottcher et al. 2009), can affect each other. So when the aliphatic glucosinolate pathway is blocked in the *cyp83a1* mutant, the pathway for indole-derived products, including indole glucosinolates and camalexin, is enhanced. Indole glucosinolates are known to contribute to plant immunity

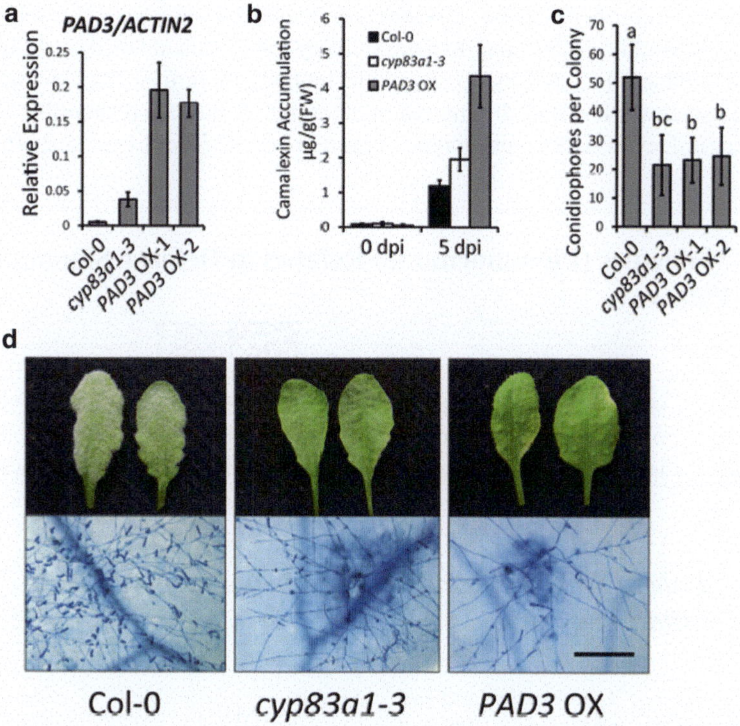

Plate 7.9 *PAD3*-overexpressing plants accumulate higher levels of camalexin and display a *cyp83a1-3*-like resistance phenotype. (**a**) The accumulation of *PAD3* transcript in 4-week-old plants was examined by quantitative real-time PCR. (**b**) Four-week-old plants were infected with *G. cichoracearum,* and camalexin accumulation was determined at 0 and 5 dpi. Three *PAD3*-OX-independent lines were tested, and similar results were obtained in these lines. One representative *PAD3*-OX line (*PAD3*-OX-*1*) is shown. (**c**) Quantification of fungal growth of wild-type, *cyp83a1-3*, and *PAD3*-OX plants at 5 dpi by counting the number of conidiophores per colony. Different letters represent statistically significant differences ($P < 0.01$, nested ANOVA). (**d**) Four-week-old wild-type, *cyp83a1-3*, and *PAD3*-OX-1 plants were infected with *G. cichoracearum*. Three *PAD3*-OX-independent lines were examined, and similar results were obtained in these three lines. Representative leaves were removed and stained with trypan blue at 8 dpi, bar = 200 µm. (Liu et al. 2016)

(Bednarek 2012). *Arabidopsis* penetration2 (*PEN2*), which initiates indole glucosinolate metabolism, plays an important role in penetration resistance (Bednarek et al. 2009; Clay et al. 2009). In addition to increased levels of camalexin, the *cyp83a1* mutant has several other aberrant phenotypes including decreased levels of very-long-chain aldehydes and alterations in many metabolites that could also contribute to resistance, so the exact mechanism of powdery mildew resistance in *cyp83a1* needs to be further studied. Consistent with this notion, *PAD3*-over expressing plants accumulate much higher levels of camalexin than the *cyp83a1-3* mutant, but no further increase in powdery mildew resistance was observed (Plate 7.9b, c). It would be interesting to measure very-long-chain aldehydes and indole-derived products in *pad3 cyp83a1* double mutants to further examine whether altered camalexin levels or differences in other metabolites are responsible for the powdery mildew resistance phenotype in *cyp83a1* (Plate 7.9d; Liu et al. 2016).

7.5 Molecules (Phytohormone) Related to Defense Signaling Pathways

The first line of plant defense (pre-penetration) against fungal pathogens largely relies on cell surface-mediated defense signaling initiated by recognition of PAMPs; secondary (post-penetration) defense responses are often induced by the salicylic acid (SA) or jasmonic acid/ethylene (JA/ET) phytohormone signaling pathways.

7.5.1 Role of Salicylic Acid-Mediated Signaling to Powdery Mildew Resistance

The fact that *npr 1*, and *pad 4*, and *eds 5* mutants are susceptible to *G. orontii* indicates that *NPR1, PAD 4*, and *EDS 5* play key roles in defense signaling pathways. *EDS 5* plays a role in salicylate-dependent signal transduction. A model of gene induction in *Arabidopsis* by *G. orontii* has been suggested by Reuber et al. (1998) shown in Fig. 7.1. As for other biotrophic interactions, SA-mediated defense signaling plays a pivotal role in *Arabidopsis* defense against adapted powdery mildew fungi: SA-dependent gene expression and immune responses increase in *Arabidopsis* leaves upon infection with powdery mildews and contribute to restriction of colony expansion and reproduction of the fungi (Zimmerli et al. 2004; Chandran et al. 2009). Consequently, many mutants with defects in SA biosynthesis, accumulation, and signaling exhibit enhanced susceptibility (hypersusceptibility) to *Go* and *Gc* (Dewdney et al. 2000; Chandran et al. 2009; Zhang et al. 2015a). Likewise, interference with SA accumulation by transgenic expression of *NahG*, a *Pseudomonas* salicylate hydroxylase that degrades SA, increases susceptibility against *Gc* isolates (Ederli et al. 2015; Zhang et al. 2015a). Despite these genetic indications for an involvement of SA in defense against powdery mildew infection, currently there is only limited direct evidence for increased SA levels during powdery mildew infection (Fabro et al. 2008).

As the final steps of SA biosynthesis in the leaf take place in chloroplasts (Strawn et al. 2007), export of the hormone is required for elevated cytosolic and nuclear SA levels. Consequently, loss-of-function mutation of DP-E2F-like 1 (*DEL1*: At3g48160), a transcriptional repressor of the gene encoding the plastidic SA exporter *EDS5*/salicylic acid induction deficient (SID) 1 (At4g39030), results in enhanced SA-dependent resistance against *Go* (Chandran et al. 2014). This phenotype correlates with elevated basal SA levels and increased transcript abundance of SA-responsive genes in the *del1* mutant (Chandran et al. 2014). Strikingly, *DEL1* also promotes cell proliferation by repressing genes involved in the induction of endo-reduplication (Vlieghe et al. 2005; Lammens et al. 2008). Together these findings suggest that *DEL1*-mediated control of SA levels regulates the balance between growth and immunity in developing leaves. The translation of distinct SA levels into specific defense responses occurs by the action of NPR proteins in *Arabidopsis* (Pajerowska-Mukhtar et al. 2013; Seyfferth and Tsuda 2014; Yan and Dong 2014). The outcome of SA-mediated signaling depends on sub-cellular SA levels and the abundance of active *NPR1* in the nucleus. Its paralogs, the SA receptors *NPR3* (At5g45110) and *NPR4* (At4g19660), cooperatively fine-tune *NPR1* degradation by their competitive SA concentration-dependent interaction with nuclear *NPR1*. Moderately elevated SA levels, as present during systemic acquired resistance (SAR), allow the accumulation of active *NPR1* protein in the nucleus. Subsequent interaction of *NPR1* with *TGA* transcription factors (binding to a TGACG nucleotide motif) promotes gene expression of SA-responsive genes and induces SA-mediated defense. The relevance of *NPR1* for the *Arabidopsis*–powdery mildew interaction is substantiated by identification of *NPR1* as a protein–protein interaction network hub during *Go* infection (Jiang et al. 2016). This *NPR1* interaction network includes the TGA-interacting GLUTAREDOXIN 480 (*GRX480/ROXY19*: At1g28480) involved in regulating SA/JA antagonism (Zander et al. 2012) and several TGA transcription factors (*TGA1* At5g65210, *TGA2* At5g06950, *TGA3* At1g22070, *TGA7* At1g77920) that can further regulate the expression of defense-related genes. Genes that show SA-dependent transcript accumulation during powdery mildew infection encode proteins involved in redox regulation, vacuolar transport, secretion, and signaling-relevant processes such as Ca^{2+} homeostasis and SA/JA crosstalk (Chandran et al. 2009).

ROP GTPases (*Rho* (RAS homolog) of plants) are molecular switches and key regulators of immunity (Kawano et al. 2014). Mutation of the *Arabidopsis ROP-GAPs ROPGAP1* (At5g22400) and *ROPGAP4* (At3g11490), trapping their (yet unidentified) target ROPs in the active state, results in enhanced susceptibility to *E. cruciferarum* (Hoefle et al. 2011; Huesmann et al. 2011). Accordantly, expression of an inactive (dominant-negative) *ROP6* (rop6dn; At4g35020) variant, unable to interact with downstream effectors, results in reduced penetration by *Go*. This correlates with an increased transcript abundance of SA-responsive genes, such as *PR1*, and elevated SA-mediated defense responses. However, *Go* resistance of the *rop6dn* transgenic plants is uncoupled from SA signaling (Poraty-Gavra et al. 2013). Nevertheless, these results, together with previous findings in barley (Hoefle et al. 2011; Scheler et al. 2016), suggest a positive role of active ROPs in mediating susceptibility to adapted powdery mildews.

A number of proteins whose deficiency leads to enhanced powdery mildew resistance, such as *EDR1* to 4, lesion initiation 2 (*LIN2*: At1g03475), and *GSL5/PMR4*, are associated with the repression of SA-mediated defense. This is indicated by requirement of SA for the increased disease resistance phenotypes of the respective mutants (Vorwerk et al. 2007; Zhang et al. 2007; Wawrzynska et al. 2010; Guo et al. 2013; Wu et al. 2015). The MAPKKK *EDR1* gene negatively regulates SA-dependent defense responses and cell death. In consequence, *edr1* mutants are constitutively primed for SA-inducible defense which might occur via the regulation of the MAPKs *MPK3* (At3g45640) and *MPK6* (At2g43790) (Beckers et al. 2009). The role of *EDR1* in the control of SA signaling probably relies on its interaction with *MKK4* and *MKK5*, the upstream MAPKKs activating *MPK3* and *MPK6* (Zhao et al. 2014). *EDR1* negatively affects *MKK4* and *MKK5* levels, presumably resulting in the repression of the *MKK4/MKK5-MPK3/MPK6* cascade involved in the induction of SA signaling (Zhao et al. 2014; Wu et al. 2015). Mutations in *MKK4*, *MKK5*, or *MPK3* (but not in *MPK6*) suppress the *edr1* phenotype, indicating a requirement of these kinases for *edr1*-mediated powdery mildew resistance. The same holds true for *edr4*, which is in line with the need of *EDR4* for the relocation of *EDR1* to the powdery mildew penetration site (Wu et al. 2015). Strikingly, overexpression of *MKK4* or *MKK5* causes *edr1*-like resistance and powdery mildew-induced cell death, pointing to an additional role for the *MKK4/MKK5-MPK3/MPK6* kinase cascade in SA-induced cell death, parallel to its contribution to SA-regulated gene expression (Zhao et al. 2014).

NPR3-mediated *NPR1* degradation at high SA levels promotes the onset of cell death. However, while no powdery mildew phenotype has been reported for *npr3* (Liu et al. 2005), *npr1* and *npr4* mutant plants are more susceptible to *Gc* (Reuber et al. 1998; Liu et al. 2005; Humphry et al. 2010). Thus, it remains elusive to which extent NPR3-mediated cell death contributes to defense against powdery mildews. In addition to *NPR3*, also *NPR4* adds to pathogen-triggered cell death (Pajerowska-Mukhtar et al. 2013; Kumar 2014; Yan and Dong 2014). A role for SA-mediated cell death responses in powdery mildew defense is supported by correlation of enhanced SA signaling with increased powdery mildew-induced cell death in several of the above-mentioned resistant mutants (Guo et al. 2013).

Mutants impaired in autophagy further corroborate a role of cell death in powdery mildew resistance, as some of them display early leaf senescence and spontaneous cell death, which in several cases coincides with increased powdery mildew resistance (Yoshimoto et al. 2009; Wang et al. 2011a, b). Autophagy targets organelles and cytosolic proteins for vacuolar/lysosome-mediated degradation (Liu and Bassham 2012). The mutant of *autophagy-related 2 (ATG2:* At3g19190), impaired in the early steps of autophagosome biogenesis, exhibits severe powdery mildew-induced cell death and increased resistance when challenged with *Gc,* while susceptibility towards *Go* is unaltered (Yoshimoto et al. 2009; Wang et al. 2011a, b). Powdery mildew resistance of *atg2* plants depends on SA signaling, while cell death is partially independent of SA. In conclusion, autophagy contributes to suppression of cell death and defense response to powdery mildew fungi; however, the mechanisms by which autophagy controls these processes are yet unknown (Wang et al. 2011a, b).

7.5.2 *Role of Jasmonic Acid (JA)- and Ethylene (ET)-Mediated Signaling to Powdery Mildew Resistance*

Salicylic acid (SA)-mediated defense appears to act mainly against biotrophic pathogens, while the JA and ET pathways are preferentially linked to resistance against necrotrophic parasites (Thomma et al. 2001; Glazebrook 2005). Induced systemic resistance (ISR) confers JA/ET-mediated protection of shoot tissues via root-to-shoot signaling. ISR is initiated by interactions with beneficial fungi or bacteria such as arbuscular mycorrhiza or plant growth-promoting rhizobacteria in the root and has proven effective against necrotrophs and herbivores (Pieterse et al. 2012). Despite the fact that powdery mildew fungi are obligate biotrophs, root colonization with the putative plant growth-promoting basidiomycetes *Piriformospora indica* reduces *Go* conidiation in a JA signaling-dependent manner (Stein et al. 2008). This finding suggests that besides SA, JA also contributes for resistance against powdery mildew fungi (Fig. 7.4). Accordingly, *Bgh* inoculation of *Arabidopsis* induces expression of genes that are controlled by the JA/ET signaling pathways (Zimmerli et al. 2004). By contrast, although endogenous JA levels are enhanced during the formation of haustoria by *Gc*, this does not result in transcript accumulation of JA/ET-responsive genes (Reuber et al. 1998; Nishimura et al. 2003; Zimmerli et al. 2004; Glazebrook 2005; Fabro et al. 2008). Nevertheless, constitutive or ectopic activation of the JA/ET pathway due to elevated JA/ET levels in the

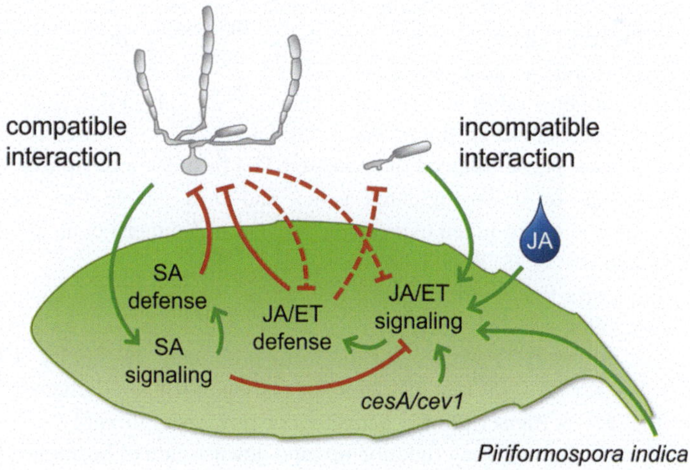

Fig. 7.4 Integration of phytohormone signaling in defense against powdery mildew fungi. Although only incompatible powdery mildew–host interactions elicit JA/ET-mediated defense, JA/ET-induced defense responses are effective against virulent powdery mildew fungi if stimulated constitutively (*cesA/cev1*), artificially (JA treatment), or systemically (*Piriformospora indica* root colonization). These findings suggest that virulent fungi suppress JA/ET signaling during compatible interactions. This suppression might involve the antagonistic action of SA signaling. Solid lines indicate experimentally supported impacts, while dashed lines indicate speculative connections (Kuhn et al. 2016)

mutant of *cellulose synthase 3/constitutive expression of VSP 1* (*CESA3/CEV1*: At5g05170) or treatment of Col-0 with methyl-JA enhances resistance against *Gc*. This effect depends on the JA receptor component *coronatine-insensitive protein 1* (*COI1*: At2g39940; Ellis et al. 2002a; Ellis et al. 2002b; Zimmerli et al. 2004). These findings suggest that, although elicitation of JA/ET-mediated defense signaling seems to be restricted to incompatible powdery mildew host interactions, JA/ET-induced defense responses are effective against virulent powdery mildew fungi if stimulated constitutively, artificially, or systemically, despite the biotrophic nature of the interaction. Consequently, JA/ET signaling must either be suppressed or failed to be elicited by adapted powdery mildews during a successful infection (Fig. 7.4; Antico et al. 2012).

7.5.3 Role of WRKY *Transcription Factors to Powdery Mildew Resistance: Expression of Resistance Genes or Defense-Related Genes*

It is also known as host transcriptional programming. Transcriptional changes in response to powdery mildew inoculation reflect a combination of both activation of defense after recognition of the pathogen and host cell manipulation by the fungal invader. *Arabidopsis* responds to powdery mildew attack with the differential regulation of defense-related genes. While many genes are induced in both host and non-host interactions, changes in gene expression interactions indicating that virulent fungi might suppress gene expression related to basal occur more rapidly and are often more pronounced in non-host interactions than in host defense (Zimmerli et al. 2004). A large subset of the powdery mildew-responsive genes is transcriptional factors (TFs). These induced or repressed TFs further transcriptionally regulate secondary up- or down-regulated genes and thus enable the coordinated expression of genes in fine-tuned expression networks (Zimmerli et al. 2004; Fabro et al. 2008; Chandran et al. 2009; Christiansen et al. 2011).

Among the genes that show altered transcript accumulation in response to *Go,* members of the plant-specific WRKY (single amino acid letter code for tryptophan-arginine-lysine-tyrosine) TFs represent the most prominent TF family (Chandran et al. 2009). WRKY TFs are key regulators of pathogen-triggered changes in gene expression that act as transcriptional activators or repressors in various homo- and heterodimer combinations. They function up- and downstream of hormone signaling pathways, are involved in the antagonistic control of SA and JA/ET signaling pathways, and can be regulated by mitogen-activated protein kinases (MAPKs) (Bakshi and Oelmuller 2014; Buscaill and Rivas 2014; Caarls et al. 2015). The involvement of WRKYs in defense of *Arabidopsis* against powdery mildews is indicated by transcriptional changes of WRKY-encoding genes in response to *Go* and an enrichment of WRKY-targeted W-box *cis*-regulatory elements in promoters of genes differentially transcribed upon powdery mildew challenge (Chandran et al. 2009).

Furthermore, expression of WRKY TFs is enhanced in powdery mildew-resistant *edr1* plants relative to the wild type in response to *Gc* and genes whose promoters contain W-boxes are likewise enriched in this data set. As powdery mildew resistance in *edr1* plants depends on the *MPKK4/5-MPK3/6* cascade, the expression of WRKY TFs might be regulated via this pathway (Christiansen et al. 2011).

In barley, the *MLA10* NB-LRR interacts with the *Bgh* AVR_{A10} (avirulence A10) effector and induces transcriptional changes by inhibition of *HvWRKY1* and *HvWRKY2*. Both TFs supposedly act as transcriptional repressors of genes involved in basal defense and effector-triggered immunity (*ETI*; Shen et al. 2007). Similarly, the closely related *Arabidopsis* TFs *WRKY18* (At4g31800) and *WRKY40* (At1g80840), whose transcription is rapidly induced during powdery mildew infection, negatively regulate defense against *Go*. Together these results indicate functional conservation of the defense-repressive role of this WRKY sub-family (Shen et al. 2007). Altered pathogen-induced transcriptional reprogramming in the *Go*-resistant *wrky18 wrky40* double mutant corroborates the negative impact of *WRKY18/40* on defense-related gene expression (Pandey et al. 2010; Schon et al. 2013). Chromatin immunoprecipitation (ChIP) experiments revealed binding of *WRKY40* to W-box containing promoter regions of *EDS1*, the AP2 (apetala 2)-type TF gene *redox responsive transcription factor 1* (*RRTF1*: At4g34410), and to *jasmonate-zim-domain protein 8* (*JAZ8*: At1g30135), a member of the JA signaling repressor gene family (Pandey et al. 2010). Thus, *WRKY18/40* TFs seem to repress the transcription of positive defense regulators such as *EDS1* and positively modulate JA signaling (Fig. 7.5 Pandey et al. 2010; Schon et al. 2013). Although the regulatory role of *HvWRKY1/2* and *WRKY18/40* is conserved between barley and *Arabidopsis*, an *Arabidopsis* R protein interfering with *WRKY18/40* function remains to be identified (Kuhn et al. 2016). The conservation of *MLA1* functionality and induction of MLA-dependent defense gene expression in response to *Bgh* might indicate the existence of a respective MLA analog in *Arabidopsis* (Maekawa et al. 2012). In contrast to *WRKY18* and *WRKY40*, *WRKY70* (At3g56400) contributes to resistance of *Arabidopsis* to *Gc,* and inactivation of the respective gene results in increased susceptibility to this pathogen (Li et al. 2006). *WRKY70* overexpression coincides with a partially *NPR1*-dependent suppression of JA-responsive genes, indicating a role of this TF in the control of SA/JA crosstalk (Li et al. 2006; Caarls et al. 2015).

7.6 Molecule (Hormone) Signaling-Induced Transcriptional Reprogramming During R to Powdery Mildew

SA signaling contributes to gene expression during the *Arabidopsis*–powdery mildew interaction. This involves *Go*-induced expression of genes related to Ca^{2+} signaling and genes coding for redox regulators that contribute to *NPR1* activation (Chandran et al. 2009). Transcript accumulation of SA-responsive TFs, SA

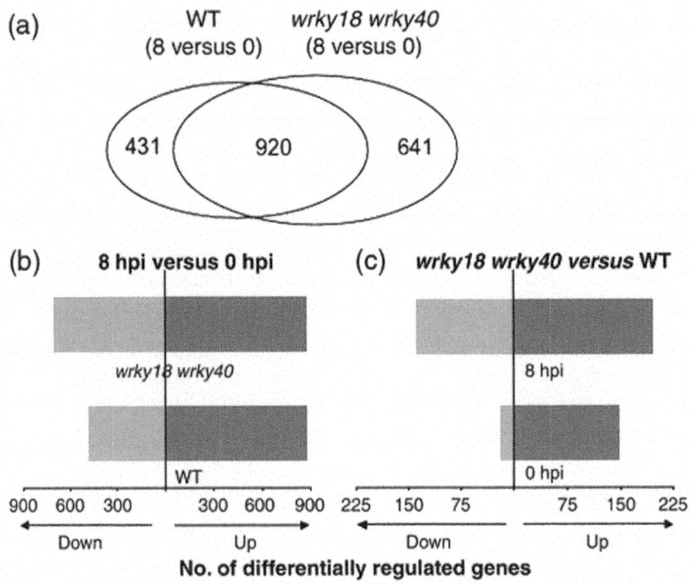

Fig. 7.5 Early transcriptional reprogramming in WT and *wrky18 wrky40* plants upon *G. orontii* infection. (**a**) Venn diagram illustrating total number and overlap of genes affected in WT and *wrky18 wrky40* 8 h post *G. orontii* infection. (**b**) Changes in the transcript profiles of *wrky18 wrky40* (upper panel) and WT (lower panel) plants 8 hpi compared with their respective profiles in the non-infected (0 hpi) states. (**c**) Differentially regulated genes between *wrky18 wrky40* and WT plants at 0 h (lower panel) and 8 hpi (upper panel). Numbers along solid lines (x-axis) indicate genes up-regulated (dark grey) and down-regulated (light grey) (Pandey et al. 2010)

biosynthesis genes (*isochorismate synthase 1* (*ICS1*)/*SID2*: At1g74710), and SA-responsive pathogenesis-related genes such as *PR1* emphasizes the predominant contribution of SA signaling to powdery mildew-induced gene expression. In line with this notion, SA signaling-dependent resistance of the *wrky18 wrky40* double mutant correlates with massive *Go*-induced transcriptional reprogramming (Pandey et al. 2010; Schon et al. 2013). The *edr1* mutation, which enhances SA-dependent powdery mildew resistance, affects accumulation of defense-related transcripts in response to *Gc* including transcripts encoding *WRKY* and *AP2*/ET-response element binding factor (ERF) TFs (Christiansen et al. 2011). Furthermore, genes encoding proteins associated with ROS production and the endo-membrane system are induced in infected *edr1* plants. Powdery mildew-induced enrichment of the latter together with transcripts associated with secretion suggests that the secretory pathway may play an important role in *edr1*-mediated immunity (Christiansen et al. 2011). This assumption is in agreement with the relocalization of *EDR1* from the ER to the plant–fungal interface during *Gc* infection (Christiansen et al. 2011; Wu et al. 2015).

TFs of the *AP2/ERF* family particularly regulate genes related to JA/ET signaling. Besides, *WRKY TFs*, *AP2/ERF TFs*, and transcripts associated with *AP2/ERF* response elements (GCC-boxes) are over-represented among genes up-regulated during powdery mildew infection in the *edr1* mutant (Christiansen et al. 2011). *Arabidopsis ERF6* (At4g17490) and *ERF104* (At5g61600) are phosphorylated by *MPK6* and/or *MPK3*, indicating a regulation of these ERFs by defense-related MAPK cascades (Bethke et al. 2009; Meng and Zhang 2013; Tsuda and Somssich 2015). Furthermore, *ERF1* (At3g23240) and *ERF2* (At5g47220) transcripts accumulate upon *Go* infection (Chandran et al. 2009). A role of *ERF1* in defense against powdery mildews is in line with the finding that *ERF1* overexpression results in enhanced resistance to *Go* (Gu et al. 2002; Chandran et al. 2009). *octadecanoid-responsive Arabidopsis AP2/ERF 59* (*ORA59*: At1g06160), a master regulator of ERF-controlled JA/ET signaling, has been identified, together with other ET/JA-responsive genes, as differentially regulated in SA-deficient mutants *versus* wild-type plants upon *Go* infection. This suggests that *ORA59* modulates the cross-talk between SA and JA/ET signaling during powdery mildew-induced defense responses (Chandran et al. 2009). In agreement with this finding, *ORA59* was identified as a major target for SA antagonism (Zander et al. 2012; Van der Does et al. 2013; Zander et al. 2014; Caarls et al. 2015). A dominant allele of *signal responsive1* (*SR1*: At2g22300), encoding a calmodulin-binding TF that regulates ET-induced senescence by binding to the *ethylene insensitive 3* (*EIN3*: At3g20770) promoter, has been identified as a gain-of-function suppressor of *edr2*-mediated resistance to *Gc* (Nie et al. 2012). SR1 binds to the promoter of *non-race-specific disease resistance1* (*NDR1*: At5g06320). *NDR1*, a membrane-associated protein, contributes to plant immunity mediated by several coiled-coil NB-LRRs and SAR (Century et al. 1997; Shapiro and Zhang 2001; Zhang and Shapiro 2002). A *sr1* null mutant conditions resistance to *Gc*, while an additional mutation in *ndr1* suppresses this phenotype. Together these data suggest that SR1 plays a critical role in powdery mildew resistance, possibly by regulating *EIN3* and *NDR1* expression (Nie et al. 2012).

Integrative analysis of protein interaction networks and transcriptomics during *Go* infection revealed a negative correlation between development-related and defense-related genes/proteins (Jiang et al. 2016). While many defense-related genes are induced after *Go* infection, the majority of genes linked to development are down-regulated. Interestingly, auxin-related genes are over-represented among nodes connecting the defense and development sub-networks. Together these findings emphasize that defense is triggered at the expense of developmental programmes and that regulation of this trade-off involves auxin (Jiang et al. 2016).

7.6.1 Harmonious Coordination Between Transcriptional Regulation and R to Powdery Mildew

Despite its compatible nature, the *Arabidopsis–Gc* interaction elicits expression of genes related to non-host resistance (NHR) (Chandran et al. 2010). Remarkably, in compatible interactions, activation neither of NHR genes nor of SA-induced defense is sufficient to confer resistance (Chandran et al. 2009, 2010). The *PEN1/ SNAP33/ VAMP722* and the *PEN2/ PEN3* pathways are important determinants of NHR and are additionally required for *mlo2*-mediated immunity. In line with their pivotal role in defense, *PEN1, PEN2, PEN3, SNAP33,* and *MLO2* share a substantial amount of co-expressed genes, with the majority of transcripts accumulating in response to biotic stresses and PAMP treatment (Humphry et al. 2010). Notably, many transcripts of this regulon accumulate in *Bgh*-inoculated *Arabidopsis* plants ectopically expressing the barley *MLA1* R protein. This implies a substantial overlap of *MLA1*-dependent transcriptional regulation and basal resistance against powdery mildew fungi (Humphry et al. 2010; Maekawa et al. 2012).

Consistent with the role of *PEN2* and *PEN3* in indole glucosinolate biosynthesis and secretion, many of the co-expressed genes are associated with the glucosinolate pathway. The gene encoding myeloblastosis *(MYB) 51 TF* (At1g18570) is a major regulator of defense-related expression of glucosinolate biosynthesis genes (Humphry et al. 2010). Transcriptomic evaluation of the powdery mildew-resistant *wrky18 wrky40* double mutant revealed that *WRKY18* and *WRKY40* suppress crucial biosynthesis genes of the indolic phytoalexin camalexin (Pandey et al. 2010). Accordingly, increased expression of camalexin and indole glucosinolate biosynthesis genes after pathogen challenge in *wrky18 wrky40* plants correlates with the enhanced accumulation of the phytoalexin camalexin and 4MI3G (4-methoxy-indol-3-ylmethylglucosinolate), an indole glucosinolate intermediate relevant for powdery mildew resistance, in this mutant (Pandey et al. 2010; Schon et al. 2013). Loss of function of the crucial 4MI3G biosynthesis gene *CYP81F2* suppresses *wrky18 wrky40*-mediated inhibition of host cell entry, indicating that the indolic metabolite is required for penetration resistance of the double mutant against *Go* (Schon et al. 2013).

The group of genes coexpressed with *PEN3* further showed a significant over-representation of components involved in Ca^{2+} signaling such as calcium/calmodulin-dependent protein kinases (CCaMKs; Humphry et al. 2010; Campe et al. 2015). Consistently, components of the Ca^{2+} homeostasis, and signaling machinery such as *CAM9/CALMODULIN-LIKE* (*CML*) *9* (At3g51920), and calreticulin-encoding genes are rapidly induced after *Go* inoculation, and *CML38* (At1g76650) is constitutively up-regulated in the highly resistant *wrky18 wrky40* mutant (Chandran et al. 2009; Pandey et al. 2010). Co-expression of genes encoding proteins involved in secondary metabolite biosynthesis (such as *PEN2* or cytochrome) P450s like *CYP83B1* (At4g31500) together with genes whose products mediate exocytosis/extrusion (SNAREs, exocyst subunits, and ABC transporters like *PEN3*) suggests that production and secretion of antimicrobial compounds are transcriptionally attuned. Additional co-regulation of receptor-like kinase genes, transcripts of Ca^{2+} signaling components, and the heterotrimeric G-protein β and g subunits *AGB1* (At4g34460) and *AGG1* (At3g63420) indicates that also recognition and signaling components are co-expressed with key components of antifungal defense (Humphry et al. 2010; Lorek et al. 2013). Enhanced host cell entry by *Go* and *E. pisi* of Gβ-deficient mutants emphasizes the importance of second messenger signaling via heterotrimeric G-protein components for powdery mildew resistance

(Lorek et al. 2013). Finally, identification of defense-related TFs in the regulon indicates that recognition of pathogens, response initiation, and defense execution are transcriptionally coordinated to enable an efficient immune output.

7.6.2 Transcription Factors and Gene Regulation for Powdery Mildew Resistance

In plants, recognition of pathogen attack and/or its entry and initiation of defense responses result in a dramatic re-programming of gene expression (Schenk et al. 2000, Zimmerli et al. 2004). Whether mediated by direct interaction between a pathogen receptor and transcription factors (Shen et al. 2007) or, as more often observed in animal systems, by a complex signal transduction cascade (mediated by MAP kinases; Asai et al. 2002), the complete activation of defense responses relies in part on the activity of so-called WRKY transcriptional regulators (Eulgem and Somssich 2007). In *Arabidopsis*, the WRKY family is composed of 72 expressed members displaying various degrees of functional redundancy with respect to the diverse biological roles they are involved in, including plant defense. *WRKY70* is transcriptionally induced by SA signaling and mediates basal resistance to a variety of pathogens including *G. cichoracearum* (Li et al. 2006; Wang et al. 2006; Ulker et al. 2007; Plate 4.1; Chap. 4). In contrast, *WRKY18* and *WRKY40* act as negative regulators of basal defense against *G. orontii* (Plate 4.1; Chap. 4), and double mutants in these genes are fully resistant to fungal attack (Shen et al. 2007). In the analogous interaction of the monocot barley with its cognate powdery mildew pathogen, *Bgh, WRKY18*, and 40 counterparts *WRKY1* and *WRKY2* were shown to interact with the CC domain of MLA immune receptors (Shen et al. 2007). Under normal (pathogen-free) conditions, *WRKY1* and *WRKY2* are proposed to suppress the transcription of genes involved in basal defense. Upon pathogen challenge, and activation by a matching fungal avirulence protein, the MLA receptor is re-localized to the nucleus where it recruits *WRKY1* and *WRKY2* away from the promoters of defense genes. As expected, silencing of these WRKY transcription factors results in enhanced basal defense against Bgh (Shen et al. 2007, Eckey et al. 2004).

In addition to WRKY transcription factors, the NAC transcription factors, barley *NAC6*, and its *Arabidopsis* homolog *ATAF1* were recently identified as positive regulators of transcription involved in early defense against powdery mildews. Silencing (barley) or knockout (*Arabidopsis*) of the respective gene resulted in decreased penetration resistance upon inoculation with *Bgh*, while overexpression (in barley) resulted in increased resistance (Plate 4.1; Chap. 4). These findings suggest a conserved contribution of NAC transcription factors in pre-haustorial defense (Jensen et al. 2007). A member of another group of transcription factors, the wheat ethylene response factor ERF3, has also been suggested to act as a positive regulator of early defense against powdery mildews (Zhang et al. 2007; Plate 4.1; Chap. 4). *ERF3* expression is induced by pathogen challenge (*Bgh*), and the gene product operates as a GCC box-binding transcriptional activator of SA signaling genes (Zhang et al. 2007). Similarly, heterologous overexpression of other members of this family from tomato in *Arabidopsis* resulted in enhanced resistance to the adapted powdery mildew, *G. orontii*, further corroborating their role as positive regulators of transcription in pathogen defense (Gu et al. 2002).

7.6.3 Transcriptional (Genes) Regulation and Expression in Response to Powdery Mildew Infection

The broad range of transcriptional changes that occur in *Arabidopsis* during powdery mildew infection (Zimmerli et al. 2004; Fabro et al. 2008) reflect both defense responses mounted by the plant and possible manipulation of the host by the fungal pathogen. With the sequencing of the *Arabidopsis* genome, and availability of novel technologies for transcript profiling (oligonucleotide-based microarray chips), it has been possible to peek into the gene expression changes that occur during the infection process. Several dedicated websites such as Genevestigator https://www.genevestigator.ethz.ch/andATGenExpress(http://www.arabidopsis.org/info/expression/ATGenExpress.jsp) provide publicly available collections of microarray data on *Arabidopsis* challenged with a variety of pathogens including powdery mildews, which represent a largely unexplored treasure of information. However, the spatial resolution of microarray data is limiting, since most microarray experiments are performed on whole leaf samples and as such only provide a gross overview about changes in transcript abundance in pathogen-challenged leaf tissue. Since powdery mildews exclusively colonize the epidermal layer of the leaf, many tissue-specific transcriptional changes are likely to be masked by proportionately larger amounts of mesophyll transcripts. Even within the epidermis, attacked cells and successfully colonized cells each differ in their transcript profile from non-attacked cells (Gjetting et al. 2004, 2007). Owing to these limitations, expression levels of numerous genes in infected host cells might be considerably under-estimated. Employment of techniques that allow transcript profiling of powdery mildew-challenged *Arabidopsis* leaves at cellular resolution, e.g. based on RNA extraction via micro-capillaries or by laser micro-dissection (Kehr 2003), is thus urgently required to refine the currently existing data sets (Micali et al. 2008).

Earlier transcriptional changes in *Arabidopsis* compared the response observed in host (*G. cichoracearum*) and non-host interactions (Bgh; Zimmerli et al. 2004) at 8, 18, and 24 hours postinoculation. Many of the differentially regulated genes displayed a similar response in both host and non-host combinations. However, in the non-host interaction, the response was more rapid and of greater amplitude than in the compatible interaction, supporting the hypothesis that common defense machinery is activated upon attack by adapted and non-adapted pathogens and reinforcing the notion that adapted pathogens are able to partially suppress basal defense in their host species. Several genes that exhibit lower transcript abundance after pathogen challenge, a majority encode proteins that are involved in photosynthesis, metabolism, transport, and transcription/translation. Consistent with these functions, half of the gene products are either known or predicted to be chloroplast residents. This observation corroborates further studies indicating that *Arabidopsis* suffers up to 30% reduction in the gain of dry weight following *Bgh* infections (Zimmerli et al. 2004). This decrease is more modest in compatible infections with *G. cichoracearum*. An inverse relationship between defense responses and plant growth has been also observed in interactions between cereals and powdery mildews and underscores the metabolic costs associated with activation of defense pathways (Swarbrick et al. 2006; Wright et al. 1995a, b). Intriguingly, a recent microarray analysis of the compatible interaction between *Arabidopsis* and *G. cichoracearum* revealed that host photosynthesis is likely stimulated by the fungus, through the induction of genes involved in chlorophyll binding and chlorophyll a oxygenase (Fabro et al. 2008). Though it is conceivable that an increase in photosynthetic rate is required, as the metabolic demands of the infected epidermal tissue increase, these findings are somewhat contradictory to those obtained in the study of Zimmerli et al. (2004). Manipulation of the host metabolism by the fungal parasite thus likely involves up-as well as down-regulation of photosynthetic genes.

However, many gene transcripts related to defense responses accumulate upon both host and non-host powdery mildew attack (transcripts encoding glycosyl hydrolases and β-1,glucanase). Consistently, many of the respective genes contain in their upstream promoter region the cis-regulatory element OCS, which is known to be responsive to auxin, SA, and hydrogen peroxide. Transcripts specifically accumulating during non-host interactions encode defensins and derive from genes that are regulated by JA and ethylene. SA signaling therefore does not appear to play an important role in non-host defense against *Bgh*. Alternatively, SA-controlled genes may only become activated once the penetration defenses have failed. However, analysis of the transcriptional response of *Arabidopsis* leaves to chitin (the main fungal cell wall polysaccharide) and chitin derivatives identified several genes involved in defense responses (Ramonell et al. 2002). Genes encoding plant-derived chitinases exhibit higher transcript accumulation upon pathogen challenge compared to control conditions, and chitin oligomers obtained as fungal cell wall degradation products constitute potent signaling molecules responsible for induced resistance to *G. cichoracearum*. This resistance response appears to be activated through signaling routes that are distinct from the SA, JA, and ET pathways (Ramonell et al. 2002). Genes encoding *PR5*, two *WRKY* transcription factors, and several disease resistance proteins belonging to the TIR-NB-LRR family were induced upon chitin treatment (Ramonell et al. 2002). Genes with reduced transcript abundance included *ETR2*, encoding a transmembrane protein kinase with a LRR domain, the defensin *PDF2.5, RAR1* (encoding a protein with two zinc-finger-like domains that is required for the accumulation of many R proteins), and several other genes coding for zinc-finger proteins. To functionally validate the microarray experiments, T-DNA insertion mutants within seven of the differentially expressed genes were tested for susceptibility to *G. cichoracearum*. Three of the seven mutants, encoding a putative RING zinc-finger, and two putative disease resistance proteins displayed enhanced susceptibility to the pathogen, further reinforcing the role of chitin sensing and signaling in plant defense responses (Ramonell et al. 2002). The critical role played by chitin perception in plant defense responses was recently strengthened by the cloning of the presumptive *Arabidopsis* chitin (co-)receptor LysM Receptor-Like Kinase1 gene (*LysM RLK1 or CERK1*) by two independent groups (Miya et al. 2007, Wan et al. 2008). The gene codes for a plasma membrane-resident protein with a single transmembrane domain, three extracellular LysM domains, and a cytoplasmic serine/threonine kinase domain (Plate 4.1; Chap. 4). Mutations in this gene abolish virtually all changes in gene transcription normally observed upon chitin perception by the plant, without affecting other elicitor perception such as flagellin sensing, suggesting that the protein is part of the chitin receptor complex (Wan et al. 2008). It is conceivable that in analogy to the *FLS2* receptor kinase, which is responsible for flagellin-triggered signaling (Asai et al. 2002), *LysM RLK1/CERK* feeds into MAP kinase signaling and defense gene activation (Plate 4.1; Chap. 4). Interestingly, although loss of *LysM RLK1* caused enhanced susceptibility to the host fungus *G. cichoracearum* at the post-penetration stage (Wan et al. 2008), it did not affect the plant response to the adapted hemibiotrophic fungus *Colletotrichum higginsianum* and resulted in only a marginal increase in susceptibility to the non-adapted necrotrophic fungus *Alternaria brassicicola*. These observations suggest that pathogen perception by the plant is microbe-specific and implies the existence of more elicitor-specific receptors yet to be discovered (Micali et al. 2008).

In addition to genes involved in defense responses, genes that are induced because they are likely required for pathogen growth have also been identified (Fotopoulos et al. 2003). The observation that *Arabidopsis* leaves infected with *G. cichoracearum* have an enhanced sugar uptake compared to non-infected leaves prompted an investigation of sugar transporter expression upon powdery mildew challenge. *G. cichoracearum* induces expression of *STP4*, encoding a glucose transporter as well as βfruct1, coding for a cell wall invertase, in several cell types of the leaf (most notably the epidermis and vascular bundle) irrespective of whether they are penetrated by the fungus or not. This transcriptional induction is sustained for up to 6 days postinoculation, and is not reproducible in a non-host interaction with *Bgh*. A similar role for sugar transporters in fungal development was also found in the interaction between the rust fungus *Uromyces fabae* and broad bean (*Vicia faba*). However, in this case, the hexose transporter is of fungal origin and resides in the haustorial membrane (Voegele et al. 2001), suggesting that exchange of nutrients at the plant/biotroph interface requires transport facilitators on both sides. It is not surprising that sugars, and, therefore, sugar biosynthesis and transport, would be important for fungal growth and development and enhanced by successful fungal pathogens. The mechanisms leading to this genetic and physiological manipulation remain an enigma. However, new efforts into powdery mildew studies promise to lift the veil on the mystery (Micali et al. 2008).

7.7 Role of *Trichoderma* in Systemic Resistance to Powdery Mildew

Induced resistance (IR) varies according to different signals. Systemic acquired resistance (SAR) and induced systemic resistance (ISR) are two forms of systemic resistance. In both SAR and ISR, plant defenses are released by a previous infection; thus, biotic and abiotic factors play a role in resistance; most agents reduce disease in the infected plants by 20–85% (Walters et al. 2013). ISR is mediated by the *NPR1* gene, which is a key gene involved in disease resistance and phenotypically similar to SAR. ISR can be defined by induction of defenses in plants against many pathogens via application of plant growth-promoting microorganisms in the soil, as well as direct spreading on plants (Hossain et al. 2008a, b; Maksimov et al. 2011), whereas SAR is usually caused by a pathogen attack locally. However, the molecular pathways of each systemic resistance are different; ISR depends on two pathways that respond to ethylene and jasmonic acid (Verhagen et al. 2004), whereas SAR depends on the salicylic acid (SA) responsiveness (Spoel et al. 2003). The SA pathway controls the expression of pathogenesis-related (PR) proteins, such as *PR1, PR2*, and *PR5*, whereas the JA/ET pathway regulates the expression of a different group of defense genes, i.e. *PR3* and *PDF1.2*, in *Arabidopsis*; these PR proteins have been defined as *PR2, PR5, PDF1.2*, and basic chitinase (*PR3*) (Thomma et al. 1998).

Trichoderma sp. is soil-borne, produces green spores, and is among the ascomycetes that are widespread throughout the world (Schuster and Schmoll 2010). The fungi of the genus *Trichoderma*, which comprises a group of plant growth-promoting fungi, can colonize the intercellular parts of plant roots and stimulate systemic resistance in all parts of the plant; the actions of these fungi suppress some plant diseases by direct mycoparasitism or antibiosis, as well as indirect induced resistance (IR) (Yedidia et al. 1999; Sharma and Sain 2004). The interaction between *Trichoderma* spp. and plant is correlated with transcriptome and systemic modulations of the plant proteome (Sharma and Sain 2004; Hermosa et al. 2012). *Trichoderma* sp. stimulates ISR through hormonal and molecular pathways in a JA/ET-dependent manner (Shoresh et al. 2010; Korolev et al. 2008). A recent study has demonstrated that fungi mainly affect the pathogenesis-related genes by increasing their expression levels, thereby resulting in resistance of the treated plants to the leaf pathogens (Harel et al. 2014).

T. harzianum is the most effective biocontrol agent against a wide range of plant pathogens and has been recommended for the integrated management of crucifer's diseases (Saharan et al. 2005, 2014, 2016, 2017). *T. harzianum* TH12 isolates are used to produce a cell-free culture filtrate (CF), which has been initially characterized as an elicitor agent of induced systemic resistance to *Sclerotinia* stem rot (SSR) on *B. napus* AACC; CF is proven to effectively control pathogen when sprayed onto rapeseed crops (Alkooranee et al. 2015). The effects of *E. cruciferarum* in *R. alboglabra* and *B. napus* treated or non-treated with *T. harzianum* and its CF have not been well examined earlier. Alkooranee et al. (2015) investigated the gene expression of common genes involved in plant disease resistance. The effects of powdery mildew on *B. napus* and *R. alboglabra* were also investigated in terms of three aspects: first, the effects of *E. cruciferarum* on *B. napus* and *R. alboglabra* in comparison with the non-infected samples at six different time points; second, the effects of *T. harzianum* and its CF on *B. napus* and *R. alboglabra* in comparison with non-treated samples at six different time points; and third, the effects of *E. cruciferarum* on *B. napus* and *R. alboglabra* treated with *T. harzianum* and its CF at six different time points of infection, namely, 1, 2, 4, 6, 8, and 10 days postinoculation (dpi).

Trichoderma harzianum TH12 is a microbial pesticide for certain rapeseed diseases. The mechanism of systemic resistance induced by TH12 or its cell-free culture filtrate (CF) in *B. napus* (AACC) and *Raphanus alboglabra* (RRCC) to powdery mildew disease caused by *E. cruciferarum* was investigated by Alkooranee et al. (2015). The investigation was on the cellular and molecular aspects of *B. napus* and *R. alboglabra* infected with *E. cruciferarum*. The histological study showed the resistance of *R. alboglabra* to powdery mildew disease. The growth of fungal colonies was not observed on *R. alboglabra* leaves at 1, 2, 4, 6, 8, and 10 days postinoculation (dpi), whereas this was clearly observed on *B. napus* leaves after 6 dpi. In addition, the gene expression of six plant defense-related genes, namely, *PR-1, PR-2* (a marker for SA signaling), *PR-3, PDF 1.2* (a marker for JA/ET signaling), *CHI620*, and *CHI570*, for both genotypes were analysed in the leaves of *B. napus* and *R. alboglabra* after treatment with TH12 or CF and compared with the non-treated ones. The qRT-PCR results showed that the *PR-1* and *PR-2* expression levels increased in *E. cruciferarum*-infected leaves, but decreased in the TH12-treated leaves compared with leaves treated with CF. The expression levels of *PR-3* and *PDF1.2* decreased in plants infected by *E. cruciferarum*. However, expression levels increased when the leaves were treated with TH12. It was observed that the nature of gene expression in *B. napus* and *R. alboglabra* to explore the resistance pathways in the leaves of both genotypes infected and non-infected by powdery mildew and inoculated or non-inoculated with elicitor factors. Results suggested that *R. alboglabra* exhibited resistance to powdery mildew disease and the application of *T. harzianum* and its CF are a useful tool to facilitate new protection methods for resistant or susceptible plants (Fig. 7.6).

7.8 Mechanisms of Non-host Resistance in Crucifers to Powdery Mildew

In nature, plants are constantly challenged by above- and below-ground attack by a vast array of potential pathogens that belong to distinct phyla and exhibit different infection strategies and lifestyles. Thus, it is surprising that natural plant communities normally look rather healthy. The reason is that most plants are immune to the majority of would-be pathogens and susceptible to only a relatively small number

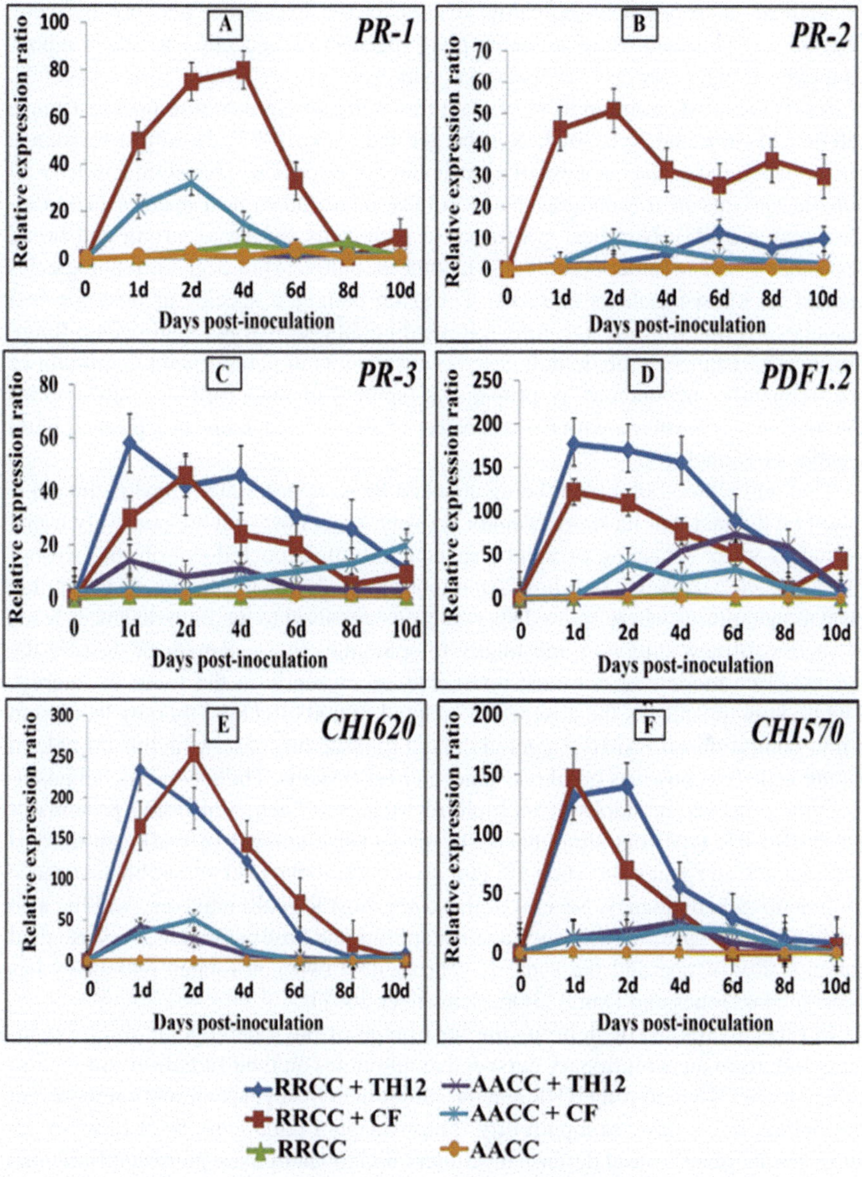

Fig. 7.6 Expression of defense-related genes of *B. napus* and *R. alboglabra* genotypes of 6 weeks old inoculated with suspensions 10 ml of TH12 and its cell-free culture filtrate (CF) separately treated by soil drenching. Three potted for each time non-inoculated served as control plants. Leaves were collected 1, 2, 4, 6, 8, and 10 days postinoculation. Total RNA was extracted, and cDNA was synthesized. Expression levels of the *PR-1, PR-2, PDF1.2* (glucanase; *BGL2*), PR-3 (basic chitinase), *CHI620*, and *CHI570* (chitinase) genes were monitored by RT q-PCR. The expression levels of genes were compared with the expression level of GAPDH (Alkooranee et al. 2015)

of adapted microbes. Consequently, disease is the exception, and not the rule. Disease resistance shown by an entire plant species to all genetic variants of a non-adapted pathogen species (or bacterial pathovar [pv] or fungal forma specialis [f.sp.]) is the most common form of plant immunity and termed non-host resistance (NHR) (Mysore and Ryu 2004; Nurnberger and Lipka 2005). Non-host resistance has been defined as the capacity of a particular plant species to resist infection by all genetic variants of a pathogen that normally infect other host species (Thordal-Christensen 2003). Non-host resistance to pathogens can be as strong and broad resistance of all members (ecotypes, landraces, cultivars) of a given plant species against all known isolates or races of a given pathogen species or host species-associated forms (*formae speciales*). Regarding attacks by phytopathogenic fungi, non-host resistance is the default case because plants in natural plant communities are frequently encountered by propagules (spores) of inappropriate (sub)-species that have not co-evolved with the respective plant host and, therefore, possess blunt weapons of attack.

The unlikelihood of a plant being attacked by an appropriate fungal pathogen is based on the fact that its spores cannot actively target hosts but are passively transported by water splashing or wind dispersal and often touch down on the 'wrong' plant species, a so-called non-host. An important exception is, of course, susceptible crop monocultures where spore dispersal even over short to medium distances leads to successful new infection and finally to epidemics that endanger or destroy the harvest. Two models of non-host resistance are currently in the focus of interest. The first one postulates the absence of adapted fungal effectors, thereby leading to a non-compromised PAMP-triggered default defense response, which is durable in nature and also known as basal resistance in host systems. The second model postulates the presence of stacks of multiple resistance (R) genes encoding proteins of the NBS-LRR type that simultaneously recognize a number of pathogen-derived avirulence (Avr) genes encoding effector proteins and that lead to durable resistance by functional redundancy. Several of these non-host R genes may very well be also mediating race-specific resistance against appropriate host pathogens, because dual functionality of one and the same R gene against two non-related pathogens has been shown (Jones and Dangl 2006; Schweizer 2007).

In other words, NHR delimits the host range of phytopathogenic microorganisms and, from an evolutionary perspective, impinges on their radiation and speciation potential. Overcoming NHR and the subsequent life cycle completion represent the hallmarks of basic compatibility. Thus, adapted pathogens need to evade or suppress the plant's basal defense machinery and to manipulate plant cell functions to their own advantage. To accomplish this mission, pathogens utilize a repertoire of effector molecules that target distinct plant mechanisms (Speth et al. 2007). Infrequent historical host range shifts are indicative of the generally robust and durable nature of NHR (Heath 2000). Conceptually, the stability of NHR has been proposed to be the consequence of several successive layers of protective mechanisms that comprise both constitutive barriers and inducible reactions (Nurnberger and Lipka 2005). Preformed physical and chemical barriers such as a rigid cell wall and toxic phytoanticipins are frequently cited as controlling invasion success of some non-adapted parasites. It is now generally accepted that inducible components of NHR and basal resistance (which defines the plant's ability to reduce the disease severity of adapted pathogens) both involve, as a first step, non-self recognition of the intruder. This is mediated by transmembrane pattern recognition receptors (PRRs) that perceive slowly evolving microbial or pathogen-associated molecular patterns (MAMPS or PAMPs) (Schweizer 2007). Downstream cell-autonomous responses of PAMP-triggered immunity include MAP-kinase signaling, production of reactive oxygen species (ROS), ethylene, ion fluxes, transcriptional induction of pathogenesis-related (PR) genes, protein phosphorylation, and callose deposits (Bittel and Robatzek 2007). Unlike cultivar-specific effector-triggered host plant immunity, which is mediated by resistance (R) proteins, NHR often (but not always) permits plants to terminate pathogenesis of invaders without a hypersensitive cell death response (HR) (Mysore and Ryu 2004). Chisholm et al. (2006) and Jones and Dangl (2006) proposed a conceptual framework for plant immunity evolution that suggests a sequential interplay of PAMP-triggered immunity (PTI), effector-triggered susceptibility (ETS), and immunity (ETI). With reference to this concept, NHR can be considered to be the consequence of ineffective microbial effectors, resulting in no suppression of PTI (and/or ETI). Alternatively, (individual) effectors might not have been selected to evade recognition and could thus be recognized in non-host plants, resulting in ETI.

Arabidopsis is a host plant for at least four distinct powdery mildews, the two closely related species *Golovinomyces cichoracearum* (*Gc*) and *G. orontii* (*Go*), as well as *Oidium neolycopersici* (*On*) and *E. cruciferarum* (*Ec*). Interestingly, *Ec* appears to be closely related to the pea pathogen *E. pisi* (*Ep*) (Saenz and Taylor 1999).

7.8.1 **Arabidopsis** ***NHR and Compatibility to Powdery Mildews***

Comparative cytological analyses of *Arabidopsis* inoculations with non-adapted and adapted powdery mildew species provided some conceptual clues: unlike compatible powdery mildews such as *Go*, which readily invades *Arabidopsis* (ca. 70% entry rate within 24 h postinoculation [hpi]), the grass and legume powdery mildews *Bgh* and *Ep* fail to penetrate approximately 95 and 75% of attacked epidermal pavement cells, respectively (Lipka et al. 2005). Entry failure correlates with timely and localized defense responses, which is reflected in dynamic cytoskeleton rearrangements, organelle transport, protein translocation, secretion processes, and focal cell wall remodeling (formation of multi-layered, callose-containing papilla) at sites of attempted fungal ingress (Hardham et al. 2007; Huckelhoven 2007). These pre-invasion defense mechanisms are backed up by post-invasion resistance: at the remaining five (25%) of successful *Bgh* (*Ep*) entry sites, haustoria become encased in callose, and the attacked epidermal cells undergo an HR-like cell death response, which is accompanied by microscopically detectable whole-cell autofluorescence. This ultimately terminates any further fungal development and prohibits colonization. By contrast, efficient invasion and cell death suppression appear to represent hallmarks of compatible and co-evolved biotrophic interactions. It is conceivable that adapted powdery mildews employ effector molecule transfer to interfere with both pre- and post-invasion defense mechanisms and to establish basic compatibility. Non-adapted isolates might entirely lack the required effector repertoire or host-specific transport machinery. Alternatively, lack of co-evolution might result in production of effector variants that are either non-functional or subject to recognition. It is plausible that the extent of taxonomic, evolutionary, and ecological separation between the interaction partners has a major impact on these alternative scenarios (O'Connell and Panstruga 2006; Lipka et al. 2008).

7.8.2 Mechanisms of Powdery Mildew Penetration Control

In order to quit descriptive phenomenology, genetic screens for chemically induced *Arabidopsis* mutants with altered non-host interactions upon *Bgh* inoculation were recently conducted. These efforts allowed the identification of three genes penetration1 (*PEN1*), *PEN2*, and *PEN3* (Lipka et al. 2005; Collins et al. 2003; Stein et al. 2006), whose products limit the entry success of non-adapted powdery mildews. Single mutants of these genes exhibit several-fold enhanced invasion frequencies of *Bgh* and *Ep*, but no increase in overall susceptibility owing to concomitant HR-like cell death of invaded epidermal cells. Systematic gene interaction analyses suggested that *PEN1* and *PEN2* act in two distinct penetration control mechanisms and that *PEN2* cooperates with *PEN3* (Lipka et al. 2005; Stein et al. 2006). These findings were further substantiated by the fact that *pen2* and *pen3* mutants, but not *pen1* mutants, also show enhanced entry of the non-adapted hemibiotrophic oomycete *Phytophthora infestans* and impaired basal resistance to the necrotrophic ascomycete *Plectosphaerella cucumerina.*

PEN1 encodes a plasma membrane-anchored syntaxin with a SNARE (soluble N-ethylmaleimide-sensitive factor attachment protein receptor) domain. SNARE domain-containing proteins are key players in vesicle-associated membrane fusion and secretion processes, including exocytosis and endocytosis. The presence of a functional homolog (required for MLO resistance 2 [*ROR2*]) in the distantly related monocot species barley supports the idea that *PEN1*-mediated defense represents an evolutionarily ancient mechanism. Cell biological analyses with functional *GFP-PEN1* fusions showed a pathogen-induced accumulation in lipid raft like plasma membrane microdomains at sites of attempted ingress by powdery mildews.

PEN1 gene product plays a role for timely secretion and cell wall remodeling processes (Plate 7.10). However, genetic evidence for *PEN1* repressor activity in multiple defense responses recently challenged simple mechanistic interpretations. Kwon et al. (2008a) could show that *PEN1* forms a pathogen-induced ternary

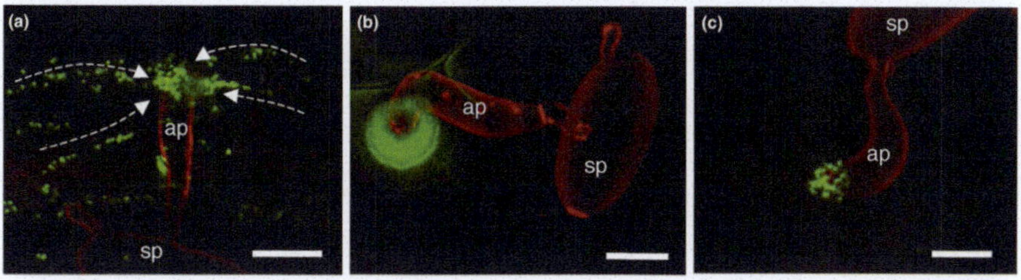

Plate 7.10 Pre-invasion resistance manifests at the cell periphery. Confocal imaging of GFP-fusion proteins in transgenic *Arabidopsis* plants inoculated with barley powdery mildew spores reveals dynamic changes in sub-cellular protein localization, secretory transport processes, and organelle relocalization at incipient fungal entry sites. (**a**) Endo-membrane compartments tagged with GFP-labeled *Arabidopsis* R-SNARE *VAMP722* move to (indicated by white dashed arrows) and accumulate at a fungal interaction site. (**b**) Focal accumulation of GFP-labeled ABC transporter *PEN3* in a lipid raft-like plasma membrane microdomain. (**c**) Peroxisome-associated *PEN2-GFP* fusions concentrate at a contact site between plant and fungal invader. Fungal structures were stained with FM 4–64. *sp* spore, *ap* appressorial germ tube. Bars = 10 μm (Lipka et al. 2008)

complex with the adaptor SNARE synaptosomal-associated protein 33 (*SNAP33*) and two endo-membrane compartment-associated SNARE proteins, vesicle-associated membrane protein *(VAMP) 721* and *VAMP722*. Moreover, transgenic expression of functional GFP *VAMP722* fusions driven by endogenous promoter sequences revealed an inducible cell autonomous expression in epidermal pavement cells that are under powdery mildew attack. The fluorescent fusion proteins tag endo-membrane compartments that move along defined routes to sites of attempted fungal penetration (Plate 7.10). These experiments provide the first evidence for SNARE complex formation in plants and, in analogy to yeast and mammal model systems, corroborate a likely role of *PEN1, SNAP33*, and *VAMP721/722* in pathogen-induced exocytotic secretion (Plate 7.11). Interestingly, *VAMP721* and *VAMP722* have a second functionally redundant role in a default secretory pathway, suggesting that an ancient transport mechanism involving these VAMP proteins was co-opted to form part of a secretory plant immune response. Homozygous/heterozygous knockout mutant combinations of *vamp721* and *vamp722* result in haplo-insufficient non-host and basal resistance to *Ep*, *Go,* and the adapted oomycete *Hyaloperonospora parasitica* (*Hp*). The experimentally challenging purification of *VAMP721/722*-tagged compartments and subsequent comparative proteomic/metabolomic cargo analyses might provide further insights into the *PEN1*-mediated defense mechanism (Lipka et al. 2005, 2007, 2008; Collins et al. 2003; Stein et al. 2006; Assaad et al. 2004; Bhat et al. 2005; Kwon et al. 2008b).

PEN2 encodes 1 out of 48 predicted *Arabidopsis* family 1 glycoside hydrolases (F1GHs). F1GHs are required for the enzymatic activation of small molecules from inactive glycosidic precursor metabolites and play diverse roles in plant biology. Catalytic activity was shown to be required for *PEN2* function in powdery mildew entry control, suggesting that the *PEN2* product(s) may have direct or indirect antifungal activity. Metabolite profiling experiments with *pen2* knockout lines suggest tryptophan-derived metabolites as potential substrates and point to a myros-

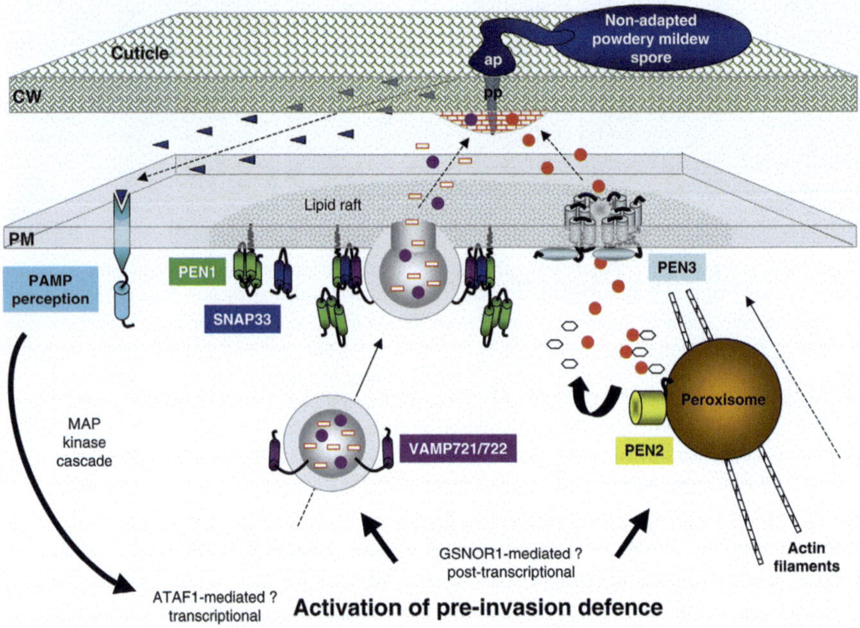

Plate 7.11 Schematic overview of pre-invasion resistance mechanisms. Barley powdery mildew spores try to penetrate the cuticle and cell wall (CW) of host and non-host plant by means of appressorium (ap) and penetration peg formation (pp). On the non-host plant *Arabidopsis*, PRR-mediated recognition of fungal *PAMPs* (blue triangles) is likely to induce MAP-kinase signaling and *ATAF1*-mediated (?) transcriptional activation of the pre-invasion defense machinery. Post-translational control (e.g. via *GSNOR1*-mediated S-nitrosylation) represents another regulatory layer. The plasma membrane (PM)-localized syntaxin *PEN1* and ABC transporter *PEN3* accumulate in a lipid raft-like microdomain. *PEN1* forms a SNARE complex with the membrane-anchored adaptor SNARE *SNAP33* and endo-membrane compartment-associated R-SNAREs *VAMP721/722*. SNARE complex formation drives secretion of cell wall precursors (red rectangles) and/or antimicrobial compounds (purple dots) at sites of attempted fungal invasion. *PEN3* discharges potentially toxic a glycons (red dots) that were catalytically released from non-toxic glycosidic (black hexagons) precursors by *PEN2* enzyme activity. *PEN2* is associated with the periphery of peroxisomes. These are known to shuttle along a focally reorganized actin cytoskeleton. Together, *PEN1/SNAP33/VAMP721/722* and *PEN2/PEN3*-mediated defense mechanisms contain the majority of fungal invasion attempts (Lipka et al. 2008)

inase-like activity of *PEN2*. Fluorescence labeling revealed an association of the *PEN2* protein with the periphery of peroxisomes, which are subject to pathogen-induced cell polarization (Plate 7.10). Intracellular transport of plant peroxisomes is known to occur along actin filaments, and genetic interference with actin dynamics has recently confirmed that pathogen-induced actin cytoskeleton reorganization plays a major role for efficient pre-haustorial NHR to powdery mildews. Therefore, peroxisomal production of *PEN2* substrate(s) and cytoskeleton-mediated peroxisome delivery to fungal entry sites represent potential underlying mechanisms (Plate 7.11). Phylogenetic analyses suggest that the *PEN2* gene represents an evolutionarily recent acquisition of the *Arabidopsis* genome. This finding demonstrates that NHR is not an exclusively ancient, fixed, and non-adaptive trait and might also explain why *pen2* mutations affect a wide range of pathogen interactions (Lipka et al. 2005, 2008; Xu et al. 2004; Miklis et al. 2007; Consonni et al. 2006).

PEN3 encodes 1 of the 15 pleiotropic drug resistance (PDR) ATP binding cassette (ABC) transporters present in the *Arabidopsis* genome. These are ubiquitous transmembrane proteins that function in the ATP-dependent transport of a wide variety of substrates across extracellular and intracellular membranes. Like *PEN1*, *PEN3* resides in the plasma membrane and accumulates at sites of attempted fungal penetration (Plate 7.10). Interestingly, lack of *PEN3* results in hyper-activation of SA-dependent defense signaling pathways, which explains the initially surprising enhanced basal resistance of *pen3* mutants to adapted powdery mildews such as *Gc* and *Go*. Intriguingly, this phenotype is significantly attenuated in *pen2pen3* double mutants.

One possible explanation is that concerted *PEN2* and *PEN3* activity drives enzymatic release and energy-dependent apoplast secretion of toxic compounds at interaction sites with invading fungi (Plate 7.11). In the absence of *PEN2*, its toxic product would no longer accumulate inside *pen3* mutant cells and thus not unintentionally activate SA-dependent hypersensitive responses (Stein et al. 2006; Crouzet et al. 2006).

The two distinct pre-haustorial non-host defense mechanisms (defined by *PEN1/SNAP33/VAMP721/VAMP722* and *PEN2/PEN3*, respectively) have in common that they are both subject to pathogen-induced cell polarization (Plate 7.10) and that they mediate the discharge of antimicrobial compounds (Plate 7.11). This highly organized battle array and the timely commitment of forces are strikingly reminiscent of the execution of immune responses at the immunological synapse in vertebrate T cells. Thus, plant innate immunity and vertebrate adaptive immunity feature mechanistic parallels that are worth further exploration (Kwon et al. 2008b; Lipka et al. 2008).

Broad-spectrum resistance to adapted powdery mildews is conferred by loss-of-function mutant alleles of mildew resistance locus O (MLO) genes in both barley and *Arabidopsis*. Powdery mildew fungi are believed to require a subset of functional MLO proteins as compatibility factors ('entry portals') for successful invasion of host epidermis cells. Lack of these potential powdery mildew effector targets results in efficient pre-invasion resistance that shares many characteristics with NHR to powdery mildews. Intriguingly, mlo resistance in *Arabidopsis* requires all PEN genes described so far. Moreover, recent experiment's showed that transient overexpression of constitutively active variants of a calcium-dependent protein kinase (CDPK) affects both mlo and non-host resistance in barley. Together, these findings imply that powdery mildew fungi utilize a common and evolutionarily old host cell entry mechanism (dependent on wild-type MLO) and that mlo resistance represents a re-establishment of pre-invasion NHR (Consonni et al. 2006; Humphry et al. 2006; Freymark et al. 2007). Other components that were recently claimed to contribute to *Arabidopsis* pre-invasion NHR against non-adapted powdery mildews include the S-nitrosoglutathione reductase *GSNOR1* and the plant transcription factor *ATAF1*, respectively. Although the entry success of *Bgh* and *Bgt* on *ataf1-1* and *atgsnor1-3* mutants was only moderately enhanced (from 9% to 14% and from 1% to 3%, respectively), these results suggest both transcriptional and post-transcriptional regulation of penetration resistance (Feechan et al. 2005; Jensen et al. 2007).

7.8.3 Mechanism of Post-penetration Defense

Arabidopsis mutants showing enhanced entry are still non-host plants for non-adapted powdery mildews owing to effective post-entry cell death. Systematic analyses with multiple mutant combinations revealed that post-haustorial NHR requires enhanced susceptibility 1 (*EDS1*), phytoalexin-deficient 4 (*PAD4*), and senescence-associated gene 101 (*SAG101*). These lipase-like proteins constitute a regulatory node that is essential for basal defense against *Hp*, SA-mediated signaling, and a subset of R gene-mediated resistance pathways. Single mutants for *EDS1*, *PAD4*, and *SAG101* have no or only a minor effect on pre-invasion resistance to powdery mildews. However, at successful invasion sites, HR is less frequent and allows ecto-parasitic secondary hyphal growth and microcolony formation of *Bgh* and *Ep*. Intriguingly, microscopic inspection revealed that NHR to *Ep*, but not to *Bgh*, was fully compromised on *pen2pad4* and *pen3eds1* double mutants, indicated by occasional conidiophore formation. On *pen2pad4sag101* triple mutants, *Bgh* was capable of completing its life cycle, while *Ep* growth and reproduction were even macroscopically detectable. Thus, removal of only three genes is sufficient to make *Arabidopsis* a fully susceptible host plant for the non-adapted pea powdery mildew and allows the monocot pathogen *Bgh* to establish basic compatibility. This is remarkable, as it suggests that both pathogens do not lack any principal adaptations to establish (presumably) complex biotrophic interactions and to complete their life cycles. The quantitative differences in virulence between *Ep* and *Bgh* possibly reflect the closer evolutionary species and effector complement relatedness between *Ep* and the adapted powdery mildews *Ec*. Comparative powdery mildew effector studies are urgently required to unequivocally resolve this issue. In summary, *Arabidopsis* NHR to non-adapted biotrophic powdery mildews is based upon two successive, multi-component, and independently effective defense layers: *PEN*-gene-mediated pre-invasion resistance and post-invasion immunity, controlled by *EDS1, PAD4*, and *SAG101* (Lipka et al. 2005, 2008; Stein et al. 2006; Wiermer et al. 2005).

HR-like cell death execution and its control by the *EDS1PAD4/SAG101* node suggested a possible contribution of R gene-mediated *ETI* to post-invasion resistance against *Bgh* and *Ep*. The proteins required for *MLA12* resistance (*RAR1*) and suppressor of *G2* allele of *SKP1* (*SGT1*) function as cofactors in heat shock protein 90 (*HSP90*)-mediated stabilization of R protein complexes and represent a genetic convergence point for *EDS1/PAD4/SAG101* and other R gene signaling pathways. Double and triple mutant combinations of either *pen2* or *pen3* with *rar1* and *sgt1b* had a significantly weaker effect on colonization success of *Bgt* or *Ep* if compared with combinations with *eds1*, *pad4,* and *sag101* and did not result in life cycle completion. Hence, R genes appear to play a minor if any role in post-haustorial *Arabidopsis* NHR to the tested non-adapted powdery mildews. Perhaps such a scenario is more likely in interactions where would-be pathogen and non-host plants share a more recent compatible ancestry, e.g. in barley/wheat interactions with the wheat/barley and powdery mildew, or where a brisk intra-specific or inter-specific exchange of effector molecules determines a pathogen's host range (Stein et al. 2006; Azevedo et al. 2006; Noel et al. 2007; Schweizer 2007).

7.8.4 Components of Non-host Resistance

Powdery mildew species either can have a wide or narrow host range or might even be specialized to a single host plant species. *Golovinomyces orontii* (*Go)* is virulent on *Arabidopsis*, on other Brassicaceae species, as well as on Solanaceae and Cucurbitaceae species, but does not infect Rosaceae or Asteraceae (Plotnikova et al. 1998). In contrast to *Go*, the pea powdery mildew pathogen *E. pisi* and the barley powdery mildew *Bgh* are not able to cause disease on *Arabidopsis*, as they show little penetration success and no completion of their asexual life cycle. Consequently, they do not give rise to any visible epiphytic colonization and symptom formation (Lipka et al. 2005). This is essentially due to NHR of plants against pathogens, which by definition is resistance of an entire plant species against all genetic variants of a microbial species (Lipka et al. 2005; Nurnberger and Lipka 2005). Mechanistically, NHR seems to be equivalent to basal defense or innate immunity and supposedly relies mainly on preformed defenses and MAMP-triggered immune responses (Thordal-Christensen 2003; Nurnberger and Lipka 2005). Accordingly, components involved in NHR often contribute to defense not only against non-adapted but also against adapted pathogens. In the case of filamentous phytopathogens, NHR can be subdivided in pre- and post-invasive resistance. Pre-invasive NHR restricts the penetration of fungal and oomycete pathogens, including powdery mildews, whereas post-invasive NHR eventuates if the non-adapted pathogen succeeds in host cell entry and frequently results in an HR associated with local host cell death. This response is mainly effective against biotrophic pathogens as it deprives the invader of nutrients (Glazebrook 2005).

Several main components from two distinct pathways of pre-invasive NHR have been identified so far. The first pathway relies on *PEN1*, which is believed to mediate exocytosis of potentially harmful cargo upon pathogen attack by forming ternary SNARE complexes with *SNAP33* and *VAMP721/722* (Kwon et al. 2008b; Kwaaitaal et al. 2010). The second pathway includes *PEN2* (At2g44490) and *PEN3. PEN2* is a tail-anchored β-thioglucoside glucohydrolase that synthesizes indole glucosinolates from a tryptophan-derived precursor (Bednarek et al. 2009; Clay et al. 2009; Fuchs et al. 2016). It contains a carboxy terminal tail anchor that targets the protein to peroxisomal and outer mitochondrial membranes. Requirement of the cytochrome P450 monooxygenase *CYP81F2* (At5g57220) for *PEN2*-mediated resistance to powdery mildew penetration indicates its involvement in pathogen-induced production of 4-substituted indol-3-ylmethyl-glucosinolate (I3G) substrates of *PEN2* (Bednarek et al. 2009; Clay et al. 2009). Consistently, *CYP81F2*-RFP localizes to the ER membrane, focally accumulating at sites of *Bgh* attack. This suggests that *PEN2* substrate production occurs in close proximity to *PEN2*-decorated mitochondrial subpopulations that are recruited to sites of attempted fungal invasion (Fuchs et al. 2016). The products of *PEN2* are thought to be exported by the *PEN3* ABC transporter (Stein et al. 2006). Mutants of *PEN3* result in pathogen-inducible, *PEN2*-dependent over-accumulation of an indole compound (4-*O*-β-D-glucosyl-indol-3-yl formamide) in leaves. This suggests that PEN3 is involved in the transport of this indole or a precursor during pathogen defense (Lu et al. 2015). *PEN3* interacts with calmodulin *(CAM)7* (At3g43810), and *cam7* mutant plants are more susceptible to the non-adapted fungal pathogens *Bgh* and the Asian soybean rust fungus *Phakopsora pachyrhizi*, demonstrating that *CAM7* and therefore Ca^{2+} sensing/transmission of Ca^{2+} signals is an important factor of NHR in *Arabidopsis* (Campe et al. 2015). Fluorophore-tagged *PEN1, PEN2*, and *PEN3* focally accumulate at powdery mildew penetration sites (Assaad et al. 2004; Lipka et al. 2005; Stein et al. 2006). Recruitment of *PEN1* and *PEN3* fusions with GFP to infection sites can be triggered by MAMPs, but distinct mechanisms contribute to the transport of the two proteins (Underwood and Somerville 2013). Interestingly, *PEN1* and *PEN3* accumulate in the apoplast at sites of papilla formation (Meyer et al. 2009; Underwood and Somerville 2013). Importantly, although the *pen* and *cam7* mutants allow increased host cell entry, subsequent host cell death due to post-penetration NHR restricts infection success. Key components of post-invasive NHR are *EDS1, PAD4*, and *SAG101*, all of which are required, to different degrees, for full resistance against various pathogens (Feys et al. 2005; Wiermer et al. 2005; Rietz et al. 2011; Wagner et al. 2013). Single mutations in *EDS1*, *SAG101,* and *PAD4* and the respective double mutants in combination with *pen2* are insufficient to allow sporulation of *Bgh* and *E. pisi* on *Arabidopsis*. However, the pre- and post-invasive NHR-deficient mutant *pen2 pad4 sag101* enables these non-adapted pathogens to form secondary hyphae. *E. pisi* even causes macroscopically visible powdery mildew symptoms resulting from moderate conidiation, and *Bgh* occasionally forms conidiospores (Lipka et al. 2005).

Besides extracellular papilla formation, a prominent aspect of pre-invasive NHR is the focal accumulation of various cellular components and organelles towards sites of attempted pathogen invasion. These structures include secretory vesicles, peroxisomes, mitochondria, and the ER (Koh et al. 2005; Bohlenius et al. 2010). In *Arabidopsis* these rearrangements further comprise the accumulation of proteins with defense functions, e.g. the *PEN* proteins (Assaad et al. 2004; Lipka et al. 2005; Stein et al. 2006) and the callose synthase *PMR4/GSL5* (Ellinger et al. 2013), at *Bgh* attack sites. This focal aggregation requires reorganization of the actin cytoskeleton towards sites of fungal ingress, emphasizing the central role of actin-based transport processes in pre-invasive NHR and plant antifungal immunity in general

(Takemoto et al. 2006; Underwood and Somerville 2008; Feechan et al. 2011; Underwood and Somerville 2013; Yang et al. 2014).

Another component of the NHR to powdery mildew is the *Arabidopsis* phospholipase Dδ (*PLDδ*: At4g35790), which is involved in the biosynthesis of phosphatidic acid (Wang 2004). Phosphatidic acid can serve as a precursor for membrane phospholipids or as a signaling molecule and may play a role in plant defense, as its levels increase after MAMP perception or recognition of various pathogen effectors (van der Luit et al. 2000; de Jong et al. 2004; Andersson et al. 2006; Kirik and Mudgett 2009). A *PLDδ* fusion with GFP accumulates around papillae at sites of attempted *Bgh* penetration. Additionally, the *pldδ* mutant allows increased cell entry by *Bgh,* and *E. pisi*, and shows delayed up-regulation of early MAMP-responsive genes after chitin treatment (Pinosa et al. 2013). Finally, the phytohormone abscisic acid (ABA) seems to be involved in NHR against powdery mildews. The NAC transcription factor (TF) *Arabidopsis thaliana* activating factor 1 (*ATAF1*: At1g01720) contributes to defense against *Bgh*. Loss of *ATAF1* partially compromises *Bgh* penetration resistance, which correlates with the induction of ABA biosynthesis and transcript accumulation of ABA-responsive genes (Jensen et al. 2007; Jensen et al. 2008). By contrast, endogenous ABA levels are decreased after inoculation of wild-type plants with *Bgh*. *ATAF1*-dependent suppression of ABA levels after pathogen challenge suggests that *ATAF1* acts as attenuator of ABA signaling in order to mediate efficient penetration resistance against *Bgh* (Jensen et al. 2008). A better understanding of the genetics and molecular mechanism understanding plant non-host resistance which is strong, and durable, bears the potential for targeted employment of the valuable trait to control host pathogens. The use of mutants and gene-silencing approaches in *A. thaliana* opens up new possibilities to discover genes involved in non-host resistance and to unravel the underlying mechanisms.

7.9 Genetics of Host–Parasite Interactions

7.9.1 Inheritance of Resistance in Brassica *to Powdery Mildew*

Inheritance of powdery mildew resistance in crucifers is the result of interaction of the genetic system of two distinct organisms, a host and powdery mildew pathogen. In cabbage, resistance to powdery mildew is controlled by a single dominant gene with modifiers (Walker and Williams 1965; Williams et al. 1968).

The six generations of two inter-specific crosses, *B. juncea* x *B. carinata* (RH 30 x HC-1; Varuna x PCC 2), were evaluated to understand the nature of resistance against powdery mildew. The F_1 hybrids of both crosses, RH 30 x HC-1 and Varuna x PCC 2, recorded 1.54 and 0.62 disease score, respectively, indicating the complete dominance of resistance (Table 7.1). The F_2 population of both the crosses segregated in the ratio of 3R:1S to *E. cruciferarum*, indicating that the single resistance gene present in HC-1, and PCC-2, and controlled the resistance with complete dom-

Table 7.1 Reaction of *B. juncea* and *B. carinata* parents and their F1 hybrids (*B. juncea* x *B. carinata*) to powdery mildew (*E. cruciferarum*) (Saharan and Krishnia 2001)

Parents/F_1 hybrids	Host reaction: Resistant	Host reaction: Susceptible	Total no. of plants	DSI	DR
Parents					
RH 30	0	24	24	3.78	S
Varuna	0	28	28	3.25	S
HC-1	23	0	23	0.65	R
PCC 2	20	0	20	0.25	R
F_2 hybrids					
RH 30 x HC-1	18	0	18	1.54	R
Varuna x PCC 2	15	0	15	0.62	R

DSI disease severity index, *DR* disease reaction, *R* resistant, *S* susceptible

Table 7.2 Mode of segregation for *Erysiphe cruciferarum* reaction to *Brassica juncea* x *B. carinata* F_2 progenies, BC_1 and BC_2 (Saharan and Krishnia 2001)

Crosses	Host reaction: R	Host reaction: S	Total no. of plants	Expected ratio (R:S)	X^2	P value
F_2						
RH 30 x HC 1	177	57	234	3:1	0.05	0.80–0.90
Varuna x PCC2	155	57	212	3:1	0.40	0.50–0.70
BC_1 (F1 x S)						
F_1 (RH 30 x HC 1) x RH 30	45	32	77	1:1	2.19	0.10–0.20
F_1 (Varuna x PCC2) x Varuna	38	30	68	1:1	0.94	0.30–0.50
BC_2 (F_1 x R)						
F_1 (RH 30 x HC 1) x HC 1	62	0	62		No segregation	
F_1 (Varuna x PCC2) x PCC 2	58	0	58		No segregation	

inance. The backcross progenies with susceptible parents segregated in the ratio of 1R:1S, well supporting the segregation pattern of F_2 population. The backcross progenies with resistant parents had no segregation (Table 7.2; Krishnia et al. 2000; Saharan and Krishnia 2001; Kumar et al. 2002).

7.9.2 Evaluation of Different Families of Plant Progenies for Resistance to Powdery Mildew

In all the crosses, the R x R plant progenies possessed comparatively less disease score for powdery mildew infection as compared to resistant self and resistant open plant progenies, except PCR 3 x Shiva, Pusa Basant x Shiva, and Pusa Bahar x

Domo. The level of susceptibility in progenies of S x S families as compared to S self families was less in all the crosses, except PCR 3 x Shiva, Pusa Basant x EC 322092, Varuna x EC322092, and RH 30 x EC 322093. The maximum GCV and PCV were observed in progenies of R x S for powdery mildew in most of crosses, while in other crosses the R x R family (Pusa Basant x Shiva), R open family (Pusa Basant x EC 322092, RH 30 x EC 322092), and S x S families (Rajat x Shiva; RH 30 x HC-1) had highest GCV and PCV. The h^2 was observed highest in R x S families of most of the crosses. However, the R x R family of Pusa Basant x Shiva, Pusa Bahar x Domo, and RH-30 x HC-1 recorded higher h^2. The genetic gain was recorded highest in progenies of R x S families of most of the crosses. However, in other crosses, the progenies of R x R (Pusa Basant x Shiva, RH-30 x HC-1), R self families (Pusa Bahar x Domo; Pusa Basant x EC 322092), and R open families (RH 30 x EC 322093) recorded the highest genetic gain. R x R families of each cross had the lowest disease score for powdery mildew; therefore, through selective intermating of resistant plants in advance generations, the level of resistance can be increased. However, the degree or magnitude of increase will depend on parents involved in the crosses and host–pathogen–environment interaction. Two crosses, i.e. RH-30 x HC-1 and Varuna x PCC 2, showed multiple resistance to all three diseases of rapeseed-mustard (white rust, *Alternaria* blight, and powdery mildew). Therefore, these parents are the best donors for developing the multiple disease resistance cultivars in oilseed *Brassica*. The positive genetic correlations obtained here indicate that selection for resistance to a single pathogen will also bring increasing resistance to other pathogens (Krishnia et al. 2000; Saharan and Krishnia 2001; Kumar et al. 2002). Similar results were reported by Mitchell-Olds et al. (1995) in *B. rapa*.

7.9.3 Inheritance of Resistance in Arabidopsis *to Powdery Mildew*

In most of the *Arabidopsis* accessions, resistance is either polygenic, based on the atypical resistance (R) gene [resistance to powdery mildew 8 (RPW8)], or combinations thereof, with RPW8 representing the major quantitative trait locus (Adam and Somerville 1996; Xiao et al. 1997; Adam et al. 1999; Schiff et al. 2001; Wilson et al. 2001; Orgil et al. 2007; Gollner et al. 2008). In general, more accessions are resistant to *E. cruciferarum* than to *G. cichoracearum* and *G. orontii* (Adam et al. 1999; Gollner et al. 2008). *RPW1, RPW2, RPW4*, and *RPW5* were described as semi-dominant R loci and localized to chromosomes II, III, IV, and V, respectively (Fig. 7.7; Adam and Somerville 1996). To determine the genetic basis of powdery mildew resistance in *Arabidopsis*, segregation analysis of progeny from crosses between each of the R accessions and Columbia (gl1), which is S to powdery mildew pathogen, was performed by Adam and Somerville (1996). For all accessions except S1-0, resistance was conferred by a single locus. S1-0 was unique in that two unlinked loci controlled the disease reaction phenotype. In accessions,

(a)

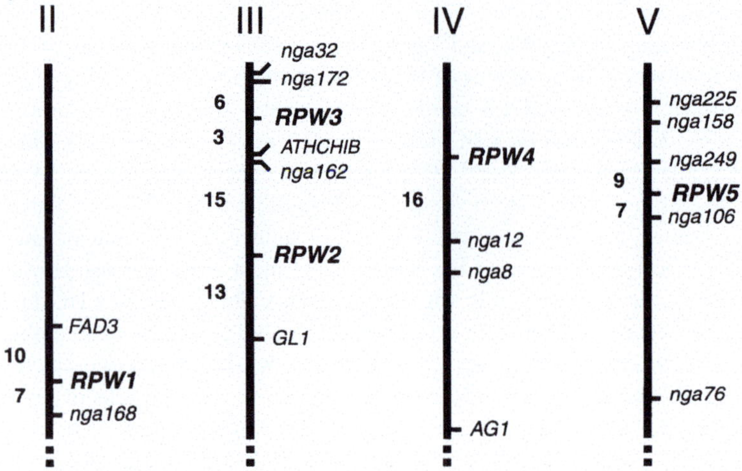

(b)

	I					II		III						
	nga59 1.9[a]	*nga63* 14.2	*nga280* 114.6	*nga111* 149.2	*ATHATPase* 153.6	*FAD3* 60	*nga168* 78.5	*nga32* 1.2	*nga172* 2.2	*GAPC* 3.8	*nga126* 13.1	*ATHCHIB* 19.4	*nga162* 21.2
RPW1	-[b]	U[c]	U	-	U	10[d]	7	U	-	-	-	-	U
RPW2	-	U	-	#[e]	-	-	U	U	-	U	-	19	17
RPW3	U	U	-	U	-	-	U	10	7	-	5	4	8
RPW4	U	-	U	-	U	-	U	-	U	-	-	-	U
RPW5	-	U	U	-	U	-	U	-	-	-	U	-	U

	III			IV				V						
	GL1 64.6	*nga6* 123.3	*nga112* 125.3	*nga12* 29.8	*nga8* 35.0	*AG1* 72.6	*SC5* 80	*nga225* 11.0	*nga158* 13.7	*nga249* 22.3	*nga106* 33.7	*nga139* 61.0	*nga76* 75.8	*nga129* 118.9
RPW1	U	U	-	-	U	-	U	U	-	U	-	U	-	U
RPW2	14	-	#	-	U	-	U	#	-	#	-	U	-	U
RPW3	U	-	U	-	U	-	U	U	-	-	-	U	-	U
RPW4	U	-	U	19	20	U	-	-	-	U	-	U	-	U
RPW5	U	-	U	-	-	-	U	13	14	9	7	22	U	U

[a] Position on map generated from recombinant inbred lines developed by Lister and Dean (version released on network June 29, 1995).
[b] -, linkage not determined
[c] U, unlinked
[d] Genetic distance in cM as determined by two-point linkage analysis.
[e] #, abnormal segregation ration ($P<0.01$)

Fig. 7.7 Map locations of five RPW loci. (**a**) Graphical representation of the map positions of the RPW loci. The most likely map order and the genetic distances (in cM) were derived via multipoint analysis of data from about 54 F_3 families for each cross. (**b**) Two point linkage data for the *RPW* loci and various SSLP and CAPS markers. Genetic distances are given in cM (Adam and Somerville 1996)

Wa-1, Kas-1, Stw-0, and su-0, powdery mildew resistance was encoded by a semi-dominant allele. Mapping studies revealed that powdery mildew resistance in Kas-1, Wa-1, Te-0, Su-0, and Stw-0 were controlled by five independent loci (Fig. 7.7).

Powdery mildew resistance in *Arabidopsis* accessions Kas-1, Wa 1, Te-0, Su-0, and Stw-0 to *E. cichoracearum* was controlled by five independent loci/genes (Table 7.3). In Kas-1, resistance to *E. cruciferarum* was conferred by *RPW1*, which

Table 7.3 F_2 segregation and test cross data illustrating the inheritance of powdery mildew resistance in six *Arabidopsis* accessions. (Adam and Somerville 1996)

			F_2 segregation analysis[a]										Test cross [b]	
	DR score[c]		# F_2 plants in each DR group										Ratio	
Accessions (Locus)	Parent	F_1	3.1	3.0	2.1	2.0	1.1	1.0	0.1	0.0	Ratio[d] (R:1:S)	X^2	(R:S)	X^2
Kas-1 (*RPW1*)	0,0-0,1	1,0-2,1	2	52	10	25	38	41	37	9	46:114:54 (1:2:1)	1.51 ($0.50 < p < 0.25$)	19.14 (1:1)	0.76 ($0.50 < p < 0.25$)
Wa-1 (*RPW2*)	0,1	1,0-2,1	19	118	94	34	113	22	125	3	128:263:137 (1:2:1)	0.31 ($0.90 < p < 0.75$)	34.29 (1:1)	0.40 ($p = 0.50$)
Te-0 (*RPW3*)	0,0-1,1	2,0-3,1	12	145	7	49	29	11	17	4	61:213 (1:3)	1.10 ($0.75 < p < 0.50$)	17.16 (1:1)	0.03 ($p = 0.90$)
Su-0 (*RPW4*)	0,0-1,1	2,0-2,1	10	136	108	158	46	31	40	30	147:266:146 (1:2:1)	1.31 ($p = 0.50$)	18.17 (1:1)	0.03 ($p = 0.90$)
Stw-0 (*RPW5*)	0,1	1,0-2,1	11	62	45	12	73	9	81	0	81:139:73 (1:2:1)	1.20 ($0.75 < p < 0.50$)	27.20 (1:1)	1.04 ($p = 0.75$)
SI-0	0,0	3,0-3,1	46	210	48	61	39	12	64	31	31:480 (1:15)	0.03 ($0.90 < p < 0.75$)	20.57 (1:3)	0.04 ($0.90 < p < 0.75$)

[a]Data from F_2 of crosses between the susceptible accession Col-5 and each resistant accession

[b]Data from test crosses between the F_1 Col-5 X (resistant accession) and the recessive parent

[c]Criteria for assessing the disease reaction phenotypes are given in the experimental procedures

[d]For F_2 progeny from crosses with Kas-1, Wa-1, Su-0, and Stw-0, plants were placed into groups consisting of those with phenotypes like the resistant parent (R), those with fully susceptible phenotypes like Col-5 (DR 3,0 or 3,1) (S), and those with intermediate phenotypes distinct from either parent (I). F_2 progeny of crosses with Te-0 and Sl-0 were grouped into two groups with one group consisting of progeny that resembled the resistant parent (R) and the second group (S) consisting of all other progeny. See text for a complete description

mapped to chromosome 2, between *nga 168* and *FAD 3* (Fig. 7.7a). Resistance in Wa-1 (*RPW2*) and in Te-0 (*RPW3*) was encoded by loci on the same arm of chromosome 3. *RPW2* was placed between markers *nga 162* and *GL1*, while *RPW3* mapped to a 9 cM interval defined by *nga 172* and *ATHCHIB*. Although SSLP marker *nga 126* also maps in this interval, it is a dominant marker in the Col-5 by Te-0 cross and thus is of limited usefulness. Resistances in Stw-0 and in Su-0 were conferred by *RPW4* (chromosome 4) and *RPW5* (chromosome 5), respectively. *RPW4* was located about 16 cM telomeric from marker nga 12. No polymorphic CAPS or SSLP marker telomeric to *RPW4* was found for the Col-5 by Su-0 cross. However, the absence of linkage to *AG-1* and slightly closer linkage to *nga 12* than *nga 8* place *RPW4* to the telomeric side rather than the centromeric side of *nga 12*. Multipoint linkage analysis placed *RPW5* to the *nga 249-nga 106* interval on chromosome 5 (Adam and Somerville 1996).

The resistance of *A. thaliana* to the powdery mildew pathogen *E. cruciferarum UEA1* is conferred by two dominant alleles, *RPW6* on chromosome 5 and *RPW7* on chromosome 3. Neither *RPW6* nor *RPW7* maps in the close vicinity of the other resistance genes that have been located on *A. thaliana* chromosomes (Kunkel 1996). *RPW7* lies more than 10 cM below the bottom of a 25 cM interval on chromosome 3 containing four loci for resistance to *Peronospora parasitica*, and *RPW6* maps more than 20 cM below *RPW5* for resistance to *E. cichoracearum*, on chromosome 5 (Fig. 7.8). The resistance of Ms-0 to *E. cruciferarum* was associated with the early arrest of the pathogen, collapse of epidermal cells underlying appressoria, and, occasionally, necrotic palisade mesophyll cells at the infection site. Resistance of Ms-0 to *E. cichoracearum* was consistently associated with localized necrosis visible to the naked eye. These host responses are comparable to the hypersensitive reaction to cereal powdery mildews (Jorgensen 1994) and therefore distinguish the resistance of Ms-0 to *E. cruciferarum UEA1* and *E. cichoracearum*, from the resistance of other *A. thaliana* accessions to *E. cichoracearum*, which was not consistently associated with host cell collapse (Adam and Somerville 1996).

Papillae are formed in *A. thaliana* at the site of penetration pegs of *E. cichoracearum* UCSC1 (Adam and Somerville 1996) and *E. cruciferarum* (Hall 1994), but could not be associated with the expression of resistance. Xiao et al. (1997) have defined resistant plants as those which do not support vegetative reproduction of the powdery mildew pathogen. Thus, *RPW6* and *RPW7* were both required for disease resistance to *E. cruciferarum* UEAI. However, plants with dominant alleles at one of these loci and recessive alleles at the other exhibited a weak resistance reaction characterized by reduced fungal growth and conidiospore production, which were referred to as an intermediate reaction. Other features of the intermediate reaction were its dependence upon high humidity, and the extensive collapse of host epidermal cells (Plate 7.12d), and necrosis of mesophyll cells, in infected tissues (Plate 7.12d). These characteristics bear some similarity to the phenomenon of partial resistance, which is well-known in barley (Jorgensen 1994). Some forms of partial resistance are associated with delayed necrosis (Carver 1986), and the pathogen grows and sporulates in the infected tissues. Like the intermediate resistance reported in *A. thaliana*, partial resistance is strongly influenced by the environment

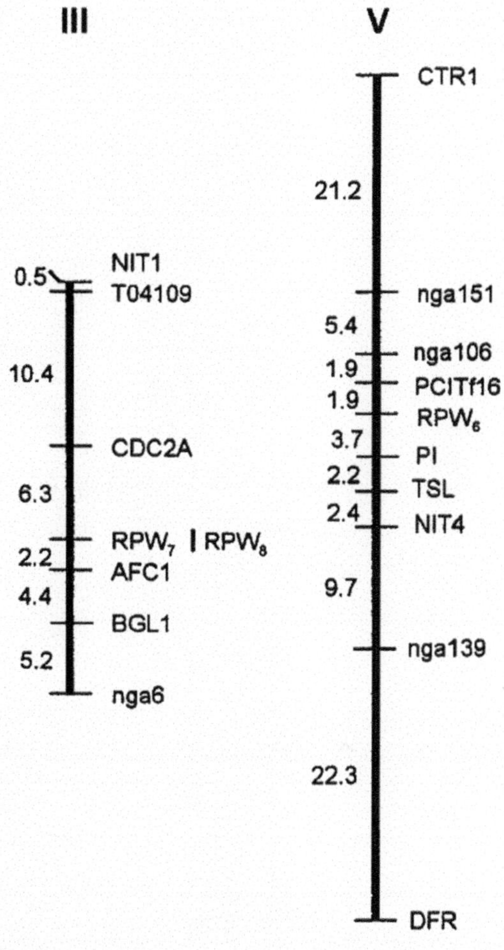

Fig. 7.8 Genetic map positions of powdery mildew resistance loci *RPW6, RPWT*, and *RPW8*, from *A. thaliana* accession Ms-0. *RPW6* and *RPW7* were both required for resistance of *A. thaliana* to *E. cruciferarum* and were mapped simultaneously on 224 F 3 families. *RPW8* conferred resistance to *E. cichoracearum* UCSC1 and was mapped on 81 F_3 families from a cross between accessions La-*er* (susceptible) × Ms-0 (resistant). Horizontal bars indicate position of locus; numbers are centimorgans (Xiao et al. 1997)

and is notoriously difficult to score, usually requiring laborious laboratory methods (Newton 1993), and individual genes have not been traced in barley. However, unlike intermediate resistance conferred by either RPW6 or RPW7, partial resistance in barley is thought to be controlled by many genes (Geiger and Heun 1989). *RPW6* and *RPW7* act independently as weak resistance genes, but in combination, they cooperate synergistically to elicit effective disease resistance against *E. cruciferarum* UEA1, apparently involving a rapid hypersensitive reaction. The genetics of resistance of *A. thaliana* to *E. cruciferarum* might therefore be similar to the resistance of mung bean to *Erysiphe polygoni* DC, which segregates as two unlinked genes with additive effects (Reddy et al. 1994). However, the independent action of the mung bean powdery mildew resistance loci was indicated by a distinctive resistance reaction which could be scored in F_2 populations.

Resistance of *A. thaliana* accession Ms-0 to *E. cichoracearum* UCSC1 was controlled by a single locus, *RPW8*, which mapped to the same position as *RPW7*.

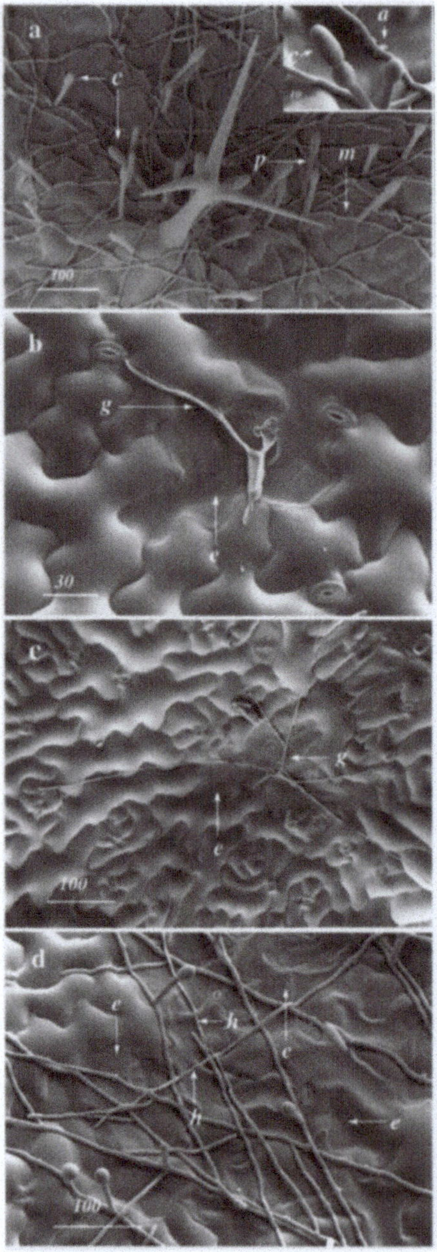

Plate 7.12 Scanning electromicrographs of *E. cruciferarum* on *A. thaliana* leaves, 8 days after inoculation. (**a**) Leaves of the susceptible *A. thaliana* accession, La-*er*, supported an extensive mycelial (*m*) network bearing conidiophores (*p*) and conidia (*c*), shown here surrounding a branched leaf trichome. Insert: Lobed appressoria (a) attached to epidermal cells of La-*er* leaves, and large terminal conidium (*c*) was characteristic features of *E. cruciferarum*. (**b**) Resistance of *A. thaliana* accession

Possibly, *RPW7* and *RPW8* are linked genes in a complex resistance locus similar to Mlo in barley (Jorgensen 1994) and *Pm3* in wheat (Zeller et al. 1993). Alternatively, *RPW7* and *RPW8* may define a single resistance gene, which recognizes products of both *E. cruciferarum* UEA1 and *E. cichoracearum* UCSC1. A precedent for a dual-specificity resistance gene is *RPM1*, which interacts with two avirulence genes from different bacterial pathogens of *A. thaliana* (Bisgrove et al. 1994; Grant et al. 1995).

7.9.4 Inheritance of Enhanced R in Arabidopsis

Genetic analysis of *edr1* mutant was made by Frye and Innes (1998) to determine the inheritance of the enhanced resistance phenotype. The *edr1* mutant was crossed with *Arabidopsis* ecotype Lands berg *erecta* (L *err*), which is susceptible to *E. cichoracearum.* The F_2 progeny was inoculated with *E. cichoracearum* conidia and scored 7 to 9 days later for development of necrotic lesions and lack of visible powdery mildew. These two traits co-segregated and behaved as a recessive mutation, producing approximately 1:3 ratios of resistant-to-susceptible plants (85:266; 250.115; P. 0.1). To obtain a chromosomal map position for the mutation in *edr1* plants, a total of 1223 F_2 plants from the La-*er* cross were scored for *E. cichoracearum* resistance, and 235 plants displaying resistance to *E. cichoracearum* were selected for mapping. DNA was isolated from the resistant F_2 plants and analysed for linkage to simple sequence length polymorphism (SSLP) and co-dominant amplified polymorphic sequence (CAPS) markers (Konieczny and Ausubel 1993; Bell and Ecker 1994). The *edr1* mutation mapped 3.2 centimorgans centromeric from the SSLP marker *ATEAT1* (15 recombinant chromosomes) and 0.85 centimorgans telomeric from the CAPS marker *NCC1* (4 recombinant chromosomes) on chromosome 1. Multiple defenses are induced more rapidly in *edr1* plants than in wild-type plants when infected with a virulent strain of powdery mildew.

Plate 7.12 (continued) Ms-0 was associated with collapse of germ tubes (*g*), shown here arising from a conidium. The underlying host epidermal cells (*e*) have also collapsed. (**c**) Resistance of F_1 plants from a cross La-*err* x Ms-0 was associated with partial collapse of hyphal germ tubes (*g*) from germinating conidium and collapse of several host epidermal cells (*e*) around the germinated conidium. (**d**) The intermediate reaction on plants from a F_3 line was characterized by a mycelial network, with conidiophores, and conidia, similar to the susceptible reaction shown in plate (*a*). However, in the intermediate reaction, many host epidermal cells underlying hyphae had collapsed (*e*). Some hyphae had also collapsed (*h*). Bars represent indicated length in Iμm (Xiao et al. 1997)

7.9.5 *Inheritance of Resistance in* Arabidopsis *Mutants to Powdery Mildew*

From a screen of approximately 26,000 M2 *Arabidopsis* plants, 32 confirmed powdery mildew-resistant mutants that did not exhibit spontaneous macroscopic lesions or constitutively express *PR1* were identified. Twelve of these mutants had obvious lesions after inoculation, reminiscent of *edr1*, and were set aside. The remaining 20 mutants showed no macroscopic lesions after inoculation, and they were crossed for inheritance pattern. Crosses between the 20 mutants defined 4 complementation groups, designated powdery mildew resistant 1–4 (*pmr1–4*) (Plate 7.13). Multiple, independent alleles were identified for *pmr1* (7), *pmr2* (6), and *pmr4* (2). No obvious phenotypic differences were noted among the alleles at each locus. The number of alleles is less than 20 because multiple mutants from the same lot of mutagenized seed were counted as one allele. The mutants were backcrossed to the wild type, Col, two times before characterization, excluding mapping. Segregation of powdery mildew resistance in the F_1 and F_2 generations indicated that all mutations are recessive and that *pmr2*, *pmr3*, and *pmr4* segregate in a simple 3:1 Mendelian fashion

Plate 7.13 Phenotype of *pmr* mutants. (**a**) Plants were inoculated with *E. cichoracearum* 8 days before being photographed. Note the extensive fungal growth on Col. (**b**) Plants was inoculated with *E. orontii* 10 days before being photographed. (**c**) Effect of light levels on *pmr3*. Twenty-two-day-old plants grown under high, 150 mEym^2 per sec, or low, 45 mEym^2 per sec, light conditions (Vogel and Somerville 2000)

Table 7.4 Genetic analysis of powdery mildew-resistant mutants (Vogel and Somerville 2000)

Cross (female x male)	Type	Total	Disease response		X^2
			Susceptible	Resistant	
PMR1/PMR1 X pmr1/pmr1	F_1	28	28	0	
PMR1/PMR1 X pmr1/pmr1	F_2	1841	1742	99	378∗; $p < 0.05$
PMR1/pmr1 X pmr1/pmr1	Testcross	124	66	58	0.47 t; $p < 0.05$
PMR1/pmr1 X PMR1/pmr1	Testcross	199	182	17	137 t; $p < 0.05$
PMR2/PMR2 X pmr2/pmr2	F_1	71	71	0	
PMR2/PMR2 X pmr2/pmr2	F_2	193	140	53	0.43∗; $p < 0.05$
PMR3/PMR3 X pmr3/pmr3	F_1	11	11	0	
PMR3/PMR3 X pmr3/pmr3	F_2	98	71	27	0.56∗; $p < 0.05$
PMR4/PMR4 X pmr4/pmr4	F_1	22	22	0	
PMR4/PMR4 X pmr4/pmr4	F_2	443	340	103	0.4∗; $p < 0.05$

∗X^2 = calculated for an expected 3:1, wild type-to-mutant ratio
t X^2 = calculated for an expected 1:1, wild type-to-mutant ratio

(Table 7.4). By contrast, all seven alleles of *pmr1* had a more complex segregation pattern (Table 7.4). Reciprocal crosses between plants heterozygous and homozygous for *pmr1* indicated that transmission of *pmr1* through the pollen is reduced (Table 7.4). To further characterize the decreased transmission of *pmr1*, a pollen competition experiment was conducted. Pollen from heterozygous plants was placed on stigmas of homozygous *pmr1* plants. When the siliques were fully ripe, but not yet brittle, they were cut into three sections: top, middle, and bottom. The segregation of powdery mildew resistance among the progeny and, by inference, the genotype of the pollen fertilizing the ovules derived from the different sections were determined. Plants originating from the top sections segregated in a 2.9:1 ratio (55 susceptible; 19 resistant), plants from middle sections segregated in a 9.3:1 ratio (93 susceptible; 10 resistant), and plants from the bottom sections segregated 52:1 (103 susceptible; 2 resistant). Thus, *pmr1* pollen fertilizes ovules close to the stigmatic surface more efficiently than distal ovules. Slower pollen tube growth or an inability of the pollen tubes to sense and grow towards unfertilized ovules could explain these results. It is important to note that homozygous *pmr1* plants are fertile, indicating that *pmr1* pollen can fertilize all ovules; it is just less efficient than wild-type pollen. *pmr1* and *pmr2* mapped to chromosome 1, and *pmr4* mapped to chromosome 4 (Fig. 7.9). *pmr3* mapped to chromosome 5, close to the position of the dwarf mutant *lu-1* (lutescens). Because the description of the *lu-1* phenotype was similar to *pmr3*, the mutants were crossed. The F_1 progeny was not dwarf, indicating that *pmr3* and *lu-1* complement one another and presumably affect different genes (Vogel and Somerville 2000).

In addition to powdery mildew resistance, *pmr3* plants are much smaller and paler than wild type at light intensities 150 mEym^2 per sec. However, at light intensities 75 mEym2 per sec, *pmr3* plants are nearly wild type in stature and are susceptible to powdery mildew. A careful analysis reveals that the size of *pmr3* plants under high- and low-light conditions is similar, but wild-type plants are much larger

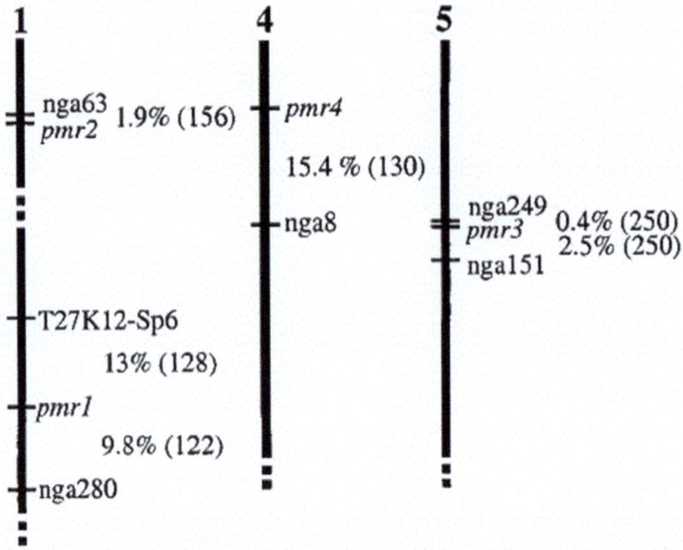

Fig. 7.9 Map positions of *pmr* mutants. Vertical bars represent chromosomes. Per cent recombination between markers and the *pmr* mutants is indicated. The number of chromosomes scored is in parentheses (Vogel and Somerville 2000)

under high light (Plate 7.13c). The dwarf phenotype co-segregated 100% with powdery mildew resistance through two backcrosses and among 598 F_2 plants. In addition, the fact that both the dwarf phenotype and the resistance phenotype are dependent on light intensity strongly suggests that the same mutation is responsible for both phenotypes. Leaves from *pmr4* plants are epinastic, especially when grown under short-day conditions. Both alleles of *pmr4* display this phenotype, indicating that the same mutations are responsible for both the powdery mildew resistance and the epinasty. Adult *pmr1* and *pmr2* plants are indistinguishable from wild type.

To determine whether fungal growth on the mutants was blocked at a defined step in the infection cycle, the growth of hyphae and level of conidiation were measured over time. The same general trend was observed for all mutants. Conidia germinate and begin growth in a normal fashion, but colonies on mutant plants grow more slowly than colonies on wild-type plants. By 3 dpi, hyphal length was significantly (ANOVA, *P*, 0.05) less on all the mutants than on wild-type plants (Fig. 7.10a). Eventually, most colonies on the mutants appeared to be dead or dying and produced few, if any, conidiophores, the aerial stalks bearing asexual conidia. In contrast, colonies on wild-type plants appeared healthy and produced many conidiophores (Fig. 7.10b). However, on all mutants a small subset of colonies does produce some conidiophores, indicating that none of the mutations result in a complete block at any stage of fungal development (Vogel and Somerville 2000).

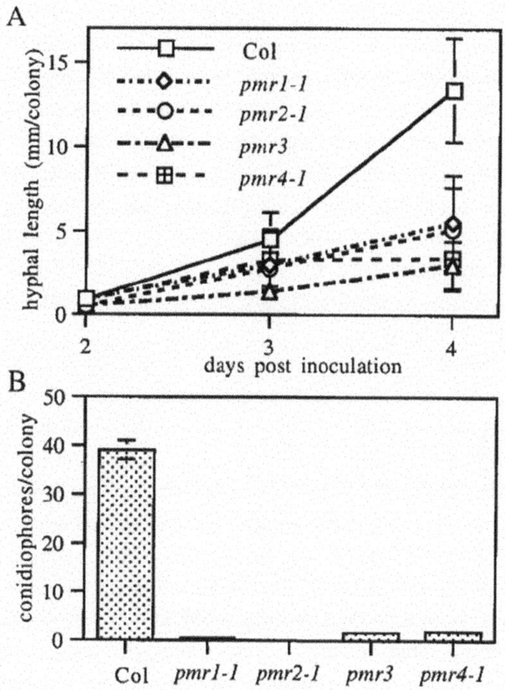

Fig. 7.10 Quantification of *E. cichoracearum* growth on *pmr* mutants. (**a**) Hyphal length per colony. (**b**) Conidiophores per colony at 6 dpi. The means 6 SD based on at least 15 replicates are plotted in *A* and *B* (Vogel and Somerville 2000)

7.10 Mutagenic Resistance

The method of induced mutagenesis is a powerful tool for creating new genotypes with an increased yield, quality, and resistance to diseases and pests (Ahloowalia et al. 2004). Upon irradiation with gamma rays, the variation of the resulting genetic alterations has high frequency and wider range of morphological and biochemical changes (Skoric et al. 2008). There is evidence that the mutations may also change the expression of certain enzymes in the biosynthetic pathway of fatty acids (Auld et al. 1992; Pleines and Friendt 1989). The irradiation of rapeseed with low doses of gamma rays of 100 Gy and 150Gy has a stimulating effect on the synthesis of linoleic and linolenic acids. Conversely, at higher doses of the mutagen factor, the concentration of these fatty acids decreases, while the amount of oleic acid increases (Rahimi and Bahrani 2011). The fatty acids synthesized in plants play an important role by participating in the composition of tissue hormones and in cell and nuclear membrane structures (Kolattukudy 1974). It is assumed that unsaturated fatty acids in plant cells are involved in cutin and suberin synthesis (Pinot et al. 1999). Cutin, which acts as first barrier and protects plant organs (leaves and fruits) from dehydration providing protection against pathogens and pests, is composed of C16 and C18 fatty acids. Suberin, which is predominantly composed of hydroxyl and epoxy unsaturated fatty acids, is a structural component of the plant underground part roots and tubers (Kolattukudy 1980). C18 fatty acids in plants play a role in building the defense mechanism against diseases and pests (Kachroo and Kachroo 2009). It has been found that hydroxyl fatty acids also inhibit the development of *Erysiphe graminis, Puccinia recondita,* and *Phytophthora infestans* (Hou and Forman 2000). The powdery mildew development is favoured by meteorological factors, such as temperature and high humidity (Penaud 1999; Enright and Cipollini 2007). Fungicide application is often not efficient enough, due to the pathogen's rapid spread in favourable conditions. The effects of gamma radiation on the fatty acid composition of rapeseed oil and the increased resistance to *E. cruciferarum* have been determined by Petkova et al. (2014). Biochemical changes in the oil after gamma-ray irradiation are also associated with changes in the susceptibility to powdery mildew.

Petkova et al. (2014) assessed through the gas chromatography the difference in the content of saturated and unsaturated fatty acids in the seeds of the non-irradiated variants and the ones irradiated with two doses of gamma radiation. A summary of the data for all 3 years of the fatty acid composition of the *Abacus* hybrid oil and its M0, M1, and M2 was decomposing the populations. The chemical analysis showed generally that the irradiated plants showed a lower content of oleic acid and a higher content of linoleic acid and linolenic acid compared with the non-irradiated variants. The data in Table 7.5 showed a tendency towards increase in the total content of raw fats in mutant populations of M1 and M2 of *Abacus* as compared with non-irradiated plants. This, in turn, is associated with a reduction in the total content of saturated fatty acids and an increase in the content of unsaturated fatty acids.

There is an inverse relationship between the values of the total fat content and the content of saturated and unsaturated fatty acids (Table 7.5). The rapeseed oil of the *Abacus* hybrid showed higher content of saturated fatty acids as compared with the M2 and M3 generations of the irradiated variants. In contrast, the percentage content of unsaturated fatty acids is influenced positively by irradiation with gamma rays and is comparatively high in M2 A1-100 and M2 A-150 92.7% and 93.7%, compared with the baseline 89.7%.

The data in Table 7.6 shows the extent of damage by the powdery mildew agent on plants of the *Abacus* variety and its gamma-ray irradiation-treated variants. It also shows the arithmetic mean values, and their errors, the accuracy indicators, and the variation coefficients. The data show that the highest extent of damage on the

Table 7.5 Total fat, saturated, and unsaturated fatty acids in the seeds of *Abacus* and M0, M1, and M2 segregating generations of *Abacus*, treated with absorbed dose of 100 Gy and 150 Gy gamma rays (Petkova et al. 2014)

Fats and fatty acids %)	*Abacus* control	M0 A-100	M0 A-150	M1 A-100	M1 A-150	M2 A-100	M2 A-150
Content of crude fat	35.9	40.7	39.5	43.2	45.1	42.5	47.1
Saturated fatty acids	10.1	9.3	7.5	6.3	6.8	6.3	6.9
Unsaturated fatty acids	89.7	82.3	83.5	85.2	86.2	92.7	93.7

Table 7.6 Fatty acid composition of rapeseed oil from the seeds of *Abacus* and M0, M1, and M2 generation of irradiated plants with 100 Gy and 150 Gy gamma rays (Petkova et al. 2014)

Fats and fatty acids (%)	*Abacus* control	M0 A-100	M0 A-150	M1 A-100	M1 A-150	M2 A-100	M2 A-150
C12:0 lauric acid	–	0.1	0.1	0.1	0.1	0.1	0.2
C14:0 myristic acid	0.1	0.1	0.2	0.2	0.2	0.1	0.1
C16:0 palmic acid	7.5	6.7	6.3	6.5	6.7	5.9	5.7
C18:0 stearic acid	2.3	1.8	2.0	1.9	1.2	0.3	0.3
C18:1 oleic acid	66.3	69.3	67.0	67.8	68.5	67.5	73.0
C18:2 linoleic acid	17.2	15.5	17.4	17.2	18.3	19.1	15.9
C18:3 linoleic acid	4.4	3.1	3.5	4.2	5.9	6.9	4.6

stems and branches in the range of 98.9%–99% ($p = 0.1\%$) is observed with the M0 *Abacus* generation, irradiated with absorbed dose 100 Gy and 150 Gy gamma rays, compared with the non-irradiated variant (Table 7.7). The irradiation did not have a positive impact on the sustainability of pods to *E. cruciferarum.* The differences found are insignificant compared with the corresponding baseline group. In contrast, the M1 and M2 generations were shown to have a proven lower extent of damage by the phytopathogen as compared with the baseline. It is assumed that the increased resistance demonstrated is due to the higher content of unsaturated fatty acids in the seeds of the irradiated plants. Unsaturated fatty acids can be used as a means of biological control of plant diseases on global scales. Increasing the dose of the mutagen leads to an increase in the concentration of unsaturated fatty acids with 18 carbon atoms and to a decrease in the extent of damage by *E. cruciferarum* (Petkova et al. 2014).

7.11 Biochemical Basis of Resistance

An adapted pathogen can recognize specific host-derived signals that trigger pathogen differentiation and expression of virulence, or both. Therefore, pathogens must adapt to the chemical composition of their hosts. This leads to the evolution of an enzymatic toolbox, which the pathogen uses to overcome structural and chemical barriers of its hosts or to metabolize host-derived substrates. In addition, plants possess robust innate immunity that involves defense responses triggered after the recognition of pathogen-derived elicitors (pathogen-associated molecular patterns or pathogen effectors) or host-derived elicitors (damage-associated molecular patterns) (Boller and Felix 2009; Maekawa et al. 2011). Consequently, pathogens require host-specific effector molecules that match host targets for the suppression of immunity and reprogramming of the host for the demands of the pathogen. Loss of susceptibility may therefore result from altered host immunity (gain of resistance

Table 7.7 Assessing the extent of powdery mildew damage on stems, branches, and pods as a percentage of the infected plant area on rapeseed plants from *Abacus* and M0, M1, and M2 generation of gamma-irradiated plants (Petkova et al. 2014)

	Extent of powdery mildew damage %											
	Stems				Branches				Pods			
Variants	Mean ± SD	D	t	Sign.	Mean ± SD	D	t	Sign.	Mean ± SD	D	t	Sign.
Abacus control	79.3 ± 1.31				90.5 ± 1.07				97.6 ± 0.49			
M0-Abascus-100 Gy	99.3 ± 0.18	20.1	17.6	+++	98.9 ± 0.22	8.4	16.4	+++	99.1 ± 0.2	1.52	1.36	Ns
M0-Abascus-150 Gy	99 ± 0.21	19.7	17.3	+++	99 ± 0.24	8.5	15.7	+++	98.9 ± 0.31	1.3	1.04	Ns
M1-Abascus-100 Gy	28 ± 1.52	−51.3	19.2	− − −	40.5 ± 1.34	−50	32.5	− − −	49.9 ± 1.17	−47.7	26.2	− − −
M1-Abascus-150 Gy	26.7 ± 1.26	−52.6	34.1	− − −	34.8 ± 1.23	−55.7	29.5	− − −	47.4 ± 1.02	−50.2	29.0	− − −
M2-Abascus-100 Gy	10.3 ± 0.58	−79.3	74.6	− − −	21.1 ± 0.77	−69	42.4	− − −	27.1 ± 1.04	−70.5	40.5	− − −
M2-Abascus-150 Gy	11.5 ± 0.56	−67.8	39.8	− − −	22.7 ± 1.31	67.8	35.1	− − −	27.5 ± 1.11	−70.1	39.8	− − −

Mean average value in % from 30 repletion, *D* difference against control, *t* Student's t distribution, *ns* non-significant

P > 5%+, P > 1% ++, P > 0.1% +++, P < 5% −, P < 1% − −, P < 0.1% − − −

functions) or, in a stricter sense, from changes in host components that are required by the pathogen for pathogenesis, but do not directly operate in the regulation of defense (de Almeida Engler et al. 2005; Huckelhoven, 2005; Pavan et al. 2010; Huckelhoven et al. 2013). In contrast with wild-type *Arabidopsis* plants, loss-of-function mutants of the cytochrome P450 monooxygenase *CYP83A1* gene are barely susceptible to the biotrophic ascomycete *E. cruciferarum* (Weis et al., 2013), which causes powdery mildew on many Brassicaceae (Adam et al. 1999; Saharan et al. 2005). The cytochrome P450 gene family in *Arabidopsis* comprises 244 genes (and 28 pseudogenes) and constitutes one of the largest gene families in plants. P450 enzymes function as monooxygenases in the biosynthesis of diverse metabolites, including pigments, phytohormones, and lignin, or defense compounds, such as flavonoids, alkaloids, or glucosinolates (Bak et al. 2011). The synthesis of aliphatic glucosinolates is divided into three stages: the chain elongation of the amino acid, formation of the glucosinolate core structure, and, finally, secondary modifications (Wittstock and Halkier 2002). The two *CYP83* proteins, *CYP83A1* and *CYP83B1*, phylogenetically belong to the *CYP71* clade (Hansen et al. 2001; Bak et al. 2011). They non-redundantly function in the core structure synthesis of glucosinolates by catalysing the initial conversion of aldoximes to thiohydroximates (Bak and Feyereisen 2001; Naur et al. 2003). *CYP83A1* has higher substrate specificity for methionine-derived aldoximes in the synthesis of aliphatic glucosinolates, whereas *CYP83B1* preferentially converts tryptophan-derived aldoximes in the synthesis of indole glucosinolates (Bak and Feyereisen 2001; Naur et al. 2003).

In plant–herbivore interactions, glucosinolates and their hydrolysis products, such as isothiocyanate, nitriles, or epithionitriles, function as deterrents against generalist herbivores, but also as attractants for specialized insects. The knowledge about the function of glucosinolates in plant–fungus interactions are more limited (Bednarek and Osbourn 2009). However, in *Arabidopsis*, the peroxisome-associated myrosinase penetration 2 (*PEN2*) hydrolyses 4-methoxy-indol-3-ylmethylglucosinolate to bioactive products involved in resistance to non-adapted powdery mildew fungi (Lipka et al. 2005; Bednarek et al. 2009). Indole glucosinolates are further important to balance the mutualistic interaction of *Arabidopsis* with the beneficial root endophyte *Piriformospora indica* (Jacobs et al. 2011; Nongbri et al. 2012). Aliphatic glucosinolates are important in resistance to lepidopteran larvae, to non-adapted bacterial pathogens, and to the necrotrophic fungus *Sclerotinia sclerotiorum* (Beekwilder et al. 2008; Fan et al. 2011; Stotz et al. 2011). However, little information is available about the role of aliphatic glucosinolates in interactions with haustorium-forming fungi. Powdery mildew fungi are obligate biotrophic ascomycete pathogens with a high degree of specialization to a limited range of hosts. They form haustoria from appressoria that sense and directly penetrate the host cuticle and cell wall (Green et al. 2002; Huckelhoven and Panstruga 2011). *Erysiphe cruciferarum* is a typical powdery mildew fungus adapted to Brassicaceae (Adam et al. 1999). Hence, it can normally cope with glucosinolates as a result of an unknown mechanism. Weis et al. (2014) have observed a loss of *CYP83A1* which changes the metabolic composition of *Arabidopsis* in a manner that greatly influences the interaction with powdery mildew pathogen.

Aliphatic glucosinolates function in the chemical defense of Capparales. The cytochrome *P450 83A1* monooxygenase (*CYP83A1*) catalyses the initial conversion of methionine-derived aldoximes to thiohydroximates in the biosynthesis of glucosinolates, and thus *cyp83a1* mutants have reduced levels of aliphatic glucosinolates. Loss of *CYP83A1* function leads to dramatically reduced parasitic growth of powdery mildew fungus *E. cruciferarum* on *Arabidopsis thaliana*. The *cyp83a1* mutants support less conidial germination, and appressorium formation of *E. cruciferarum* on the leaf surface, and post-penetration conidiophore formation by the fungus. By contrast, a *myb28-1 myb29-1* double mutant, which totally lacks aliphatic glucosinolates, shows a high level of susceptibility to *E. cruciferarum*. The *cyp83a1* mutants also lack very-long-chain aldehydes on their leaf surface. Such aldehydes support appressorium formation by *E. cruciferarum* in vitro. In addition, when chemically complemented with the C26 aldehyde *n*-hexacosanal, *cyp83a1* mutants can again support appressorium formation. The mutants further accumulate 5-methylthiopentanaldoxime, the potentially toxic substrate of *CYP83A1*. Loss of powdery mildew susceptibility by *cyp83a1* may be explained by a reduced supply of the fungus with inductive signals from the host and an accumulation of potentially fungitoxic metabolites (Weis et al. 2014).

7.11.1 Induction of Biochemical Metabolites

Powdery mildew fungi are biotrophic pathogens that form a complex interface, the haustorium, between the host plant and the pathogens. The pathogen acts as an additional sink, competing with host sinks, resulting in considerable modification of photo assimilate production, and partitioning within the host tissue. The biotrophic interaction of *Arabidopsis*/*E. cichoracearum* induces a number of responses: Glc uptake in host tissues is enhanced after fungal infection; this coincides with the induction of expression of the monosaccharide transporter gene, *Arabidopsis sugar transport protein 4* (*AtSTP4*), in infected leaves; invertase activity and transcript levels for a cell wall invertase, *Atβfruct1*, increase substantially in *Arabidopsis* during infection of pathogen. Before infection, *Arabidopsis* plants transformed with an *AtSTP4* promoter-β-glucuronidase construct show expression mainly in sink tissues such as roots; after infection, *AtSTP4* expression is induced in the mature leaves and increases over the 6-d time period. Sections of infected leaves stained for β-glucuronidase show that *AtSTP4* expression is not confined to infected epidermal cells but is also evident in a wider range of cells, including those of the vascular tissue (Fotopoulos et al. 2003).

The finding that infection by powdery mildew induces *AtSTP4* expression in a wide range of host cells despite the fungus being confined to the epidermis implicates the role of signaling mechanisms in this response. There is evidence that defense response genes are induced by elevated sugar levels (Ehness et al. 1997; Herbers et al. 2000) and hexose transport into the host cells may be enhanced to cope with this increased energy demand and/or to reduce availability of sugars to the pathogen. Enhanced cell wall invertase reduces Suc loading (Stitt et al. 1990; von Schaewen et al. 1990), and an increase in hexose availability as a consequence could increase the concentration gradient of hexoses to the fungus. The transfer of carbon to the fungus, primarily as Glc, is thought to occur down a concentration gradient created by fungal solute uptake and use. The nature of the sugar transporters at the haustorial membrane in the powdery mildew fungi is unknown, but by analogy with rust, carriers may exist at this membrane to facilitate transport into the mycelium. A major goal will be to dissect the signaling pathway in a compatible interaction that results in the diversion of host resources to the fungus (Fotopoulos et al. 2003).

7.11.2 Role of Camalexin in Powdery Mildew Resistance

To protect themselves against pathogens, plants have evolved intricate immune responses that include accumulation of reactive oxygen species, deposition of callose, enhanced expression of *pathogenesis-related* (*PR*) genes, and biosynthesis of phytoalexins. Phytoalexins are low molecular mass secondary metabolites that are induced by both biotic and abiotic stress. During pathogen infection, plants synthesize a wide variety of structurally different phytoalexins to defend against pathogen invasion (Hammerschmidt 1999; Pedras et al. 2011). Camalexin, 3-thiazol-20-yl-indole, is one of the major phytoalexins of *Arabidopsis thaliana* and was long considered to be the only phytoalexin of *Arabidopsis*, until the discovery of rapalexin A (Pedras and Adio 2008). The biosynthesis and regulation of camalexin in *Arabidopsis* remain only partially understood, and the full scope of camalexin functions also remains to be defined. Camalexin is derived from tryptophan and requires many cytochrome P450s, including *CYP79B2, CYP71A13*, and *CYP71B15* [which corresponds to the camalexin-deficient mutant *phytoalex-indeficient 3* (*PAD3*); Nafisi et al. 2007; Schuhegger et al. 2007a, b]. The *Arabidopsis* transcription factors *WRKY33, WRKY18*, and *WRKY40* appear to be involved in the regulation of camalexin biosynthesis (Qiu et al. 2008; Pandey et al. 2010; Mao et al. 2011). Camalexin plays an important role in the response to necrotrophic pathogens *Alternaria brassicicola* and *Botrytis cinerea* (Thomma et al. 1999; Ferrari et al. 2003; Kliebenstein et al. 2005; Nafisi et al. 2007) and the oomycete *Phytophthora brassicae* (Schlaeppi et al. 2010), as well as the biotrophic fungus *Golovinomyces orontii* (Consonni et al. 2010; Pandey et al. 2010). Although camalexin produces broad-spectrum resistance to many species of plant pathogens, how it functions remains unclear (Liu et al. 2016).

In addition to camalexin, plants synthesize other related secondary metabolites, such as glucosinolates, that participate in the defense response. Plant cells usually store glucosinolates in stable forms; during insect and/or pathogen attack, myrosinases hydrolyse these stable forms in to active compounds (Halkier and Gershenzon 2006). According to their side-chain radical, glucosinolates can be divided into aliphatic glucosinolates, indole glucosinolates, and aromatic glucosinolates. Many cytochrome P450s function in glucosinolate synthesis, including *CYP83A1* and *CYP83B1*. The *Arabidopsis* cytochrome P450 monooxygenase *CYP83A1* participates in the biosynthesis of aliphatic glucosinolates from aliphatic oximes, whereas *CYP83B1*, the *Arabidopsis* protein most similar to *CYP83A1*, functions in the biosynthesis of indole glucosinolates. The biosynthetic pathways of alkyl glucosinolates and indole glucosinolates affect each other; for instance, the *cyp83a1-2* (also called *ref2-1*) mutant produces slower levels of aliphatic glucosinolates, but accumulates higher levels of indole-derived glucosinolates compared with wild type (Hemm et al. 2003; Naur et al. 2003; Sonderby et al. 2010). The biosynthesis of glucosinolates, especially indole glucosinolates, shares the intermediate product indole-3-acetaldoxime (IAOx) with biosynthetic pathways that produce many other secondary metabolites or hormones like camalexin and indole-3-acetic acid (IAA), respectively (Hemm et al. 2003; Grubb and Abel 2006; Nafisi et al. 2007). Although the regulation of callose biosynthesis, in response to bacterial elicitors of *Arabidopsis* immunity, requires 4-methoxy-indol-3-ylmethylglucosinolate (4MI3G; Bednarek et al. 2009; Clay et al. 2009), how 4MI3G and related metabolites participate in the immune response is not well-understood. Powdery mildew fungi, as biotrophic pathogens, infect many plant species and cause huge agricultural losses worldwide. The plant hormone salicylic acid (SA) plays an important role in resistance to powdery mildew in *Arabidopsis*, and many mutants showing enhanced resistance to powdery mildew require SA signaling for their resistance phenol type; these mutants include *edr1* (*enhanced disease resistance1*), *edr2,* and *edr4* (Frye and Innes 1998; Frye et al. 2001; Tang et al. 2005a, b; Zhao et al. 2014; Wu et al. 2015). Liu et al. (2016) characterized an *Arabidopsis* mutant that exhibits enhanced resistance to a variety of powdery mildew species. A mutation in the gene encoding cytochrome P450 monooxygenase *CYP83A1*, a component of the glucosinolate pathway, leads to higher accumulation of camalexin and enhanced resistance to the powdery mildew fungus *Golovinomyces cichoracearum*, which is consistent with the previous finding that *cyp83a1* exhibits increased resistance to powdery mildew fungus *E. cruciferarum* (Weis et al. 2013). The *cyp83a1-3* accumulates higher levels of camalexin. The mutations in genes affecting camalexin production suppress the resistance of *cyp83a1-3*, indicating that higher accumulation of camalexin in *cyp83a1-3* mutants contributes to their enhanced powdery mildew resistance.

Small secondary metabolites, including glucosinolates and the major phytoalexin camalexin, play important roles for immunity in *Arabidopsis thaliana*. Liu et al. (2016) isolated an *Arabidopsis* mutant with increased resistance to the powdery mildew fungus *Golovinomyces cichoracearum* and identified a mutation in the gene encoding cytochrome *P450 83A1* monooxygenase (*CYP83A1*), which functions in glucosinolate biosynthesis. The *cyp83a1-3* mutant exhibited enhanced defense responses to *G. cichoracearum,* and double mutant analysis showed that this enhanced resistance requires *NPR1, EDS1*, and *PAD4*, but not *SID2* or *EDS5*. In *cyp83a1-3* mutants, the expression of genes related to camalexin synthesis increased upon *G. cichoracearum* infection. Significantly, the *cyp83a1-3* mutant also accumulated higher levels of camalexin. Decreasing camalexin levels by mutation of the camalexin synthetase gene *PAD3* or the camalexin synthesis regulator *AtWRKY33* compromised the powdery mildew resistance in these mutants. Consistent with these observations, overexpression of *PAD3* increased camalexin levels and enhanced resistance to *G. cichoracearum.* It indicates that accumulation of higher levels of camalexin contributes to increased resistance to powdery mildew of *A. thaliana*.

Camalexin biosynthesis and accumulation are affected by wrky18 wrky40 transcription factor of *Arabidopsis*. Camalexin accumulation enforces defenses to penetrating pathogens and is the main *Arabidopsis* phytoalexin induced at infection sites (Rauhut and Glawischnig 2009). Genes encoding all three enzymes of camalexin biosynthesis were up-regulated in microarrays. Conversion of indole-3-acetaldoxime (IAOx) to indole-3-acetonitrile (IAN), catalysed by *CYP71A13*, marks a committed step (Nafisi et al. 2007), whereas *PAD3* catalyses the final step in camalexin biosynthesis (Schuhegger et al. 2006). Pandey et al. (2010) examined expression changes for both of these genes over a time course in WT and *wrky18 wrky40* plants by qPCR. Compared with uninfected plants, *G. orontii*-infected *wrky18 wrky40* plants had >15-fold elevated *CYP71A13* transcript levels, whereas infected WT plants showed only a 4-fold increase. Similarly, the *wrky18 wrky40* mutant accumulated overall more *PAD3* transcripts than WT plants (Fig. 7.11b). The phytoalexin camalexins levels prior to, and 24 hpi with *G. orontii* was observed compared with WT plants, the *wrky18 wrky40* mutant already had elevated levels of camalexin in uninfected tissue (Fig. 7.11a). Nevertheless, these levels increased significantly in both genotypes upon infection, with the mutant accumulating 18-fold higher concentrations of camalexin than WT. These findings substantiated microarray studies and revealed that loss of *WRKY18* and *WRKY40* functions resulted in increased biosynthesis and accumulation of camalexin, which is further strongly enhanced upon *G. orontii* infection. The pre-existing higher camalexin levels found in uninfected *wrky18 wrky40* plants may in part be due to the nearly twofold elevated transcript levels observed for *CYP79B2, CYP71A13*, and *CYP71B15/PAD3* (Pandey et al. 2010).

7.12 Transfer of Powdery Mildew Resistance Through Embryo Rescue

Inter-specific hybrid plants and backcross 1 (BC1) progeny were produced through sexual crosses and embryo rescue between *B. carinata* accession PI 360883 and *B. oleracea* cvs. Titleist and Cecile to transfer resistance to powdery mildew to *B. oleracea.* Four inter-specific hybrids were obtained through application of embryo rescue from crosses with *B. carinata* as the maternal parent, and their inter-specific nature was confirmed through plant morphology and random amplified polymorphic DNA (RAPD) analysis (Plate 7.14). Twenty-one BC1 plants were obtained through sexual crosses and embryo rescue although embryo rescue was not necessary to produce first backcross generation plants between inter-specific hybrids and *B. oleracea.* All inter-specific hybrids and eight of the BC1 plants were resistant to powdery mildew (Tongue and Griffiths 2004).

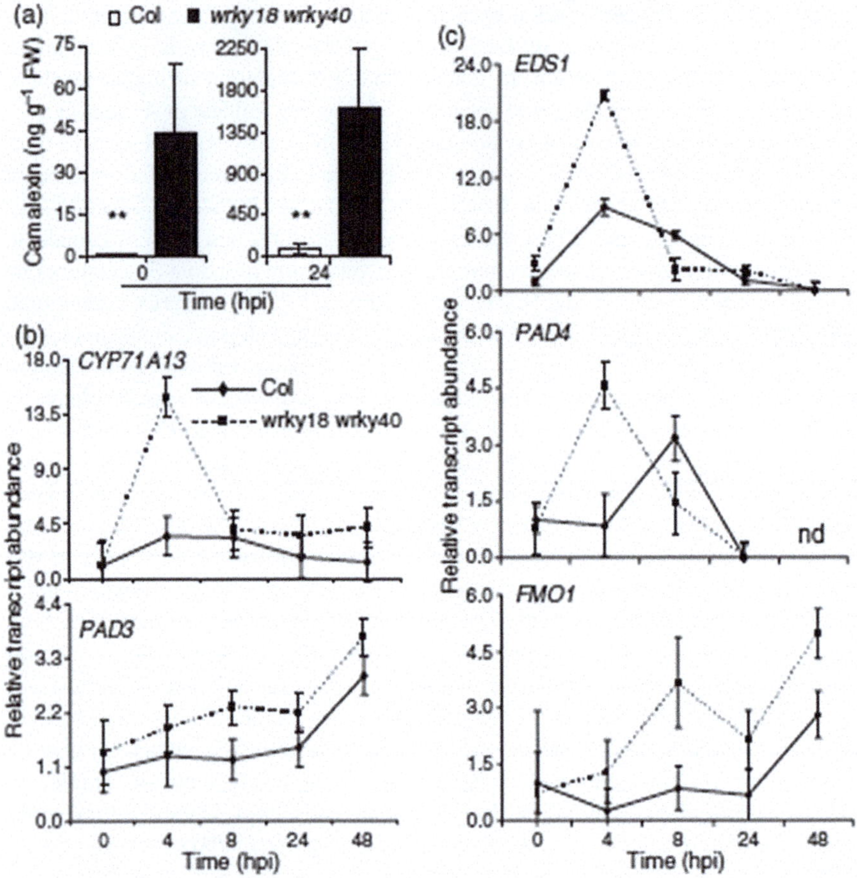

Fig. 7.11 Loss of *WRKY18/40* functions up-regulate the accumulation and biosynthesis of camalexin and the *EDS1* signaling pathway upon *G. orontii* infection. (**a**) Camalexin levels were determined in WT (open bars) and *wrky18 wrky40* (solid bars) plants before (0 hpi) and at 24 hpi with *G. orontii*. (**b**) Temporal expression of *G. orontii*-induced host genes *CYP71A13* and *PAD3* essential for camalexin biosynthesis and (**c**) of *EDS1*, *PAD4*, and *FMO1* in WT (solid lines) and *wrky18 wrky40* (broken lines) plants as determined by qPCR at the indicated time points. Samples were collected, and gene expression levels were calculated with respect to time 0. ∗∗Student's t-test, n = 10, $P < 0.05$ (Pandey et al. 2010)

7.13 Sources of Powdery Mildew Resistance

Major gene sources of resistance against powdery mildew of crucifers have been identified. These sources can be easily incorporated through conventional and biotechnological approaches to breed powdery mildew-resistant cultivars of different crucifers. The number of resistant sources identified belong to five species of oil-yielding crops (*B. alba* all available accessions, *B. carinata* 5, *B. juncea* 15, *B. napus* 5, *B. rapa* ssp. yellow sarson 2), one of fodder crops (*B. napus* ssp. *rapifera*

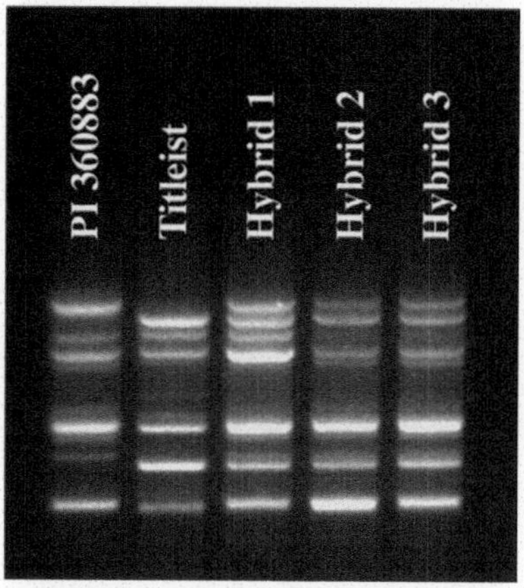

Plate 7.14 Confirmation of inter-specific hybrids using random amplified polymorphic DNA (RAPD) polymorphisms from *Brassica carinata* parent (PI 360883) and broccoli parent (Titleist) with 1.5% agarose gel electrophoresis (Tongue and Griffiths 2004)

Table 7.8 Sources of resistance in crucifers against powdery mildew

Crucifer species	Sources of resistance	References
Brassica napus ssp. *rapifera*	UG 3, UG 4	Shattuck (1993)
B. napus	GS-7027, Midas, tower, GSL-1, MNS-9605	Dang et al. (2000); Mehta et al. (2008)
B. alba, Sinapis alba	All accessions, local	Mehta et al. (2008); Meena et al. (2018)
B. carinata	HC-1, HC-2, HC-9605, HC-9603, PI 360883	Dang et al. (2000); Mehta et al. (2008); Tonquc and Griffiths (2004)
B. juncea	JMO 6014, JMO 6015, JMO 6009, JMO 6012 (from Australia), JM3, DIR-1507, DIR-1522, Kranti, DIR 621, IJWHJ 001, PCR-10, PCA 9201, RK 8602, RK 8615, RAUD 101	Dang et al. (2000); Singh et al. (2010); Singh and Singh (2003)
Arabidopsis thaliana	Accessions Su-0, Ms-0, Wa-1, Kas-1, SI-0, Te-0, Stw-0	Adam et al. (1999); Adam and Somerville (1996)
B. rapa ssp. *yellow sarson*	YSPb-24, TH-68	Mehta et al. (2008)
B. rapa ssp. *brown sarson*	BSH-1	Mehta et al. (2008)

2), and a weed (*Arabidopsis thaliana* 7; Table 7.8). These sources are being used as model through powdery mildew host pathosystem for molecular and genetical studies (Table 7.8).

In India, tolerant and resistant sources against powdery mildew of crucifers have been identified during 2000–2017 under AICRPRM programme by different researchers, and reactions of various *Brassica* species against powdery mildew with genotypes are presented in Table 7.9. However, registration of resistant source against powdery mildew of rapeseed-mustard in India is lacking so far.

Table 7.9 Sources identified from crucifers for powdery mildew disease tolerance/resistance under AICRPRM (Anonymous 2000–2017; Meena et al. 2018)

Year	Powdery mildew-tolerant/mildew-resistant sources[a]
2001	TM 18, RM 505 (Bj)
2004	EC -338997 (Bn), PBC-9221 (Bc), PBN-2001 (Bn), PBN-2001 (Bn)
2005	NPC-14, JTC-55, PBC-2002-(Bc)
2006	OCN 3 (Bn)
2007	HNS-9605 (Bn), PT-303 (Brt)
2008	EC 338997 (Bn), ONK 1 (Bn)
2009	EC-414299 (Br), EC-339000 (Bn)
2010	NPJ-143 (Bj)
2011	DRMR 243 (Bc), DRMR 261 (Bc), DLSC 1 (Bc)
2012	DRMR 312 (Bc)
2014	NPC-16 (Bc), NPC-21 (Bc)
2015	PPBN-3 (Bn), PPBR-2 (Br), PT-2006-4 (Brt), RMT-10-7 (Brt)
2016	PRD 2013-3 (Bj), DRMR-316 (Bc), DRMR-100 (Bc)
2017	DRMR1-5 (Bj)

[a]*Bj* (*Brassica juncea*), *Bc* (*B. carinata*), *Bn* (*B. napus*), Brt (*B. rapa* ssp Toria), *Br* (*B. rapa*)

7.13.1 Sources of Slow Mildewing Resistance

To assess the nature of powdery mildew resistance in *Brassica* crops, seven cvs. of *B. juncea* and one each of *B. napus* and *B. carinata* were selected for evaluation of slow mildewing components (Singh 2004; Meena et al. 2018). Various components of slow mildewing, viz. incubation period, latent period, no. of colony/speck per leaf, no. of conidia per colony/speck, progression of the disease, and disease intensity, were recorded under field conditions. The incubation period of test cvs. ranged from 3 to 4 days. However, powdery mildew was not observed on variety HC-9603 even under artificial inoculation conditions. The maximum incubation period of 4 days was recorded on the varieties RH-9304 and RH-9801, whereas in the rest of the varieties, it was 3 days. However, no significant differences in incubation period in case of *Brassica* cvs. were recorded (Table 6.7). The latent period of test cvs. ranged between 1 and 3 days. The variety HC-9603 did not contract powdery mildew. The latent period was 1–2 days in varieties belonging to the *B. juncea*. However, it was 3 days in the case of variety GSL-1 (Table 6.7). The results also revealed that there was no significant difference in the latent period in the varieties belonging to *B. juncea*. However, it differed in the case of *B. napus* where it was slightly higher. The number of powdery mildew specks/leaf was also recorded on all the nine varieties as a test of slow mildewing components. The variety GSL-1 showed minimum number of specks/leaf (5.52), whereas the variety RH-9801 contracted maximum number of the specks/leaf (39.40). It was followed by the variety RH-9304 (34.0). On the other varieties, viz. RH-8812, RH-9901, RC-781, and Purple Mutant, the number of specks ranged between 18 and 27 per leaf, which is moderate. The variety GSL-1 contracted less number of specks per leaf which significantly differed from the other varieties (less than 10 specks/leaf) which may be considered as resistant, whereas other varieties such as RH-9801, RH-9304, and RH-30 had higher number of specks/leaf (more than 30) and may be termed as susceptible. The other varieties such as RH-9901, RH-8812, RC-781, and Purple Mutant contracted the powdery mildew specks ranging between 10 and 30 specks/leaf and considered as moderately susceptible (Meena et al. 2018).

The disease intensity was recorded on all the varieties except HC-9603 on which disease did not appear till the end of the crop season to test the behaviour of varieties against powdery mildew. The maximum disease (49.5%) was recorded on the variety RH-9801 followed by the varieties RH-9304 (47.8%) and RH-30 (46.2%) though statistically at par. The minimum disease (4.1%) was observed on the variety GSL-1, whereas, on other varieties including RH-8812, RH-9901, Purple Mutant, and RC-781, the disease intensity ranged from 33.6 to 43.4 per cent. There were no significant differences in all the varieties in relation to disease intensity except GSL-1 and HC-9603 which appeared as resistant to powdery mildew disease and others as susceptible (Table 6.7).

The number of conidia produced in each speck by *E. cruciferarum* on different cultivars/varieties of mustard was examined under the compound microscope (10 X 10 x). The minimum number of conidia/speck was recorded on the variety GSL-1 (10 conidia/speck). It was followed by the variety RC-781 where it was 41.58 conidia/speck. In other varieties such as RH-30, RH-9801, RH-9304, RH-8812, RH-9901, and Purple Mutant, the conidial production ranged between 50.08 and 83.75 conidia per speck being maximum on the variety RH-30 (83.8) and minimum on RH-8812 (50.1). It was revealed that the considerable amount of conidia per speck was produced in all the susceptible varieties belonging to *B. juncea* except GSL-1 and HC-9603 which belong to *B. napus* and *B. carinata*, respectively (Table 6.7).

The progression of the powdery mildew on all the nine varieties was recorded from the appearance of the disease till the maturity of the leaves on ten randomly selected marked tagged leaves from each replication. The disease appeared on all the varieties in first week of March except that on HC-9603 (disease did not appear). The minimum speck size was recorded in the variety GSL-1 (1.98 mm) whereas maximum in the variety RH-30 (5.80 mm). It was followed by the varieties RH-9901 (5.25 mm) and RH-9304 (4.98 mm), whereas on other varieties, the speck size ranged between 3.81 and 4.81 mm which is moderate (Table 6.8). The progression of powdery mildew on different varieties/cultivars of mustard presented in Fig. 6.13 revealed that the progression of powdery mildew was maximum up to mid of March; after that, the disease was slowed down. The minimum progression was recorded in the variety GSL-1 where it was almost static after initiation of the disease. Similarly, on Purple Mutant variety also, the disease progression was slow as compared to the other varieties.

The progression of the powdery mildew in relation to weather variables was evaluated which revealed the maximum R^2 value, i.e. 0.91, on variety RH-30 followed by RH-9901 (0.86), Purple Mutant (0.86), RH-9801 (0.83), and RH-8812

(0.83). The minimum value (R^2 0.47) was recorded on variety GSL-1 which indicated that in addition to weather variables included here, other factors such as varietal resistance and some unknown factors have a significant role in the disease development (Table 6.4). The varieties GSL-1 and HC-9603 appeared as resistant to powdery mildew with the expression of slow mildewing components, whereas other varieties belonging to *B. juncea* group appeared as susceptible to the disease showing faster powdery mildew development under field conditions.

The correlation matrix for progression of the powdery mildew in relation to weather variables on all the test varieties/cultivars was analysed. It was observed that T. Max. (X_1) had significant and positive role in the progression of powdery mildew on all the varieties/cultivars except GSL-1 where it was positive but non-significant. Similarly, Avp. M (X_5) also had significant and positive role in the disease progression on all the varieties/cultivars except HC-9603 (disease did not appear). The RHE (X_4) has negative and significant correlation in the disease development on all the varieties/cultivars except GSL-1 where it was negative but non-significant. Similarly Sunshine (X_8) has negative and significant correlation in all the varieties/cultivars. Other weather variables such as T. Min. (X_2), RHM (X_3), and Avp. E (X_6) had positive but non-significant correlation in disease progression on all the varieties/cultivars (Table 6.9, Meena et al. 2018).

References

Abraham V (1993) Transfer of resistance to powdery mildew from *Brassica carinata* to Indian mustard (*B. juncea*). In: Natl Symp on Oilseed Research and Development in India. Status and Strategies Abstr. August 2–5, 1993, Hyderabad, pp 86

Acevedo-Garcia J, Kusch S, Panstruga R (2014) *Magical mystery tour:* MLO proteins in plant immunity and beyond. New Phytol 204:273–281

Acevedo-Garcia J, Gruner K, Reinstaedler A, Kemen A, Kemen E, Cao L, Takken FLW, Reitz MU, Schafer P, O'Connell RJ, Kusch S, Kuhn H, Panstruga R (2017) The powdery mildew-resistant *Arabidopsis* mlo2 mlo6 mlo12 triple mutant displays altered infection phenotypes with diverse types of phytopathogens. Sci Rep 7:–9319

Adam L, Somerville SC (1996) Genetic characterization of five powdery mildew disease resistance loci in *Arabidopsis thaliana*. Plant J 9:341–356

Adam L, Ellwood S, Wilson I, Saenz G, Xiao S, Oliver RP, Turner JG, Somerville S (1999) Comparison of *Erysiphe cichoracearum* and *E. cruciferarum* and a survey of 360 *Arabidopsis thaliana* accessions for resistance to these two powdery mildew pathogens. Mol Plant-Microbe Interact 12:1031–1043

Adams DJ (2004) Fungal cell wall chitinases and glucanases. Microbiology 150:2029–2035

Afzal AJ, Wood AJ, Lightfoot DA (2008) Plant receptor-like serine threonine kinases: roles in signaling and plant defense. Mol Plant-Microbe Interact 21:507–517

Ahloowalia B, Maluszynski M, Nichterlein K (2004) Global impact of mutation-derived varieties. Euphytica 135:187–204

Aist JR (1976) Papillae and related wound plugs of plant cells. Annu Rev Phytopathol 14:145–163

Ali S, Mir ZA, Tyagi A, Mehari H, Meena RP, Bhat JA, Yadav P, Papalou P, Rawat S, Grover A (2017) Over expression of NPR1 in *Brassica juncea* confers broad spectrum resistance to fungal pathogens. Front in Plant Sci 8:1693

Alkooranee JT, Yin Y, Aledan TR, Jiang Y, Lu G, Wu J, Li M (2015) Systemic resistance to powdery mildew in *Brassica napus* (AACC) and *Raphanus alboglabra* (RRCC) by *Trichoderma harzianum* TH12. PLoS One 10(11):e0142177

Alvarez ME (2000) Salicylic acid in the machinery of hypersensitive cell death and disease resistance. Plant Mol Biol 44:429–442

An Q, Huckelhoven R, Kogel KH, van Bel AJ (2006) Multi-vesicular bodies participate in a cell wall-associated defence response in barley leaves attacked by the pathogenic powdery mildew fungus. Cell Microbiol 8:1009–1019

Andersson MX, Kourtchenko O, Dangl JL, Mackey D, Ellerstrom M (2006) Phospholipase-dependent signalling during the AvrRpm1- and AvrRpt2-induced disease resistance responses in *Arabidopsis thaliana*. Plant J 47:947–959

Antico CJ, Colon C, Banks T, Ramonell KM (2012) Insights into the role of jasmonic acid-mediated defenses against necrotrophic and biotrophic fungal pathogens. Front Biol 7:48–56

Aravind L, Koonin EV (1999) Fold prediction and evolutionary analysis of the POZ domain: structural and evolutionary relationship with the potassium channel tetramerization domain. J Mol Biol 285:1353–1361

Asai T, Tena G, Plotnikova J, Willmann MR, Chiu WL, Gomez-Gomez L, Boller T, Ausubel FM, Sheen J (2002) MAP kinase signalling cascade in *Arabidopsis* innate immunity. Nature 415:977–983

Assaad FF, Qiu JL, Youngs H, Ehrhardt D, Zimmerli L, Kalde M, Wanner G, Peck SC, Edwards H, Ramonell K, Somerville CR, Thordal-Christensen H (2004) The PEN1 syntaxin defines a novel cellular compartment upon fungal attack and is required for the timely assembly of papillae. Mol Biol Cell 15:5118–5129

Auld D, Heikkinen M, Erickson D, Sernyk J, Romero J (1992) Rapeseed mutants with reduced levels of polyunsaturated fatty acids and increased levels of oleic acid. Crop Sci 32:657–662

Ausubel FM (2005) Are innate immune signaling pathways in plants and animals conserved? Nature Immunol 6:973–979

Azevedo C, Betsuyaku S, Peart J, Takahashi A, Noel L, Sadanandom A, Casais C, Parker J, Shirasu K (2006) Role of SGT1 in resistance protein accumulation in plant immunity. EMBO J 25:2007–2016

Bai YL, van der Hulst R, Bonnema G, Marcel BC, Meijer-Dekens F, Niks RE, Lindhout P (2005) Tomato defense to *Oidium neolycopersici*: dominant Ol genes confer isolate-dependent resistance via a different mechanism than recessive ol-2. Mol Plant-Microbe Interact 18:354–362

Bak S, Feyereisen R (2001) The involvement of two p450 enzymes, CYP83B1 and CYP83A1, in auxin homeostasis and glucosinolate biosynthesis. Plant Physiol 127:108–118

Bak S, Beisson F, Bishop G, Hamberger B, Hofer R, Paquette S, Werck-Reichhart D (2011) Cytochromes p450. Arabidopsis Book 9:e0144

Bakshi M, Oelmuller R (2014) WRKY transcription factors: Jack of many trades in plants. Plant Signal Behav 9:e27700

Baluska F, Bacigalova K, Oud JL, Hauskrecht M, Kubica S (1995) Rapid reorganization of microtubular cytoskeleton accompanies early changes in nuclear ploidy and chromatin structure in post-mitotic cells of barley leaves infected with powdery mildew. Protoplasma 185:140–151

Bari R, Jones JD (2009) Role of plant hormones in plant defence responses. Plant Mol Biol 69:473–488

Bayer M, Nawy T, Giglione C, Galli M, Meinnel T, Lukowitz W (2009) Paternal control of embryonic patterning in *Arabidopsis thaliana*. Science 323:1485–1488

Beckers GJM, Jaskiewicz M, Liu Y, Underwood WR, He SY, Zhang S, Conrath U (2009) Mitogen-activated protein kinases 3 and 6 are required for full priming of stress responses in *Arabidopsis thaliana*. Plant Cell 21:944–953

Bednarek P (2012) Chemical warfare or modulators of defence responses-the function of secondary metabolites in plant immunity. Curr Opinion Plant Biol 15:407–414

Bednarek P, Osbourn A (2009) Plant–microbe interactions: chemical diversity in plant defense. Science 324:746–748

Bednarek P, Pislewska-Bednarek M, Svatos A, Schneider B, Doubsky J, Mansurova M, Humphry M, Consonni C, Panstruga R, Sanchez-Vallet A, Molina A, Schulze-Lefert P (2009) A glucosinolate metabolism pathway in living plant cells mediates broad-spectrum antifungal defense. Science 323:101–106

Beekwilder J, van Leeuwen W, van Dam NM, Bertossi M, Grandi V, Mizzi L, Soloviev M, Szabados L, Molthoff JW, Schipper B, Verbocht H, de Vos RCH, Morandini P, Arts MGM, Bovy A (2008) The impact of the absence of aliphatic glucosinolates on insect herbivory in *Arabidopsis*. PLoS One 3:e2068

Beers EP, Jones AM, Dickerman AW (2004) The S8 serine, C1A cysteine and A1 aspartic protease families in *Arabidopsis*. Phytochemistry 65:43–58

Belanger RR, Bushnell WR, Dik AJ, Carver TLW (2002) The powdery mildews: a comprehensive treatise. (St. Paul: Am Phytopathol Soc (APS Press)

Belanger RR, Benhamou N, Menzies JG (2003) Cytological evidence of an active role of silicon in wheat resistance to powdery mildew (*Blumeria graminis* f. sp. *tritici*). Phytopathology 93:402–412

Bell CJ, Ecker JR (1994) Assignment of 30 microsatellite markers to the linkage map of *Arabidopsis*. Genomics 19:137–144

Berken A (2006) ROPs in the spotlight of plant signal transduction. Cell Mol Life Sci 63:2446–2459

Bethke G, Unthan T, Uhrig JF, Poschl Y, Gust AA, Scheel D, Lee J (2009) Flg22 regulates the release of an ethylene response factor substrate from MAP kinase 6 in *Arabidopsis thaliana* via ethylene signaling. Proc Natl Acad Sci U S A 106:8067–8072

Bhat RA, Miklis M, Schmelzer E, Schulze-Lefert P, Panstruga R (2005) Recruitment and interaction dynamics of plant penetration resistance components in a plasma membrane microdomain. Proc Natl Acad Sci U S A 102:3135–3140

Bisgrove SR, Simonich MT, Smith NM, Sattler A, Innes RW (1994) A disease resistance gene in *Arabidopsis* with specificity for two different pathogen avirulence genes. Plant Cell 6:927–933

Bittel P, Robatzek S (2007) Microbe-associated molecular patterns (MAMPs) probe plant immunity. Curr Opinion Plant Biol 10:335–341

Boch J, Verbsky ML, Robertson TL, Larkin JC, Kunkel BN (1998) Analysis of resistance gene-mediated defense responses in *Arabidopsis thaliana* plants carrying a mutation in CPR5. Mol Plant-Microbe Interact 11:1196–1206

Bohlenius H, Morch SM, Godfrey D, Nielsen ME, Thordal-Christensen H (2010) The multivesicular body-localized GTPase ARFA1b/1c is important for callose deposition and ROR2 syntaxin-dependent pre-invasive basal defense in barley. Plant Cell 22:3831–3844

Boller T, Felix G (2009) A renaissance of elicitors: perception of microbe-associated molecular patterns and danger signals by pattern-recognition receptors. Ann Rev Plant Biol 60:379–406

Bottcher C, Westphal L, Schmotz C, Prade E, Scheel D, Glawischnig E (2009) The multifunctional enzyme CYP71B15 (PHYTOALEXIN DEFICIENT3) converts cysteine-indole-3-acetonitrile to camalexin in the indole-3-acetonitrile metabolic network of *Arabidopsis thaliana*. Plant Cell 21:1830–1845

Bowling SA, Clarke JD, Liu Y, Klessig DF, Dong X (1997) The cpr5 mutant of *Arabidopsis* expresses both NPR1-dependent and NPR1- independent resistance. Plant Cell 9:1573–1584

Bracker CE (1968) Ultra-structure of the haustorial apparatus of *Erysiphe graminis* and its relationship to the epidermal cell of barley. Phytopathology 58:12–30

Brininstool G, Kasili R, Simmons LA, Kirik V, Hulskamp M, Larkin JC (2008) Constitutive Expressor of pathogenesis-related genes 5 affects cell wall biogenesis and trichome development. BMC Plant Biol 8:58

Burr CA, Leslie ME, Orlowski SK, Chen I, Wright CE, Daniels MJ, Liljegren SJ (2011) CAST AWAY, a membrane-associated receptor-like kinase, inhibits organ abscission in *Arabidopsis*. Plant Physiol 156:1837–1850

Buscaill P, Rivas S (2014) Transcriptional control of plant defence responses. Curr Opinion Plant Biol 20:35–46

Caarls L, Pieterse CMJ, Van Wees SCM (2015) How salicylic acid takes transcriptional control over jasmonic acid signaling. Front Plant Sci 6:170. https://doi.org/10.3389/fpls.2015.00170

Cahill D, Rookes J, Michalczyk A, McDonald K, Drake A (2002) Microtubule dynamics in compatible and incompatible interactions of soybean hypocotyl cells with *Phytophthora sojae*. Plant Pathol 51:629–640

Caillaud MC, Abad P, Favery B (2008a) Cytoskeleton reorganization, a key process in root-knot nematode-induced giant cell ontogenesis. Plant Signal Behav 3:816–818

Caillaud MC, Lecomte P, Jammes F, Quentin M, Pagnotta S, Andrio E, de Almeida Engler J, Marfaing N, Gounon P, Abad P, Favery B (2008b) MAP65-3 microtubule-associated protein is essential for nematode-induced giant cell ontogenesis in *Arabidopsis*. Plant Cell 20(2):423–437

Campe R, Langenbach C, Leissing F, Popescu GV, Popescu SC, Goellner K, Beckers GJM, Conrath U (2015) ABC transporter PEN3/PDR8/ABCG36 interacts with calmodulin that, like PEN3, is required for *Arabidopsis* non-host resistance. New Phytol 209:294–306

Cao H, Bowling SA, Gordon AS, Dong X (1994) Characterization of an *Arabidopsis* mutant that is nonresponsive to inducers of systemic acquired resistance. Plant Cell 6:1583–1592

Cao H, Glazebrook J, Clarke JD, Volko S, Dong X (1997) The Arabidopsis NPR1 gene that controls systemic acquired resistance encodes a novel protein containing ankyrin repeats. Cell 88:57–63

Cao H, Li X, Dong XN (1998) Generation of broad–spectrum disease resistance by over expression of an essential regulatory gene in systemic acquired resistance. Proc Natl Acad Sci U S A 95:6531–6536

Cao Y, Liang Y, Tanaka K, Nguyen CT, Jedrzejczak RP, Joachimiak A, Stacey G (2014) The kinase LYK5 is a major chitin receptor in *Arabidopsis* and forms a chitin-induced complex with related kinase CERK1. elife 3:e03766

Carver TLW (1986) Histology of infection by *E. graminis* f.sp. *hordei* in spring barley lines with various levels of partial resistance. Plant Pathol 35:232–240

Century KS, Shapiro AD, Repetti PP, Dahlbeck D, Holub E, Staskawicz BJ (1997) *NDR1*, a pathogen-induced component required for *Arabidopsis* disease resistance. Science 278:1963–1965

Chan J, Jensen CG, Jensen LC, Bush M, Lloyd CW (1999) The 65-kDa carrot microtubule-associated protein forms regularly arranged filamentous cross-bridges between microtubules. Proc Natl Acad Sci U S A 96:14931–14936

Chandran D, Tai YC, Hather G, Dewdney J, Denoux C, Burgess DG, Ausubel FM, Speed TP, Wildermuth MC (2009) Temporal global expression data reveal known and novel salicylate-impacted processes and regulators mediating powdery mildew growth and reproduction on *Arabidopsis*. Plant Physiol 149:1435–1451

Chandran D, Inada N, Hather G, Kleindt CK, Wildermuth MC (2010) Laser microdissection of *Arabidopsis* cells at the powdery mildew infection site reveals site-specific processes and regulators. Proc Natl Acad Sci U S A 107:460–465

Chandran D, Rickert J, Cherk C, Dotson BR, Wildermuth MC (2013) Host cell ploidy underlying the fungal feeding site is a determinant of powdery mildew growth and reproduction. Mol Plant-Microbe Interact 26:537–545

Chandran D, Rickert J, Huang Y, Steinwand MA, Marr SK, Wildermuth MC (2014) A typical E2F transcriptional repressor DEL1 acts at the intersection of plant growth and immunity by controlling the hormone salicylic acid. Cell Host Microbe 15:506–513

Chelkowski J, Tyrka M, Sobkiewicz A (2003) Resistance genes in barley (*Hordeum vulgare* L.) and their identification with molecular markers. J Appl Genet 44:291–309

Chen L, Shiotani K, Togashi T, Miki D, Aoyama M, Wong HL, Kawasaki T, Shimamoto K (2010) Analysis of the Rac/Rop small GTPase family in rice: expression, sub-cellular localization and role in disease resistance. Plant Cell Physiol 51:585–595

Chen X, Barnaby JY, Sreedharan A, Huang X, Orbovic V, Grosser JW, Wang N, Dong X, Song WY (2013) Over–expression of the citrus gene CtNH1 confers resistance to bacterial canker disease. Physiol Mol Plant Pathol 84:115–122

Chisholm ST, Coaker G, Day B, Staskawicz BJ (2006) Host–microbe interactions: shaping the evolution of the plant immune response. Cell 124:803–814

Christiansen KM, Gu Y, Rodibaugh N, Innes RW (2011) Negative regulation of defence signalling pathways by the EDR1 protein kinase. Mol Plant Pathol 12:746–758

Clarke JD, Volko SM, Ledford H, Ausubel FM, Dong X (2000) Roles of salicylic acid, jasmonic acid, and ethylene in *cpr*-induced resistance in *Arabidopsis*. Plant Cell 12:2175–2190

Clay NK, Adio AM, Denoux C, Jander G, Ausubel FM (2009) Glucosinolate metabolites required for an *Arabidopsis* innate immune response. Science 323:95–101

Collins NC, Thordal-Christensen H, Lipka V, Bau S, Kombrink E, Qiu JL, Huckelhoven R, Stein M, Freialdenhoven A, Somerville SC, Schulze-Lefert P (2003) SNARE-protein-mediated disease resistance at the plant cell wall. Nature 425:973–977

Consonni C, Humphry ME, Hartmann HA, Livaja M, Durner J, Westphal L, Vogel J, Lipka V, Kemmerling B, Schulze-Lefert P, Somerville SC, Panstruga R (2006) Conserved requirement for a plant host cell protein in powdery mildew pathogenesis. Nature Genet 38:716–720

Consonni C, Bednarek P, Humphry M, Francocci F, Ferrari S, Harzen A, Ver Loren van Themaat E, Panstruga R (2010) Tryptophan-derived metabolites are required for antifungal defense in the *Arabidopsis mlo2* mutant. Plant Physiol 152:1544–1561

Crouzet J, Trombik T, Fraysse AS, Boutry M (2006) Organization and function of the plant pleiotropic drug resistance ABC transporter family. FEBS Lett 580:1123–1130

Dang JK, Sangwan MS, Mehta N, Kaushik CD (2000) Multiple disease resistance against four fungal foliar diseases of rapeseed-mustard. Indian Phytopathol 53(4):455–458

Dangl JL, Jones JDG (2001) Plant pathogens and integrated defence responses to infection. Nature 411:826–833

Dangl JL, Horvath DM, Staskawicz BJ (2013) Pivoting the plant immune system from dissection to deployment. Science 341:746–751

De Almeida EJ, Favery B (2011) The plant cytoskeleton remodelling in nematode induced feeding sites. In: Jones J, Gheysen G, Fenoll C (eds) Genomics and molecular genetics of plant–nematode interactions. Springer, Heidelberg, pp 369–393

De Almeida EJ, Favery B, Engler G, Abad P (2005) Loss of susceptibility as an alternative for nematode resistance. Curr Opinion Biotech 16:112–117

de Jong CF, Laxalt AM, Bargmann BO, de Wit PJ, Joosten MH, Munnik T (2004) Phosphatidic acid accumulation is an early response in the *Cf-4/Avr4* interaction. Plant J 39:1–12

Delaney TP, Friedrich L, Ryals JA (1995) *Arabidopsis* signal transduction mutant defective in chemically and biologically induced disease resistance. Proc Natl Acad Sci U S A 92:6602–6606

Dempsey DA, Shah J, Klessig DF (1999) Salicylic acid and disease resistance in plants. Critical Rev Plant Sci 18:547–575

De-Vos M, Van-Oosten VR, van Poecke RMP, Van Pelt JA, Pozo MJ, Mueller MJ, Buchala AJ, Metraux JP, Van Loon LC, Dicke M, Pieterse CMJ (2005) Signal signature and transcriptome changes of *Arabidopsis* during pathogen and insect attack. Mol Plant-Microbe Interact 18:923–937

Devoto A, Hartmann HA, Piffanelli P, Elliott C, Simmons C, Taramino G, Goh CS, Cohen FE, Emerson BC, Schulze-Lefert P, Panstruga R (2003) Molecular phylogeny and evolution of the plant-specific seven-transmembrane MLO family. J Mol Evol 56:77–88

Dewdney J, Reuber TL, Mary CW, Alessandra D, Jianping C, Lisa MS, Emma PD, Frederick MA (2000) Three unique mutants of *Arabidopsis* identify eds loci required for limiting growth of a biotrophic fungal pathogen. Plant J 24:205–218

Dickman MB, Fluhr R (2013) Centrality of host cell death in plant-microbe interactions. Annu Rev Phytopathol 51:543–570

Dorjgotov D, Jurca ME, Fodor-Dunai C, Szucs A, Otvos K, Klement E, Biro J, Feher A (2009) Plant rho-type (Rop) GTPase-dependent activation of receptor-like cytoplasmic kinases *in vitro*. FEBS Lett 583:1175–1182

Dormann P, Kim H, Ott T, Schulze-Lefert P, Trujillo M, Wewer V, Huckelhoven R (2014) Cell-autonomous defense, re-organization and trafficking of membranes in plant-microbe interactions. New Phytol 204:815–822

Durrant WE, Dong X (2004) Systemic acquired resistance. Annu Rev Phytopathol 42:185–209

Dutt M, Barthe G, Irey M, Grosser J (2015) Transgenic citrus expressing an *Arabidopsis* NPR1 gene exhibit enhanced resistance against huanglongbing (HLB; citrus greening). PLoS One 10:e0137134

Eckey C, Korell M, Leib K, Biedenkopf D, Jansen C, Langen G, Kogel KH (2004) Identification of powdery mildew-induced barley genes by cDNA-AFLP: functional assessment of an early expressed MAP kinase. Plant Mol Biol 55:1–15

Ederli L, Dawe A, Pasqualini S, Quaglia M, Xiong L, Gehring C (2015) *Arabidopsis* flower specific defense gene expression patterns affect resistance to pathogens. Front Plant Sci 6:79

Eggert D, Naumann M, Reimer R, Voigt CA (2014) Nanoscale glucan polymer network causes pathogen resistance. Sci Rep 4:4159

Ehness R, Ecker M, Godt DE, Roitsch T (1997) Glucose and stress independently regulate source and sink metabolism and defence mechanisms via signal transduction pathways involving protein phosphorylation. Plant Cell 9:1825–1841

Ellinger D, Voigt CA (2014a) The use of nano scale fluorescence microscopic to decipher cell wall modifications during fungal penetration. Front Plant Sci 5:270

Ellinger D, Voigt CA (2014b) Callose biosynthesis in *Arabidopsis* with a focus on pathogen response: what we have learned within the last decade. Ann Bot 114:1349–1358

Ellinger D, Naumann M, Falter C, Zwikowics C, Jamrow T, Manisseri C, Somerville SC, Voigt CA (2013) Elevated early callose deposition results in complete penetration resistance to powdery mildew in *Arabidopsis*. Plant Physiol 161:1433–1444

Ellinger D, Glockner A, Koch J, Naumann M, Sturtz V, Schutt K, Manisseri C, Somerville SC, Voigt CA (2014) Interaction of the *Arabidopsis* GTPase RabA4c with its effector PMR4 results in complete penetration resistance to powdery mildew. Plant Cell 26:3185–3200

Ellis C, Karafyllidis I, Turner JG (2002a) Constitutive activation of jasmonate signaling in an *Arabidopsis* mutant correlates with enhanced resistance to *Erysiphe cichoracearum*, *Pseudomonas syringae*, and *Myzus persicae*. Mol Plant-Microbe Interact 15:1025–1030

Ellis C, Karafyllidis I, Wasternack C, Turner JG (2002b) The *Arabidopsis* mutant *cev1* links cell wall signaling to jasmonate and ethylene responses. Plant Cell 14:1557–1566

Enright SM, Cipollini D (2007) Infection by powdery mildew *Erysiphe cruciferarum* (*Erysiphaceae*) strongly affects growth and fitness of *Alliaria petiolata* (*Brassicaceae*). Am J Bot 94(11):1813

Eulgem T, Somssich IE (2007) Networks of WRKY transcription factors in defense signaling. Curr Opinion Plant Biol 10:366–371

Eulgem T, Rushton PJ, Robatzek S, Somssich IE (2000) The WRKY super family of plant transcription factors. Trends Plant Sci 5:199–206

Fabro G, Di Rienzo JA, Voigt CA, Savchenko T, Dehesh K, Somerville S, Alvarez ME (2008) Genome-wide expression profiling Arabidopsis at the stage of *Golovinomyces cichoracearum* haustorium formation. Plant Physiol 146:1421–1439

Falk A, Feys BJ, Frost LN, Jones JDG, Daniels MJ, Parker JE (1999) EDS1, an essential component of R genemediated disease resistance in *Arabidopsis* has homology to eukaryotic lipases. Proc Natl Acad Sci U S A 96:3292–3297

Fan J, Crooks C, Creissen G, Hill L, Fairhurst S, Doerner P, Lamb C (2011) *Pseudomonas* sax genes overcome aliphatic isothiocyanate-mediated non-host resistance in *Arabidopsis*. Science 331:1185–1188

Fauteux F, Rémus-Borel W, Menzies JG, Bélanger RR (2005) Silicon and plant disease resistance against pathogenic fungi. FEMS Microbiol Lett 249:1–6

Fauteux F, Chain F, Belzile F, Menzies JG, Bélanger RR (2006) The protective role of silicon in the *Arabidopsis*–powdery mildew pathosystem. Proc Natl Acad Sci U S A 103:17554–17559

Feechan A, Kwon E, Yun BW, Wang Y, Pallas JA, Loake GJ (2005) A central role for S-nitrosothiols in plant disease resistance. Proc Natl Acad Sci U S A 102:8054–8059

Feechan A, Kabbara S, Dry IB (2011) Mechanisms of powdery mildew resistance in the vitaceae family. Mol Plant Pathol 12:263–274

Ferrari S, Plotnikova JM, DeLorenzo G, Ausubel FM (2003) *Arabidopsis* local resistance to *Botrytis cinerea* involves salicylic acid and camalexin and requires EDS4 and PAD2, but not SID2, EDS5 or PAD4. Plant J 35:193–205

Feys BJ, Moisan LJ, Newman MA, Parker JE (2001) Direct interaction between the *Arabidopsis* disease resistance signaling proteins, EDS1 and PAD4. EMBO J 20:5400–5411

Feys BJ, Wiermer M, Bhat RA, Moisan LJ, Medina-Escobar N, Neu C, Cabral A, Parker JE (2005) *Arabidopsis* senescence-associated gene101 stabilizes and signals within an enhanced disease susceptibility1 complex in plant innate immunity. Plant Cell 17:2601–2613

Fotopoulos V, Gilbert MJ, Pittman JK, Marvier AC, Buchanan AJ, Sauer N, Hall JL, Williams LE (2003) The monosaccharide transporter gene, *AtSTP4,* and the Cell-Wall invertase, *Atfruct1*, are induced in *Arabidopsis* during infection with the fungal biotroph *Erysiphe cichoracearum.* Plant Physiol 132:

Freymark G, Diehl T, Miklis M, Romeis T, Panstruga R (2007) Antagonistic control of powdery mildew host cell entry by barley calcium-dependent protein kinases (CDPKs). Mol Plant-Microbe Interact 20:1213–1221

Friedrich L, Lawton K, Dietrich R, Willits M, Cade R, Ryals J (2001) NIM1 overexpression in Arabidopsis potentiates plant disease resistance and results in enhanced effectiveness of fungicides. Mol Plant-Microbe Interact 14:1114–1124

Frye CA, Innes RW (1998) An *Arabidopsis* mutant with enhanced resistance to powdery mildew. Plant Cell 10:947–956

Frye CA, Tang D, Innes RW (2001) Negative regulation of defense responses in plants by a conserved MAPKK kinase. Proc Natl Acad Sci U S A 98:373–378

Fu Y, Gu Y, Zheng Z, Wasteneys G, Yang Z (2005) *Arabidopsis* interdigitating cell growth requires two antagonistic pathways with opposing action on cell morphogenesis. Cell 120:687–700

Fu Y, Xu T, Zhu L, Wen M, Yang Z (2009) A ROP GTPase signaling pathway controls cortical microtubule ordering and cell expansion in *Arabidopsis*. Curr Biol 19:1827–1832

Fu ZQ, Yan S, Saleh A, Wang W, Ruble J, Oka N, Mohan R, Spoel SH, Tada Y, Zheng N, Dong X (2012) NPR3 and NPR4 are receptors for the immune signal salicylic acid in plants. Nature 486:228–232

Fuchs R, Kopischke M, Klapprodt C, Hause G, Meyer AJ, Schwarzlander M, Fricker MD, Lipka V (2016) Immobilized subpopulations of leaf epidermal mitochondria mediate PEN2-dependent pathogen entry control in *Arabidopsis*. Plant Cell 28:130–145

Galun M, Braun A, Frensdorff A, Galun E (1976) Hyphal walls of isolated lichen fungi: autoradiographic localization of precursor incorporation and binding of fluorescein-conjugated lections. Acta Microbiol 108:9–16

Gardiner J (2013) The evolution and diversification of plant microtubule associated proteins. Plant J 75:219–229

Geiger HH, Heun M (1989) Genetics of quantitative resistance to fungal diseases. Annu Rev Phytopathol 27:317–341

Genre A, Chabaud M, Timmers T (2005) Arbuscular mycorrhizal fungi elicit a novel intracellular apparatus in *Medicago truncatula* root epidermal cells before infection. Plant Cell 17:3489–3499

Ghanmi D, McNally DJ, Benhamou N, Menzies JG, Belanger RR (2004) Powdery mildew of *Arabidopsis thaliana*: a pathosystem for exploring the role of silicon in plant–microbe interactions. Physiol Mol Plant Pathol 64:189–199

Gil F, Gay JL (1977) Ultra-structural and physiological properties of the host interfacial components of haustoria of *Erysiphe pisi in vivo* and *in vitro*. Physiol Plant Pathol 10:1–4

Gillmor CS, Poindexter P, Lorieau J, Sujino K, Palcic M, Somerville CR (2002) The a-glucosidase I encoded by the KNOPF gene is required for cellulose biosynthesis and embryo morphogenesis in *Arabidopsis*. J Cell Biol 256:1003–1013

Gish LA, Clark SE (2011) The RLK/Pelle family of kinases. Plant J 66:117–127

Gjetting T, Carver TL, Skot L, Lyngkjaer MF (2004) Differential gene expression in individual papilla-resistant and powdery mildew-infected barley epidermal cells. Mol Plant-Microbe Interact 17:729–738

Gjetting T, Hagedorn PH, Schweizer P, Thordal-Christensen H, Carver TL, Lyngkjaer MF (2007) Single-cell transcript profiling of barley attacked by the powdery mildew fungus. Mol Plant-Microbe Interact 20:235–246

Glazebrook J (2005) Contrasting mechanisms of defense against biotrophic and necrotrophic pathogens. Annu Rev Phytopathol 43:205–227

Glazebrook J, Rogers EE, Ausubel FM (1996) Isolation of *Arabidopsis* mutants with enhanced disease susceptibility by direct screening. Genetics 143:973–982

Gollner K, Schweizer P, Bai Y, Panstruga R (2008) Natural genetic resources of *Arabidopsis thaliana* reveals a high prevalence and unexpected phenotypic plasticity of *RPW8*-mediated powdery mildew resistance. New Phytol 177:725–742

Gomez-Gómez L, Felix G, Boller T (1999) A single locus determines sensitivity to bacterial flagellin in *Arabidopsis thaliana*. Plant J 18:277–284

Grant M, Lamb C (2006) Systemic immunity. Curr Opinion Plant Biol 9:414–420

Grant MR, Godiard L, Straube E, Ashfield T, Lewald J, Sattler A, Innes RW, Dangl JL (1995) Structure of the *Arabidopsis* RPM7 gene enabling dual specificity disease resistance. Science 269:843–846

Green JR, Carver TL, Gurr SJ (2002) The formation and function of infection and feeding structures. In: Belanger RR, Bushnell WR, Dik AJ, Carver TL (eds.) The powdery mildews: a comprehensive treatise. St Paul: APS Press, 66–82

Grubb CD, Abel S (2006) Glucosinolate metabolism and its control. Trends Plant Sci 11:89–100

Gu Y, Innes RW (2011) The keep on going protein of Arabidopsis recruits the enhanced disease resistance1 protein to trans-golgi network/early endosome vesicles. Plant Physiol 155:1827–1838

Gu Y, Innes RW (2012) The KEEP ON GOING protein of *Arabidopsis* regulates intracellular protein trafficking and is degraded during fungal infection. Plant Cell 24:4717–4730

Gu YQ, Wildermuth MC, Chakravarthy S, Loh YT, Yang C, He X, Han Y, Martin GB (2002) Tomato transcription factors Pti4, Pti5, and Pti6 activate defense responses when expressed in *Arabidopsis*. Plant Cell 14:817–831

Guo CY, Wu GH, Xing J, Li WQ, Tang DZ, Cui BM (2013) A mutation in a coproporphyrinogen III oxidase gene confers growth inhibition, enhanced powdery mildew resistance and powdery mildew-induced cell death in *Arabidopsis*. Plant Cell Report 32:687–702

Gus-Meyer S, Naton B, Hahlbrock K, Schmelzer E (1998) Local mechanical stimulation induces components of the pathogen defense response in parsley. Proc Natl Acad Sci U S A 146(1):8398–8403

Halkier BA, Gershenzon J (2006) Biology and biochemistry of glucosinolates. Ann Rev Plant Biol 57:303–333

Hall D (1994) Interactions of Arabidopsis with fungal pathogens. PhD thesis. Norwich: University of East Anglia

Hamada T (2014) Microtubule organization and microtubule-associated proteins in plant cells. Intl Rev Cell Mol Biol 312:1–52

Hammerschmidt R (1999) Phytoalexins: what have we learned after 60 years? Annu Rev Phytopathol 37:285–306

Hansen CH, Du L, Naur P, Olsen CE, Axelsen KB, Hick AJ, Pickett JA, Halkier BA (2001) CYP83b1 is the oxime-metabolizing enzyme in the glucosinolate pathway in *Arabidopsis*. J Biol Chem 276:24790–24796

Hardham AR (2013) Microtubules and biotic interactions. Plant J 75:278–289

Hardham A, Jones DA, Takemoto D (2007) Cytoskeleton and cell wall function in penetration resistance. Curr Opinion Plant Biol 106(1):342–348

Hardham AR, Takemoto D, White RG (2008) Rapid and dynamic sub-cellular reorganization following mechanical stimulation of *Arabidopsis* epidermal cells mimics responses to fungal and oomycete attack. BMC Plant Biol 8:63

Harel TM, Mehar ZHI, Rav-David D, Elad Y (2014) Systemic resistance to gray Mold induced in tomato by Benzothiadiazole and *Trichoderma harzianum* T39. Phytopathology 104:150–157

Heath MC (2000) Non-host resistance and nonspecific plant defences. Curr Opinion Plant Biol 3:315–319

Helm M, Schmid M, Hierl G, Terneus K, Tan L, Lottspeich F, Kieliszewski MJ, Gietl C (2008) KDEL-tailed cysteine end peptidases involved in programmed cell death, inter-calation of new cells and dismantling of extension scaffolds. Am J Bot 95:1049–1062

Hemm MR, Ruegger MO, Chapple C (2003) The *Arabidopsis* ref2 mutant is defective in the gene encoding CYP83A1 and shows both phenylpropanoid and glucosinolate phenotypes. Plant Cell 15:179–194

Hepworth SR, Zhang Y, McKim S, Li X, Haughn GW (2005) BLADE-ON-PETIOLE–dependent signaling controls leaf and floral patterning in *Arabidopsis*. Plant Cell 17:1434–1448

Herbers K, Takahata Y, Melzer M, Mock H-P, Hajirezaei M, Sonnewald U (2000) Regulation of carbohydrate partitioning during the interaction of potato virus Y with tobacco. Mol Plant Pathol 1:51–59

Hermosa R, Viterbo A, Chet I, Monte E (2012) Plant-beneficial effects of *Trichoderma* and of its genes. Microbiology 158:17–25

Hierl G, Vothknecht U, Gietl C (2012) Programmed cell death in *Ricinus* and *Arabidopsis*: the function of KDEL cysteine peptidases in development. Physiol Plant 145:103–113

Hierl G, Howing T, Isono E, Lottspeich F, Gietl C (2014) *Ex vivo* processing for maturation of Arabidopsis KDEL-tailed cytokine endopeptidase2 (AtCEP2) pro-enzyme and its storage in endoplasmic reticulum derived organelles. Plant Mol Biol 84:605–620

Ho CMK, Hotta T, Guo F, Roberson RW, Lee YRJ, Liu B (2011) Interaction of antiparallel microtubules in the phragmoplast is mediated by the microtubule-associated protein MAP65-3 in *Arabidopsis*. Plant Cell 23:2909–2923

Hoefle C, Huesmann C, Schultheiss H, Bornke F, Hensel G, Kumlehn J, Huckelhoven R (2011) A barley ROP GTPase ACTIVATING PROTEIN associates with microtubules and regulates entry of the barley powdery mildew fungus into leaf epidermal cells. Plant Cell 23:2422–2439

Hossain MY, Jasmine S, Ibrahim AHM, Ahmed ZF, Rahman MM, Ohtomi J (2008a) Length-weight and length-length relationships of ten small fish species from the Ganges (Bangladesh). J Appl Ichthyol 25:117–119

Hossain MY, Leunda PM, Ohtomi J, Ahmed ZF, Oscoz J, Miranda R (2008b) Biological aspects of the ganges river sprat *Corica soborna* (Clupeidae) in the Mathabhanga river (SW Bangladesh). Cybium 32:241–246

Hou CT, Forman RJ (2000) Growth inhibition of plant pathogenic fungi by hydroxy fatty acids. J Indust Microbiol Biochem 24:275–276

Howing T, Huesmann C, Hoefle C, Nagel M-K, Isono E, Hueckelhoven R, Gietl C (2014) Endoplasmic reticulum KDEL-tailed cysteine endo-peptidases 1of *Arabidopsis* (AtCEP1) is involved in pathogen defense. Front Plant Sci 5:58. https://doi.org/10.3389/fpls.2014.00058

Howing T, Dann M, Hoefle C, Huckelhoven R, Gietl C (2017) Involvement of *Arabidopsis thaliana* endoplasmic reticulum KDEL- tailed cysteine endopeptidase 1 (AtCEP1) in powdery mildew-induced and AtCPR5-controlled cell death. PLoS One 12(8):e0183870

Huang YY, Shi Y, Lei Y, Li Y, Fan J, Xu YJ, Ma XF, Zhao JQ, Xiao S, Wang WM (2014) Functional identification of multiple nucleocytoplasmic trafficking signals in the broad-spectrum resistance protein RPW8.2. Planta 239:455–468

Huckelhoven R (2005) Powdery mildew susceptibility and biotrophic infection strategies. FEMS Microbiol Lett 245:9–17

Huckelhoven R (2007) Cell wall-associated mechanisms of disease resistance and susceptibility. Annu Rev Phytopathol 45:101–127

Huckelhoven R, Panstruga R (2011) Cell biology of the plant-powdery mildew interaction. Curr Opinion Plant Biol 14:738–746

Huckelhoven R, Eichmann R, Weis C, Hoefle C, Proels RK (2013) Genetic loss of susceptibility: a costly route to disease resistance? Plant Pathol 62(Suppl 1):56–62

Huesmann C, Hoefle C, Huckelhoven R (2011) ROPGAPs of *Arabidopsis* limit susceptibility to powdery mildew. Plant Signal Behav 6:1691–1694

Huesmann C, Reiner T, Hoefle C, Preuss J, Jurca ME, Domoki M, Fehér A, Huckelhoven R (2012) Barley ROP binding kinase1 is involved in microtubule organization and in basal penetration resistance to the barley powdery mildew fungus. Plant Physiol 159:311–320

Humphry M, Consonni C, Panstruga R (2006) *mlo*-based powdery mildew immunity: silver bullet or simply non-host resistance? Mol Plant Pathol 7:605–610

Humphry M, Bednarek P, Kemmerling B, Koh S, Stein M, Gobel U, Stuber K, Pislewska-Bednarek M, Loraine A, Schulze-Lefert P, Somerville S, Panstruga R (2010) A regulon conserved in monocot and dicot plants defines a functional module in antifungal plant immunity. Proc Natl Acad Sci U S A 107:21896–21901

Humphry M, Reinstadler A, Ivanov S, Bisseling T, Panstruga R (2011) Durable broad-spectrum powdery mildew resistance in pea *er1* plants is conferred by natural loss-of-function mutations in *PsMLO1*. Mol Plant Pathol 12:866–878

Hussey PJ, Hawkins TJ, Igarashi H, Kaloriti D, Smertenko A (2002) The plant cytoskeleton: recent advances in the study of the plant microtubule-associated proteins MAP-65, MAP-190 and the Xenopus MAP215-like protein, MOR1. Plant Mol Biol 50:915–924

Inada N, Ueda T (2014) Membrane trafficking pathways and their roles in plant-microbe interactions. Plant Cell Physiol 55:672–686

Inada N, Higaki T, Hasezawa S (2016) Nuclear function of subclass I actin depolymerizing factor contributes to susceptibility in *Arabidopsis* to an adapted powdery mildew fungus. Plant Physiol 170:1420–1434

Jacobs AK, Lipka V, Burton RA, Panstruga R, Strizhov N, Schulze-Lefert P, Fincher GB (2003) An *Arabidopsis* callose synthase, GSL5, is required for wound and papillary callose formation. Plant Cell 15:2503–2513

Jacobs S, Zechmann B, Molitor A, Trujillo M, Petutschnig E, Lipka V, Kogel KH, Schcafer P (2011) Broad-spectrum suppression of innate immunity is required for colonization of Arabidopsis roots by the fungus *Piriformospora indica*. Plant Physiol 156:726–740

Jensen MK, Rung JH, Gregersen PL, Gjetting T, Fuglsang AT, Hansen M, Joehnk N, Lyngkjaer MF, Collinge DB (2007) The *Hv*NAC6 transcription factor: a positive regulator of penetration resistance in barley and *Arabidopsis*. Plant Mol Biol 65:137–150

Jensen MK, Hagedorn PH, de Torres-Zabala M, Grant MR, Rung JH, Collinge DB, Lyngkjaer MF (2008) Transcriptional regulation by an NAC (NAM-ATAF1, 2-CUC2) transcription factor attenuates ABA signalling for efficient basal defence towards *Blumeria graminis* f. sp. *hordei* in *Arabidopsis*. Plant J 56:867–880

Jiang Z, Dong X, Zhang Z (2016) Network-based comparative analysis of *Arabidopsis* immune responses to *Golovinomyces orontii* and *Botrytis cinerea* infections. Sci Rep 6:19149

Jirage D, Tootle TL, Reuber TL, Frost LN, Feys BJ, Parker JE, Ausubel FM, Glazebrook J (1999) *Arabidopsis thaliana* PAD4 encodes a lipase-like gene that is important for salicylic acid signaling. Proc Natal Accad Sci, USA 96:13583–13588

Jones JDG, Dangl JL (2006) The plant immune system. Nature 444(7117):323–329

Jorgensen JH (1992) Multigene families of powdery mildew resistance genes in locus *Mlo* on barley chromosome 5. Plant Breed 108:53–59

Jorgensen JH (1994) Genetics of powdery mildew resistance in barley. Critical Rev Plant Sci 13:97–119

Jurca ME, Bottka S, Feher A (2008) Characterization of a family of *Arabidopsis* receptor-like cytoplasmic kinases (RLCK class VI). Plant Cell Report 27:739–748

Kachroo A, Kachroo P (2009) Fatty acid-derived signals in plant defense. Annu Rev Phytopathol 47:153–176

Kawano Y, Kaneko-Kawano T, Shimamoto K (2014) Rho family GTPase-dependent immunity in plants and animals. Front Plant Sci 5:522

Kehr J (2003) Single cell technology. Curr Opinion Plant Biol 6:617–621

Kim H, O'Connell R, Maekawa-Yoshikawa M, Uemura T, Neumann U, Schulze-Lefert P (2014) The powdery mildew resistance protein RPW8.2 is carried on VAMP721/722 vesicles to the extrahaustorial membrane of haustorial complexes. Plant J 79:835–847

Kirik A, Mudgett MB (2009) SOBER1 phospholipase activity suppresses phosphatidic acid accumulation and plant immunity in response to bacterial effector AvrBsT. Proc Natl Acad Sci U S A 106:20532–20537

Kirik V, Bouyer D, Schobinger U, Bechtold N, Herzog M, Bonneville JM, Hulskamp M (2001) CPR5 is involved in cell proliferation and cell death control and encodes a novel transmembrane protein. Curr Biol 11:1891–1895

Kliebenstein DJ, Rowe HC, Denby KJ (2005) Secondary metabolites influence *Arabidopsis/Botrytis* interactions: variation in host production and pathogen sensitivity. Plant J 44:25–36

Kobayashi I, Kobayashi Y, Hardham AR (1994) Dynamic reorganization of microtubules and microfilaments in flax cells during the resistance response to flax rust infection. Planta 61:237–247

Kobayashi Y, Kobayashi I, Funaki Y, Fujimoto S, Takemoto T, Kunoh H (1997a) Dynamic reorganization of microfilaments and microtubules is necessary for the expression of non-host resistance in barley coleoptile cells. Plant J 11:525–537

Kobayashi Y, Yamada M, Kobayashi I, Kunoh H (1997b) Actin microfilaments are required for the expression of non-host resistance in higher plants. Plant Cell Physiol 38:725–733

Koh S, Andre A, Edwards H, Ehrhardt D, Somerville S (2005) *Arabidopsis thaliana* sub-cellular responses to compatible *Erysiphe cichoracearum* infections. Plant J 44:516–529

Kolattukudy PE (1974) Biosynthesis of a hydroxy fatty acid polymer, cutin. Identification and biosynthesis of 16-oxo-9- or 10- hydroxypalmitic acid, a novel compound in *Vicia faba*. Biochemistry 13:1354–1363

Kolattukudy PE (1980) Biopolyester membranes of plants: cutin and suberin. Science 208:990–1000

Konieczny A, Ausubel FM (1993) A procedure for mapping *Arabidopsis* mutations using co-dominant ecotype-specific PCR based markers. Plant J 4:403–410

Korolev N, David DR, Elad Y (2008) The role of phytohormones in basal resistance and *Trichoderma*-induced systemic resistance to *Botrytis cinerea* in *Arabidopsis thaliana*. BioControl 53:667–683

Kragler F, Curin M, Trutnyeva K, Gansch A, Waigmann E (2003) MPB2C, a microtubule-associated plant protein binds to and interferes with cell-to-cell transport of tobacco mosaic virus movement protein. Plant Physiol 132:1870–1883

Krishnia SK, Saharan GS, Singh D (2000) Genetic variation for multiple disease resistance in the families of interspecific cross of *Brassica juncea* x *Brassica carinata*. Cruciferae Newslett 22:51–53

Kuhn H, Kwaaitaal M, Kusch S, Acevedo-Garcia J, Wu H, Panstruga R (2016) Biotrophy at its best: novel findings and unsolved mysteries of the *Arabidopsis*-powdery mildew pathosystem. Arabidopsis Book. https://doi.org/10.1199/tab.0184

Kumar D (2014) Salicylic acid signaling in disease resistance. Plant Sci 228:127–134

Kumar S, Saharan GS, Singh D (2002) Inheritance of resistance in inter and intraspecific crosses of *Brassica juncea* and *Brassica carinata* to *Albugo candida* and *Erysiphe cruciferarum*. J Mycol Pl Pathol 32(1):59–63

Kunkel BN (1996) A useful weed put to work: genetic analysis of disease resistance in *Arabidopsis thaliana*. Trends Genet 12:63–69

Kusch S, Panstruga R (2017) *mlo*-based resistance: An apparently universal "weapon" to defeat powdery mildew disease. Mol Plant-Microbe Interact 30:179–189

Kusch S, Pesch L, Panstruga R (2016) Comprehensive phylogenetic analysis sheds light on the diversity and origin of the MLO family of integral membrane proteins. Genom Biol Evol 8:878–895

Kwaaitaal M, Keinath NF, Pajonk S, Biskup C, Panstruga R (2010) Combined bimolecular fluorescence complementation and forster resonance energy transfer reveals ternary SNARE complex formation in living plant cells. Plant Physiol 152:1135–1147

Kwon C, Panstruga R, Schulze-Lefert P (2008a) Les liaisons dangereuses: immunological synapse formation in animals and plants. Trends Immunol 29:159–166

Kwon C, Neu C, Pajonk S, Yun HS, Lipka U, Humphry M, Bau S, Straus M, Kwaaitaal M, Rampelt H, El Kasmi F, Jurgens G, Parker J, Panstruga R, Lipka V, Schulze-Lefert P (2008b) Co-option of a default secretory pathway for plant immune responses. Nature 451:835–840

Laluk K, Luo H, Chai M, Dhawan R, Lai Z, Mengiste T (2011) Biochemical and genetic requirements for function of the immune response regulator BOTRYTIS-INDUCED KINASE1 in plant growth, ethylene signaling, and PAMP-triggered immunity in *Arabidopsis*. Plant Cell 23:2831–2849

Lammens T, Boudolf V, Kheibarshekan L, Zalmas LP, Gaamouche T, Maes S, Vanstraelen M, Kondorosi E, La Thangue NB, Govaerts W, Inzé D, De Veylder L (2008) Atypical E2F activity restrains APC/CCCS52A2 function obligatory for endocycle onset. Proc Natl Acad Sci U S A 105:14721–14726

Lapin D, Van den Ackerveken G (2013) Susceptibility to plant disease: more than a failure of host immunity. Trends Plant Sci 18:546–554

Lebel E, Heifetz P, Thorne L, Uknes S, Ryals J, Ward E (1998) Functional analysis of regulatory sequences controlling PR-1 gene expression in *Arabidopsis*. Plant J 16:223–233

Leborgne-Castel N, Bouhidel K (2014) Plasma membrane protein trafficking in plant-microbe interactions: a plant cell point of view. Front Plant Sci 5:735

Li H, Shen JJ, Zheng ZL, Lin Y, Yang Z (2001) The Rop GTPase switch controls multiple developmental processes in Arabidopsis. Plant Physiol 126:670–684

Li J, Brader G, Kariola T, Palva ET (2006) WRKY70 modulates the selection of signaling pathways in plant defense. Plant J 46:477–491

Lin W, Ma X, Shan L, He P (2013) Big roles of small kinases: the complex functions of receptor-like cytoplasmic kinases in plant immunity and development. J Integ Plant Biol 55:1188–1197

Lipka V, Dittgen J, Bednarek P, Bhat R, Wiermer M, Stein M, Landtag J, Brandt W, Rosahl S, Scheel D, Llorente F, Molina A, Parker J, Somerville S, Schulze-Lefert P (2005) Pre- and post invasion defenses both contribute to non-host resistance in *Arabidopsis*. Science 310:1180–1183

Lipka V, Kwon C, Panstruga R (2007) SNARE-ware: the role of SNARE-domain proteins in plant biology. Ann Rev Cell Develop Biol 23:147–174

Lipka U, Fuchs R, Lipka V (2008) *Arabidopsis* non-host resistance to powdery mildews. Curr Opinion Plant Biol 11:404–411

Liu Y, Bassham DC (2012) Autophagy: pathways for self-eating in plant cells. Ann Rev Plant Biol 63:215–237

Liu H, Stone SL (2013) Cytoplasmic degradation of the *Arabidopsis* transcription factor ABSCISIC ACID INSENSITIVE 5 is mediated by the RING-type E3 ligase KEEP ON GOING. J Biol Chem 288:20267–20279

Liu G, Holub EB, Alonso JM, Ecker JR, Fobert PR (2005) An *Arabidopsis NPR1*-like gene, *NPR4*, is required for disease resistance. Plant J 41:304–318

Liu S, Bartnikas LM, Volko SM, Ausubel FM, Tang D (2016) Mutation of the glucosinolate biosynthesis enzyme cytochrome P450 83A1 monooxygenase increases camalexin accumulation and powdery mildew resistance. Front Plant Sci 7:227

Loake G, Grant M (2007) Salicylic acid in plant defence-the players and protagonists. Curr Opinion Plant Biol 10:466–472

Loapez-Fernaandez MP, Maldonado S (2015) Programmed cell death in seeds of angiosperms. J Integ Plant Biol 57:996–1002

Lorek J, Griebel T, Jones AM, Kuhn H, Panstruga R (2013) The role of *Arabidopsis* heterotrimeric G-protein subunits in MLO2 function and MAMP-triggered immunity. Mol Plant-Microbe Interact 26:991–1003

Lu D, Wu S, Gao X, Zhang Y, Shan L, He P (2010) A receptor-like cytoplasmic kinase, BIK1, associates with a flagellin receptor complex to initiate plant innate immunity. Proc Natl Acad Sci U S A 107:496–501

Lu X, Dittgen J, Piślewska-Bednarek M, Molina A, Schneider B, Svatos A, Doubský J, Schneeberger K, Weigel D, Bednarek P, Schulze-Lefert P (2015) Mutant allele-specific uncoupling of PENETRATION3 functions reveals engagement of the ATP-binding cassette transporter in distinct tryptophan metabolic pathways. Plant Physiol 168:814–827

Luna E, Pastor V, Robert J, Flors V, Mauch-Mani B, Ton J (2011) Callose deposition: a multifaceted plant defense response. Mol Plant-Microbe Interact 24:183–193

Lyngkjær MF, Newton AC, Atzema JL, Baker SJ (2000) The barley *mlo* -gene: an important powdery mildew resistance source. Agronomie 20:745–756

Ma XF, Li Y, Sun JL, Wang TT, Fan J, Lei Y, Huang YY, Xu YJ, Zhao JQ, Xiao S, Wang WM (2014) Ectopic expression of *RESISTANCE TO POWDERY MILDEW8.1* confers resistance to fungal and oomycete pathogens in *Arabidopsis*. Plant Cell Physiol 55:1484–1496

Maeda K, Houjyou Y, Komatsu T, Hori H, Kodaira T, Ishikawa A (2009) AGB1 and PMR5 contribute to PEN2-mediated pre-invasion resistance to *Magnaporthe oryzae* in *Arabidopsis thaliana*. Mol Plant-Microbe Interact 22:1331–1340

Maekawa T, Kufer TA, Schulze-Lefert P (2011) NLR functions in plant and animal immune systems: so far and yet so close. Nat Immun 12:817–826

Maekawa T, Kracher B, Vernaldi S, Ver Loren van Themaat E, Schulze-Lefert P (2012) Conservation of NLR-triggered immunity across plant lineages. Proc Natl Acad Sci U S A 109:20119–20123

Maekawa S, Inada N, Yasuda S, Fukao Y, Fujiwara M, Sato T, Yamaguchi J (2014) The carbon/nitrogen regulator ARABIDOPSIS TOXICOS EN LEVADURA31 controls papilla formation in response to powdery mildew fungi penetration by interacting with SYNTAXIN OF PLANTS121 in *Arabidopsis*. Plant Physiol 164:879–887

Makandar R, Essig JS, Schapaugh MA, Trick HN, Shah J (2006) Genetically engineered resistance to Fusarium head blight in wheat by expression of *Arabidopsis* NPR1. Mol Plant-Microbe Interact 19:123–129

Maksimov V, Abizgil'dina RR, Pusenkova LI (2011) Plant growth promoting rhizobacteria as alternative to chemical crop protectors from pathogens (review). Appl Biochem Microbiol 47(4):373–385

Malamy J, Klessig DF (1992) Salicylic acid and plant disease resistance. Plant J 2:643–654

Malinovsky FG, Fangel JU, Willats WGT (2014) The role of the cell wall in plant immunity. Front Plant Sci 5.:PMID, 24834069

Mao G, Meng X, Liu Y, Zheng Z, Chen Z, Zhang S (2011) Phosphorylation of a WRKY transcription factor by two pathogen-responsive MAPKs drives phytoalexin biosynthesis in *Arabidopsis*. Plant Cell 23:1639–1653

Marone D, Russo MA, Laido G, De Vita P, Papa R, Blanco A, Gadaleta A, Rubiales D, Mastrangelo AM (2013) Genetic basis of qualitative and quantitative resistance to powdery mildew in wheat: from consensus regions to candidate genes. BMC Genomics 14:562

Martiniere A, Gargani D, Uzest M, Lautredou N, Blanc S, Drucker M (2009) A role for plant microtubules in the formation of transmission specific inclusion bodies of cauliflower mosaic virus. Plant J 58:135–146

Meena PD, Mehta N, Rai PK, Saharan GS (2018) Geographical distribution of rapeseed-mustard powdery mildew disease in India. J Mycol Pl Pathol 48(3):284–302

Mehta N, Singh K, Sangwan MS (2008) Assessment of yield losses and evaluation of different varieties/ genotypes of mustard against powdery mildew in Haryana. Plant Dis Res 23(1):55–59

Meng X, Zhang S (2013) MAPK cascades in plant disease resistance signaling. Annu Rev Phytopathol 51:245–266

Meyer D, Pajonk S, Micali C, O'Connell R, Schulze-Lefert P (2009) Extracellular transport and integration of plant secretory proteins into pathogen-induced cell wall compartments. Plant J 57:986–999

Micali C, Gollner K, Humphry M, Consonni C, Panstruga R (2008) The powdery mildew disease of *Arabidopsis*: a paradigm for the interaction between plants and biotrophic fungi. Arabidopsis Book 6:e0115

Micali C, Neumann U, Grunewald D, Panstruga R, O'Connell R (2011) Biogenesis of a specialized plant-fungal interface during host cell internalization of *Golovinomyces orontii* haustoria. Cell Microbiol 13:210–226

Miklis M, Consonni C, Bhat RA, Lipka V, Schulze-Lefert P, Panstruga R (2007) Barley mlo modulates actin-dependent and actin-independent antifungal defense pathways at the cell periphery. Plant Physiol 144(2):1132–1143

Misas-Villamil JC, van der Hoorn RA, Doehlemann G (2016) Papain-like cysteine proteases as hubs in plant immunity. New Phytol 212:902–907

Mitchell-Olds T, James RV, Palmer MJ, Williams PH (1995) Genetics of *Brassica rapa* (Syn. *campestris*) L. multiple disease resistance to three fungal pathogens: *Peronospora parasitica*, *Albugo candida* and *Leptosphaeria maculans*. Heredity 75:362–369

Miya A, Albert P, Shinya T, Desaki Y, Ichimura K, Shirasu K, Narusaka Y, Kawakami N, Kaku H, Shibuya N (2007) CERK1, a LysM receptor kinase, is essential for chitin elicitor signaling in *Arabidopsis*. Proc Natl Acad Sci U S A 104:19613–19618

Molendijk AJ, Ruperti B, Singh MK, Dovzhenko A, Ditengou FA, Milia M, Westphal L, Rosahl S, Soellick TR, Uhrig J, Weingarten L, Huber M, Palme K (2008) A cysteine rich receptor-like kinase NCRK and a pathogen-induced protein kinase RBK1 are Rop GTPase interactors. Plant J 53:909–923

Mou Z, Fan W, Dong X (2003) Inducers of plant systemic acquired resistance regulate NPR function through redox changes. Cell 113:935–944

Mucha E, Fricke I, Schaefer A, Wittinghofer A, Berken A (2011) Rho proteins of plants—functional cycle and regulation of cytoskeletal dynamics. European J Cell Biol 90:934–943

Muller S, Smertenko A, Wagner V, Heinrich M, Hussey PJ, Hauser MT (2004) The plant microtubule-associated protein AtMAP65-3/PLE is essential for cytokinetic phragmoplast function. Curr Biol 14:412–417

Murase K, Shiba H, Iwano M, Che FS, Watanabe M, Isogai A, Takayama S (2004) A membrane-anchored protein kinase involved in *Brassica* self-incompatibility signaling. Science 303:1516–1519

Mysore KS, Ryu CM (2004) Non-host resistance how much do we know? Trends Plant Sci 9:97–104

Nafisi M, Goregaoker S, Botanga CJ, Glawischnig E, Olsen CE, Halkier BA (2007) *Arabidopsis* cytochrome P450 monooxygenase 71A13 catalyzes the conversion of indole-3-acetaldoxime in camalexin synthesis. Plant Cell 19:2039–2052

Nakano RT, Yamada K, Bednarek P, Nishimura M, Hara-Nishimura I (2014) ER bodies in plants of the Brassicales order: biogenesis and association with innate immunity. Front Plant Sci 5:73

Naumann M, Somerville S, Voigt C (2013) Differences in early callose deposition during adapted and non-adapted powdery mildew infection of resistant *Arabidopsis* lines. Plant Signal Behav 8:e24408

Naur P, Petersen BL, Mikkelsen MD, Bak S, Rasmussen H, Olsen CE, Halkier BA (2003) CYP83A1 and CYP83B1, two non-redundant cytochrome P450 enzymes metabolizing oximes in the biosynthesis of glucosinolates in *Arabidopsis*. Plant Physiol 133:63–72

Newton AC (1993) The effect of humidity on the expression of partial resistance to powdery mildew in barley. Plant Pathol 42:364–367

Nibau C, Wu HM, Cheung AY (2006) RAC/ROP GTPases: 'hubs' for signal integration and diversification in plants. Trends Plant Sci 11:309–315

Nie H, Wu Y, Yao C, Tang D (2011) Suppression of *edr2*-mediated powdery mildew resistance, cell death and ethylene-induced senescence by mutations in *ALD1* in *Arabidopsis*. J Genet Genom 38:137–148

Nie H, Zhao C, Wu G, Wu Y, Chen Y, Tang D (2012) SR1, a calmodulin binding transcription factor, modulates plant defense and ethylene-induced senescence by directly regulating *NDR1* and *EIN3*. Plant Physiol 158:1847–1859

Nielsen ME, Feechan A, Bohlenius H, Ueda T, Thordal-Christensen H (2012) *Arabidopsis* ARF-GTP exchange factor, GNOM, mediates transport required for innate immunity and focal accumulation of syntaxin PEN1. Proc Natl Acad Sci U S A 109:11443–11448

Nishimura MT, Stein M, Hou BH, Vogel JP, Edwards H, Somerville SC (2003) Loss of a callose synthase results in salicylic acid-dependent disease resistance. Science 301:969–972

Noel LD, Cagna G, Stuttmann J, Wirthmuller L, Betsuyaku S, Witte C-P, Bhat R, Pochon N, Colby T, Parker JE (2007) Interaction between SGT1 and cytosolic/nuclear HSC70 chaperones regulates *Arabidopsis* immune responses. Plant Cell 19:4061–4076

Nongbri PL, Johnson JM, Sherameti I, Glawischnig E, Halkier BA, Oelm€uller R (2012) Indole-3-acetaldoxime-derived compounds restrict root colonization in the beneficial interaction between *Arabidopsis* roots and the endophyte *Piriformospora indica*. Mol Plant-Microbe Interact 25:1186–1197

Nurnberger T, Lipka V (2005) Non-host resistance in plants: new insights into an old phenomenon. Mol Plant Pathol 6:335–345

O'Connell RJ, Panstruga R (2006) Tete a Tete inside a plant cell: establishing compatibility between plants and biotrophic fungi and oomycetes. New Phytol 171:699–718

Opalski KS, Schultheiss H, Kogel KH, Huckelhoven R (2005) The receptor-like MLO protein and the RAC/ROP family G-protein RACB modulate actin reorganization in barley attacked by the biotrophic powdery mildew fungus *Blumeria graminis* f.sp. *hordei*. Plant J 41:291–303

Orgil U, Araki H, Tangchaiburana S, Berkey R, Xiao S (2007) Intraspecific genetic variations, fitness cost and benefit of *RPW8*, a disease resistance locus in *Arabidopsis thaliana*. Genetics 176:2317–2333

Ouko MO, Sambade A, Brandner K, Niehl A, Pena E, Ahad A, Heinlein M, Nick P (2010) Tobacco mutants with reduced microtubule dynamics are less susceptible to TMV. Plant J 62:829–839

Pajerowska-Mukhtar KM, Emerine DK, Mukhtar MS (2013) Tell me more: roles of NPRs in plant immunity. Trends Plant Sci 18:402–411

Pandey SP, Roccaro M, Schon M, Logemann E, Somssich IE (2010) Transcriptional reprogramming regulated by WRKY18 and WRKY40 facilitates powdery mildew infection of *Arabidopsis*. Plant J 64:912–923

Parkhi V, Kumar V, Campbell LM, Bell AA, Shah J, Rathore KS (2010) Resistance against various fungal pathogens and reniform nematode in transgenic cotton plants expressing *Arabidopsis* NPR1. Transgenic Res 19:959–975

Pavan S, Jacobsen E, Visser RG, Bai Y (2010) Loss of susceptibility as a novel breeding strategy for durable and broad-spectrum resistance. Mol Breed 25:1–12

Pedras MSC, Adio AM (2008) Phytoalexins and phytoanticipins from the wild crucifers *Thellungiella halophila* and *Arabidopsis thaliana*: rapalexin a, was alexins and camalexin. Phytochemistry 69:889–893

Pedras MSC, Yaya EE, Glawischnig E (2011) The phytoalexins from cultivated and wild crucifers: chemistry and biology. Nat Prod Rep 28:1381–1405

Penaud A (1999) Chemical control and yield losses caused by *Erysiphe cruciferarum* on oilseed rape in France. In: Wratten N, Salisbury PA (eds) Proceedings of the 10th international Rapeseed Congress. The Regional Institute Ltd., Canberra, Australia CD-ROM, Doc. No. 327:1–8

Perrin RM, Jia X, Wagner TA, O'Neill MA, Sarria R, York WS, Raikhel NV, Keegstra K (2003) Analysis of xyloglucan fucosylation in *Arabidopsis*. Plant Physiol 132:768–778

Peters W, Latka I (1986) Electron microscopic localization of chitin using colloidal gold labeled with wheat germ agglutinin. Histochem 84:155–160

Petkova M, Dimova M, Dimova D, Bistrichanov S (2014) Effect of gamma-irradiation on the fatty acid composition and susceptibility to powdery mildew (*Erysiphe cruciferarum*) of oilseed rape plants. Agril Sci Technol 6(4):413–416

Pieterse CMJ, Van der Does D, Zamioudis C, Leon-Reyes A, Van Wees SCM (2012) Hormonal modulation of plant immunity. Ann Rev Cell Dev Biol 28:489–521

Pinosa F, Buhot N, Kwaaitaal M, Fahlberg P, Thordal-Christensen H, Ellerstrom M, Andersson MX (2013) *Arabidopsis* phospholipase Dδ is involved in basal defense and non-host resistance to powdery mildew fungi. Plant Physiol 163:896–906

Pinot F, Benveniste IJ, Salau P, Loreau O, Noe J, Schreiber L, Durst F (1999) Production in vitro by the cytochrome P450 CYP94A1 of major C18 cutin monomers and potential messengers in plant–pathogen interactions: enantioselectivity studies. Biochem J 342:27–32

Pleines S, Friendt W (1989) Genetic control of linolenic acid concentrations in seed oil of rapeseed (*Brassica napus* L.). Theor Appl Genet 78:793–797

Plotnikova JM, Reuber TL, Ausubel FM, Pfister DH (1998) Powdery mildew pathogenesis of *Arabidopsis thaliana*. Mycologia 90:1009–1016

Poraty-Gavra L, Zimmermann P, Haigis S, Bednarek P, Hazak O, Stelmakh OR, Sadot E, Schulze-Lefert P, Gruissem W, Yalovsky S (2013) The *Arabidopsis* rho of plants GTPase AtROP6 functions in developmental and pathogen response pathways. Plant Physiol 161:1172–1188

Qiu JL, Fiil BK, Petersen K, Nielsen HB, Botanga CJ, Thorgrimsen S (2008) *Arabidopsis* MAP kinase 4 regulates gene expression through transcription factor release in the nucleus. EMBO J 27:2214–2221

Quentin M, Baures I, Hoefle C, Caillaud MC, Allasia V, Panabieres F, Abad P, Huckelhoven R, Keller H, Favery B (2016) The *Arabidopsis* microtubule-associated protein MAP65-3 supports infection by filamentous biotrophic pathogens by down-regulating salicylic acid-dependent defences. J Exp Bot 67(6):1731–1743

Quilis J, Penas G, Messeguer J, Brugidou C, Segundo BS (2008) The *Arabidopsis* AtNPR1 inversely modulates defense responses against fungal, bacterial, or viral pathogens while conferring hypersensitivity to abiotic stresses in transgenic rice. Mol Plant-Microbe Interact 21:1215–1231

Rahimi M, Bahrani A (2011) Effect of gamma irradiation on qualitative and quantitative characteristics of canola. Middle East J Sci Res 8:519–525

Ramonell KM, Zhang B, Ewing RM, Chen Y, Xu D, Stacey G, Somerville S (2002) Microarray analysis of chitin elicitation in *Arabidopsis thaliana*. Mol Plant Pathol 3:301–311

Ramonell K, Berrocal-Lobo M, Koh S, Wan J, Edwards H, Stacey G, Somerville S (2005) Loss-of-function mutations in chitin responsive genes show increased susceptibility to the powdery mildew pathogen *Erysiphe cichoracearum*. Plant Physiol:138

Rauhut T, Glawischnig E (2009) Evolution of camalexin and structurally related indolic compounds. Phytochemistry 70:1638–1644

Reddy KS, Pawar SE, Bhatia CR (1994) Inheritance of powdery mildew resistance (*Erysiphe polygoni* DC) resistance in mungbean. Theor Appl Genet 88:945–948

Reiner T, Hoefle C, Huesmann C, Menesi D, Feher A, Huckelhoven R (2014) The *Arabidopsis* ROP-activated receptor-like cytoplasmic kinase RLCK VI-A3 is involved in control of basal resistance to powdery mildew and trichome branching. Plant Cell Reptr 34(3):457–468

Reuber TL, Plotnikova JM, Dewdney J, Rogers EE, Wood W, Ausubel FM (1998) Correlation of defense gene induction defects with powdery mildew susceptibility in *Arabidopsis* enhanced disease susceptibility mutants. Plant J 16:473–485

Rietz S, Stamm A, Malonek S, Wagner S, Becker D, Medina-Escobar N, Vlot AC, Feys BJ, Niefind K, Parker JE (2011) Different roles of enhanced disease susceptibility1 (EDS1) bound to and dissociated from phytoalexin deficient4 (PAD4) in *Arabidopsis* immunity. New Phytol 191:107–119

Roux F, Voisin D, Badet T, Balague C, Barlet X, Huard-Chauveau C, Roby D, Raffaele S (2014) Resistance to phytopathogens *e tutti quanti*: placing plant quantitative disease resistance on the map. Mol Plant Pathol 15(5):427–432

Rushton PJ, Somssich IE (1998) Transcriptional control of plant genes responsive to pathogens. Curr Opinion Plant Biol 1:311–315

Russo VM, Bushnell WR (1989) Responses of barley cells to puncture by micro needles and to attempted penetration by *Erysiphe graminis* f. sp. *hordei*. Can J Bot 67:2912–2921

Rusterucci C, Aviv DH, Holt BF, Dangl JL, Parker JE (2001) The disease resistance signaling components EDS1 and PAD4 are essential regulators of the cell death pathway controlled by LSD1 in *Arabidopsis*. Plant Cell 13:2211–2224
Saenz GS, Taylor JW (1999) Phylogeny of the Erysiphales (powdery mildews) inferred from internal transcribed spacer (ITS) ribosomal DNA sequences. Can J Bot 77:150–169
Saharan GS, Krishnia SK (2001) Multiple disease resistance in rapeseed and mustard. In: Nagarajan S, Singh DP (eds) Role of resistance in intensive agriculture, vol 10. Kalyani Publishers, New Delhi, pp 98–100
Saharan GS, Mehta N, Sangwan MS (2005) Development of disease resistance in rapeseed-mustard pp 561-617. In: Saharan GS, Mehta N, Sangwan MS (eds) Diseases of oilseed crops. Indus Publishing Co., New Delhi, 643p
Saharan GS, Verma PR, Meena PD, Kumar A (2014) White rust of crucifers: biology, ecology and management. Springer, New Delhi. 244pp
Saharan GS, Mehta N, Meena PD (2016) Alternaria blight of crucifers: biology, ecology and disease management. Springer Verlag, Singapore 326. ISBN 978-981-10-0019-5
Saharan GS, Mehta N Meena PD (2017) Downy mildew disease of crucifers: biology, ecology and disease management. Springer Verlag Singapore LVI: 357. ISBN 978-981-10-7499-8
Schaller A (2004) A cut above the rest: the regulatory function of plant proteases. Planta 220:183–197
Scheler B, Schnepf V, Galgenmuller C, Ranf S, Huckelhoven R (2016) Barley disease susceptibility factor RACB acts in epidermal cell polarity and positioning of the nucleus. J Exp Bot. https://doi.org/10.1093/jxb/erw141
Schenk PM, Kazan K, Wilson I, Anderson JP, Richmond T, Somerville SC, Manners JM (2000) Coordinated plant defense responses in *Arabidopsis* revealed by microarray analysis. Proc Natl Acad Sci U S A 97:11655–11660
Schiff CL, Wilson IW, Somerville SC (2001) Polygenic powdery mildew disease resistance in *Arabidopsis thaliana*: quantitative trait analysis of the accession Warschau-1. Plant Pathol 50:690–701
Schlaeppi K, Abou-Mansour E, Buchala A, Mauch F (2010) Disease resistance of *Arabidopsis* to *Phytophthora brassicae* is established by the sequential action of indole glucosinolates and camalexin. Plant J 62:840–851. https://doi.org/10.1111/j.1365-313X.2010.04197.x
Schmelzer E (2002) Cell polarization, a crucial process in fungal defence. Trends Plant Sci 7:411–415
Schmid M, Simpson D, Kalousek F, Gietl C (1998) A cysteine endopeptidase with a C-terminal KDEL motif isolated from castor bean endosperm is a marker enzyme for the ricinosome, a putative lytic compartment. Planta 206:466–475
Schmidt SM, Panstruga R (2007) Cytoskeleton functions in plant–microbe interactions. Physiol Mol Plant Pathol 71:135–148
Schmidt SM, Kuhn H, Micali C, Liller C, Kwaaitaal M, Panstruga R (2014) Interaction of a *Blumeria graminis* f. sp. *hordei* effector candidate with a barley ARF-GAP suggests that host vesicle trafficking is a fungal pathogenicity target. Mol Plant Pathol 15:535–549
Schon M, Toller A, Diezel C, Roth C, Westphal L, Wiermer M, Somssich IE (2013) Analyses of *wrky18 wrky40* plants reveal critical roles of SA/EDS1 signaling and indole-glucosinolate biosynthesis for *Golovinomyces orontii* resistance and a loss-of resistance towards *Pseudomonas syringae* pv. *tomato* AvrRPS4. Mol Plant-Microbe Interact 26:758–767
Schuhegger R, Nafisi M, Mansourova M, Petersen BL, Olsen CE, Svatos A, Halkier BA, Glawischnig E (2006) CYP71B15 (PAD3) catalyzes the final step in camalexin biosynthesis. Plant Physiol 141:1248–1254
Schuhegger R, Nafisi M, Mansourova M, Petersen BL, Olsen CE, Svatos A, Halkier BA, Glawischnig E (2007a) CYP71 B15 (PAD3) catalyzes the final step in camalexin biosynthesis-correction. Plant Physiol 145:1086–1086. https://doi.org/10.1104/pp.104.900240
Schuhegger R, Rauhut T, Glawischnig E (2007b) Regulatory variability of camalexin biosynthesis. J. Plant Physiol 164:636–644. https://doi.org/10.1016/j.jplph.2006.04.012

Schultheiss H, Dechert C, Kogel KH, Huckelhoven R (2002) A small GTP-binding host protein is required for entry of powdery mildew fungus into epidermal cells of barley. Plant Physiol 128:1447–1454

Schultheiss H, Dechert C, Kogel KH, Huckelhoven R (2003) Functional analysis of barley RAC/ROP G-protein family members in susceptibility to the powdery mildew fungus. Plant J:589–601

Schuster A, Schmoll M (2010) Biology and biotechnology of *Trichoderma*. Appl Microbiol Biotech 87:787–799

Schweizer P (2007) Non-host resistance of plants to powdery mildew-new opportunities to unravel the mystery. Physiol Mol Plant Pathol 70:3–7

Schweizer P, Kmecl A, Carpita N, Dudler R (2000) A soluble carbohydrate elicitor from *Blumeria graminis* f. sp *tritici* is recognized by a broad range of cereals. Physiol Mol Plant Pathol 56:157–167

Serrano I, Gu Y, Qi D, Dubiella U, Innes RW (2014) The *Arabidopsis* EDR1 protein kinase negatively regulates the ATL1 E3 ubiquitin ligase to suppress cell death. Plant Cell 26:4532–4546

Seyfferth C, Tsuda K (2014) Salicylic acid signal transduction: the initiation of biosynthesis, perception and transcriptional reprogramming. Front Plant Sci 5:697

Shapiro AD, Zhang C (2001) The role of *NDR1* in avirulence gene-directed signaling and control of programmed cell death in *Arabidopsis*. Plant Physiol 127:1089–1101

Sharma P, Sain SK (2004) Induction of systemic resistance in tomato and cauliflower by *Trichoderma* spp. against stalk rot pathogen, *Sclerotinia sclerotiorum* Lib de Bary. J Biol Control 18(1):21–27

Shattuck VI (1993) Powdery mildew-resistant UG3 and UG4 rutabaga germplasm. Can J Plant Sci 73:301–302

Shen QH, Saijo Y, Mauch S, Biskup C, Bieri S, Keller B, Seki H, Ulker B, Somssich IE, Schulze-Lefert P (2007) Nuclear activity of MLA immune receptors links isolate-specific and basal disease-resistance responses. Science 315:1098–1103

Shirasu K, Nakajima H, Rajasekhar VK, Dixon RA, Lamb C (1999) Salicylic acid potentiates an agonist-dependent gain control that amplifies pathogen signals in the activation of defense mechanisms. Plant Cell 9:261–270

Shiu SH, Karlowski WM, Pan R, Tzeng YH, Mayer KF, Li WH (2004) Comparative analysis of the receptor-like kinase family in *Arabidopsis* and rice. Plant Cell 16:1220–1234

Shoresh M, Harman GE, Mastouri F (2010) Induced systemic resistance and plant responses to fungal biocontrol agents. Annu Rev Phytopathol 48:21–43

Singh RB, Singh RN (2003) Management of powdery mildew of mustard Department of Plant Pathology, N. D. University of Agriculture and Technology, Kumarganj, Faizabad 224 229

Singh K (2004) Studies on the ecofriendly management of powdery mildew (*Erysiphe cruciferarum* Opiz ex. Junell) of mustard [*Brassica juncea* (Linn.) Czern & Coss]. M. Sc Thesis, CCS HAU, Hisar 108 + xix

Singh D, Chandra N, Gupta PP (1997) Inheritance of powdery mildew resistance in interspecific crosses of Indian and Ethiopian mustard. Ann Biol 13:73–77

Singh R, Singh D, Salisbury P, Barbetti MJ (2010) Field evaluation of indigenous and exotic *Brassica juncea* genotypes against Alternaria blight, white rust, downy mildew and powdery mildew diseases in India. Indian J Agril Sci 80(2):155–159

Skoric D, Jocic S, Sakac Z, Lecic N (2008) Genetic possibilities for altering sunflower oil quality to obtain novel oils. Can J Physio Pharmacology 86:215–221

Smits GJ, van den Ende H, Klis FM (2001) Differential regulation of cell wall biogenesis during growth and development in yeast. Microbiol 147:781–794

Sonderby IE, Geu-Flores F, Halkier BA (2010) Biosynthesis of glucosinolates–gene discovery and beyond. Trends Plant Sci 15:283–290. https://doi.org/10.1016/j.tplants.2010.02.005

Spencer-Phillips PTN, Gay JL (1981) Domains of ATPase in plasma membrane and transport through infected plant cells. New Phytol 89:393–400

Speth EB, Lee YN, He SY (2007) (2007). Pathogen virulence factors as molecular probes of basic plant cellular functions. Curr Opinion Plant Biol 10:580–586

Spoel SH, Koornneef A, Claessens SM, Korzelius JP, Van-Pelt JA, Mueller MJ, Buchala AJ, Metraux J-P, Brown R, Kazan K, Van Loon LC, Dong X, Pieterse CMJ (2003) NPR1 modulates cross–talk between salicylate and jasmonate dependent defense pathways through a novel function in the cytosol. Plant Cell 15:760–770. https://doi.org/10.1105/tpc.009159

Sreeramulu S, Mostizky Y, Sunitha S, Shani E, Nahum H, Salomon D, Hayun LB, Gruetter C, Rauh D, Ori N, Sessa G (2013) BSKs are partially redundant positive regulators of brassinosteroid signaling in *Arabidopsis*. Plant J 74:905–919

Stein M, Dittgen J, Sánchez-Rodríguez C, Hou BH, Molina A, Schulze-Lefert P, Lipka V, Somerville S (2006) *Arabidopsis* PEN3/PDR8, an ATP binding cassette transporter, contributes to non-host resistance to inappropriate pathogens that enter by direct penetration. Plant Cell 18:731–746

Stein E, Molitor A, Kogel KH, Waller F (2008) Systemic resistance in *Arabidopsis* conferred by the mycorrhizal fungus *Piriformospora indica* requires jasmonic acid signaling and the cytoplasmic function of NPR1. Plant Cell Physiol 49:1747–1751

Stitt M, von Schaewen A, Willmitzer L (1990) “Sink” regulation of photosynthetic metabolism in transgenic tobacco plants expressing yeast invertase in their cell wall involves a decrease of the Calvin-cycle enzymes and an increase in glycolytic enzymes. Planta 183:40–50

Stotz HU, Sawada Y, Shimada Y, Hirai MY, Sasaki E, Krischke M, Brown PD, Saito K, Kamiya Y (2011) Role of camalexin, indole glucosinolates, and side chain modification of glucosinolate-derived isothiocyanates in defense of *Arabidopsis* against *Sclerotinia sclerotiorum*. Plant J 67:81–93

Strack D, Vogt T, Schliemann W (2003) Recent advances in betalain research. Phytochem 62:247Ð269

Strawn MA, Marr SK, Inoue K, Inada N, Zubieta C, Wildermuth MC (2007) *Arabidopsis* isochorismate synthase functional in pathogen-induced salicylate biosynthesis exhibits properties consistent with a role in diverse stress responses. J Biol Chem 282:5919–5933

Sundaresha S, Rohini S, Appanna V K, Arthikala M K, Shanmugam N B, Shashibhushan N B, Kishore C M , Pannerselvam R , Kirti P B , Udayakumar M (2016). Co-overexpression of *Brassica juncea* NPR1 (BjNPR1) and *Trigonella foenum-graecum* defensin (Tfgd) in transgenic peanut provides comprehensive but varied protection against *Aspergillus flavus* and *Cercospora arachidicola*. Plant Cell Rep 35:1189–1203. https://doi.org/10.1007/s00299-016-1945-7

Swarbrick PJ, Schulze-Lefert P, Scholes JD (2006) Metabolic consequences of susceptibility and resistance (race-specific and broad-spectrum) in barley leaves challenged with powdery mildew. Plant Cell Environ 29:1061–1076

Swiderski MR, Innes RW (2001) The *Arabidopsis* PBS1 resistance gene encodes a member of a novel protein kinase subfamily. Plant J 26:101–112

Tada Y, Spoel SH, Pajerowska-Mukhtar K, Mou Z, Song J, Wang C, Zuo J, Dong X (2008) Plant immunity requires conformational charges of NPR1 via S-nitrosylation and thioredoxins. Science 321:952–956

Takemoto D, Jones DA, Hardham AR (2003) GFP-tagging of cell components reveals the dynamics of sub-cellular re-organization in response to infection of *Arabidopsis* by oomycete pathogens. Plant J 33:775–792

Takemoto D, Jones DA, Hardham AR (2006) Re-organization of the cytoskeleton and endoplasmic reticulum in the *Arabidopsis pen1*-1 mutant inoculated with the non-adapted powdery mildew pathogen, *Blumeria graminis* f. sp. *hordei*. Mol Plant Pathol 7:553–563

Takken FL, Goverse A (2012) How to build a pathogen detector: structural basis of NB-LRR function. Curr Opinion Plant Biol 15:375–384

Tang D, Innes RW (2002) Overexpression of a kinase-deficient form of the *EDR1* gene enhances powdery mildew resistance and ethylene-induced senescence in *Arabidopsis*. Plant J 32:975–983

Tang XY, Xie MT, Kim YJ, Zhou JM, Klessig DF, Martin GB (1999) Overexpression of Pto activates defense responses and confers broad resistance. Plant Cell 11:15–29

Tang D, Christiansen KM, Innes RW (2005a) Regulation of plant disease resistance, stress responses, cell death, and ethylene signaling in *Arabidopsis* by the EDR1 protein kinase. Plant Physiol 138:1018–1026

Tang D, Ade J, Frye CA, Innes RW (2005b) Regulation of plant defense responses in *Arabidopsis* by EDR2, a PH and START domain-containing protein. Plant J 44:245–257

Tang D, Ade J, Frye CA, Innes RW (2006) A mutation in the GTP hydrolysis site of *Arabidopsis* dynamin-related protein 1E confers enhanced cell death in response to powdery mildew infection. Plant J 47:75–84

Tang W, Kim TW, Oses-Prieto JA, Sun Y, Deng Z, Zhu S, Wang R, Burlingame AL, Wang ZY (2008) BSKs mediate signal transduction from the receptor kinase BRI1 in *Arabidopsis*. Science 321:557–560

Teh OK, Hofius D (2014) Membrane trafficking and autophagy in pathogen-triggered cell death and immunity. J Exp Bot 65:1297–1312

Thaler JS, Humphrey PT, Whiteman NK (2012) Evolution of jasmonate and salicylate signal crosstalk. Trends Plant Sci 17:260–270

Than ME, Helm M, Simpson DJ, Lottspeich F, Huber R, Gietl C (2004) The2.0-Å crystal structure of the KDEL-tailed cysteine endopeptidase from germinating endosperm of *Ricinus communis* confirms its function in the final stage of programmed cell death. J Mol Biol 336:1103–1116

Thomma BPHJ, Eggermont K, Penninckx IA, Mauch-Mani B, Vogelsang R, Cammue BP, Broekaert WF (1998) Separate jasmonate-dependent and salicylate-dependent defense-response pathways in *Arabidopsis* are essential for resistance to distinct microbial pathogens. Proc Natl Acad Sci U S A 95:15107–15111

Thomma BPHJ, Nelissen I, Eggermont K, Broekaert WF (1999) Deficiency in phytoalexin production causes enhanced susceptibility of *Arabidopsis thaliana* to the fungus *Alternaria brassicicola*. Plant J 19:163–171

Thomma BPHJ, Penninckx IAMA, Broekaert WF, Cammue BPA (2001) The complexity of disease signaling in *Arabidopsis*. Curr Opinion Immun 13:63–68

Thordal-Christensen H (2003) Fresh insights into processes of non-host resistance. Curr Opinion Plant Biol 6:351–357

Timonen KL, Vanninen E, de Hartog J, Ibald-Mulli A, Brunekreef B, Gold DR, Heinrich J, Hoek G, Lanki T, Peters A, Tarkiainen P, Kreyling W, Pekkanen J (2006) Effects of ultrafine and fine particulate and gaseous air pollution on cardiac autonomic control in subjects with coronary artery disease-the ULTRA study. J Expo Sci Environ Epidemiol 16(4):332–341

Tongue M, Griffiths PD (2004) Transfer of powdery mildew resistance from *Brassica carinata* to *Brassica oleracea* through embryo rescue. Plant Breed 123:587–589

Tronchin G, Poulain D, Herbaut J, Biguet J (1981) Localization of chitin in the cell wall of *Candida albicans* by means of wheat germ agglutinin. Fluorescence and ultra-structural studies. Eur J Cell Biol 26:121–128

Tsuda K, Somssich IE (2015) Transcriptional networks in plant immunity. New Phytol 206:932–947

Uemura T, Kim H, Saito C, Ebine K, Ueda T, Schulze-Lefert P, Nakano A (2012) Qa-SNAREs localized to the *trans*-Golgi network regulate multiple transport pathways and extracellular disease resistance in plants. Proc Natl Acad Sci U S A 109:1784–1789

Ulker B, Shahid Mukhtar M, Somssich IE (2007) The WRKY70 transcription factor of Arabidopsis influences both the plant senescence and defense signaling pathways. Planta 226:125–137

Underwood W, Somerville SC (2008) Focal accumulation of defences at sites of fungal pathogen attack. J Exp Bot 59:3501–3508

Underwood W, Somerville SC (2013) Perception of conserved pathogen elicitors at the plasma membrane leads to relocalization of the *Arabidopsis* PEN3 transporter proc Natl Acad Sci. USA 110:12492–12497

Van der Does D, Leon-Reyes A, Koornneef A, Van Verk MC, Rodenburg N, Pauwels L, Goossens A, Korbes AP, Memelink J, Ritsema T, Van Wees SCM, Pieterse CMJ (2013) Salicylic acid suppresses jasmonic acid signaling downstream of SCFCOI1-JAZ by targeting GCC promoter motifs via transcription factor ORA59. Plant Cell 25:744–761

Van der Luit AH, Piatti T, van Doorn A, Musgrave A, Felix G, Boller T, Munnik T (2000) Elicitation of suspension-cultured tomato cells triggers the formation of phosphatidic acid and diacylglycerol pyrophosphate. Plant Physiol 123:1507–1516

Van Schie CC, Takken FL (2014) Susceptibility genes: how to be a good host. Annu Rev Phytopathol 52:551–581

Vassileva VN, Kouchi H, Ridge RW (2005) Microtubule dynamics in living root hairs: transient slowing by lipochitin oligosaccharide nodulation signals. Plant Cell 17:1777–1787

Verhagen BWM, Glazebrook J, Zhu T, Chang HS, van Loon LC, Pieterse CMJ (2004) The transcriptome of rhizobacteria-induced systemic resistance in *Arabidopsis*. Mol Plant-Microbe Interact 17:895–908

Veronese P, Nakagami H, Bluhm B, AbuQamar S, Chen X, Salmeron J, Dietrich RA, Hirt H, Mengiste T (2006) The membrane-anchored *BOTRYTIS-INDUCED KINASE1* plays distinct roles in *Arabidopsis* resistance to necrotrophic and biotrophic pathogens. Plant Cell 18:257–273

Vivancos J, Labbé C, Menzies JG, Bélanger RR (2015) Silicon-mediated resistance of Arabidopsis against powdery mildew involves mechanisms other than the salicylic acid (SA)-dependent defence pathway. Mol Plant Pathol 16:572–582

Vlieghe K, Boudolf V, Beemster GTS, Maes S, Magyar Z, Atanassova A, de Almeida EJ, De Groodt R, Inzé D, De Veylder L (2005) The DP-E2F-like gene *DEL1* controls the endocycle in *Arabidopsis thaliana*. Curr Biol 15:59–63

Voegele RT, Stuck C, Hahn M, Mendgen K (2001) The role of haustoria in sugar supply during infection of broad bean by the rust fungus *Uromyces fabae*. Proc Natl Acad Sci U S A 98:8133–8138

Vogel J, Somerville S (2000) Isolation and characterization of powdery mildew-resistant *Arabidopsis* mutants. Proc Natl Acad Sci U S A 97:1897–1902

Vogel JP, Raab TK, Schiff C, Somerville SC (2002) *PMR6*, a pectate lyase–like gene required for powdery mildew susceptibility in *Arabidopsis*. Plant Cell 14:2095–2106

Vogel JP, Raab TK, Somerville CR, Somerville SC (2004) Mutations in *PMR5* result in powdery mildew resistance and altered cell wall composition. Plant J 40:968–978

von Schaewen A, Stitt M, Schmidt R, Sonnewald U, Willmitzer L (1990) Expression of a yeast-derived invertase in the cell wall of tobacco and *Arabidopsis* plants leads to accumulation of carbohydrate and inhibition of photosynthesis and strongly influences growth and phenotype of transgenic tobacco plants. EMBO J 9:3033–3044

Vorwerk S, Schiff C, Santamaria M, Koh S, Nishimura M, Vogel J, Somerville C, Somerville S (2007) *EDR2* negatively regulates salicylic acid-based defenses and cell death during powdery mildew infections of *Arabidopsis thaliana*. BMC Plant Biol 7:35

Wagner S, Stuttmann J, Rietz S, Guerois R, Brunstein E, Bautor J, Niefind K, Parker JE (2013) Structural basis for signaling by exclusive EDS1 heteromeric complexes with SAG101 or PAD4 in plant innate immunity. Cell Host Microbe 14:619–630

Walker JC, Williams PH (1965) The inheritance of powdery mildew resistance in cabbage. Plant Dis Reptr 49:198–201

Wally O, Jayaraj J, Punja ZK (2009) Comparative resistance to foliar fungal pathogens in transgenic carrot plants expressing genes encoding for chitinase, beta–1, 3– glucanase and peroxidase. Eur J Plant Pathol 123:331–342

Walters DR, Ratsep J, Havis Neil D (2013) Controlling crop diseases using induced resistance: challenges for the future. J Exp Bot 64:1263–1280

Wan J, Zhang XC, Neece D, Ramonell KM, Clough S, Kim SY, Stacey MG, Stacey G (2008) A LysM receptor-like kinase plays a critical role in chitin signaling and fungal resistance in *Arabidopsis*. Plant Cell 20:471–481

Wang X (2004) Lipid signaling. Curr Opinion Plant Biol 7:329–336

Wang D, Amornsiripanitch N, Dong X (2006) A genomic approach to identify regulatory nodes in the transcriptional network of systemic acquired resistance in plants. PLOS Pathol 2:e123

Wang W, Devoto A, Turner JG, Xiao S (2007) Expression of the membrane-associated resistance protein RPW8 enhances basal defense against biotrophic pathogens. Mol Plant-Microbe Interact 20:966–976
Wang W, Wen Y, Berkey R, Xiao S (2009) Specific targeting of the *Arabidopsis* resistance protein RPW8.2 to the interfacial membrane encasing the fungal haustorium renders broad-spectrum resistance to powdery mildew. Plant Cell 21:2898–2913
Wang W, Berkey R, Wen Y, Xiao S (2010) Accurate and adequate spatiotemporal expression and localization of RPW8.2 is key to activation of resistance at the host-pathogen interface. Plant Signalling Behav 5:1002–1005
Wang Y, Nishimura MT, Zhao T, Tang D (2011a) ATG2, an autophagy-related protein, negatively affects powdery mildew resistance and mildew-induced cell death in *Arabidopsis*. Plant J 68:74–87
Wang Y, Wu Y, Tang D (2011b) The autophagy gene, *ATG18a*, plays a negative role in powdery mildew resistance and mildew-induced cell death in *Arabidopsis*. Plant Signalling Behav 6:1408–1410
Wang WM, Ma XF, Zhang Y, Luo MC, Wang GL, Bellizzi M, Xiong XY, Xiao SY (2012) PAPP2C interacts with the atypical disease resistance protein RPW8.2 and negatively regulates salicylic acid-dependent defense responses in *Arabidopsis*. Mol Plant 5:1125–1137
Wang W, Zhang Y, Wen Y, Berkey R, Ma X, Pan Z, Bendigeri D, King H, Zhang Q, Xiao S (2013) A comprehensive mutational analysis of the *Arabidopsis* resistance protein RPW8.2 reveals key amino acids for defense activation and protein targeting. Plant Cell 25:4242–4261
Wasteneys GO (2004) Progress in understanding the role of microtubules in plant cells. Curr Opinion Plant Biol 7:651–660
Wasteneys GO, Galway ME (2003) Remodeling the cytoskeleton for growth and form: an overview with some new views. Ann Rev Plant Biol 54:691–722
Wawrzynska A, Rodibaugh NL, Innes RW (2010) Synergistic activation of defense responses in *Arabidopsis* by simultaneous loss of the GSL5 callose synthase and the EDR1 protein kinase. Mol Plant-Microbe Interact 23:578–584
Weis C, Huckelhoven R, Eichmann R (2013) LIFEGUARD proteins support plant colonization by biotrophic powdery mildew fungi. J Exp Bot 64:3855–3867
Weis C, Hildebrandt U, Hoffmann T, Hemetsberger C, Pfeilmeier S, Konig C, Schwab W, Eichmann R, Huckelhoven R (2014) CYP83A1 is required for metabolic compatibility of *Arabidopsis* with the adapted powdery mildew fungus *Erysiphe cruciferarum*. New Phytol 202:1310–1319
Weymann K, Hunt M, Uknes S, Neuenschwander U, Lawton K, Steiner HY, Ryals J (1995) Suppression and restoration of lesion formation in *Arabidopsis* lsd mutants. Plant Cell 7:2013–2022
Wiermer M, Feys BJ, Parker JE (2005) Plant immunity-the EDS1 regulatory node. Curr Opinion Plant Biol 8:383–389
Williams PH, Walker JC, Pound GS (1968) Hybelle and Sanibel, multiple disease-resistant F1 hybrid cabbages. Phytopathology 58:791–796
Wilson IW, Schiff CL, Hughes DE, Somerville SC (2001) Quantitative trait loci analysis of powdery mildew disease resistance in the *Arabidopsis thaliana* accession Kashmir-1. Genetics 158:1301–1309
Winge P, Brembu T, Kristensen R, Bones AM (2000) Genetic structure and evolution of RAC-GTPases in *Arabidopsis thaliana*. Genetics 156:1959–1971
Wittstock U, Halkier BA (2002) Glucosinolate research in the *Arabidopsis* era. Trends Plant Sci 7:263–270
Wright DP, Baldwin BC, Shephard MC, Scholes JD (1995a) Source-sink relationships in wheat leaves infected with powdery mildew. II. Changes in the regulation of the Calvin cycle. Physiol Mol Plant Pathol 47:255–267
Wright DP, Baldwin BC, Shephard MC, Scholes JD (1995b) Source-sink relationships in wheat leaves infected with powdery mildew. I. Alterations in carbohydrate metabolism. Physiol Mol Plant Pathol 47:237–253

Wu Y, Zhang D, Chu JY, Boyle P, Wang Y, Brindle ID, De Luca V, Despres C (2012) The *Arabidopsis* NPR1 protein is a receptor for the plant defense hormone salicylic acid. Cell Rep 1:639–647

Wu G, Liu S, Zhao Y, Wang W, Kong Z, Tang D (2015) Enhanced disease resistance 4 associates with clathrin heavy chain 2 and modulates plant immunity by regulating relocation of edr1 in *Arabidopsis*. Plant Cell 27:857–873

Xiao S, Ellwood S, Findlay K, Oliver RP, Turner JG (1997) Characterization of three loci controlling resistance of *Arabidopsis thaliana* accession Ms-0 to two powdery mildew diseases. Plant J 12(4):757–768

Xiao S, Ellwood S, Calis O, Patrick E, Li T, Coleman M, Turner JG (2001) Broad-spectrum mildew resistance in *Arabidopsis thaliana* mediated by *RPW8*. Science 291:118–120

Xiao S, Brown S, Patrick E, Brearley C, Turner JG (2003) Enhanced transcription of the *Arabidopsis* disease resistance genes RPW8.1 and RPW8.2 via a salicylic acid–dependent amplification circuit is required for hypersensitive cell death. Plant Cell 15:33–45

Xiao S, Emerson B, Ratanasut K, Patrick E, O'Neill C, Bancroft I, Turner JG (2004) Origin and maintenance of a broad-spectrum disease resistance locus in *Arabidopsis*. Mol Biol Evol 21:1661–1672

Xiao S, Calis O, Patrick E, Zhang G, Charoenwattana P, Muskett P, Parker JE, Turner JG (2005) The atypical resistance gene, *RPW8*, recruits components of basal defence for powdery mildew resistance in *Arabidopsis*. Plant J 42:95–110

Xu X, Kanbara K, Azakami H, Kato A (2004) Expression and characterization of *Saccharomyces cerevisiae* Cne1p, a calnexin homologue. J Biochem 135(5):615–618

Xu T, Wen M, Nagawa S, Fu Y, Chen JG, Wu MJ, Perrot-Rechenmann C, Friml J, Jones AM, Yang Z (2010) Cell surface- and rho GTPase-based auxin signaling controls cellular interdigitation in *Arabidopsis*. Cell 143:99–110

Yan S, Dong X (2014) Perception of the plant immune signal salicylic acid. Curr Opinion Plant Biol 20:64–68

Yang Y, Shah J, Klessig DF (1997) Signal perception and transduction in plant defense responses. Genes Dev 11:1621–1639

Yang X, Wang W, Coleman M, Orgil U, Feng J, Ma X, Ferl R, Turner JG, Xiao S (2009) *Arabidopsis* 14-3-3 lambda is a positive regulator of RPW8-mediated disease resistance. Plant J 60:539–550

Yang L, Qin L, Liu G, Peremyslov VV, Dolja VV, Wei Y (2014) Myosins XI modulate host cellular responses and penetration resistance to fungal pathogens. Proc Natl Acad Sci U S A 111:13996–14001

Yedidia I, Benhamou N, Chet I (1999) Induction of defense responses in cucumber plants (*Cucumis sativus* L.) by the biocontrol agent *Trichoderma harzianum*. Appl Env Microbiol 65:1061–1070

Yoshimoto K, Jikumaru Y, Kamiya Y, Kusano M, Consonni C, Panstruga R, Ohsumi Y, Shirasu K (2009) Autophagy negatively regulates cell death by controlling NPR1-dependent salicylic acid signaling during senescence and the innate immune response in *Arabidopsis*. Plant Cell 21:2914–2927

Yu D, Chen C, Chen Z (2001) Evidence for an important role of WRKY DNA binding proteins in the regulation of NPR1 gene expression. Plant Cell 13:1527–1540

Yun BW, Atkinson HA, Gaborit C, Greenland A, Read ND, Pallas JA, Loake GJ (2003) Loss of actin cytoskeletal function and EDS1 activity, in combination, severely compromises non-host resistance in *Arabidopsis* against wheat powdery mildew. Plant J 34:768–777

Zablackis E, Huang J, Muller B, Darvill AG, Albersheim P (1995) Characterization of the cell wall polysaccharides of *Arabidopsis thaliana* leaves. Plant Physiol 107:1129–1138

Zander M, Chen S, Imkampe J, Thurow C, Gatz C (2012) Repression of the *Arabidopsis thaliana* jasmonic acid/ethylene-induced defense pathway by TGA-interacting glutaredoxins depends on their C-terminal ALWL motif. Mol Plant 5:831–840

Zander M, Thurow C, Gatz C (2014) TGA transcription factors activate the salicylic acid-suppressible branch of the ethylene-induced defense program by regulating *ORA59* expression. Plant Physiol 165:1671–1683

Zeller FJ, Lutz J, Reimlein E, Limpert E, Koenig J (1993) Identification of powdery mildew resistance genes in common wheat (*Triticum aestivum* L.) II. French cultivars. Agronomie 13:201–207

Zeyen RJ, Kruger WM, Lyngkjær MF, Carver TLW (2002) Differential effects of D-mannose and 2-deoxym-D-glucose on attempted powdery mildew fungal infection of inappropriate and appropriate gramineae. Physiol Mol Plant Pathol 61:315–323

Zhang C, Shapiro AD (2002) Two pathways act in an additive rather than obligatorily synergistic fashion to induce systemic acquired resistance and *PR* gene expression. BMC Plant Biol 2:9

Zhang Y, Tessaro MJ, Lassner M, Li X (2003) Knockout analysis of *Arabidopsis* transcription factors TGA2, TGA5, and TGA6 reveals their redundant and essential roles in systemic acquired resistance. Plant Cell 15:2647–2653

Zhang Z, Feechan A, Pedersen C, Newman MA, Qiu JL, Olesen KL, Thordal-Christensen H (2007) A SNARE-protein has opposing functions in penetration resistance and defence signalling pathways. Plant J 49:302–312

Zhang X, Francis MI, Dawson WO, Graham JH, Orbovic V (2010) Over-expression of the *Arabidopsis* NPR1 gene in citrus increases resistance to citrus canker. Eur J Plant Pathol 128:91–100

Zhang D, Liu D, Lv X, Wang Y, Xun Z, Liu Z, Li F, Lu H (2014) The cysteine protease CEP1, a key executor involved in tapetal programmed cell death, regulates pollen development in *Arabidopsis*. Plant Cell 26:2939–2961

Zhang LL, Ma XF, Zhou BB, Zhao JQ, Fan J, Huang F, Li Y, Wang WM (2015a) EDS1-mediated basal defense and SA-signaling contribute to post-invasion resistance against tobacco powdery mildew in *Arabidopsis*. Physiol Mol Plant Pathol 91:120–130

Zhang Q, Berkey R, Pan Z, Wang W, Zhang Y, Ma X, King H, Xiao S (2015b) Dominant negative RPW8.2 fusion proteins reveal the importance of haustorium-oriented protein trafficking for resistance against powdery mildew in *Arabidopsis*. Plant Signalling Behav 10:e989766

Zhao J, Last RL (1996) Coordinate regulation of the tryptophan biosynthetic pathway and indolic phytoalexin accumulation in *Arabidopsis*. Plant Cell 8:2235–2244

Zhao C, Nie H, Shen Q, Zhang S, Lukowitz W, Tang D (2014) EDR1 physically interacts with MKK4/MKK5 and negatively regulates a MAP kinase cascade to modulate plant innate immunity. PLoS Genet 10:e1004389

Zhong X, Xi L, Lian Q, Luo X, Wu Z, Seng S, Yuan X, Yi M (2015) The NPR1 homolog GhNPR1 plays an important role in the defense response of *Gladiolus hybridus*. Plant Cell Reptr 34:1063–1074

Zhou N, Tootle TL, Tsui F, Klessig DF, Glazebrook J (1998) PAD4 functions upstream from salicylic acid to control defense responses in *Arabidopsis*. Plant Cell 10:1021–1030

Zhou L-Z, Howing T, Muller B, Hammes UZ, Getl C, Dresselhaus T (2016) Expression analysis of KDEL CysEPs programmed cell death markers during reproduction in *Arabidopsis*. Plant Reprod 29:265–272

Zimmerli L, Stein M, Lipka V, Schulze-Lefert P, Somerville S (2004) Host and non-host pathogens elicit different jasmonate/ethylene responses in *Arabidopsis*. Plant J 40:633–646

Zimmermann P, Hirsch-Hoffmann M, Hennig L, Gruissem W (2004) Genevestigator. *Arabidopsis* microarray data base and analysis toolbox. Plant Physiol 136:2621–2632

Zipfel C, Robatzek S, Navarro L, Oakeley EJ, Jones JDG, Felix G, Boller T (2004) Bacterial disease resistance in Arabidopsis through flagellin perception. Nat Intl J Sci 428:764–767

Chapter 8
Disease Management

8.1 Introduction

In a comprehensive review, Yarwood (1957) has emphasized the control of powdery mildew by exclusion, eradication, protection, immunization, and therapy. Out of these methods, chemicals and genetic manipulation are perhaps the most important at present time considering the nature and mechanisms of the pathogen to take epidemic forms within a very short period of time after initiation of the disease. The future predictions of climate change worldwide with rise in temperature and dry spell may aggravate the situation to develop powdery mildews in epidemic forms. Under such conditions, chemical control is the best option to manage the disease right from initiation stage. However, growing of resistant cvs. takes the priority wherever they have been developed or available to protect from environmental pollution and save the cost of cultivation which may be imposed in the form of cost of chemicals and labour. Alternatives like cultural control based on sowing time to avoid the favourable period for multiplication of the pathogen has also been advocated and effective. Some micronutrients and biological control agents are also helpful in managing powdery mildew of crucifers. Hyperparasites of powdery mildews may be exploited as a biological control agent. Cultural practices like mixed cropping and intercropping may also be a useful preposition.

8.2 Chemical (Fungicidal) Control

Several chemicals have been tested against powdery mildews including powdery mildew of crucifers. Some of them have shown their efficacy in managing the disease of crucifers and increasing the yield of protected crops (Tables 8.1 and 8.2). For efficient management of the disease, fungicides should be sprayed soon after the appearance of the disease since it spreads very fast after its initiation. All the

G. S. Saharan et al., *Powdery Mildew Disease of Crucifers: Biology, Ecology and Disease Management*, https://doi.org/10.1007/978-981-13-9853-7_8

Table 8.1 Chemicals (fungicides) tested against powdery mildews

Chemicals[a]	References
Acrex	Singh et al. (1980)
Actidione	Felber and Hammer (1948) and Yarwood (1957)
Ammoniacal copper carbonate	Bewley (1923) and Yarwood (1957)
Ammonium sulphide	Eyre et al. (1919) and Yarwood (1957)
Apron 35 SD	Meena et al. (2011, 2013)
Arsenates	Szembel (1930) and Yarwood (1957)
Bavistin	Singh et al. (1980) and Ashraf and Yadav (2009)
Baycor	Singh et al. (1980)
Bayleton	Singh et al. (1980)
Baytan	Singh et al. (1980)
Benomyl	Agrios (1978)
Blitox-50	Ashraf and Yadav (2009)
Borax	Kumar et al. (2015)
Bordeaux mixture	Yarwood (1942, 1957)
Burgundy mixture	Horne (1918) and Yarwood (1957)
Ca Co_3	Singh (2004) and Singh et al. (2010)
Calixin	Singh and Solanki (1974), Sharma et al. (1990), Saharan and Sheoran (1986), Jani et al. (1991), Saharan et al. (2005), Ashraf and Yadav (2009), Saharan (1992, 1998) and Saharan and Mehta (2002)
Carbendazim	Penaud (1998), Meena et al. (2011, 2013) and Kumar et al. (2015)
Chloranil	Middleton (1943) and Yarwood (1957)
Chlorothalonil	Lomate et al. (2014)
Copper-lime dust	Davis and Blair (1938) and Yarwood (1957)
Copper sulphate	Radclyffe (1861) and Yarwood (1945)
Cycloheximide	Agrios (1978)
Dichloronaphthoquinone	Yarwood (1951, 1957)
Difenoconazole	Penaud (1998) and Patel and Patel (2008)
Dinitro capryl phenyl crotonate	Yarwood (1951, 1957)
Dinocap	Agrios (1978)
Dinocap	Patel et al. (1995a, b), Singh and Singh (2003), Shete et al. (2008) and Lomate et al. (2014)
Elgetol	Rowe (1944) and Yarwood (1957)
Elosal	Singh and Solanki (1974)
Fermate	Middleton (1943) and Yarwood (1957)
Flusilazole	Penaud (1998)
Formalin	Wenzi (1949) and Yarwood (1957)
Hexaconazole	Ashraf and Yadav (2009)
Iprodione	Penaud (1998) and Kumar et al. (2015)

(continued)

Table 8.1 (continued)

Chemicals[a]	References
Karathane	Singh and Solanki (1974), Kaur (1981), Saharan and Sheoran (1986), Sharma et al. (1990), Jani et al. (1991), Singh (2004), Saharan (1992, 1998), Saharan and Mehta (2002) and Saharan et al. (2005)
Liquid lime sulphur	Fisher (1938) and Yarwood (1957)
Lithium carbonate	Vidali (1951) and Yarwood (1957)
Malachite green	Mcllellan (1947) and Yarwood (1957)
Mancozeb	Kumar et al. (2015)
Maneb	Lomate et al. (2014)
Manzate	Thomas and Holtzmann (1951) and Yarwood (1957)
Metalaxyl	Meena et al. (2011)
Morestan	Singh and Solanki (1974)
Morocide	Singh and Solanki (1974)
Nimbicidine	Singh (2004) and Singh et al. (2010)
Penconazole	Shete et al. (2008) and Lomate et al. (2014)
Potassium permanganate	Guillon (1903)
Propiconazole (tilt)	Kumar et al. (2015)
Ridomil MZ-72	Meena et al. (2013) and Kumar et al. (2015)
Rizolex	Jani et al. (1991)
Salicylic acid	Singh (2004) and Singh et al. (2010)
Sodium bicarbonate	Curry (1924) and Yarwood (1957)
Sodium chloride	Jorstad (1925) and Yarwood (1957)
Sodium thiosulphate	Yarwood (1951, 1957)
Spergen	Middleton (1943) and Yarwood (1957)
Sulfex	Singh et al. (1980), Saharan and Sheoran (1986), Singh (2004), Saharan et al. (2005), Patel and Patel (2008) and Saharan (1992, 1998)
Sulphuric acid	Guba (1928); Yarwood (1957)
Sulphur dust	Kyle (1841); Yarwood (1957); Agrios (1978)
Sulphur WP	Kaur (1981); Jani et al. (1991); Patel et al. (1995a, b); Singh and Singh (2003); Shete et al. (2008); Kurowski et al. (2010); Ashraf and Yadav (2009); Lomate et al. (2014); Kumar et al. (2015)
Tebuconazole (Folicur 48 EC)	Patel and Patel (2008)
Thiovit	Singh and Solanki (1974) and Jani et al. (1991)
Topsin	Jani et al. (1991), Saharan and Sheoran (1986), Ashraf and Yadav (2009) and Saharan (1992, 1998)
Triadimefon	Singh and Singh (2003), Shete et al. (2008), Patel and Patel (2008) and Lomate et al. (2014)
Tridemorph	Patel et al. (1995a, b), Singh and Singh (2003), Shete et al. (2008) and Lomate et al. (2014)
Vegetable oils	Martin and Salmon (1930) and Yarwood (1957)
Vulcanized rubber	Brodie and Neufeld (1942) and Yarwood (1957)
Zinc sulphate	Kumar et al. (2015)

[a]Name of the chemical is as reported in original publications

Table 8.2 Effect of time and number of fungicidal sprays on powdery mildew of mustard (Saharan and Sheoran 1986)

Fungicide	Concentration (%)	Disease intensity (%) (T_1)	Reduction in disease (%)	Disease intensity (%) (T_2)	Reduction in disease (%)	Disease intensity (%) (T_3)	Reduction in disease (%)
Sulfex	0.20	32.0	65.2	70.0	30.0	7.5	92.5
	0.30	30.0	67.3	70.0	30.0	5.2	94.8
Topsin M	0.05	63.0	31.5	82.0	18.0	14.5	85.5
	0.10	62.5	32.0	80.5	19.5	12.5	87.5
Karathane	0.05	26.6	71.0	63.4	36.6	6.2	93.8
	0.10	24.7	73.1	60.2	39.8	4.3	95.7
Calixin	0.05	36.1	60.7	72.3	27.7	8.4	91.6
	0.10	35.0	61.9	70.4	39.6	5.2	94.8
Check	–	92.0	–	100.0	–	100.0	–

T_1 = Only one spray was given at the time of appearance of the disease
T_2 = Only one spray was given after 10 days of appearance of the disease
T_3 = Two sprays were given, first at the time of appearance of the disease and second at the intervals of 10 days

four fungicides tested by Saharan and Sheoran (1986) against powdery mildew of mustard were found effective in reducing the disease intensity at both the concentrations used (Table 8.2). However, higher concentration is more effective in controlling the disease. Karathane (0.1%) was found most effective followed by Sulfex (0.3%) and Calixin (0.1%) in reducing the disease intensity and increasing the yield. Singh and Solanki (1974) also reported that powdery mildew in *Brassica juncea* under Rajasthan, India, conditions can be controlled with three sprays of Karathane (0.2%). However, under Haryana (India) conditions, two sprays of Karathane (0.1%) at 10-day intervals are sufficient to manage the disease. Similar results have been obtained by Sharma et al. (1990) who found that Karathane (0.1%) gave maximum control of the disease closely followed by Calixin (0.1%) with higher yields. Time and number of sprays are very important for effective and economical control of the disease. There is 95.7% reduction in disease when Karathane (0.1%) is sprayed twice at an interval of 10 days immediately after appearance of the disease. However, when first spray is delayed by 10 days after appearance of the disease, only 39.8% reduction in disease intensity is obtained. Moreover, when only one spray is given at the initiation of the disease, it reduces the disease to the extent of 73.1% (Saharan and Sheoran 1986). Karathane and wettable sulphur can persist in the plant for 2 weeks to control the disease (Kaur 1981).

According to Singh and Singh (2003), spray of triadimefon 25 WP (0.1%) suspension thrice at fortnightly intervals, beginning at the initiation of the symptoms, @ 1000 l/ha resulted in lowest disease severity and highest yield. Taking it as a measure of protection, 42.4% avoidable yield loss due to this disease and 17.5% loss in 1000 seed weight have been recorded. Lower loss in test weight indicated role of other yield contributing characters in the overall higher yield loss due to the disease. All the fungicidal treatments reduced powdery mildew, and increased yield over check. During 1995–1996, even water sprays alone also reduced powdery mildew and increased yield over unsprayed check. In triadimefon, treatment, PDI 15.1 and 11.9% and yield 13.7 and 14.2 q/ha were recorded during the years 1995–1996, respectively. It gave significantly superior performance over all other treatments. Tridemorph, wettable sulphur, and dinocap were at par in managing the disease, but wettable sulphur with 12.2 and 12.9q/ha yield during the 2 years out yielded dinocap and tridemorph significantly. The highest 1000 seed weight was recorded in triadimefon-treated plants.

Triadimefon gave the highest net additional income of Rs. 5618.0/ha followed by wettable sulphur of Rs. 4264.3. However, the latter gave the most favourable cost-benefit ratio. Period of spray of wettable sulphur was increased from 10 days (Singh and Solanki 1974) to 15 days due to persistence of wettable sulphur for more than 2 weeks to control powdery mildew (Kaur 1981).

Patel and Patel (2008) studied health status of mustard under protected conditions compared to powdery mildew-affected crop. Out of five fungicides, five plant extract, and one biocontrol agent (*Trichoderma viride*) used as spray to control powdery mildew, all the five fungicides reduced abnormality significantly in the seeds. However, among the plant extracts, and biocontrol agents, only *T. viride* was found to be better in reducing abnormality. Significant better germination, seedling length, and vigour index were recorded in fungicides and *T. viride*-sprayed plots (Table 8.3).

The fungicide dinocap showed significantly least incidence (16.70%) and intensity (16.86%) of powdery mildew of mustard with maximum disease control of 79.14%. It was followed by tridemorph in which the per cent incidence, intensity, and disease control were 19.05%, 19.78%, and 75.52%, respectively. The next treatments in order of superiority were penconazole, triadimefon, and wettable sulphur with 71.25%, 72:28%, and 72.51% powdery mildew control, respectively (Shete et al. 2008). The efficacy of dinocap and tridemorph against powdery mildew has also been recorded by Singh and Chauhan (1998) and Khodke and Kakade (2004).

The application of chemical fertilizers showed significantly higher incidence (31.20%) and intensity (32.06%) of powdery mildew as compared to FYM application alone. The interaction between nutrients and fungicides (Table 8.4) was statistically significant. The studies showed that interaction between FYM and dinocap showed significantly least incidence (14.87%) and intensity (14.91%) of powdery mildew and thus highest disease control of 81.13%. The interaction between FYM and tridemorph was superior, which recorded 76.83% disease control. Fungicides and nutrients have positive correlation with all yield contributing parameters. The highest 1000 seed weight (3.15 g), plant yield (17.50 g/plant), and total yield (1551 kg/ha) were obtained in dinocap treatment, wherein a maximum yield increase of 92.67% was noticed (Table 8.5). The next treatments in order of superiority were wettable sulphur, triadimefon, and penconazole. Dange et al. (2002) also reported the reduction in powdery mildew incidence and increase in yield of mustard with fungicidal control.

Among the nutrients, chemical fertilizers recorded significantly higher seed weight (15.75 g/plant) and total yield (1356.50 kg/ha) as against FYM application. The interaction between dinocap and chemical fertilizers produced maximum 1000

Table 8.3 Effect of powdery mildew severity on seed health status of mustard (Patel and Patel 2008)

Treatments	Conc. (%)	PDI	No. of abnormal seeds/100 seeds	Percent reduction over control	Seed germination (%)	Seedling length (cm)	Seedling vigour index
Hexaconazole (Contaf 5EC)	0.05	35.00	18.66	45.65	92.75	12.86	1192.8
Wettable sulphur (Sulfex 80WP)	0.20	38.67	18.33	46.61	91.75	12.53	1149.6
Tebuconazole (Folicur 25 EC)	0.05	35.33	19.00	44.65	90.75	12.74	1156.2
Tridemorph (Calixin 48EC)	0.04	23.33	18.00	47.57	92.75	12.81	1188.1
Difenoconazole (Score 25EC)	0.05	51.67	19.66	42.73	90.50	12.70	1149.4
Onion bulb extract (*Allium cepa* L.)	5.00	78.33	32.66	4.86	86.00	12.50	1075.0
Neem leaf extracts (*Azadirachta indica* A. Juss.)	5.00	74.00	30.66	10.69	85.25	11.99	1022.1
Eucalyptus leaf extracts (*Eucalyptus globulus* L.)	5.00	74.67	29.00	15.53	85.75	12.36	1059.9
Karan leaf extracts (*Nerium indicum* L.)	5.00	73.33	30.00	12.61	86.00	12.25	1053.5
Karanj leaf extracts (*Pongamia pinnata* L.)	5.00	76.33	33.00	3.87	84.75	12.45	1055.1
Trichoderma viride suspension	3.00	65.67	23.33	32.04	87.75	12.51	1097.8
Untreated check	–	80.67	34.33	–	85.50	12.18	1041.4
S Em±		2.55	1.02		00.80	00.17	
CD 0.05		7.48	3.01		02.49	00.51	
CV%		7.50	6.96		01.96	02.42	

seed weight (3.16 g), plant yield (19.75 g/plant), and total yield (1685.75 kg/ha) (Shete et al. 2008). Cost-benefit ratio calculation showed the maximum net income of Rs. 13,437/ha with additional income of Rs. 9374.87/ha with fungicide dinocap, which was followed by tridemorph in respect of net income. However, the maximum B:C ratio of 2.34 was obtained in wettable sulphur treatments followed by tridemorph (2.13) and dinocap (2.10). Furthermore, net income (Rs. 12,348.50/ha), additional income (Rs. 7408.80/ha), and B:C ratio (1.75) was obtained in chemical fertilizers as compared to FYM. The interaction between wettable sulphur and

Table 8.4 Effect of nutrients and fungicides on powdery mildew of mustard (Shete et al. 2008)

		Percent powdery mildew								
		Incidence			Intensity			Control		
Fungicides	Conc. %	M1	M2	Mean	M1	M2	Mean	M1	M2	Mean
Sulphur WP	0.25	18.43	22.52	20.47	22.73	21.71	22.22	71.23	73.72	72.51
Triadimefon	0.10	22.41	22.13	22.27	20.38	24.41	22.40	74.21	70.45	72.28
Penconazole	0.10	20.85	20.08	20.46	22.13	23.53	22.83	71.99	71.52	71.75
Tridemorph	0.10	18.08	20.02	19.05	18.31	21.24	19.78	76.83	74.29	75.52
Dinocap	0.10	14.87	18.53	16.70	14.91	18.81	16.86	81.13	77.23	79.14
Control	–	82.70	83.96	83.33	79.03	82.63	80.83	–	–	–
Mean	–	29.56	31.20	–	29.58	32.06	–	–	–	–

Source	SE±	CD 5%	SE±	CD 5%
Nutrients (N)	0.33	1.49	0.20	0.93
Fungicides (F)	0.72	2.09	0.47	1.38
N × F	1.02	2.96	0.67	1.95

M1 FYM as source of nutrients, *M2*: fungicidal application

chemical fertilizer yielded the highest net income (Rs. 17,833/ha) and B:C ratio (2.90). These higher values of economic returns in the above interaction are due to low cost of wettable sulphur; similarly higher yields in chemical fertilizers as compared to FYM were obtained (Table 8.6). Out of six fungicides (Table 8.7) tested by Ashraf and Yadav (2009), hexaconazole at the rate of 0.2% was found most effective with 66% reduction in disease intensity along with higher seed yield of Indian mustard.

During 1999–2005, Chattopadhyay et al. (2007) tested garlic (*Allium sativum*) bulb aqueous extract (1%) as seed treatment and foliar spray alone or in combination with fungicides for the management of powdery mildew of mustard along with other diseases. Treatment with Apron 35 SD + carbendazim was found most effective in reducing powdery mildew followed by carbendazim treatment (Table 8.8). In an experiment conducted at 8 and 11 locations during 2006–2009, Meena et al. (2011, 2013) found that ecofriendly treatments with garlic bulb extract, *Trichoderma harzianum* as seed treatment alone or in combination with foliar spray by garlic aqueous extract, *T. harzianum*, and *Pseudomonas fluorescens* are effective for powdery mildew control.

Foliar sprays with chemical fungicides did significantly better than the non-chemical fungicides against powdery mildew. Similar results were observed when seed treated with garlic bulb extract followed by foliar spray with either garlic bulb extract or spray with *P. fluorescens* or *T. harzianum*. *T. harzianum* as seed treatment in combination with *Pseudomonas fluorescens* spraying was significantly superior for powdery mildew control (Table 8.9). The combination of treatment with *T. harzianum* followed by its use as a foliar spray and seed treatment with foliar spraying of *A. sativum* bulb extract gave higher plant stand, higher yield, and higher cost-benefit ratio (Tables 8.10, 8.11, and 8.12).

Lomate et al. (2014) evaluated the efficacy of fungicides and bioagents on percent disease infection, percent disease control, and percent disease intensity of pow-

Table 8.5 Effect of nutrients and fungicides on yield and yield contributing parameters of mustard (Shete et al. 2008)

		Yield contributing parameters											
		1000 seed weight (g)			Yield/plant (g)			Yield kg/ha			Yield increase (%)		
Fungicides	Conc. %	M1	M2	Mean	M1	M2	Mean	M1	M2	Mean	M1	M2	Mean
Sulphur WP	0.25	3.09	3.11	3.10	14.75	16.00	15.37	1148.0	1599.0	1373.5	**50.26**	88.78	70.62
Triadimefon	0.10	3.02	3.10	3.06	13.25	16.00	14.62	1231.5	1449.0	1340.1	61.19	71.07	66.47
Penconazole	0.10	3.08	3.10	3.09	12.75	14.00	13.37	1037.0	1169.2	1103.1	35.73	38.04	37.03
Tridemorph	0.10	3.07	3.14	3.11	15.25	18.50	17.37	1416.2	1389.0	1438.0	85.37	63.99	78.63
Dinocap	0.10	3.15	3.16	3.15	16.25	19.75	17.50	1487.0	1685.7	1551.0	94.63	99.02	92.67
Control	–	2.45	2.49	2.47	9.75	10.25	10.00	764.0	847.0	805.0	–	–	–
Mean	–	2.98	3.01	–	13.66	15.75	–	1180.6	1356.5	–	–	–	–

Source	SE±	CD 5%	SE±	CD 5%	SE±	CD 5%
Nutrients (N)	0.01	0.07	0.07	0.34	15.11	68.02
Fungicides (F)	0.02	0.06	0.36	1.04	39.90	115.24
N × F	0.03	0.09	0.50	1.47	56.43	162.98

M1 FYM as source of nutrients, *M2* fungicidal application

Table 8.6 Cost-benefit ratio as influenced by nutrients and fungicides in management of powdery mildew of mustard (Shete et al. 2008)

		Income/ha								
		Net income (Rs./ha)			Additional income (Rs./ha)			BC ratio		
Fungicides	Conc. %	M1	M2	Mean	M1	M2	Mean	M1	M2	Mean
Sulphur WP	0.25	8556.0	17,833.0	13,194.0	5403.0	11,659.0	8531.0	1.78	2.90	2.34
Triadimefon	0.10	7125.5	12,433.0	9729.2	3972.5	6259.0	5115.5	1.51	2.01	1.76
Penconazole	0.10	4044.0	7902.2	5973.0	891.0	1728.2	1309.6	1.29	1.65	1.47
Tridemorph	0.10	12,066.2	13,216.0	13,241.0	8913.5	7039.0	7976.1	2.00	2.27	2.13
Dinocap	0.10	11,544.0	16,532.7	13,437.0	8391.0	10,358.7	9374.9	1.84	2.36	2.10
Control	–	3153.0	6174.0	4663.0	–	–	–	1.32	1.75	1.53
Mean	–	7748.1	12,348.5	–	5514.2	7408.8	–	–	–	–

M1 FYM as source of nutrients, *M2*: fungicidal application

Table 8.7 Effect of different fungicides on powdery mildew of mustard (Ashraf and Yadav 2009)

Treatments	Conc.	Disease intensity (%)	Disease reduction over control (%)
Calixin	0.2	33.4	62.88
Hexaconazole	0.2	30.6	66.0
Bavistin	0.1	46.5	48.3
Wettable sulphur	0.2	40.3	55.2
Topsin M	0.2	63.7	29.2
Blitox-50	0.2	49.9	44.5
Control	–	100.0	–
CD at 0.05%	–	0.69	–

Table 8.8 Effect of different treatments on powdery mildew of Indian mustard (Chattopadhyay et al. 2007)

Treatment	Disease reduction over control (%)
Garlic (ST)	50.6
Apron 35 SD (ST)	22.0
Carbendazim (ST)	16.0
Apron 35 SD (ST) + carbendazim (ST)	13.0
Garlic (ST + FS)	67.7
Apron 35 SD (ST) + Ridomil (FS)	29.0
Mancozeb (FS)	59.6
Carbendazim (FS)	DUA
Carbendazim (ST + FS)	DUA
Carbendazim (ST) + Ridomil (FS)	34.0

ST treatment, *FS* foliar spray DUA: data unavailable mean of 7 locations × 2 years

dery mildew of Indian mustard after first and second spraying. Two sprays of dinocap minimized the powdery mildew (76.29%) significantly as compared to all other treatments followed by tridemorph (74.18%), wettable sulphur (73.80%), and triadimefon (72.72%). With bioagents, maximum disease was reduced with *Ampelomyces quisqualis* (65.53%) which proved better than *T. harzianum* (61.65%). Highest seed yield (1023.62 kg/ha) and 1000 seed weight (5.5 g) of mustard was recorded by dinocap treatment. The bioagent *Ampelomyces quisqualis* gave higher seed yield (810.13 kg/ha) and 1000 seed weight (3.9 g) as compared to *T. harzianum* (Tables 8.13, 8.14, and 8.15).

Kumar et al. (2015) found treatment with iprodione + carbendazim (1:1) at the rate of 2 g/kg as most effective in reducing powdery mildew intensity with increase in yield and 1000 seed weight. Removal of three lower leaves also helps in reducing powdery mildew. Use of Zn SO_4 at the rate of 15 kg/ha + sulphur as per recommendations was most effective for the management of powdery mildew.

Powdery mildew caused by *E. cruciferarum* is a common disease of oilseed rape crops grown in the south of France. Field experiments were carried out to assess the

Table 8.9 Effect of different treatments on powdery mildew of Indian mustard during 2006–2009 over eight locations (Meena et al. 2013)

Treatments	Powdery mildew severity (%)
Aqueous garlic extract @1% (w/v)- ST[a]	28.4 (27.0)
Metalaxyl @ 1 or 6 g/kg Apron 35SD-ST	29.3 (28.3)
Carbendazim@1 g a.i. or 2 g/kg Bavistin 50 WP-ST	31.0 (31.0)
Apron 35SD@6 g/kg + carbendazim @1 g a.i. or 2 g/kg Bavistin 50 WP-ST	30.3 (30.1)
Trichoderma harzianum @ 10 g/kg -ST	31.2 (31.2)
Trichoderma harzianum @ 10 g/kg –ST + *Pseudomonas fluorescens* (oil based) @ 10 ml/l water -FS	28.5 (27.0)
Aqueous garlic extract @1% (w/v)- ST+ aqueous garlic extract @1% (w/v)- FS[a]	27.9 (27.2)
Metalaxyl @ 1 or 6 g/kg Apron 35 SD -ST+ Metalaxyl + Mancozeb or Ridomil MZ-72 WP @2 g/l-FS	27.6 (26.2)
Carbendazim@1 g a.i. or 2 g/kg Bavistin 50 WP-ST+ Ridomil MZ -72WP@2 g/l FS	25.8 (23.7)
Trichoderma harzianum @ 10 g/kg – ST+ *Trichoderma harzianum* @ 10 ml/l -FS	29.1 (26.5)
Trichoderma harzianum @ 10 g/kg – ST + *T. harzianum* + *T. viride* @ 2 g/l –FS at 50 days after sowing	35.0 (33.4)
Oxydematon-methyl@ 1 ml/L-FS when aphid population reaches ETL	30.4 (30.5)
Control	35.4 (38.5)
CD ($P < 0.05$)	3.6

[a]Pooled data of 4 years and eight locations. Figures in parenthesis are arc sin transformed values

yield losses and to evaluate the efficacy of fungicides. During the 1989–1993 periods, and in 1998, yield losses were quantified using an effective chemical control and an untreated check in 20 field trials. Powdery mildew has resulted in crop losses, estimated to be 0.5 tonnes per hectare. Several fungicides were evaluated; the treatment flusilazole + carbendazim (200 + 100 g a.i./ha) was the most effective one. These results are considered in developing spraying recommendations for farmers. Over several years, powdery mildew caused by the obligate fungus *E. cruciferarum* was regularly observed on winter oilseed rape grown in the south of France (Penaud 1991). Disease symptoms appear in autumn on leaf surface as small, white powdery spots which as they increase in size coalesce to cover the whole leaf surface. In spring, the powdery mildew covers first the lower leaves, and gradually it develops on stems and on pods. Severely diseased pods produce small siliquae. The development of the disease is influenced by weather factors such as temperature and rain which directly affect growth of the fungus on leaf or pod surface. All the cultivars are currently susceptible to powdery mildew infection. There is no information on the effect of the disease on yield, but some increases in production were reported from plots where powdery mildew was controlled by fungicides.

Table 8.10 The effect of different treatments on the powdery mildew severity (percent) in Indian mustard 90 days after sowing (Meena et al. 2013)

Treatment	2006–2007[a]	2007–2008[b]	2008–2009[c]	Pooled mean
Garlic bulb extract 1% w/v (ST)	23.6 (17.1)	36.8 ^{rs}(35.2)	20.4 ^{pq}(12.1)	27.6 ^{pq} (21.5)
Apron 35 SD 6 g/kg (ST)	28.5 (24.5)	35.8 ^{qrs}(33.3)	23.1 ^{rst}(15.4)	29.6 ^{pq} (24.4)
Carbendazim 1 g a.i. (ST)	31.5 (29.5)	36.6 ^{qrs}(36.0)	24.1^{stu} (16.7)	31.6 ^{qr}(27.4)
Apron 35 SD 6 g/kg + carbendazim 1 g a.i. (ST)	30.0 (26.9)	35.3 ^{qr} (34.3)	24.5 ^{tu}(17.2)	30.7^{pqr} (26.1)
Trichoderma harzianum 10 g/kg (ST)	29.7 (26.4)	35.9 ^{qrs} (34.7)	23.8 ^{s}(16.3)	30.5^{pqr} (25.8)
T. harzianum (ST) + *P. fluorescens* 10 ml/l (FS)	26.8 (22.1)	30.6^{qrs} (26.3)	24.0 ^{t}(16.2)	27.6 ^{pq} (21.5)
T. harzianum (ST) + *T. harzianum* 10 ml/l (FS)	29.6 (26.8)	29.3^{pq} (24.1)	18.9 ^{P}(10.5)	26.9 ^{pq} (20.5)
Garlic bulb extract (ST) + garlic bulb extract (FS)	30.4 (28.4)	29.5^{qrs} (24.3)	21.1^{pqr} 13.0)	27.9 ^{pq} (21.9)
Apron 35 SD (ST) + Ridomil MZ 72 WP 2 g/l (FS)	27.0 (22.5)	27.6 ^{P}(22.2)	21.9^{qrs} 14.0)	26.2 ^{p} (19.6)
Carbendazim (ST) + Ridomil MZ 72 WP 2 g/l (FS)	28.2 (24.6)	32.3^{qrs} (28.9)	21.1^{pqr} 13.0)	28.1 ^{pq}(22.2)
Control	34.9 (34.9)	42.9 ^{s}(46.3)	26.4^{u} (19.8)	35.5 ^{r} (33.7)
LSD ($P < 0.05$)	NS	7.5	2.4	5.1

The figures in parentheses are the actual means of disease severity and the other figures are angular transformed values; same letter in superscript indicates no significant difference among data within the column. *ST* treatment, *FS* foliar spray, *cv*. Varuna, *NS* not significant

[a]Mean of three replications at each of the following locations: Faizabad, Morena, Ludhiana, and Navgaon

[b]Mean of three replications at each of the following locations: Faizabad, Jagdalpur, and Kanpur

[c]Mean of three replications at Jagdalpur

In France four fungicides, viz. flusilazole + carbendazim, difenoconazole + carbendazim, carbendazim, and iprodione + carbendazim, were tested against powdery mildew of oil seed rape by Penaud (1999). All the fungicides in combination or alone provided substantial disease control compared to untreated plots. However, flusilazole + carbendazim (200 + 100 g a.i./ha) has been regularly observed as the most effective one against powdery mildew. Difenoconazole + carbendazim and iprodione + carbendazim have also given good control, but both appeared slightly less effective than flusilazole + carbendazim (200 + 100 g a. i./ha). Carbendazim used alone showed large variability in its efficacy (Table 8.16). Fungicide treatments have shown marked increase in yield of oilseed rape (Table 8.17 and Fig. 8.1). For a satisfactory disease control, two applications of fungicides are necessary, the first one when powdery mildew is appearing, mainly at the flowering stage, and the second after 2–4 weeks to control pod infections (Penaud 1999).

Table 8.11 The effect of different treatments on the yield (q/ha) of Indian mustard (Meena et al. 2013)

Treatment	2006–2007[a]	2007–2008[b]	2008–2009[c]	Pooled mean
Garlic bulb extract 1% w/v (ST)	11.6^{qrs}	12.9^{t}	16.1^{pqrs}	13.5^{rs}
Apron 35 SD 6 g/kg (ST)	11.1^{rs}	13.0^{rst}	15.2^{rst}	13.1st
Carbendazim 1 g a.i. (ST)	11.3^{rs}	13.1^{rst}	15.0st	13.1st
Apron 35 SD 6 g/kg + carbendazim 1 g a.i. (ST)	11.7^{qrs}	13.7^{rst}	15.4^{qr}	13.6^{qrs}
Trichoderma harzianum 10 g/kg (ST)	11.6^{qrs}	12.9st	15.2^{rst}	13.2^{rst}
T. harzianum (ST) + *P. fluorescens* 10 ml/l (FS)	12.2^{qr}	13.5^{rst}	16.2^{Pqr}	14.0^{Pqr}
T. harzianum (ST) + *T. harzianum* 10 ml/l (FS)	12.2^{qr}	14.2^{Pqr}	16.9^{P}	14.4^{pqr}
Garlic bulb extract (ST) + garlic bulb extract (FS)	13.4^{P}	15.4^{P}	16.8^{P}	15.2^{P}
Apron 35 SD (ST) + Ridomil MZ 72 WP 2 g/l (FS)	12.5^{Pq}	15.1^{Pq}	16.9^{P}	14.8^{Pq}
Carbendazim (ST) + Ridomil MZ 72 WP 2 g/l (FS)	11.4^{qrs}	13.9^{qrs}	16.4^{Pq}	13.9^{qrs}
Control	10.6^{s}	11.2^{u}	14.2^{t}	12.0^{t}
LSD ($P < 0.05$)	1.2	1.3	1.2	1.3

ST treatment, *FS* foliar spray, *cv.* Varuna; same letter in superscript indicates no significant difference among data within the column

[a]Mean of three replications at each of the following locations: Faizabad, Morena, Pantnagar, Ludhiana, and Kanpur

[b]Mean of three replications at each of the following locations: Sri Ganganagar, Faizabad, Morena, Pantnagar, Navgaon, Jagdalpur, Kanpur, Hisar, Dholi, and SK Nagar

[c]Mean of three replications at each of the following locations: Sri Ganganagar, Faizabad, Morena, Pantnagar, Ludhiana, Navgaon, and Jagdalpur

8.3 Effect of Nitrogen and Fungicides on Quality of Oilseed Rape

Nitrogen fertilization increased the protein, but lowered the oil content of the seed of oilseed rape under Turkey conditions. Fungicidal treatments significantly increased oil contents in all the five varieties tested, but it reduced protein levels in fertilized and non-fertilized plots. The level of linolenic acid did not change significantly in any plots of any treatment combinations, and similar result was observed in the level of oleic acid in most of the genotypes. Nitrogen fertilization increased GSL and SAE levels, whereas fungicidal treatment had no effect (Table 8.18). These findings demonstrated that nitrogen fertilization can markedly influence some quality parameters in oilseed rape. However, the application of fungicides reduced side effects of nitrogen fertilizer and resulted in a reduction on GSL, SAE, and protein contents but an increase in total oil and oleic acid contents (Table 8.19) (Mert-Turk et al. 2008). Disease severity was the highest in fertilized plots with no fungicide application. Fungicide application significantly reduced disease on overall plant

Table 8.12 The effect of different treatments on the yield (q/ha) and B:C ratio of Indian mustard (Meena et al. 2013)

Treatment	Yield[a]	Additional cost (INR)	Total cost (INR)	Total returns (INR)	Cost-benefit ratio	% change in cost-benefit ratio over control
Garlic bulb extract 1% w/v (ST)	13.5	250	18,400	23,531	1:28	10.97
Apron 35 SD 6 g/kg (ST)	13.1	105	18,255	22,833	1:25	8.54
Carbendazim 1 g a.i. (ST)	13.1	7	18,157	22,833	1:26	9.12
Apron 35 SD 6 g/kg + carbendazim 1 g a.i. (ST)	13.6	112	18,262	23,705	1:30	12.64
Trichoderma harzianum 10 g/kg (ST)	13.2	75	18,225	23,008	1:26	9.55
T. harzianum (ST) + *P. fluorescens* 10 ml/l (FS)	14.0	430	18,580	24,402	1:31	13.97
T. harzianum (ST) + *T. harzianum* 10 ml/l (FS)	14.4	430	18,580	25,099	1:35	17.22
Garlic bulb extract (ST) + garlic bulb extract (FS)	15.2	1470	19,620	26,494	1:35	17.18
Apron 35 SD (ST) + Ridomil MZ 72 WP 2 g/l (FS)	14.8	3285	21,435	25,796	1:20	4.43
Carbendazim (ST) + Ridomil MZ 72 WP 2 g/l (FS)	13.9	3187	21,337	24,228	1:14	−1.47
Control	12.0	0	18,150	20,916	1:15	NA

[a]The aggregated mean of the yield over all locations and years listed in Table 8.11; *ST* treatment, *FS* foliar spray, *INR* Indian rupees

appearance compared to the untreated control. The use of fungicide resulted in the least disease occurrence in unfertilized but fungicide-applied plots (Plate 8.1 and Table 8.20).

8.4 Cultural Control

Common cultural practices like (a) use of clean, bold, and healthy seeds of improved cvs., (b) destruction of weeds and crop residues, (c) planting on recommended time, (d) long crop rotation, (e) proper plant population, and (f) judicious use of nutrition, though look very simple, pay unaware dividends by way of curb-

Table 8.13 Efficacy of fungicides and bioagents on per cent disease infection of powdery mildew of mustard (Lomate et al. 2014)

Treatments	PDI before spraying	PDI after spraying	PDI after second spraying	PDC after first spraying	PDC after second spraying
Chlorothalonil 75%WP @ 0.15%	65.30	41.05	29.68	50.96	70.06
Maneb 75%WP @ 0.2%	65.60	41.62	30.48	50.28	69.25
Wettable sulphur 75% WP @ 0.25%	65.30	39.22	25.97	53.15	73.80
Triadimefon 25% WP @ 0.02%	65.97	39.76	27.04	52.20	72.72
Penconazole 10% EC @ 0.10%	65.97	40.13	28.12	52.24	71.63
Dinocap 48%EC @ 0.10%	66.09	38.50	23.50	54.10	76.29
Tridemorph 80% EC @ 0.05%	64.44	38.93	25.70	53.49	74.78
Ampelomyces quisqualis 10^8CFU/ml	65.21	42.71	34.17	48.98	65.53
Trichoderma harzianum, 10^8CFU/ml	65.40	43.06	38.02	48.56	61.65
Control	66.00	83.72	99.15	00	00
F test	NS	Sig	Sig		
SE (m) ±	0.39	0.78	0.82		
C.D. (P = 0.05)	1.11	2.33	2.43		

ing primary sources of inoculum and secondary spread of powdery mildew disease in the field. Cultural control is based on the principle of avoidance and escape of host plant from the favourable period of pathogen infection and multiplication. Some cultural practices like intercropping and mixed cropping may also be helpful to curb the powdery mildew progress. In some areas (Indian conditions), timely sown crop (before last week of October) escapes infection since it matures by the time pathogen assumes an epidemic form (Mukerji et al. 1999). However, it may not work under the conditions where disease appears very early in the areas like Jharkhand and Bihar or in the areas where temperature does not go below 20 °C, nontraditional area of India. Under Gujarat, India, conditions, lower severity of powdery mildew was observed in October-sown crop than late sown crop of November. A maximum powdery mildew severities of 63.0 and 71.5% were recorded on GM-1 and Varuna cvs. when they were sown on 25 November, whereas only 17.5% powdery mildew severity was recorded on GM-1 on 5 October sowing and 20.5% on Varuna on 15 October sowing (Table 2.4, Chap. 2). Planting time has shown great bearing on yield of mustard cvs. Significantly higher seed yield of 1696.51 kg/ha was recorded on GM-1 cv. sown on 25 October. The lowest seed

Table 8.14 Efficacy of fungicides and bioagents on percent disease intensity of powdery mildew of mustard (Lomate et al. 2014)

Treatments	PDI before spraying	PDI after spraying	PDI after second spraying
Chlorothalonil 75%WP @ 0.15%	18.52 (4.35)	14.07 (3.81)	6.90 (2.72)
Maneb 75%WP @ 0.2%	19.00 (4.47)	14.31 (3.84)	7.15 (2.76)
Wettable sulphur 75% WP @ 0.25%	17.77 (4.27)	12.71 (3.63)	5.67 (2.48)
Triadimefon 25% WP @ 0.02%	17.15 (4.20)	13.08 (3.68)	6.04 (2.55)
Penconazole 10% EC @ 0.10%	18.14 (4.31)	13.33 (3.71)	6.53 (2.65)
Dinocap 48%EC @ 0.10%	19.00 (4.41)	12.09 (3.54)	5.53 (2.45)
Tridemorph 80% EC @ 0.05%	19.13 (4.42)	12.46 (3.60)	5.40 (2.43)
Ampelomyces quisqualis 10^8CFU/ml	18.63 (4.37)	14.68 (3.89)	7.64 (2.85)
Trichoderma harzianum, 10^8CFU/ml	19.65 (4.48)	15.18 (3.95)	8.63 (3.02)
Control	17.64 (4.25)	32.22 (5.71)	44.68 (6.72)
F test	NS	Sig.	Sig.
SE (m) ±	0.08	0.04	0.05
C.D. (P = 0.05)	0.23	0.13	0.16

Figures in parenthesis are square root transformed value

Table 8.15 Efficacy of fungicides and bioagents on yield (kg/ha) of mustard (Lomate et al. 2014)

Treatments	Yield (kg/ha)	1000 weight (g)
Chlorothalonil 75%WP @ 0.15%	864.93	4.1
Maneb 75%WP @ 0.2%	845.77	4.0
Wettable sulphur 75% WP @ 0.25%	968.06	5.1
Triadimefon 25% WP @ 0.02%	946.03	4.9
Penconazole 10% EC @ 0.10%	907.08	4.4
Dinocap 48%EC @ 0.10%	1023.62	5.5
Tridemorph 80% EC @ 0.05%	979.23	5.3
Ampelomyces quisqualis 10^8CFU/ml	810.13	3.9
Trichoderma harzianum, 10^8CFU/ml	737.22	3.5
Control	708.73	3.2
F test	Sig.	Sig.
SE (m) ±	10.82	0.24
C.D. (P = 0.05)	32.15	0.72

yield of 392.66 kg/ha of GM-1 was recorded when crop was sown on 25 November (Dange et al. 2003; Table 2.5, Chap. 2). The effect of intercropping, plant density, and date of sowing on severity of powdery mildew of rapeseed-mustard was observed under organic farming system in Manipur, India, by Devi and Chhetry (2017). Intercropping of rapeseed-mustard with pea has significant effects on disease severity compared to sole crop. With intercropping minimum disease

Table 8.16 Effect of fungicides for controlling powdery mildew on pods of rapeseed (Penaud 1999)

Locations/treatments	Chambon (17)	Juzes (31)	Beziers (34)	Portiragnes (34)	Mean rank
Healthy check	6.3 (1.5)	1.9 (1.0)	3.7 (1.0)	5.5 (3.0)	1
Flusilazole + carbendazim	5.7 (1.5)	1.3 (2.0)	3.9 (2.0)	5.0 (1.0)	2
Difenoconazole + carbendazim	6.6 (3.0)	4.2 (4.0)	5.5 (3.0)	5.8 (4.0)	3
Carbendazim	6.7 (5.0)	5.4 (5.0)	5.5 (5.0)	5.0 (2.0)	5
Iprodione + carbendazim	6.5 (4.0)	3.6 (3.0)	6.0 (4.0)	6.3 (5.0)	4
Untreated control	6.9 (6.0)	6.4 (6.0)	8.3 (6.0)	6.5 (6.0)	6

Table 8.17 Effect of fungicide treatments on the yield of rapeseed (Penaud 1999)

Locations/treatments	Chambon (17)	Juzes (31)	Beziers (34)	Portiragnes (34)
	Three treatments on healthy check			
Healthy check	35.4a	37.4a	18.6a	24.8
Flusilazole + carbendazim	35.7a	33.3ab	17.4ab	25.4
Difenoconazole + carbendazim	32.0b	30.8abc	16.8ab	23.9
Carbendazim	30.1bc	30.8abc	16.1ab	23.2
Iprodione+ carbendazim	31.2bc	28.3bc	17.8ab	23.0
Untreated control	28.8c	24.7c	13.9b	21.6
CV	4.5	10.9	10.4	9.1
Pr > F	0.0001	0.003	0.016	0.23

Significant difference between one treatment with another treatment is presented as a, b, and c. Same letter on two data indicates that both are similar or insignificant

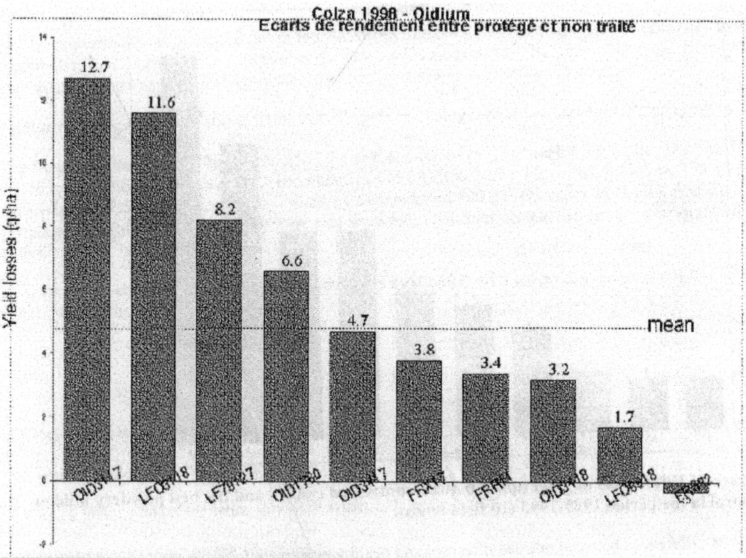

Fig. 8.1 Differences of yield (q/ha) between untreated control, and the best powdery mildew control (ten field experiments) (Penaud 1999)

Table 8.18 The effect of the nitrogen fertilization, and fungicidal treatment on the levels of quality components of five varieties of oilseed rape (Mert-Turk et al. 2008)

Rape varieties	Oil (%)	Protein (%)	GSL (μmol/g)	SAE (g/kg)	Oleic acid (% oil)	Linolenic acid (% oil)
Traingle	40.81*c	19.90a	21.30a	0.43a	59.31d	10.64a
H602016	42.67ab	18.93b	19.93a	0.41a	60.99c	10.40a
H602004	43.16ab	18.29bc	13.68b	0.36b	62.70b	9.95b
Titan	43.55a	17.96c	11.22c	0.33c	64.30a	9.88b
Mendel	42.38b	19.01ab	14.16b	0.36b	60.42 cd	10.71a
LSD (5%)	1.02	0.94	1.59	0.024	1.37	0.39

*Mean value followed by a different letter are significantly different

Table 8.19 Effects of fungicide and nitrogen applications on the total oil, protein, GSL, SAE, oleic acid, and linolenic acid contents of five varieties of oilseed rape naturally infected by powdery mildew in four different treatment combinations (Mert-Turk et al. 2008)

Varieties	Treatments*	Oil (%)	Protein (%)	GSL (μmol/g)	SAE (g/kg)	Oleic acid (% oil)	Linolenic acid (% oil)
Traingle	N0–F0	42.34a	18.42bc	20.32ab	0.40b	60.45a	10.61
	N0–F1	42.50a	18.19c	21.58ab	0.40b	60.79a	10.60
	N1–F0	37.05b	22.34a	24.42a	0.50a	55.19b	11.01
	N1–F1	41.36a	20.65ab	18.86b	0.43b	60.82a	10.33
	LSD (5%)	2.24	2.28	4.85	0.07	2.72	0.91
H602016	N0–F0	44.43a	16.95bc	18.30	0.37b	62.07a	10.07
	N0–F1	44.87a	16.92bc	18.85	0.36b	62.49a	10.27
	N1–F0	39.56c	21.32a	21.89	0.48a	57.55b	10.59
	N1–F1	41.84b	20.51a	20.70	0.42b	61.87a	10.68
	LSD (5%)	2.00	2.13	3.95	0.06	1.93	0.75
H602004	N0–F0	44.86a	16.12c	13.82	0.32b	62.05	9.95
	N0–F1	44.51a	17.39bc	13.59	0.33b	64.50	10.18
	N1–F0	40.11b	20.57a	15.30	0.42a	61.08	9.69
	N1–F1	43.18a	19.09ab	12.03	0.36b	63.19	9.97
	LSD (5%)	2.94	2.49	3.95	0.05	4.46	1.08
Titan	N0–F0	44.56a	16.41b	11.86a	0.31b	64.28b	9.77
	N0–F1	45.28a	16.84b	11.35 ab	0.32b	67.24a	9.76
	N1–F0	40.77b	20.28a	12.25a	0.37a	62.09b	10.13
	N1–F1	43.59a	18.31b	9.41b	0.31b	63.60b	9.86
	LSD (5%)	1.88	1.96	2.10	0.05	2.77	0.94
Mendel	N0–F0	44.14a	17.35b	12.91b	0.33bc	60.96ab	10.63ab
	N0–F1	45.09a	16.81b	13.39b	0.31c	63.98a	10.04b
	N1–F0	38.65c	21.71a	16.79a	0.42a	56.07c	11.12a
	N1–F1	41.64b	20.17a	13.54b	0.36b	60.65b	11.07a
	LSD (5%)	2.47	1.93	2.35	0.04	3.18	0.73

Mean values followed by a different letter are significantly different. The legend to the four treatment groups: N0–F0, no nitrogen fertilization and no fungicidal treatment; *treatments were N0–F1, no nitrogen fertilization but fungicidal treatment; N1–F0, nitrogen fertilization but no fungicidal treatment; N1–F1, nitrogen fertilization and fungicidal treatment

Plate 8.1 Pods and stems of oilseed rape infected with *Erysiphe cruciferarum* in fungicide-treated (F1) plants (left) and plants not treated with fungicide (F0) (right) (Mert-Turk et al. 2008)

Table 8.20 Visual assessment of the powdery mildew disease severity of oilseed rape varieties in four treatment combinations of nitrogen and fungicide during two growing seasons (2004–2005 and 2005–2006) (Mert-Turk et al. 2008)

	N0–F0[a]		N0–F1		N1–F0		N1–F1		
Rape varieties	2004–2005	2005–2006	2004–2005	2005–2006	2004–2005	2005–2006	2004–2005	2005–2006	Mean
Traingle	5.03[b]	4.90	1.16	1.20	6.00	5.90	2.16	1.90	3.38d
H602016	5.06	5.76	1.80	1.23	5.93	5.96	2.70	2.90	2.92b
H602004	5.09	5.13	1.30	1.23	5.93	5.93	2.03	2.06	2.60c
Titan	5.03	5.70	1.16	1.16	5.96	5.96	3.90	3.00	3.99a
Mendel	5.03	5.06	1.90	1.13	5.96	5.96	1.80	2.06	3.62c
Mean of years	5.05	5.31	1.46	1.19	5.96	5.95	2.52	2.38	–
Mean of treatment[c]	5.18b		1.33d		5.96a		2.45c		

[a]N0–F0, no nitrogen fertilization and no fungicidal treatment; N0–F1, no nitrogen fertilization but fungicidal treatment; N1–F0, nitrogen fertilization but no fungicidal treatment; N1–F1, nitrogen fertilization and fungicidal treatment

[b]1 = no powdery mildew present; 2 = occurrence of some white spots on stems; 3 = powdery mildew present on stems and lower leaves; 4 = powdery mildew present on stems, lower leaves, and upper leaves; 5 = powdery mildew present on stems, lower leaves, upper leaves, and partly pods; 6 = powdery mildew covers all the plants

[c]LSD = 0.06 among treatment, LSD = 0.07 among varieties (P = 0.05). Mean values followed by a different letter are significantly different

severity (15.68%) of powdery mildew was observed on *B. rapa* cv. M-27 compared to maximum powdery mildew of 40.46% on *B. juncea* cv. Lamtachabi (Table 8.21). Plant density has significant effect on powdery mildew severity of all the four cvs. of rapeseed-mustard. With the increase in spacing from 20 × 5 cm^2 through 30 × 10 cm^2, there is significant reduction in powdery mildew severity on all the cvs.

Table 8.21 Effect of intercropping with pea on disease severity of powdery mildew of rapeseed-mustard (Devi and Chhetry 2017)

	Powdery mildew (%)			
Treatments	*B. juncea* Local Yella	*B. juncea* Lamtachabi	*B. rapa* M-27	*B. rapa* Ragini
Intercropping	33.12	37.79	15.68	16.50
Control	36.66	40.46	16.54	19.19
t-values (5%)*	03.32	13.06	13.53	17.66

Treatment and control data are mean of six observations in each year; *significant at 5% level of significance

Table 8.22 Effect of plant density on disease severity of powdery mildew of rapeseed-mustard (Devi and Chhetry 2017)

	Powdery mildew (%)			
Spacing (cm^2)	*B. juncea* Local Yella	*B. juncea* Lamtachabi	*B. rapa* M-27	*B. rapa* Ragini
20 × 5	37.59	41.28	16.92	20.06
30 × 10	35.32	37.84	16.06	18.00
40 × 15	33.21	35.81	15.08	17.10
CD (5%)*	0.37	0.74	0.35	0.60

*Significant at 5% level of significance

Table 8.23 Effect of date of sowing on disease severity of powdery mildew of rapeseed-mustard (Devi and Chhetry 2017)

	Powdery mildew (%)			
Date of sowing	*B. juncea* Local Yella	*B. juncea* Lamtachabi	*B. rapa* M-27	*B. rapa* Ragini
30.09.14	34.73	36.99	15.99	15.73
15.10.14	35.54	38.41	16.77	17.94
30.10.14	36.17	38.97	17.66	18.75
14.11.14	38.19	39.68	18.44	19.31
29.11.14	38.84	40.43	18.75	20.02
C.D (5%)*	0.51	0.56	0.38	0.38

*Significant at 5% level of significance

with lower disease severity values under 40 × 15 cm^2 spacing (Table 8.22). Under Manipur, India, conditions, sowing of rapeseed-mustard earlier than 30 October contracts less disease severities of powdery mildew. With the delay in sowing time, there is sharp increase in powdery mildew severity values of rapeseed-mustard (Table 8.23). The comparison of powdery mildew severity values under intercropping, spacing (pant density), and date of sowing on all the four cvs. of rapeseed-mustard revealed that *B. rapa* cv M-27 has performed better with significantly lower powdery mildew severity under all situations. It indicates its inbuilt (genetic) resistance or tolerance against powdery mildew.

8.5 Biological Control

Plant extracts and some biological control agents of fungal and bacterial origin have shown their antagonistic effects against powdery mildew of crucifers (Table 8.24). Garlic bulb aqueous extract (1%) and *Trichoderma harzianum* @ 10 g/kg as seed treatment individually or in combination with foliar sprays with garlic aqueous extract, *T. harzianum*, and *Pseudomonas fluorescens* @ 10 ml/l of water controlled powdery mildew of mustard. These treatments have also shown higher cost-benefit ratio in comparison to chemical control (Meena et al. 2011, 2013). However, two sprays of bioagents *Ampelomyces quisqualis* @ 10^8 CFU/ml has been found better than *T. harzianum* @ 10^8 CFU/ml in reducing powdery mildew of mustard along with higher seed yield and 1000 seed weight by Lomate et al. (2014). The antagonistic effects of *P. fluorescens* @ 10^8 cells/ml on *E. cruciferarum* radical growth and spore germination have been observed by Prasad et al. (2015). The *Pseudomonas* spp. produces diffusible metabolites like pyoluteorin (PLT), phenazine-1-carboxylic acid, 2–4-diacetylphloroglucinol (DAPG), and 2-4-de-epoxy-2-2-di-dehydrorhizoxin, which are associated with antimicrobial effects. In addition, the antifungal activity of metabolites like siderophore, HCN, ammonia, lipase, and chitinase also plays a major role. Leaf extracts of eucalyptus, neem (*Azadirachta indica*), and datura (*Datura stramonium*) at 2% were found very effective for the control of powdery mildew of mustard by Singh (2004). Patel and Patel (2008) found *T. viride* (3%) was very helpful in reducing seed abnormality caused by powdery mildew of mustard under Gujarat conditions.

Table 8.24 Bioagents tested against powdery mildew of crucifers

Bioagents	References
Allium sativum bulb aqueous extracts (1%)	Chattopadhyay et al. (2007), Meena et al. (2011, 2013) and Patel and Patel (2008)
Azadirachta indica leaf extracts (2%)	Singh (2004) and Patel and Patel (2008)
Datura stramonium leaf extract 2%	Singh (2004)
Eucalyptus leaf extracts (2%)	Singh (2004) and Patel and Patel (2008)
Trichoderma harzianum (10 g/kg)	Meena et al. (2011, 2013)
Trichoderma harzianum (10^8CFU/kg)	Lomate et al. (2014)
T. viride suspension	Patel and Patel (2008)
Ampelomyces quisqualis (10^8CFU/ml)	Lomate et al. (2014)
Nerium indicum L. (Karan) leaf extracts (5%)	Patel and Patel (2008)
Pseudomonas fluorescens oil based (10 g/l water)	Meena et al. (2011, 2013)
Pongamia pinnata L. (Karanj) leaf extracts (5%)	Patel and Patel (2008)
Pseudomonas fluorescens (10^8 cells /ml)	Prasad et al. (2015)

8.6 Host Resistance

Under natural conditions, the powdery mildew of crucifers is caused by *E. cruciferarum* which is very widespread and economically most important, causing very heavy yield losses under epidemic conditions (see Chap. 2). The other non-adopted powdery mildew species, viz. *E. cichoracearum* (GC), *E. orontii*, and *Oidium neolycopersici*, are of academic interest and have been used to unravel molecular mechanisms of host resistance using *Arabidopsis*-powdery mildew host pathosystem. The occurrence of these powdery mildew species on crucifers under natural ecosystem has not been recorded. The number of R genes has been identified using this host pathosystem which governs molecular and genetical mechanisms of pre- and post-penetration powdery mildew pathogenesis and expression of resistance by the host. However, the application of this knowledge is under the process to manage the powdery mildews of crucifers under field conditions (see Chap. 7). Major gene host-specific resistance sources have been identified in *B. juncea*, *B. rapa*, *B. napus*, *B. carinata*, *B. alba*, *A. thaliana*, and *R. sativus* from different countries (Table 7.13, Chap. 7). Some slow mildewing sources of resistance from *B. juncea* have also been identified. These sources are being utilized to breed powdery mildew resistance cvs. through conventional and biotechnological techniques. The most effective, economical, and environmental friendly powdery mildew disease management can be achieved by the use of resistant cvs., which can be easily developed by the use of sources of resistance by transferring cloned durable R genes in the agronomically superior susceptible cvs.

8.7 Integrated Disease Management

Integrated disease management (IDM), integrated pest management (IPM), and integrated crop management (ICM) are the strategies to manage a crop for better yield production and high economic returns with holistic approach using all available technology and effective control measures. It includes all cultural, nutritional, biological, biochemical, biotechnological, chemical, host resistance, and molecular or genetic engineering techniques in an integrated way. Cultural practices like manipulating date of sowing to escape congenial environmental conditions, wider spacing (40×15 cm^2), and mixed cropping with pea and mustard have been helpful to manage powdery mildew with high yield of mustard crop. Chemicals/ fungicides like Sulfex (0.2%), Karathane (0.05%), Calixin (0.05%), dinocap (0.1%), and flusilazole + carbendazim (200 + 100 g a.i. /ha) are most effective to control powdery mildew of crucifers. Fungicidal spray is essential immediately after appearance of the disease with second spray at 10-day interval. Biological control agents like *Allium sativum* bulb aqueous extracts (1%), *Trichoderma harzianum* (10 g/kg or 10^3 CFU/kg), and *Pseudomonas fluorescens* (10 g/l) have shown promise after seed treatment and spray to control powdery mildew of crucifers. Fungicidal application

Table 8.25 Priorities of crucifers' powdery mildew control strategies[a]

Priorities	Order of importance
Resistant cultivars	1
Fungicidal spray	2
Seed treatment	3
Sanitation	4
Crop rotation	5
Mixed cropping	6
Spacing of plants	7
Date of sowing	2
Balanced nutrition	4
Barrier crop	?

[a]Repetition of number indicates useful combination.? – No information

to control powdery mildew also increases total oil and oleic acid content of oil-yielding crops. Balanced nutrition in the form of organic (FYM) and inorganic (fertilizers) sources with fungicidal protection cover to the crop is always beneficial for getting healthy crop and higher yields (Saharan and Sheoran 1986; Patel and Patel 2008; Shete et al. 2008; Meena et al. 2013; Lomate et al. 2014; Penaud 1999; Mert-Turk et al. 2008; Devi and Chhetry 2017; Saharan et al. 2005; Saharan 1992, 1998; Saharan and Chand 1988; Saharan and Mehta 2002; Mukerji et al. 1999). There is a need to fix priorities of powdery mildew control strategies according to the appearance of the disease in an area with extent and amount of disease intensity visualizing environmental conditions to become favourable for epidemic development (Table 8.25). The order of priorities indicated in Table 8.25 may be manipulated as per the time of occurrence of the disease with different combinations of disease control strategies for which historical data on the occurrence of the disease will be useful measure to take the decision.

References

Agrios GN (1978) Plant pathology, 2nd edn. Academic, London

Ashraf S, Yadav B (2009) Studies on the anamorph characters and management of powdery mildew of mustard. Trends Biosci 2(2):79–80

Bewley WF (1923) Diseases of glasshouse plants. E. Benn, London, p 208

Brodie HJ, Neufeld CC (1942) The development and structure of the conidia of *Erysiphe polygoni* DC and their germination at low humidity. Can J Res C 20:41–62

Chattopadhyay C, Kumar VR, Meena PD (2007) Biomanagement of Sclerotinia rot of *Brassica juncea* in India – a case study. Phytomorphology 57:71–83

Curry JA (1924) Bicarbonate of soda spray effective. Am Rose Ann:67–90

Dange SRS, Patel S, Mishra RL, Saxsena CM (2002) Assessment of losses in yield due to powdery mildew disease in mustard under north Gujarat conditions. J Mycol Plant Pathol 32(2):249–250

Dange SRS, Patel RL, Patel SI, Patel KK (2003) Effect of planting time on the appearance and severity of white rust and powdery mildew diseases of mustard. Indian J Agric Res 37(2):154–156

Davis MB, Blair DS (1938) The strawberry and its cultivation in Canada. Can Dep Agric Farmers Bull 63:43
Devi YP, Chhetry GKN (2017) Effect of traditional agronomic practices on fungal diseases of rape-mustard under organic farming system in Manipur. Intl J Adv Res 5(6):1256–1260
Eyre JV, Salmon ES, Wormald LK (1919) Further notes on powdery mildews and the ammonium polysulphide wash. J Breed Agric Great Brit 25:1494–1497
Felber IM, Hammer CL (1948) Control of mildew on bean plants by means of an antibiotic. Bot Gaz 110:324–325
Fisher DF (1938) Control of apple powdery mildew. U S Dep Agric Farm Bull 1120:10
Guba EF (1928) Control of cucumber powdery mildew in greenhouses. Phytopathology 18:847–860
Guillon GM (1903) Le permanganate de potasse et I'Oidium. Rev Vitic 20:357–358
Horne EV (1918) American gooseberry mildew. J Royal Hortic Soc 43:32–33
Jani SM, Dange SR, Desai SG, Patel KK (1991) Chemical control of powdery mildew of mustard. Indian Phytopathol 44(4):535–536
Jorstad I (1925) The Erysiphaceae of Norway. Schriften Utgitt av det Norske Videnskaps Akademi i Oslo I Maten Naturvid, 116p
Kaur J (1981) Persistence of different fungitoxicants in wheat and *raya* against *Erysiphe graminis* f sp. *tritici* and *E. polygoni.* Third international symposium on plant pathology, Indian Phytopathological Society, December 14–18, IARI, New Delhi 183 (Abstr.)
Khodke SW, Kakade SU (2004) Effect of chemicals and botanicals on powdery mildew of mustard (*Brassica juncea* L.). PKV Res J 28:61–63
Kumar S, Singh D, Yadav SP, Prasad R (2015) Studies on powdery mildew of rape-mustard (*Brassica juncea* L.) caused by *Erysiphe cruciferarum* and its management. J Pure Appl Microbiol 9(2):1481–1486
Kurowski TP, Majchrzak B, Jankowski K (2010) Effect of sulfur fertilization on the sanitary state of plants of the family Brassicaceae. Acta Agrobot 63(1):171–178
Kyle J (1841) Cure for mildew and scale on peach trees. Gard Chron 1:69–70
Lomate CB, Mate GD, Kalaskar RR (2014) Management of powdery mildew of mustard with chemicals and biogents. Int J Plant Prot 7:122–124
Martin H, Salmon ES (1930) Vegetable oils as fungicides. Nature 126:58
Mclellan WD (1947) Insecticides and their relationship to mildew on roses. Florists Rev 101:39
Meena PD, Awasthi RP, Godika S, Gupta JC, Kumar A, Sandhu PS, Sharma P, Rai PK, Singh YP, Rathi AS, Prasad R, Rai D, Kolte SJ (2011) Eco-friendly approaches managing major diseases of Indian mustard. World Appl Sci J 12(8):1192–1195
Meena PD, Gour RB, Gupta JC, Singh HK, Awasthi RP, Netam RS, Godika S, Sandhu PS, Prasad R, Rathi AS, Rai D, Thomas L, Patel GA, Chattopadhyay C (2013) Non-chemical agents provide tenable, eco-friendly alternatives for the management of the major diseases devastating Indian mustard (*Brassica juncea*) in India. Crop Prot 53:169–174
Mert-Turk F, Gul MK, Egesel CO (2008) Nitrogen and fungicide applications against *Erysiphe cruciferarum* affect quality components of oil rape. Mycopathologia 165(1):27–35
Middleton JT (1943) Disease control with fermate and spergon. Plant Dis Rep 27:169–170
Mukerji KG, Upadhyay RK, Saharan GS, Sokhi SS, Khangura RK (1999) Diseases of rapeseed-mustard and their integrated management. In: Upadhyay RK, Mukerji KG, Jalak RL (eds) IPM system in agriculture, vol 5. Aditya Books Pvt Ltd, New Delhi, pp 91–135
Patel JS, Patel SJ (2008) Health status of mustard as influenced by foliar sprays against powdery mildew disease. Indian J Agric Sci 21(3):464–465
Patel VA, Vaishnav KA, Dhedhi BM, Kakani BK (1995a) Efficacy and economy of different fungicides against powdery mildew of mustard. J Oil Res 12(1):126–127
Patel VA, Vaishnav KA, Dhedhi BM (1995b) Chemical control of powdery mildew of mustard. Indian J Mycol Plant Pathol 25(12):115
Penaud A (1991) Les nouvelles maladies du colza. Phytoma 427:15–18
Penaud A (1998) Oidium du colza, la protection du colza est maintenant possible. Oleoscope 50:36–38

Penaud A (1999) Chemical control and yield losses caused by *Erysiphe cruciferarum* on oilseed rape in France. In: Wratten N, Salisbury PA (eds) Proceedings of the 10th international rapeseed congress, vol 327. The Regional Institute Ltd, Canberra. CD-ROM, pp 1–8
Prasad R, Chandra H, Sinha BK, Srivastava J (2015) Antagonistic effect of *Pseudomonas fluorescens* isolated from soil of Doon Valley (Dehradun–India) on certain phyto-pathogenic fungi. Octa J Biosci 3(2):92–95
Radclyffe WF (1861) Rose mildew. Gard Chron 21:967
Rowe CW (1944) My experience in the use of elgetol for thinning Newtown apples. Wash State Hortic Assoc Proc Annu Meet 40:41
Saharan GS (1992) Management of rape and mustard diseases. In: Kumar D, Rai M (eds) Advances in oils research (vol. 1) Rape and mustard. Scientific Publishers, Jodhpur, pp 155–188
Saharan GS (1998) Diseases of rapeseed and mustard and their management. In: Thind TS (ed) Diseases of field crops and their management. National Agricultural Technology Information Centre, Ludhiana, pp 95–113
Saharan GS, Chand JN (1988) Diseases of oilseeds crops (in Hindi) (chap. 3). Directorate of Publication, CCS HAU, Hisar, pp 84–146
Saharan GS, Mehta N (2002) Fungal diseases of rape-mustard. In: Gupta VK, Paul YS (eds) Diseases of field crops. Indus Pub Co, New Delhi, pp 193–228
Saharan GS, Sheoran BS (1986) Management of powdery mildew of mustard with fungicides. Oil Crops Newsl 31:63–65
Saharan GS, Mehta N, Sangwan MS (2005) Diseases of oil crops. Indus Pub Co, New Delhi, p 643
Sharma KB, Bhargava AK, Chhipa HP (1990) Fungicidal control of powdery mildew of mustard. Indian J Mycol Plant Pathol 10:49
Shete MH, Dake GN, Gaikwad AP, Pawar NB (2008) Chemical management of powdery mildew of mustard. Plant Dis Sci 3(1):46–48
Singh K (2004) Studies on the ecofriendly management of powdery mildew (*Erysiphe cruciferarum* Opiz ex. Junell) of mustard [*Brassica juncea* (L.) Czern & Coss]. MSc thesis, Department of Plant Pathology, CCSHAU, Hisar,. 106p
Singh VK, Chauhan VB (1998) Management of powdery mildew of mustard. Ann Plant Prot Sci 6(1):93–99
Singh RB, Singh RN (2003) Management of mildew of mustard. Indian Phytopathol 56:147–150
Singh RR, Solanki JS (1974) Fungicidal control of powdery mildew of *Brassica juncea*. Indian J Mycol Plant Pathol 4(2):210–211
Singh RB, Lal Thakore BB, Mathur S (1980) Fungicidal control of white rust, downy mildew and powdery mildews of *Brassica juncea*. Indian J Mycol Plant Pathol (Abstr) 10:XLVII
Singh R, Singh D, Salisbury P, Barbetti MJ (2010) Field evaluation of indigenous and exotic *Brassica juncea* genotypes against Alternaria blight, white rust, downy mildew and powdery mildew diseases in India. Indian J Agric Sci 80(2):155–159
Szembel SJ (1930) Control of powdery mildew of cucumbers by means of disodium hydrogen ortho-arsenate. Comment Inst Astrachanensis ad defensionern Plantarium 2:21–31. [In Russian Abstr Rev Appl Mycol 10: 500–501]
Thomas WD, Holtzmann DW (1951) Studies on the control of powdery mildew of roses. Colo Flower Grs Assoc Bull 25 [Mimen]
Vidali A (1951) Esperienze di lotto eontro l'oidio del Tobacco *(Eryslphe cichoracearum)* effectuate in campo a mezzo di carbonato di litio. Notiz Mal Piante 16:35–39. [Abstr Rev Appl Mycol 31:461–462]
Wenzi H (1949) Die Grenzen der therapeutischen Sommerbekampfung des amerikanischen. Stachelbeermeltaus PflSch Bet Wien 3:10–16
Yarwood CE (1942) Stimulatory and toxic effect of copper sprays on powdery mildews. Am J Bot 29:132–135
Yarwood CE (1945) Copper sulphate as an eradicant spray for powdery mildews. Phytopathology 35:895–909
Yarwood CE (1951) Fungicides for powdery mildews. Proc IInd Int Cong Crop Prot 2:500–521
Yarwood CE (1957) Powdery mildews. Bot Rev 23:235–312

Chapter 9
Techniques

9.1 Introduction

Powdery mildew of crucifers is caused by biotrophic fungal pathogens. So far, no protocol has been developed to culture the pathogen under laboratory conditions on artificial culture medium. Some standardized, reproducible, widely acceptable techniques to validate the research results on various aspects of crucifers–powdery mildew pathosystem have been developed by crucifer scientists of the world. These techniques are very essential tools, procedures, and methods to carry out research work by students for their graduate and postgraduate research, for teachers to demonstrate research results with standard techniques, and for research scientists to refine their research pursuits.

9.2 Collection, Preservation, and Cultivation of Crucifer's Powdery Mildew

Symptoms of powdery mildew on the infected host parts (leaves, stem, branches, and siliquae) are conspicuous in the form of white floury patches consisting of mycelium, conidia, and conidiophores. To locate formation of chasmothecia, careful search is needed especially at the maturity stage of the crop. Chasmothecia are visible in the form of brown to dark brown pinhead size bodies on the infected leaves, stem, and siliquae. Special care should be taken to properly identify the host plant. It is always good to collect flowering and/or fruiting parts of the host along with infected leaves. For the systematic study of powdery mildew, dry material is very satisfactory. For further study, infected parts of plants are collected and placed between driers for a few days. The material is then transferred to suitable envelops,

G. S. Saharan et al., *Powdery Mildew Disease of Crucifers: Biology, Ecology and Disease Management*, https://doi.org/10.1007/978-981-13-9853-7_9

properly labeled, and laid aside for future study. To maintain the culture, plants should be grown under growth room/greenhouse for continuous availability of inoculum and pathogen to use in subsequent studies. In general, a succession of crops of young host must be ready to maintain powdery mildew on the host.

9.3 Artificial Inoculation

Conflicting ideas of the relation of water to infection with powdery mildews have led to a variety of contrasting methods of making artificial inoculations. Suspending of conidia in water, spraying of plants with water before inoculation, and incubating the plants in a moist chamber after inoculation are probably all of no value in obtaining infection in most cases, though they may incidentally serve such useful purposes as confinement of strains, reduction of light, and reduction of temperature (Yarwood 1957). Dusting dry conidia onto dry leaves and leaving the inoculated plants in a dry environment (Yarwood 1936) will give heavy infection with most powdery mildews. Use of detached leaves (Yarwood 1946) has facilitated certain experiments with powdery mildews, especially in securing infection under conditions unfavorable to infection of the entire plants. Growing powdery mildews on tissue cultures will probably be useful for specific purposes. Single-spore inoculations are commonly successful (Homma 1937).

9.4 Molecular Identification of Anamorphic Powdery Mildews (Erysiphales)

The powdery mildew fungi (*Erysiphales*) are a common group of obligate plant pathogens that can be extremely difficult to identify, depending on the reproductive stage encountered. The ribosomal DNA internal transcribed spacer (ITS) region was assessed for its usefulness to link anamorphic *Erysiphales* with their respective teleomorphs. PCR primers for the rDNA ITS region were designed and found to have enhanced specificity for the 12 genera tested, even in the presence of contaminating fungi. Recent herbarium specimens of both anamorphic and teleomorphic materials yielded sufficient DNA for amplification using a Chelex-based extraction method. The ITS regions from 25 anamorphic specimens were sequenced and compared with the ITS sequences of their suspected teleomorphs. In most cases, an ITS region sequence similarity of above 99% indicated that a correct match had been made. Although this technique will not always unambiguously identify an anamorphic specimen, it will provide valuable information to use in conjunction with morphological and host range data to aid in the final identification (Cunnington et al. 2003).

9.4.1 Primer Design and Test Specimen Methods

Erysiphales specific primers for the ITS region were designed from 18S and 28S sequences (Mori et al. 2000). The forward primer, PMITS1 (5´-TCGGACTGGCC(T/C)AGGGAGA-3′), was designed from 18S sequences of *Blumeria graminis* (GenBank Accession AB022361), *Erysiphe heraclei* (AB022390), *Golovinomyces cichoracearum* var. *cichoracearum* (AB022359), *Phyllactinia moricola* (AB0022400), *Podosphaera xanthii* (as *Sphaerotheca cucurbitae*, AB022409), and *Pleochaeta shiraiana* (AB022402), whereas the reverse primer, PMITS2 (5´-TCACTCGCCGTTACTGAGGT-3′), was designed from 28S sequences of *Erysiphe heraclei* (AB022391), *Erysiphe* (*Uncinula*) *mori* (AB022418), *Phyllactinia kakicola* (AB022372), *Golovinomyces cichoracearum* var. *cichoracearum* (AB022360), *Sawadaea polyfida* var. *japonica* (AB022364), and *Blumeria graminis* (AB022362). The primer sites were chosen by comparing these sequences with those from a range of other fungi available in GenBank. In particular, 26 ascomycete sequences from species in the Pezizales, Ophiostomatales, Microascales, Saccharomycetales, Chaetothyriales, Eurotiales, Hypocreales, Phyllachorales, Cyttariales, Rhytismatales, Dothidales, Sordariales, Onygenales, Heliotales, and Schizosaccharomycetales were used. To gain the specificity required, particular attention was paid to ensure that the 3′ end of both primers consisted of a sequence unique to the powdery mildews (Innis and Gelfand 1990; Kwok et al. 1990).

Fourteen test specimens were chosen to represent species from as many genera as possible and also to include specimens with common contaminating fungi growing in the powdery mildew colony. Although most of these specimens were purely anamorphic, they were species that could be confidently identified by a combination of their characteristic anamorph morphology and host. The only specimen not identified to species level was the *Sawadaea* species on *Acer*.

9.4.2 DNA Extraction and PCR

DNA was extracted using a protocol modified from Hirata and Takamatsu (1996). A small amount (<1 mm^3) of the fungus was scraped from the host plant and placed into 50 μl of 5% Chelex 100 (Bio-Rad) containing 0.01% Triton X-100. This was incubated at 56 °C for 2 h (vortexing briefly after 1 h), vortexed, incubated in boiling water for 8 min, vortexed, and spun in a benchtop centrifuge at 20,000 g for 5 min. The top 40 μl was removed, added to 40 μl isopropanol, mixed, left on ice for 10 min, and centrifuged for 10 min at 20,000 g. The supernatant was removed, and the sample was dried and resuspended in 40 μl of water.

The initial PCR was performed in 25 μl containing 5 μl DNA, 200 μM of each dNTP (Pharmacia Biotech), 1.5 m M $MgCl_2$, 2.5 μl 10× buffer, 4 ng each of primers PMITS1 and PMITS2, 6% Tween 20 (Lab chem.), 0.5 units of Amplitaq Gold (PerkinElmer), and 100 ng BSA (Bresatec). Reactions were overlaid with 30 μl of light paraffin and subjected to 10 min at 94 °C, 35 cycles of 1 min at 94 °C, 1 min at 65 °C, 2 min at 72 °C, and a final extension of 10 min at 72 °C. PCR products were detected by running 4 μl on a 1.4% agarose gel in TBE buffer. A nested PCR was performed in 50 μl as outlined above, but without Tween 20, using primers PMITS1 and ITS4 (White et al. 1990) at twice the previous concentration and containing 1 μl of a 1:100 dilution of the first round product or, if no product was visible, then an undiluted 1 μl. Cycling times were the same, but with an annealing temperature of 60 °C. Nested PCR products were purified using a QIA quick PCR Purification Kit (Qiagen). To ensure that only the powdery mildew ITS region had been amplified, the products were sequenced directly using primers ITS5 (White et al. 1990) and ITS4, with an ABI PRISM BigDye Terminator Cycle Sequencing Kit (PerkinElmer). These sequences were used in similarity searches on the GenBank database using the computer program BLAST (Altschul et al. 1990) to confirm that they did belong to the *Erysiphales*.

9.4.3 Anamorph–Teleomorph Connection

The method described in the previous section was used to sequence the ITS region of several anamorphic specimens collected on a variety of Australian host plants. Additionally, some specimens from previous list were used. Cleistothecial material of their suspected teleomorphs was obtained from either Northern Hemisphere localities or from Australia and the ITS region sequenced. Sequence data for several species were obtained from GenBank. In each case, to determine if a match had been made, a percentage DNA similarity was calculated by aligning the two sequences using Clustal W (Thompson et al. 1994) and using the programme Homologies (Leunissen, J. A. M., CAOS/CAMM Center, University of Nijmegen, The Netherlands).

9.5 Light and Scanning Electron Microscopy

Growth of plant material and *Erysiphe* isolate MGH–*Arabidopsis thaliana* was grown in Metro-Mix2 000 substrate in either a climate-controlled greenhouse at 19 C under a 12 h light/dark cycle with supplemental fluorescent illumination or in a Percival AR-60 L growth chamber at 20 °C and 80% relative humidity. The MGH isolate was grown on *A. thaliana* ecotype Col-0 or a mutant of Col-0 called *pad4* (Glaze brook et al. 1996) which showed enhanced fungal susceptibility and gave greater yields of conidia. The MGH isolate was propagated by inoculating 4–5-week-old *A. thaliana* plants with dry conidia from 2- to 3-week-old infections using modified settling towers (Adam and Somerville 1996), which consisted of a square metal tower 71 cm high, which fits over a 13 cm square pot, covered with nylon mesh with

95 μm openings. Conidia from two heavily infected *A. thaliana* leaves were brushed onto the mesh and passed through to break up the conidial chains. Seeds of *Capsella bursa-pastoris* were harvested from wild plants growing in Medford, Massachusetts. These were cold-treated for 7–14 days in 0.1% agar at 4 C, then germinated in Metro-Mix2 00 (Scotts-Sierra Horticultural Products, Marysville, Ohio), and grown in the greenhouse as described above for *A. thaliana*.

9.5.1 Examination of Fungal Development

Fungal development was examined *in vitro* and *in vivo* on *A. thaliana* leaves. Microscopic analysis of powdery mildew development was performed 1, 2, 5, 24, 48, 72, 96, and 120 h after inoculation. Fungal development of the MGH and UCSC isolates was also compared on leaves of squash cultivar Kuta. Individual leaves of Kuta (first fully expanded leaf) were cut in half at the mid-vein, and each half was incubated in a Petri plate containing 1.5% water agar. One half of each leaf was infected with the MGH isolate, and one half was infected with the UCSC isolate. The Petri plates were placed in a growth chamber (20 °C, 12 h day–night cycle). Fungal development was assessed at 6–8 days postinoculation (dpi). The MGH isolate was tested for growth on other hosts by transferring conidia from infected *A. thaliana* leaves to fully expanded leaves of the plants being tested, either by tapping conidia from the infected leaves onto the recipient leaf surface or by use of a camel's-hairbrush. Leaves were examined under a dissecting microscope for visible fungal growth from approximately 5 dpi until leaf senescence.

9.5.2 Light Microscopy

Infected leaves were cleared in a solution of lacto-phenol/ethanol (1 vol. of lacto-phenol and 2 vol. ethanol) for a period of 12–24 h with 1 change of solution. Lacto-phenol was prepared of equal vol. of phenol, lactic acid, glycerol, and water. Cleared *Arabidopsis* leaves were moved to a fresh lacto-phenol solution containing 1 mg/ml trypan blue. Leaves were stained for 10 min before mounting on slides (Shipton and Brown 1962). Samples were stained overnight for visualization of haustoria. An Axioscope (Zeiss) with bright field and Nomarski optics were used for the study of pathogen development on the leaf and stem surface and for analysis of host responses to the pathogen. Photomicrographs were taken with an MC 80 automatic camera. Measurements of conidia, conidiophores, hyphae, and haustoria were made using object and ocular micrometers. Freshly harvested samples were examined in water, because accurate measurements of the curved conidiophores of the UCSC isolate were difficult to obtain.

9.5.3 Scanning Electron Microscopy

Pieces of infected *A. thaliana* leaves at various stages of disease development were fixed in 4% paraformaldehyde and passed through an ethanol series: 30, 50, 70, 96, and 100%. The fixed plant material was dried in a critical point drying apparatus (Samdri-PVT-3B, Tousimis). Dried leaf pieces were mounted on stubs and covered with 20–25 μm of gold–palladium in a Hummer II sputter coater (Techniques). Coated samples were studied under an AMRAY 1000 scanning electron microscope (Plotnikova et al. 1998).

9.6 DNA Sequence Analysis

DNA from the MGH isolate was prepared by a slight modification of the method of Pfister (1997). Briefly, conidia (about 0.1 ml dry vol) were ground in a microcentrifuge tube with a plastic pestle. The ground conidia were resuspended in 0.4 ml lysis buffer (200 mM Tris/HCl, pH 8.5, 250 mM NaCl, 25 mM EDTA, 0.5% SDS), and the suspension was treated as described by Pfister (1997). One μl of the resulting DNA preparation was used for amplification by the polymerase chain reaction (PCR). The region of the ribosomal DNA containing the internal transcribed pacers (ITS) 1 and 2 was amplified using the primers ITS4 and NS7 (White et al. 1990) and the reaction conditions recommended by White et al. (1990) with an annealing temperature of 55 °C. The PCR products were sequenced without cloning utilizing ABI PRISM dye terminator chemistries and ABI 377 DNA Sequencers. The primers ITS4, ITS2, NS8, NS7, ITS5, and ITS3 were used for DNA sequencing (White et al. 1990). Products from three independent PCR amplifications, one of which was performed on an independent DNA preparation, were sequenced to ensure that no PCR errors were obtained in the sequence. Analysis of the sequence data was carried out using Genetics Computer Group software, version 8 (Genetics Computer Group, Madison, Wisconsin), and the sequences of the ITS regions of the MGH isolate and the *E. cichoracearum* UCSC isolate were compared using the GAOP algorithm (Plotnikova et al. 1998).

9.7 Use of qPCR and Spore Count Assays to Quantify Powdery Mildew

Two protocols for the quantification of the powdery mildew disease on *Arabidopsis* with reproducible results over a broad range of infection phenotypes have been developed (Weßling and Panstruga 2012) (Fig. 9.1 and Table 9.1). The techniques are also useful for the quantification of more subtle differences, although such instances may require a higher number of experimental replications. For the

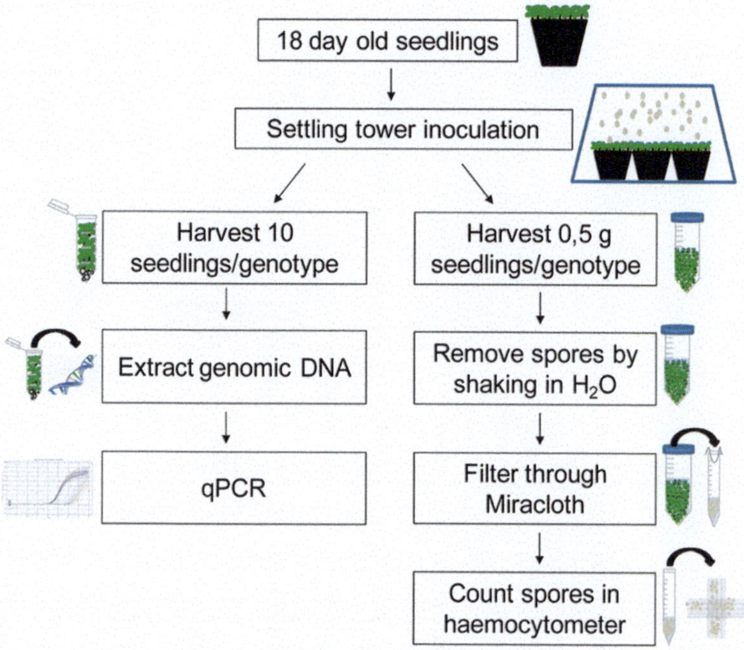

Fig. 9.1 Schematic overview of developed methods. Simplified view of the workflow of the qPCR- and spore count-based powdery mildew quantification procedures (Weßling and Panstruga 2012)

qPCR-based method, efficient and specific primer combinations for the amplification of both host and pathogen DNA have been used. The quantified amount of *G. orontii* relative to *Arabidopsis* genomic DNA correlated well with the amount of fungal biomass. Fungal development was reflected accurately across both time and different genotypes and could also be determined upon the occurrence of localized powdery mildew-induced host cell death. 5 dpi was determined to be the optimal time point for comparative analysis among genotypes, but the method can also be used to resolve kinetics of powdery mildew pathogenesis (Plate 9.1).

A spore count assay can also be useful for powdery mildew research. This assay has a wide dynamic range and accurately and reproducibly determined differences in conidiation by the pathogen (Plate 9.2). Both assays yielded comparable results for the susceptibility of the genotypes tested and can easily be adapted to other *Arabidopsis*-infecting powdery mildew species and other plant-powdery mildew pathosystems. Both methods require similar amounts of time for data generation. The qPCR method is intrinsically more cost-intensive but allows simultaneous DNA extraction and PCR analysis of many samples. Spore counts are more cost-effective and, owing to the lower number of practical steps involved, offer less possibilities for experimental error (Fig. 9.1). Currently, powdery mildew infection is either assessed semi-quantitatively by coarse categorization, and thus macroscopic

Table 9.1 Comparison of methods to assess powdery mildew infection (Weßling and Panstruga 2012)

	Macroscopic categorization	Host cell entry counts	Conidiophore counts	Hyphal area quantification	qPCR	Conidia counts
Assay type	Semi-quantitative	Quantitative	Quantitative	Quantitative	Quantitative	Quantitative
Staining required	No	Yes	Yes	Yes	No	No
Microscopy required	No	Yes (hundreds of interaction sites)	Yes (multiple colonies)	Yes (multiple colonies)	No	Yes
Stage of pathogenesis scored	Late (conidiation)	Early (host cell entry)	Late (conidiation)	Early to late (host cell entry to conidiation)	Middle to late (hyphal expansion to conidiation)	Late (conidiation)
Importance of equal inoculation density	High (no normalization)	Low (internal normalization)	Low (internal normalization)	Low (internal normalization)	Low (internal normalization)	Medium (averaging effect of scoring multiple plants at once)
Suitable for high through put analysis	Yes	No	No	No	Yes	Yes

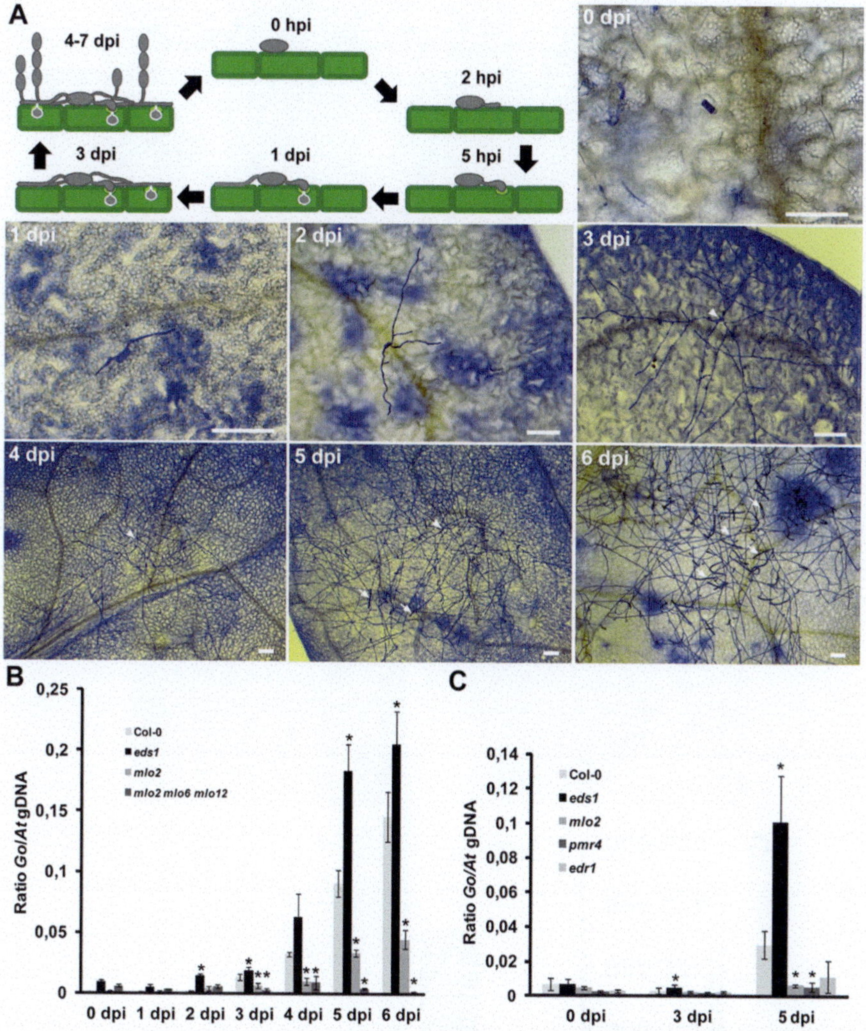

Plate 9.1 Powdery mildew disease progression on *Arabidopsis* seedlings. (**a**) Schematic overview and microscopic images of powdery mildew disease progression on Col-0 seedlings. Samples were harvested at indicated time points and stained with Coomassie Brilliant Blue. Arrows indicate conidiospore chains, and arrowheads point to the initial spore. hpi, hours post inoculation (**b**) qPCR analysis of a time series of powdery mildew infection on Col-0 wild type, *eds1*, *mlo2*, and *mlo2 mlo6 mlo2* seedlings. Ratios of *G. orontii* to *Arabidopsis* gDNA were determined by qPCR with primers R189/R192 and R193/R194, respectively. Bars represent the mean ± standard deviation of three technical replicates from a DNA sample of ten pooled seedlings grown in five different pots (two seedlings/pot used). (**c**) qPCR analysis of powdery mildew infection on *Arabidopsis* mutants that show powdery mildew-induced cell death. Representative time points of infection on Col-0 wild type *eds1*, *mlo2*, *pmr4*, and *edr1* seedlings were used. Ratios of *G. orontii* to *Arabidopsis* gDNA were determined by qPCR with primers R189/R192 and R193/R194, respectively. Bars represent the mean ± standard deviation of three DNA samples (each derived from ten pooled seedlings grown in five different pots) with three technical replicates each. Asterisks indicate statistically significant differences to Col-0 in two-tailed Student's *t*-test ($p < 0.05$). Schematic overview in (**a**) is courtesy of Justine Lorek. Scale bars in (**b**) are 100 μm (Weßling and Panstruga 2012)

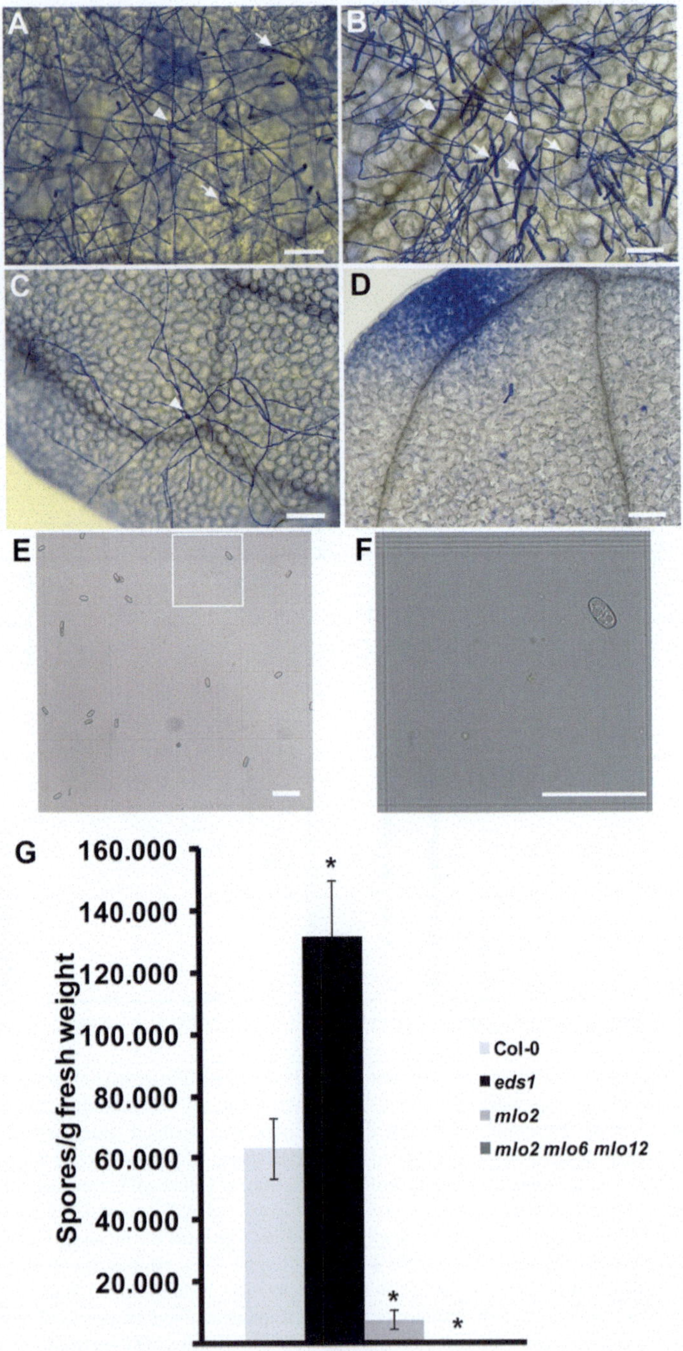

Plate 9.2 Analysis of powdery mildew infection by spore counts. *G. orontii*-infected leaves were harvested at 5 dpi from Col-0 wild type (**a**), *eds1* (**b**), *mlo2* (**c**), and *mlo2 mlo6 mlo2* (**d**) seedlings,

symptoms, or quantitatively by microscopic penetration rate counts at early and/or conidiophore counts at late stages (Table 9.1). The latter methods require staining, mounting, and microscopy and are therefore time-consuming and labor-intensive. Both assays introduced here are quantitative and are applied at late stages of infection when differences between genotypes are more pronounced. They do not require staining, are easy to perform, and can generate quantitative data for larger numbers of genotypes (Table 9.1).

In addition, the spore count assay is based on many seedlings rather than few mature plants, which offers the advantage of an averaging effect, lowering the possibility of experimental outliers. These assays can overcome limitations of currently used methods of powdery mildew quantification. They might be particularly suitable for the assessment of larger numbers of different mutant lines, e.g. in the course of follow-up analyses of genes identified by omics approaches.

9.7.1 Plant Material and Inoculation Methods

This technique has been developed using the *Arabidopsis thaliana* Col-0 genotype and the *edr1*-1, *eds1*-2, *pmr4*-1, *mlo2*-6 single, and *mlo2*-5 *mlo6*-2 *mlo12*-1 triple mutants in the Col-0 genetic background. Approximately 100 seeds were sown per pot of soil substrate, and five pots were used per genotype. After stratification for 2 days at 4 °C in darkness, plants were grown for 18 days at a day/night cycle of 10/14 h in a light chamber with 22 °C/20 °C day/night temperature and a relative humidity of 60%. The *G. orontii* isolate MPIPZ was propagated on 4-week-old *eds1*-2 plants, and conidia were used at 14–21 dpi. Inoculations were performed in a simple 80-cm-high cardboard settling tower whose opening was covered with 80 μm nylon mesh (Reuber et al. 1998). The tower contained up to nine pots per inoculation. Three consecutive rounds of infection have been used. In each round, pots of all genotypes were included. A fine paint brush was used to harvest conidia from four heavily infected leaves and to separate the conidia by brushing them through the nylon mesh. Inoculation density was approximately 750 spores/cm^2. Newly inoculated seedlings were then returned to the growth chamber.

Plate 9.2 (continued) and stained with Coomassie Brilliant Blue. Arrows indicate conidiospore chains, and arrowheads point to the initial spore. (**e, f**) Bright field image of isolated spores in the haemocytometer. (**f**) is a close-up of the indicated area in (**e**). (**g**) Spore counts of indicated genotypes at 6 dpi normalized to seedling fresh weight. Bars represent the mean ± standard deviation of three samples (500 mg of seedlings each) from one experiment counting eight fields/sample. Asterisks indicate statistically significant differences to Col-0 in two-tailed Student's *t*-test ($p < 0.05$). Scale bars in (**a–f**) are 100 μm (Weßling and Panstruga 2012)

9.7.2 *Staining and Microscopy*

For the visualization of fungal structures, seedlings were harvested at indicated time points and destained and stored in ethanol/glacial acetic acid 3:1 (v/v). Fungal structures were stained with Coomassie Brilliant Blue as described previously by Gollner et al. (2008), and bright field images were obtained using a AxioImager. A2 system with an AxioCamHRc (Zeiss, Jena, Germany). The experiment was repeated twice, and 5–10 images were analysed per replicate, genotype, and time point.

9.7.3 *Genomic DNA Extraction*

Ten seedlings per genotype were harvested across pots and frozen in liquid nitrogen. Genomic DNA was extracted essentially as previously described by Brouwer et al. (2003). Approximately 151 mm and 100 mg 0.2 mm diameter glass beads were added, and the frozen material was disrupted in a MM400 mixer mill (Retsch, Haan, Germany) for 2 × 1 min at 30 Hz. Subsequently, 300 µl lysis buffer (2.5 M LiCl, 50 mM Tris–HCl, 62.5 mM Na2-ethylenediamine tetra-acetic acid (EDTA), and 4.0% Triton X100, pH 8.0) and an equal volume of phenol/chloroform/isoamyl alcohol (25:24:1 v/v, Carl Roth, Karlsruhe, Germany) were added, and samples were homogenized for 30s at 30 Hz in the mixer mill. After centrifugation (5 min, 16.000 g), the supernatant was recovered, and the genomic DNA was precipitated by the addition of two volumes of 100% ethanol, incubation for 15 min at −20 °C, and another round of centrifugation. The DNA pellet was washed with 70% ethanol, air-dried, and resuspended in Millipore water. DNA quality and concentration were inspected on a Nano Drop system (Thermo Scientific, Bonn, Germany).

9.7.4 *Quantitative Real-Time PCR*

For qPCR, 15 µl samples were prepared using the Brilliant Sybr Green QPCR Reagent Kit (Stratagene, Waldbronn, Germany) according to the manufacturer's protocol. The Taq DNA polymerase provided was replaced by another Taq polymerase (Ampliqon, Odense, Denmark) and the corresponding standard buffer. A final primer concentration of 0.4 µM and three technical replicates per sample were used. qPCR was carried out according to the following protocol: denaturation at 95 °C for 3 min, 40 repeats of 95 °C for 20 s, 61 °C for 20 s, and 72 °C for 15 s. A melting curve analysis was conducted from 55 to 95 °C in 0.5 °C steps and 10 s dwell time to confirm the amplification of single amplicons. Additionally, amplicon size and identity were confirmed on a 2% agarose gel and by DNA sequencing, respectively. The ratio of *G. orontii* to *Arabidopsis* genomic DNA was calculated using the ΔΔCt method (Pfaffi 2001).

9.7.5 *Spore Counts*

At 6 dpi, three samples of approximately 500 mg of seedlings were harvested per genotype. Five ml H_2O was added, and spores were liberated by vortexing for 30 s at maximum speed. The spore solution was filtered through Miracloth (Merck, Darmstadt, Germany) to remove large debris, and spores were fourfold concentrated by centrifugation (5 min, 4000 g). For each sample, spores were counted in eight 1 mm^2 fields of a Neubauer-improved haemocytometer (Marienfeld Lauda-Konigshofen, Germany), and results were averaged. Finally, spore counts were normalized to the initial weight of seedlings (Weßling and Panstruga 2012).

9.8 Embryo Rescue Technique to Transfer Powdery Mildew Resistance

Fully developed flower buds were emasculated prior to anthesis and allowed to bloom. Fresh pollen grains from newly opened flowers were used to pollinate emasculated flowers. For embryo rescue, pistils were harvested 5–7 days after pollination, washed in 70% ethanol for 1 min, washed in 10% sodium hypochlorite solution for 10 min, and then rinsed twice with sterile distilled water. Ovaries were cultured for 7 days in hormone-free MS medium (Murashiga and Skoog 1962) supplemented with 5% sucrose, 0.8% agar, and 500 mg/l casein hydrolysate (MS1) under continuous illumination with fluorescent lights (115 lmol/m2/s). Pistils were cut open aseptically, and ovules were transferred to MS medium with 3% sucrose and 0.8% agar without casein hydrolysate (MS2). All ovules were incubated under fluorescent lights with 16 h photoperiod, and developed ovules were transferred to individual plant boxes. Well-established seedlings were transferred to plastic pots containing Cornell mix (Boodley and Sheldrake 1972), covered with plastic bags, and placed in growth chambers for acclimatization with 6 h photoperiod day/night for two cycles per day with fluorescent lights. Fully developed plants were transferred to a greenhouse where they were grown to maturity (Tonguc and Griffiths 2004).

9.9 DNA Marker Analysis

DNA for PCR was extracted from F_2 plants using a mini-prep procedure. Two or three young leaves were placed in a 1.5 ml minifuge tube, frozen in liquid nitrogen, and quickly ground to powder with a small pestle. The powder was thawed in 700 μl of extraction buffer containing 100 mM of Tris–HCI (pH 8.0), 500 mM of NaCI, 50 mM EDTA, 1.5% SDS, and 10 μM of 2-mercaptoethanol. The homogenate was mixed by vortex and incubated at 65 °C for 10 min. Homogenates were mixed with 700 μl of phenol, and the upper phase after centrifugation for 5 min at 13,000 g was transferred to a fresh tube and extracted with 700 μl of phenol/chloroform (1:1). DNA in the aqueous phase was precipitated with 50 μl of 3 M NaAc (pH 5.2) and 450 μl of cold isopropanol and collected by centrifugation. The pellets were dissolved in 100 μl of TE (10 mM Tris, 1 mM EDTA, pH 8.0) and stored at −20 °C. Larger-scale DNA isolation for RFLP analysis of F_2 plants used 1–2 g of fresh leaves from 20 to 30 young F_3 plants, and the extraction was, as for the miniprep, scaled up proportionately. The DNA pellets were dissolved in 200–300 μl of TE. The primers of microsatellite markers (Bell and Ecker 1994) and CAPS markers (Konieczny and Ausubel 1993) were either obtained from Research Genetics (Huntsville, AL, USA) or synthesized in-house. Amplification reactions contained 0.3 μM of each primer, 0.2 μl of Taq polymerase (1 unit; Boehringer), 1 μl of DNA solution (about 5–10 ng), and 20 μl of PCR mixture containing 50 mM NH_4CI, 10 mM Tris, pH 8.0, 2–3 mM $MgCI_2$, and 200 μM dNTPs.

Amplification was for 45 cycles: for microsatellite markers, each cycle was 94 °C for 20 s, 55 °C for 20 s, and 72 °C for 40 s; and for CAPS markers, each cycle was 94 °C for 30 s, 55 °C or higher, according to GC content and primer length, for 30 sec, and 72 °C for 60 s. Amplified products were resolved on a 1–3% agarose (w/v) gel, depending upon the size of DNA fragments. New CAPS markers between La-*er* and Ms-0 were developed from TSL and PI using the following primers: TSL, forward primer 5' CTGCCGCTGTCAAAGTCGTCAC 3′ reverse primer 5' CCGTCCTCCCATGTCTCAGCTATG 3′; polymorphism revealed by BclI; and PI, forward primer 5' GGGTAGAGGAAAGATCGAGATA 3′, reverse primer CACAAATAGCTTATCACAGAAT, polymorphism revealed by *DraI* and *AluI*. The RFLP probe T01409 was amplified with the forward primer, 5' CGGAGCTTATCAGATCCAGACA 3′, and reverse primer, 5' CCTCAATTTTCCCCAAAGAAGA 3′. Other RFLP probes were amplified (BGL1) or cloned (pCITf16, obtained from Ohio DNA Stock Center, ABRC). After digestion, 1.5–2.0 μg of DNA was separated on a 1.0% agarose gel with 1 × TAE buffer and then transferred to a nylon membrane (Hybond N, Amersham). Probes were labeled using the Pharmacia Oligo labeling kit, and hybridization was conducted according to manufacturer's instruction (Xiao et al. 1997).

9.10 Histological Assessment of *E. cruciferarum* Growth

Similar to other pathogenic powdery mildew and biotrophic fungi, *E. cruciferarum* is noncultural on standard laboratory media and reproduces and proliferates only on susceptible crucifer hosts. After inoculating the upper leaf surface with conidia of *E. cruciferarum*, samples were collected at different time periods for histological examination. Approximately four leaves from all infected plants of each genotype were examined for each time point. At 6 dpi, symptoms of infection were observed in white powdery mildew on the upper surface of *B. napus* leaves. Symptoms can be observed with the naked eye and increased with the progress of time (Plate 9.3d^1–f^1).

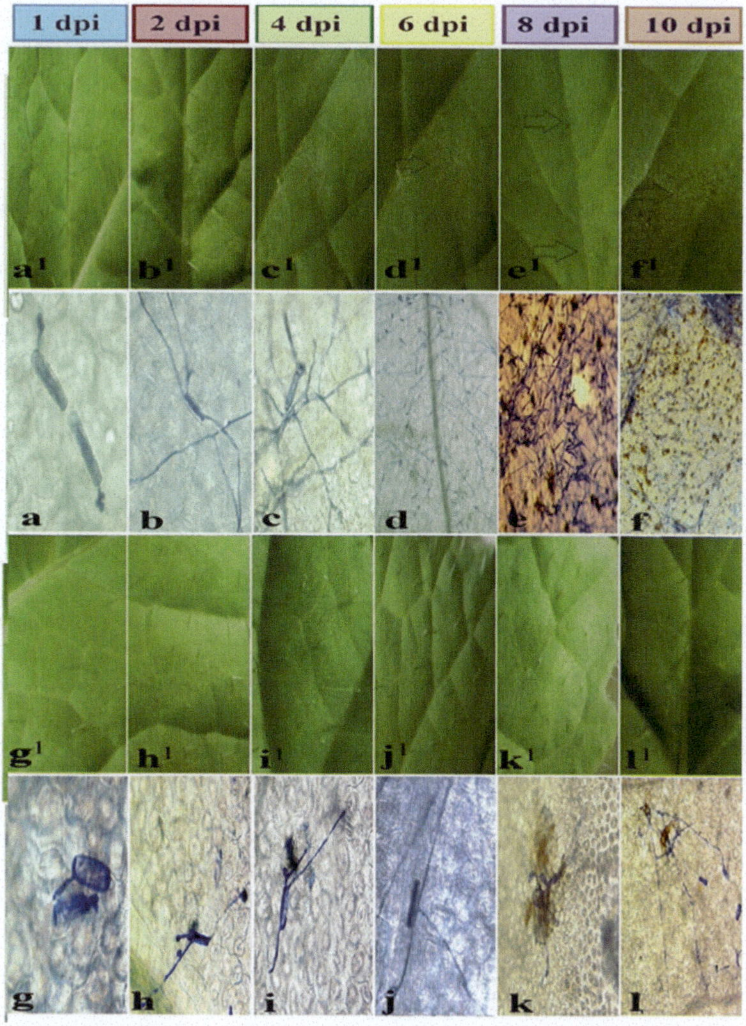

Plate 9.3 Symptoms of *Erysiphe cruciferarum* on leaves infected of *B. napus* of six time points. The images of the symptoms were (a^1, b^1, c^1, d^1, e^1, f^1), and light micrograph were (a, b, c, d, e, f) for leaves of *R. alboglabra* infected by pathogen were (g^1, h^1, i^1, j^1, k^1, l^1), and light micrograph were (g, h, i, j, k, l) for leaves of RRCC infected by pathogen at 1, 2, 4, 6, 8, and 10 days postinoculation (dpi), respectively. Stocks indicate to colonies and the growth of pathogenic fungus. Scale bars for light micrograph at 8 and 10 dpi are 25 μm (Alkooranee et al. 2015)

The observable signs of the pathogen included hyphae, conidia, conidiophores, and dead cells on the infected leaf surface, whereas visual symptoms were not observed in the *R. alboglabra* genotype (Plate 9.3g^1–i^1). At 1 dpi, the structures of pathogenic fungi appeared on the securities under microscopic examination but lower compared with *B. napus* (Plate 9.3g). Microscopic examination showed the conidia of the pathogenic fungus produced germination tubes on the leaves of both genotypes, where conidia produced primary germ tube, aspersorium, and primary hyphae at 2 dpi, followed by hyphae in *B. napus* and to lesser extent in *R. alboglabra* (Plate 9.3b). The rapid growth and widespread of the pathogen and hyphae were also observed. Several conidia were observed on the upper surface of the infected leaves of *B. napus*. The fungus growth increased with time (Plate 9.3c–f). On the contrary, the microscopic examination on *R. alboglabra* showed that most of the conidia produced germination tubes but failed to format the hyphae. These conidia did not develop intensively and rapidly compared with *B. napus* during the same time period (Plate 9.3i–l). The results also showed the presence of dead cells in *B. napus*, and *R. alboglabra* increased. However, fungus growth was densely spread and increased with time. The microscopic examination clearly emerged after 6, 8, and 10 dpi (Plate 9.3f, i). The percentage of spores that successfully germinated on the leaf surface did not differ between the genotypes and time points. However, the further development of the fungus differed during cell penetration; development was lower for *R. alboglabra* (5%) compared with *B. napus* (75%) at 2 dpi. This continued penetration resistance reached 14% successful penetration at 4 dpi compared with 68% of *B. napus* (Plate 9.3g, h). Thus, no or very little colonies were formed on *R. alboglabra* at 4, 6, 8, and 10 dpi. Finally, the histological study confirmed that the *R. alboglabra* genotype was excluded from infection because it was able to prevent the pathogenic fungus *E. cruciferarum* from penetrating the cell wall of host successfully and inhibited the growth of the pathogen at different stages. This penetration resistance was activated at 1 dpi and allowed complete resistance under certain conditions (Alkooranee et al. 2015).

9.11 Maintenance of *Erysiphe cruciferarum* Isolates

A culture of a powdery mildew pathogen, later identified as *Erysiphe cruciferarum*, was isolated from an infected *B. napus* cv. Cobra plant growing in greenhouses at the University of East Anglia. To obtain a pure culture, conidiospores from a single colony were sub-cultured five times on *B. napus* cv. Cobra plants. This pure isolate, designated UEA1, was thereafter propagated on *B. napus* cv. Cobra plants growing in the controlled environment. *A. thaliana* plants with 6–8 true leaves (about 3–4 weeks old) were inoculated with *E. cruciferarum* UEA1 in a wind-still chamber, using heavily infected young leaves of *B. napus* cv. Cobra as the source of inoculum. For inoculation, an artists, fine brush was used to gently dislodge conidia from infected leaves onto *A. thaliana* plants approximately 20 cm below. The inoculated plants were kept under transparent plastic lids for 24 h, to give high humidity (>90% RH), after which the lids were removed (Adam and Somerville 1996).

9.12 Characterization of the Disease Reaction Phenotypes

The disease reaction scores were evaluated by comparison with the level of infection observed on the fully susceptible Col-5. Two measures of infection were used. The level of fungal growth at 7 dpi was ranked on a scale from 0 to 3, with a score of 0 indicating that no fungal growth was apparent by eye and no conidiophores developed. A score of 1 indicated that fungal growth was restricted and no conidiation occurred. An intermediate density of fungal growth accompanied by the development of conidiophores was scored as a 2, and a score of 3 indicated dense fungal growth and abundant conidiation (Le. a fully compatible interaction). The host response to inoculation was scored as either a 1 or a 0 indicating the presence or absence, respectively, of chlorotic or necrotic flecks at the point of inoculation. For example, a disease reaction score of 0.1 (with the first number referring to the extent of fungal growth and the second number referring to the presence of chlorotic/necrotic lesions) was given to highly resistant *Arabidopsis* accessions that supported no visible fungal growth and that did produce chlorotic or necrotic lesions at the point of inoculation. A disease reaction score of 3.0 was assigned to accessions supporting dense fungal growth with no chlorotic or necrotic flecks (Adam and Somerville 1996).

9.13 Disease Scoring Scales

Number of disease scoring scales has been used by researchers to assess disease intensity with respect to evaluation of germplasm of crucifers for sources of resistance and efficacy of disease control measures (cultural, chemical, biological, host resistance, etc.) and to assess damages/losses caused by powdery mildews. The details are given in Chap. 2 at Sect. 2.6.

References

Adam L, Somerville SC (1996) Genetic characterization of five powdery mildew disease resistance loci in *Arabidopsis thaliana*. Plant J 9:341–356

Alkooranee JT, Yin Y, Aledan TR, Jiang Y, Lu G, Wu J, Li M (2015) Systemic resistance to powdery mildew in *Brassica napus* (AACC) and *Raphanus alboglabra* (RRCC) by *Trichoderma harzianum* TH12. PLoS One 10(11):e0142177

Altschul SF, Gish W, Miller W, Myers E, Lipmann DJ (1990) Basic local alignment search tool. J Mol Biol 219:403

Bell CJ, Ecker JR (1994) Assignment of 30 microsatellite loci to the linkage map of *Arabidopsis*. Genomics 19:137–144

Boodley JW, Sheldrake R Jr (1972) Cornell peat-lite mixes for commercial plant growing. Cornell Univ Coop Exten Info Bull 43:1–8

Brouwer M, Lievens B, Van Hemelrijck W, Van den Ackerveken G, Cammue BPA, Thomma BPHJ (2003) Quantification of disease progression of several microbial pathogens on *Arabidopsis thaliana* using real-time fluorescence PCR. FEMS Microbiol Lett 228:241–248

Cunnington JH, Takamatsu S, Lawrie AC, Pascoe IG (2003) Molecular identification of anamorphic powdery mildews (*Erysiphales*). Aust Plant Pathol 32:421–428
Glaze brook J, Rogers EE, Ausubel FM (1996) Isolation of *Arabidopsis* mutants with enhanced disease susceptibility by direct screening. Genetics 143:973–982
Gollner K, Schweizer P, Bai Y, Panstruga R (2008) Natural genetic resources of *Arabidopsis thaliana* reveals a high prevalence and unexpected phenotypic plasticity of RPW8-mediated powdery mildew resistance. New Phytol 177:725–742
Hirata T, Takamatsu S (1996) Nucleotide sequence diversity of rDNA internal transcribed spacers extracted from conidia and cleistothecia from several powdery mildew fungi. Mycoscience 37:283–288
Homma Y (1937) Erysiphaceae of Japan. J Faculty Agric Hokkaido Univ 88:183–461
Innis MA, Gelfand DH (1990) Optimization of PCRs. In: Innis MA, Gelfand DH, Sninsky JJ, White TJ (eds) PCR protocols: a guide to methods and applications. Academic, San Diego, pp 3–12
Konieczny A, Ausubel FM (1993) A procedure for mapping Arabidopsis mutations using co-dominant ecotype-specific PCR based markers. Plant J 4:403–410
Kwok S, Kellogg DE, Mc Kinnery N, Spasic D, Goda L, Levenson C, Sninsky JJ (1990) Effects of primer-template mismatches on the polymerase chain reaction: human immunodeficiency virus type 1 model studies. Nucleic Acids Res 18:999–1005
Mori Y, Sato Y, Takamatsu S (2000) Evolutionary analysis of the powdery mildew fungi using nucleotide sequences of the nuclear ribosomal DNA. Mycologia 92:74–93
Murashiga T, Skoog F (1962) A revised medium for rapid growth and bioassays with tobacco tissue cultures. Physiol Plant 15:473–497
Pfaffi MW (2001) A new mathematical model for relative quantification in real time RT–PCR. Nucleic Acids Res 29:45
Pfister DH (1997) Castor, pollux, and life histories of fungi. Mycologia 89:1–23
Plotnikova JM, Reuber TL, Ausubel FM, Pfister DH (1998) Powdery mildew pathogenesis of *Arabidopsis thaliana*. Mycologia 90:1009–1016
Reuber TL, Plotnikova JM, Dewdney J, Rogers EE, Wood W, Ausubel Frederick M (1998) Correlation of defense gene induction defects with powdery mildew susceptibility in *Arabidopsis* enhanced disease susceptibility mutants. Plant J 16:473–485
Shipton WA, Brown JF (1962) A whole-leaf clearing and staining technique to demonstrate host-pathogen relationships of wheat stem rust. Phytopathology 52:1313
Thompson JD, Higgins DG, Gibson TJ (1994) Clustal W: improving the sensitivity of progressive multiple progressive sequence alignment through sequence weighting, position specific gap penalties and weight matrix choice. Nucleic Acids Res 22:4673–4680
Tonguc M, Griffiths PD (2004) Transfer of powdery mildew resistance from *Brassica carinata* to *Brassica oleracea* through embryo rescue. Plant Breed 123:587–589
Weßling R, Panstruga R (2012) Rapid quantification of plant-powdery mildew interactions by qPCR and conidiospores counts. Plant Methods 8:35
White TJ, Bruns T, Lee S, Taylor J (1990) Amplification and direct sequencing of fungal ribosomal RNA genes for phylogenetics. In: Innis MA, Gelfand DH, Sninsky JJ, White TJ (eds) PCR protocols: a guide to methods and applications. Academic, New York, pp 315–322
Xiao S, Ellwood S, Findlay K, Oliver RP, Turner JG (1997) Characterization of three loci controlling resistance of *Arabidopsis thaliana* accession Ms-0 to two powdery mildew diseases. The Plant J 12(4):757–768
Yarwood CE (1936) The diurnal cycle of the powdery mildew *Erysiphe polygoni*. J Agric Res 52:645–657
Yarwood CE (1946) Detached leaf culture. The Bot Rev 12:1–56
Yarwood CE (1957) Powdery mildews. Bot Rev 23:235–312

Chapter 10
Powdery Mildew Epilog

10.1 Introduction

Crucifers include very large group of oilseed *Brassica* species and vegetable crops grown all over the world for quality vegetable oil and as vegetables apart from a source of fodder crops. The major oil-yielding *Brassica* crops are *B. juncea* (mustard); *B. napus* (rapeseed); *B. carinata* (Ethiopian mustard); *B. rapa* subsp. *oleifera* (turnip rape); var. brown sarson, var. yellow sarson, and var. toria; and *B. nigra* (black mustard). Main vegetable crops are *B. oleracea* var. *acephala* (kale), var. *capitata* (cabbage), var. *sabauda* (Savoy cabbage), var. *gemmifera* (brussels sprouts), var. *botrytis* (cauliflower), var. gongylodes (kohlrabi), var. *italica* (broccoli), and var. *alboglabra* (Chinese kale) and *B. rapa* subsp. *chinensis* (bok choy), subsp. *pekinensis* (Chinese cabbage), subsp. *nipposinica*, and subsp. *pamchinensis*. Fodder crops are *B. oleracea* var. *fruticosa* (branching bush kale), *B. rapa* subsp. *rapifera* (turnip), and *B. napus* subsp. *rapifera* (rutabaga, swede). In addition to these crops, there are several weeds including *Arabidopsis thaliana* which is drosophila or ecoli for molecular studies using microbes as a host pathosystem. Out of more than 40 biotic and abiotic stress reported on crucifers (Saharan et al. this volume), powdery mildew is the fourth important biotic disease causing heavy losses and widely distributed over more than 25 countries of the world on crucifers. More so, powdery mildew is at very alarming stage on oil-yielding *Brassica* crops, viz. *B. juncea* and *B. napus*, all over the world. These two *Brassica* species are widely cultivated throughout oil-yielding countries as a source of quality edible oil.

Powdery mildews (*Erysiphales*) are a very large group of biotrophic pathogens which grow principally on the aerial parts of angiosperms including crucifers and cause qualitative and quantitative damage on a wide variety of crops. Throughout the more than 250 years since Linnaeus first gave a scientific name to a powdery mildew, they have figured prominently in the history of plant pathology. The white, floury superficial mycelium, conidiophores, conidia, and pinhead dark brown chasmothecia (cleistothecia) of powdery mildews make them among the most easily

G. S. Saharan et al., *Powdery Mildew Disease of Crucifers: Biology, Ecology and Disease Management*, https://doi.org/10.1007/978-981-13-9853-7_10

recognized and best known plant pathogens. The taxonomy and identification of powdery mildews are based on the characteristics of teleomorph (cleistothecial appendages, number of asci per cleistothecium) and anamorphs morphology of conidial stages on host including host range data, etc. With the use of scanning electron microscopy (SEM), light microscopy (LM), host range, and phylogenetic analysis, identification keys have been constructed for different powdery mildews genera and species infecting different crops including crucifers. The major powdery mildew of crucifers is caused by *E. cruciferarum* under field conditions all over the world. However, on *Arabidopsis* under artificial inoculation conditions, non-adapted species of powdery mildew (not reported under natural conditions) like *Golovinomyces* (syn. *Erysiphe*) *cichoracearum*, *G. orontii,* and *Oidium neolycopersici* have been used for molecular and genetical studies. Powdery mildews can take epidemic form under field conditions within a very short period of time if congenial weather conditions prevail after landing of primary inoculum on the host surface.

10.2 The Disease: Powdery Mildew

The powdery mildew of crucifers shows its symptoms on aerial parts of host plants in the form of white to dirty white circular floury patches on leaves, stems, inflorescence, and siliquae. These floury patches increase in size and coalesce to cover entire plant with the increase in atmospheric temperature. Later in the season, on infected leaves and siliquae, powdery growth may be embedded with brown to dark brown pinhead size cleistothecia or chasmothecia especially in case of *B. juncea*. Severely diseased leaves exhibit chlorotic or necrotic brown lesions and leaf distortion followed by senescence. Severely diseased siliquae remain small in size and produce small-sized shriveled few seeds. The disease is distributed in more than 25 countries of the world infecting more than 120 crucifers host plants including vegetables and oil-yielding economically important crops. In India, powdery mildew is widely distributed from north to south and east to west with more severity in central states on crucifers vegetables and oil-yielding crops. Powdery mildew of crucifers causes 10–90% loss in yield of rapeseed-mustard crops with reduction in oil quality and quantity up to 7%. The disease severity on different crucifers have been assessed using different scoring scales with respect to assessment of losses, efficacy of chemicals, influence of cultural operations (sowing time, spacing, intercropping, nitrogen and micronutrients application, cultivars grown), inbuilt resistance of genotype, efficacy of inoculation techniques, and yield potential of genotypes/cvs under protected, and non-protected environments.

10.3 The Pathogen

On the basis of morphological, pathological, host range, anatomical, phylogenetical, and developmental features, and characteristics of the powdery mildew fungus at anamorph and teleomorph states, the major causal organism of crucifer's powdery mildew has been identified as *Erysiphe cruciferarum*. Under artificial inoculation conditions, three other powdery mildew fungi, viz. *Golovinomyces cichoracearum*, *G. orontii,* and *Oidium neolycopersici*, are able to complete their anamorph state on a crucifers weed *Arabidopsis thaliana*. *Arabidopsis*–powdery mildew host pathosystem has been used recently as a model system for molecular and genetical studies all over the world. Morphological features of all the four powdery mildew species pathogenic to crucifers, *Brassica* species, and *Arabidopsis* have been described as observed on the host species. Classification, taxonomy, and nomenclature of powdery mildew pathogens of crucifers have been updated on the basis of features on the surfaces of powdery mildew conidia revealed through scanning electron microscopy and light microscopy at anamorph state where teleomorph state was absent. The four powdery mildew species pathogenic on crucifers have been distinguished on the basis of conidial size, shape, appearance, number of conidia per conidiophore, conidiophores shape, and size, haustoria, hyphal branch angles, conidial germination with respect to temperature, relative humidity, light, and substrate, size of chasmothecia, asci, ascospores, number of ascospores per chasmothecium, and type of appendages. Phylogenetics relationship of crucifers powdery mildew with other powdery mildew species has been determined through sequence analysis of the ITS rDNA and sequence comparison of DNA encoding the 5.8 s, rRNA, and ITS2. The powdery mildew pathogen genomes and transcriptomes have been determined. Powdery mildew fungi have sizeable genomes, which are about four times larger than those of other ascomycetes. The number of coding genes in the powdery mildew genomes is comparatively low. The structure of powdery mildew family, *Erysiphaceae*; the position of genes *Erysiphe*, its sections, and subsections *Golovinomyces*, key to the recognized genera; and species of the family, and relationship within the family have been exploited.

10.4 Infection, Pathogenesis, and Disease Cycle

Powdery mildew fungi of crucifers are ectoparasites on their respective host plant species. Primary infection takes place through ascospores or conidia after landing on the host surface through germ tube which develops an appressorium at the tip, a thickened infection structure. After cell wall penetration, specialized hyphae feeding structure called haustorium is formed which gets necessary nutrients from the host tissues for the growth and development of the pathogen to produce mycelium, conidiophores, and conidia to continue its asexual state. Secondary spread of the disease is through airborne conidia causing infection and pathogenesis under congenial environmental conditions. At the maturity stage of the crop, sexual state of the pathogen is visible in the form of small, pinhead, dark brown spherical chasmothecia over summer or over winter to serve as a source of primary inoculum for the next season. Chasmothecia imbibe to burst and release asci and ascospores which germinate after landing on host surface and cause primary infection. The pathogenesis of powdery mildew fungi is achieved through the suppression of the hosts preformed and inducible penetration resistance mechanisms as well as maintenance of the compatible interaction after host cell penetration. After host cell penetration, the haustorial complex is established, which consists of the fungal haustorium, an extra-haustorial modified plant plasma membrane, and the haustorial matrix. The haustorial complex is the site of the most intimate contact of the pathogen with its host cell, and functions in nutrient uptake, and provides a platform for the delivery of pathogens effector genes for pathogenesis. The expression of hundreds of host genes are altered within 8 hpi indicating that signaling between the pathogen and individual host cells already occurs a few hours post contact of the powdery mildew conidia or ascospores with the leaf surface of about the stage of penetration but certainly prior to establishment of the extra-haustorial matrix. For the successful pathogenesis of crucifer's powdery mildew, there is great role of genetic factors (genes), MLO proteins, and host transcriptional changes. Powdery mildew of crucifers is activated by SA signaling pathways at the early stage of infection and by the JA/ET signaling pathways at the later stage of pathogenesis. Infection is highly activated by chitinase in the genotypes. Adjustment of the plant host metabolism to support the growth of the powdery mildew pathogen is consistent with an increased ploidy level of the mesophyll cells underlying infected epidermal cells. Powdery mildew disease cycle operates at both asexual and sexual stages depending on its host plant species infected under prevailing environmental conditions.

10.5 Fine Structures and Electron Microscopy

The formation of fine structures of powdery mildew pathogens during infection and pathogenesis after artificial inoculation of *Arabidopsis* have been revealed through light and scanning electron microscopy (SEM). The sequence of events has been observed from time of inoculation to production of asexual states and symptoms production on the host leaf and stem. The comparative observations on the morphological characteristics of different structures formed by three pathogens, viz. *E. cruciferarum, E. orontii,* and *E. cichoracearum* of *Arabidopsis*, have been distinguished as a compatible host–parasite interaction system for molecular and genetical studies. The observations have been recorded on pathogens hypha, its branches; angles, mycelium growth; conidiophore shape, size; number of conidia formed; conidial shape, size, germ tubes number, shape, its tips shape; haustoria, its shape, number of infection pegs, their shape; appressoria, their shape, and formation of papillae. For each event, time period has been recorded including incubation period of powdery mildew pathogens on *Arabidopsis*. The fine structures revealed through SEM such as septa, haustorial sac, extra-haustorial membrane, vacuolar membrane of the host cell, host cytoplasm, granular structures in the extra-haustorial matrix, space between the haustorial sac and the haustorium, and haustorial lobes are very crucial to understand the mechanisms of pathogenesis by powdery mildews of crucifers in susceptible and resistant hosts. The observation of encapsulation of haustoria, formation of papilla-like structures, deposition of granular material adjacent to haustoria, and necrosis of host cells are potential resistant reactions of the invaded host plant. Now a days, SEM observations have become important and essential criteria for all taxon of powdery mildews especially when only anamorph states are produced like in powdery mildews of *Arabidopsis*.

10.6 Epidemiology and Disease Forecasting

Epidemiology of crucifier's powdery mildews is influenced by numerous factors of host and pathogen origin under the prevailing environmental and physical factors of soil, nutrition, and cultivation practices. After landing of inoculum in the form of pathogens conidia/ascospores, their germination and formation of appressoria to cause infection are influenced by environmental factors like temperature (minimum, maximum, averages), relative humidity (morning, evening, averages), sunshine hours, and wind velocity along with critical host growth stages. Host factors like age of the leaf and stem under the influence of ambient temperature determines the infection and spread of disease under field conditions. Interaction of pathogen with host at right growth stage coinciding with congenial environmental conditions favours disease development at faster rate. Stem infection is maximum, if they are exposed to the inoculum for longer duration especially at maturity stage of the crop. Infection on younger plants and newly emerging leaves is rare. Disease development is at faster rate if host–pathogen interaction coincides with favourable crop growth stages and crucial environmental variables. Late planting is associated with high disease intensity. Host resistance curbs the area under disease progress curve (AUDPC), less disease intensity, higher incubation, and latent periods, low infection rate, less number of powdery mildew colony with less number of conidiophores, and conidial production. Disease forecasting system has been developed using historical and current data on maximum severity of disease, crop age at first appearance of the disease, and crop age at peak severity of disease for two susceptible cultivars sown at different locations. Weather data have been analysed to develop disease forecasting models.

10.7 Host Resistance

Host resistance in crucifers to powdery mildews is multilayered and multicomponents both at pre- and post-penetration stages. The intricate immune responses are evolved through accumulation of ROS, H_2O_2, deposition of callose, pectin, cellulose, waxes, silicon, ion fluxes, formation of papilla, cell wall apposition, phenolic compounds, overexpression of R genes, PR proteins, protein phosphorylation, biosynthesis of phytoalexins, fungal enzymes inhibiters, chito-octamers, triggering of HR induction of SAR, and non-host resistance mechanisms. These cytoskeleton components have very important and crucial functional and structural roles in host resistance to powdery mildew pathogens of crucifers. Defenses are activated either through SA signaling or simultaneous perception of ethylene and jasmonic acid (JA). The overexpression of several R genes in crucifers – powdery mildew host pathosystem induces host resistance. MLO genes encoding seven-transmembrane, calmodulin-binding protein confer broad-spectrum resistance to adapted powdery mildews of *Arabidopsis*. *edr* mutants of *Arabidopsis* have a general link between SA-mediated resistance, mitochondrial function, and programmed cell death. Powdery mildews mutants confer resistance to powdery mildew through altered cell wall composition of host. Increased SA enhances the expression of RPW 8.1 and RPW 8.2 leading to HR or SHL and resistance. BjNPR1 gene activates SAR to confer broad-spectrum resistance to powdery mildew of *B. juncea*. AtROP-regulated AtRLCK V1A3 has a role in basal resistance to powdery mildews. The *AtMLO2*, *AtMLO6*, and *AtMLO12* triple mutants are resistant to powdery mildews and induce PCD in response to infection by *E. cruciferarum*. There is a role of *WRKY* transcription factors and overexpression of R genes like *PMR, MLO, PEN, EDR, MAPK, MAPK 65-3, MPR1, PAD3, PAD4, ED5, SNARE, RLCKs*, and *KDL* (At *CEP1*) to confer R to powdery mildews of crucifers. Higher levels of camalexin contribute to the enhanced R to powdery mildew in Cyp83 a1-3 mutants of *Arabidopsis*. *SR1* plays a critical role in powdery mildew resistance by regulating *EIN3* and *NDR1* expression. There is harmonious condition between transcriptional regulation and resistance to powdery mildews. The application of *Trichoderma harzianum* and its CF induces (ISR) resistance in crucifers. Mechanisms of non-host R in crucifers to powdery mildews have been unrevealed which is strong and durable. Non-host resistance is *PEV*-gene-mediated at preinvasion and controlled by genes *EDS1, PAD4*, and *SAG* (101) at post-invasion of powdery mildew pathogens. In cabbage, R to powdery mildew is controlled by a single dominant gene with modifiers. A single R gene controls R to powdery mildew in HC-1 and PCC-2 with complete dominance. In *Arabidopsis*, R to PM is polygenic and based on R gene *RPW8* or on combination of *RPW* 8 gene complex locus. Powdery mildew resistance genes of *Arabidopsis* have been mapped on chromosomes II (*RPW1*), III (*RPW2, RPW3, RPW7, RPW8*), IV (*RPW4*), and V (*RPW 5, RPW 6*). In rapeseed gamma rays mutagenic plants exhibit R to powdery mildew due to an increase in concentration of unsaturated fatty acids with 18 carbon atoms. Induction of glucosinolates and camalexin plays important roles for resistance to powdery mildew of crucifers. Camalexin biosynthesis and accumulation are affected by *WRKY 18 and WRKY 40* transcription factor of *Arabidopsis* and enhance upon *G. orontii* infection to confer resistance. Transfer of powdery mildew resistance to *B. oleracea* from *B. carinata* through embryo rescue has been confirmed. Major gene sources of resistance against powdery mildew of crucifers have been identified.

10.8 Disease Management

Powdery mildew of crucifers is a compound interest disease (polycyclic) whose epidemic development within the growth cycle of the host is determined by the initial amount of inoculum or primary inoculum and the rate at which the disease/

inoculum increases, i.e. the apparent infection rate. To reduce the severity of powdery mildew, the reproductive rate of the pathogen must be decreased. The major contributing factors which affect the reproduction rate of pathogen are reduction in the infection frequency or lesion, colony numbers, and size, longer latent period of the pathogen, and decreased spore production. These can be controlled by spraying of fungicides soon after the appearance of the disease symptoms and repeating of second spray at an interval of 10 days. To escape the crop from powdery mildew infection date of planting should be manipulated to avoid congenial weather conditions and susceptible growth stage of the crop. Out of several chemicals tested, fungicides like sulphur (wettable sulphur), Karathane, Calixin, dinocap, flusilazole, and carbendazim have shown their efficacy against powdery mildew with positive effect on total yield and yield-contributing attributes. Some biological control agents have shown promise to control powdery mildew when used as seed treatment and foliar sprays (*Allium sativum*, *Trichoderma harzianum*, and *Pseudomonas fluorescens*). Cultural practices like mixed cropping with pea and wider spacing of crop reduce powdery mildew intensity under field conditions. Crop nutrition in the form of FYM is better than chemical fertilizers to manage powdery mildew of crucifers, but higher yields are obtained with chemical fertilization. Integrated disease management using all means to control powdery mildew gives better results for high yield and better quality of crucifers.

10.9 Techniques

Techniques, methods, and standardized, reproducible procedures are essential parts of biological research to carry out research by students and researchers for wider acceptability. This chapter includes a number of standardized and reproducible techniques to help crucifer researchers to validate, refine, and update their research pursuits using crucifer – powdery mildew pathosystem.

10.10 Research Gaps

After critical analysis of the information on powdery mildew of crucifers, some research gaps have been observed which are highlighted in Chap. 12 in the form of future research priorities.

Reference

Saharan GS, Mehta N, Meena PD (this volume) Techniques. In: Powdery mildew disease of crucifers: biology, ecology and disease management. Springer Nature, Singapore

Chapter 11
Future Research Priorities of Crucifer's Powdery Mildew

With the globalization of agriculture, there is a significant increase in the movement of crucifer's germplasm and in the cropping patterns of this important commodity all over the world. The absence of strict measures on the restriction of movement of germplasm and the intensive and/or extensive cultivation of these crops have resulted in large scale perpetuation, build-up, and dissemination of *Erysiphe cruciferarum* virulence on cruciferous species all over the world. The information gathered in this book indicates that some gaps still exist in the complete comprehension of this disease, which is highlighted below. Even after well-developed techniques, procedures, and protocols available using *Arabidopsis–Erysiphe* pathosystem, progress to use this information has not attained satisfactory level. The following areas of research needs to be exploited for comprehensive study of the Erysiphe–crucifers pathosystem

1. In-depth study on interaction between host plant age, developmental growth stage of host and pathogen, environmental conditions, and genotypes interaction with powdery mildew pathogen under natural as well as controlled laboratory conditions may reveal useful information on powdery mildew epidemiology to develop disease forecasting system.
2. Mechanisms of host resistance through histology of host–pathogen interaction, induction of resistance through chemical and bioagents, expression of PR proteins, and identification of QTL markers for host resistance.
3. Exploitation of hyperparasitism of powdery mildews as a biological control measure. Identification of hyperparasites which can multiply and grow under low humidity and at temperature range of 20–25 °C synchronizing with pathogens (PM) growth stages.
4. To develop a protocol of powdery mildew pathogen to grow and maintain under artificial laboratory conditions on artificial media. To study life cycle pattern of asexual and sexual states under artificial as well as natural growth conditions of the all four powdery mildew pathogens adapted to crucifer species.

G. S. Saharan et al., *Powdery Mildew Disease of Crucifers: Biology, Ecology and Disease Management*, https://doi.org/10.1007/978-981-13-9853-7_11

5. Genome sequencing of the powdery mildew pathogens of crucifers. A comparative analysis of the genomes of adapted and non-adapted pathogens will provide precious glimpses into their virulence arsenals.
6. Identification of determinants of crucifers–powdery mildew host pathosystem both for adapted and non-adapted pathogens.
7. Identification of host components that can be manipulated to increase resistance level against powdery mildew pathogens. Components should be identified at both pre- and post-infection levels.
8. Isolation and characterization of the avirulence genes (AVR). To study inheritance of virulence in the pathogens isolates.
9. Strengthening of genetical and molecular studies of powdery mildew fungi.
10. Ultrastructural and biochemical basis of infection and pathogenesis, nutrient uptake, and host manipulation strategies of powdery mildew pathogen will be helpful to plan powdery mildew management techniques.
11. Sequences of powdery mildew fungal genomes, identification of powdery mildew effectors (genes), critical host targets, adaptation of the host metabolism to the presence of the pathogen, and modulation of plant development will contribute to the identification of determinants of powdery mildew relationship with their hosts and among themselves. It will contribute to the identification of common determinants of obligate parasitism and allow identification of potential host components that can be targeted to increase host resistance to powdery mildew pathogens.
12. The mechanism of non-host resistance against powdery mildew of crucifers should be exploited which is more durable and robust in nature.
13. Factors governing disease development and consequent progression are not completely understood. Deeper insight with regard to various epidemiological aspects will help to develop strategies to curb the progression of the potential disease. There is an urgent need to carry out intensive investigation with regard to the environmental parameters responsible for causation and spread of disease under field conditions. Multilocational trials with staggering dates of planting could be helpful in analysing disease development in relation to environment by computing disease progression at regular intervals. These efforts could be undertaken to study various *Disease Prediction Models* (geophytopathological, bioclimatic, simulation system analysis, etc.) which could play a meaningful role in developing an effective disease control strategy. PCR-based procedures for detecting powdery mildew can improve capabilities of disease forecasting and management.
14. To work out cost-effective cultural, chemical, biological, and survival behaviour and host resistance to powdery mildew management strategies at each level considering prevailing environmental conditions and area under cultivation of cruciferous crop species.
15. Identification and standardization of host differentials is a fundamental prerequisite for the gathering of meaningful information on races/pathotypes. This is a high-priority area of investigation. Designation and nomenclature of pathotypes should be uniform, logical, and internationally acceptable.

16. The collection, maintenance, characterization, and description of taxa on crucifer's using morphological, molecular, and host range data.
17. Information on the ultrastructures of conidiogenesis will reveal evolution of conidial states of crucifer's powdery mildew taxa.
18. Powdery mildew host pathosystem is a well-suited model for studying the effects of climate change on plant diseases.
19. Transcriptional (genes) regulation and expression in response to powdery mildew infection through employment of techniques that allow transcript profiling of powdery mildew challenged host leaves at cellular resolution based on RNA extraction via microcapillaries or by laser microdissection can be exploited for better comprehension of mechanisms of pathogenesis and host resistance.
20. The use of mutants and gene-silencing approaches in *A. thaliana* can open up new possibilities to discover genes involved in non-host resistance and to unravel the underlying mechanisms.
21. The biosynthesis, regulation, functions, and role of camalexin (phytoalexin) in crucifers–powdery mildew host pathosystem in not fully understood, which needs attention of *Brassica* scientists.
22. Identification of common determinants of biotrophy and potential host components that can be targeted to increase resistance against powdery mildew of crucifers.
23. Characterization of the mechanisms by which the powdery mildew pathogens enforces adjustment of the host cytoskeleton system to promote its intracellular proliferation.
24. To dissect molecular pathways of pathogenesis by different powdery mildew species on crucifers host species.
25. Identification, characterization, cloning, and utilization of genes governing powdery mildew pathogenesis and crucifers host resistance to biotic and abiotic stresses.
26. Identification and characterization of signaling molecular pathways during pre- and post-penetration stages of powdery mildews and a common plate form where signaling molecules synchronizes during both biotic and abiotic stresses.

Index

G. S. Saharan et al., *Powdery Mildew Disease of Crucifers: Biology, Ecology and Disease Management*, https://doi.org/10.1007/978-981-13-9853-7

C

E

F

S

T

U

The manufacturer's authorised representative in the EU is Springer Nature Customer Service Centre GmbH, Europaplatz 3, 69115 Heidelberg, Germany. If you have any concerns regarding our products, please contact ProductSafety@springernature.com

Printed and bound by CPI Group (UK) Ltd, Croydon, CR0 4YY
30/09/2026
02225304-0015